Modal Analysis and Testing

NATO Science Series

A Series presenting the results of activities sponsored by the NATO Science Committee. The Series is published by IOS Press and Kluwer Academic Publishers, in conjunction with the NATO Scientific Affairs Division.

A. **Life Sciences**	IOS Press
B. **Physics**	Kluwer Academic Publishers
C. **Mathematical and Physical Sciences**	Kluwer Academic Publishers
D. **Behavioural and Social Sciences**	Kluwer Academic Publishers
E. **Applied Sciences**	Kluwer Academic Publishers
F. **Computer and Systems Sciences**	IOS Press

1. **Disarmament Technologies**	Kluwer Academic Publishers
2. **Environmental Security**	Kluwer Academic Publishers
3. **High Technology**	Kluwer Academic Publishers
4. **Science and Technology Policy**	IOS Press
5. **Computer Networking**	IOS Press

NATO-PCO-DATA BASE

The NATO Science Series continues the series of books published formerly in the NATO ASI Series. An electronic index to the NATO ASI Series provides full bibliographical references (with keywords and/or abstracts) to more than 50000 contributions from internatonal scientists published in all sections of the NATO ASI Series.
Access to the NATO-PCO-DATA BASE is possible via CD-ROM "NATO-PCO-DATA BASE" with user-friendly retrieval software in English, French and German (WTV GmbH and DATAWARE Technologies Inc. 1989).

The CD-ROM of the NATO ASI Series can be ordered from: PCO, Overijse, Belgium

Series E: Applied Sciences – Vol. 363

Modal Analysis and Testing

edited by

Júlio M.M. Silva

and

Nuno M.M. Maia

Instituto Superior Técnico,
Technical University of Lisbon,
Portugal

Kluwer Academic Publishers

Dordrecht / Boston / London

Published in cooperation with NATO Scientific Affairs Division

Proceedings of the NATO Advanced Study Institute on
Modal Analysis and Testing
Sesimbra, Portugal
3–15 May, 1998

A C.I.P. Catalogue record for this book is available from the Library of Congress.

ISBN 0-7923-5893-7 (HB)
ISBN 0-7923-5894-5 (PB)

Published by Kluwer Academic Publishers,
P.O. Box 17, 3300 AA Dordrecht, The Netherlands.

Sold and distributed in North, Central and South America
by Kluwer Academic Publishers,
101 Philip Drive, Norwell, MA 02061, U.S.A.

In all other countries, sold and distributed
by Kluwer Academic Publishers,
P.O. Box 322, 3300 AH Dordrecht, The Netherlands.

Printed on acid-free paper

Contents

Preface

The present book contains the revised texts of the main lectures presented at the NATO Advanced Study Institute (ASI) on Modal Analysis and Testing, held in Sesimbra, Portugal, 3-15 May 1998.

Modal Analysis is recognized nowadays as one of the most powerful tools, available to the engineer, for the dynamic analysis of structures. It can be said that the derivation of reliable models to represent the dynamics of complex structures is not possible without modal analysis. This field of study has suffered a strong development is the past few years, covering a wide range of interests and applications namely the identification and evaluation of vibration phenomena, the validation, correction and updating of analytical models, the assessment of structural integrity, etc. However, the subject, in its broad aspects, is currently considered too advanced and/or too specialised to be part of normal undergraduate or even post-graduate courses. On the other hand, the available literature is very reduced in number if one excludes papers published in specialised journals or presented at international scientific meetings.

Thus, it was thought that the organisation of an Advanced Study Institute (ASI) on the foundations, current state-of-the-art and recent developments of all essential techniques related to Modal Analysis and Testing would be an important and welcome contribution to increase research collaboration among different countries. In addition, such an ASI would provide a forum of interaction among leading scientists and young researchers from several countries and different schools of thought, assessing present directions and unifying trends for future research.

With the previous objectives in mind and taking advantage of their connections with the Modal Analysis research community, the ASI organizers managed to meet, in the U.S.A., some of the internationally recognized leading experts in the field and to agree upon their cooperation on the organisation of a comprehensive and well structured advanced course on Modal Analysis and Testing. The participation of these well known researchers insured the high scientific level of the course and its success. The venue was held in Sesimbra, Portugal, 20 miles from Lisbon.

The texts of the main lectures have been subjected to an editing and revision process, so that they could be "assembled" to form a coherent and comprehensive harmonized work. The result is the present book. After a first chapter on the fundamentals of Modal Analysis, the reader is introduced to signal processing and to the basic rules of exchange and analysis of dynamic information. The subject of the derivation of theoretical models for modal analysis is then addressed, followed by three chapters discussing the different approaches to the derivation of models based on the identification of experimental data: time domain, frequency domain and pseudo-testing. The previous subjects lead to the discussion of updating of analytical models and to model quality assessment techniques in the chapters that follow.

Further on, the reader can find more specialised subjects such as damage detection and evaluation, structural modification and damping modelling. The normalisation of complex modes is also addressed.

Finally, other seldom covered subjects are included: active control of structures, acoustic modal analysis and neural networks for modal analysis are discussed followed by advanced optimisation methods for model updating, modal analysis for rotating machinery and nonlinearity in modal analysis.

We believe that this book may constitute a fundamental reference for many people in different domains of activity, as it covers not only the basics but also a complete and wide range of subjects that embrace nowadays the most important fields of interest of the modal analysis research community.

We wish to take this opportunity to thank all contributors and participants to the ASI. Special thanks go to Professors Samir Ibrahim, Michael Link and David Ewins for their efforts in the planning and organisation of the Institute. Finally, we are grateful to the NATO Office of Scientific Affairs and Dr. Veiga da Cunha, without whom the Institute and this book would not have been possible. Their support, together with the other sponsors listed in the book, is gratefully acknowledged.

March, 1999

Júlio M. Montalvão e Silva
Nuno M. M. Maia

NATO Advanced Study Institute
Modal Analysis & Testing
Sesimbra, Portugal
May 3 - 15, 1998

Sponsors

NATO North Atlantic Treaty Organization
FCT Fundação para a Ciência e a Tecnologia
FLAD Fundação Luso-Americana para o Desenvolvimento
ADtranz
LMS International
BPA Banco Português do Atlântico
JTS Junta de Turismo de Setúbal
IDMEC Instituto de Engenharia Mecânica
IST Instituto Superior Técnico

Director

Júlio M. Montalvão e Silva, IDMEC/Instituto Superior Técnico, Portugal

Co-Director

Nuno M. M. Maia, IDMEC/Instituto Superior Técnico, Portugal

Organizing Committee

Júlio M. Montalvão e Silva, IDMEC/Instituto Superior Técnico, Portugal
Nuno M. M. Maia, IDMEC/Instituto Superior Técnico, Portugal
Samir Ibrahim, Old Dominion University, USA
Michael Link, University of Kassel, Germany
David J. Ewins, Imperial College, United Kingdom

Main Lecturers

Júlio M. Montalvão e Silva, IDMEC/Instituto Superior Técnico, Portugal
Nuno M. M. Maia, IDMEC/Instituto Superior Técnico, Portugal
Jimin He, Victoria University of Technology, Australia
Paul Sas, Katholieke Universiteit Leuven, Belgium
Paulo Varoto, Universidade de São Paulo, Brasil
Gérard Lallement, Université de Franche-Comté, France

Michael Link, Universität Kassel, Germany
Lothar Gaul, University of Stuttgart, Germany
David J. Ewins, Imperial College, United Kingdom
J. K. Hammond, University of Southampton, United Kingdom
Nicholas A. J. Lieven, University of Bristol, United Kingdom
G. R. Tomlinson, University of Sheffield, United Kingdom
Samir Ibrahim, Old Dominion University, USA
Kenneth McConnell, Iowa State University, USA
Daniel Inman, Virginia Polytechnic Institute and State University, USA
Charles R. Farrar, Los Alamos National Laboratory, USA

Participants

Hal Gurgenci, University of Queensland, Australia
Vincent Lenaerts, University of Liège, Belgium
Luc Hermans, LMS International, Belgium
Alexandre Mesquita, Federal University of Pará, Brasil
Petia Georgieva, Bulgaria Academy of Sciences, Bulgaria
Ludmila Mihaylova, Bulgaria Academy of Sciences, Bulgaria
Rune Brincker, Aalborg University, Denmark
Marie Reynier, Université Paris X, CNRS, France
Ina Taralova-Roux, GESNLA, DGEI – INSA, France
Walter Ponge-Ferreira, University of Kassel, Germany
Bruno Piombo, Politecnico di Torino, Italy
Paolo Pasqua, University of Rome "La Sapienza", Italy
Rosario Ceravolo, Politecnico di Torino, Italy
Alessandro De Stefano, Politecnico di Torino, Italy
Michele Arturo Caponero, ENEA, Italy
Massimo Ruzzene, Politecnico di Torino, Italy
Wojciech Lisowski, University of Mining and Metallurgy, Poland
Piotr Kurowski, University of Mining and Metallurgy, Poland
Paula Silva, Instituto Politécnico de Setúbal, Portugal
Teresa Morgado, Instituto Politécnico de Setúbal, Portugal
José Viriato Araújo Santos, Instituto Politécnico de Setúbal, Portugal
Relógio Ribeiro, IDMEC/Instituto Superior Técnico, Portugal
Paulo Costa, Instituto Politécnico da Guarda, Portugal
Miguel Moreira, Instituto Politécnico de Setúbal, Portugal
António Urgueira, FCT da Universidade Nova de Lisboa, Portugal
João Paulo Martins, FCT da Universidade Nova de Lisboa, Portugal
Raquel Almeida, FCT da Universidade Nova de Lisboa, Portugal
Aurélio Araújo, IDMEC/Instituto Superior Técnico, Portugal
José Negreiros de Carvalho, Universidade Independente, Portugal
António J. Araújo Gomes, Escola Sup. de Tecn. de Castelo Branco, Portugal
Rui Chedas Sampaio, Escola Náutica Infante D. Henrique, Portugal
Miguel Eurico Lisboa, ADTRANZ - Divisão Sorefame, Portugal
Afzal Suleman, IDMEC/Instituto Superior Técnico, Portugal

Paulo Piloto, Instituto Politécnico de Bragança, Portugal
Nuno Nunes, Instituto Politécnico de Setúbal, Portugal
Mircea Rades, University Politehnica of Bucharest, Romania
Titus Gh. Cioara, Technical University of Timisoara, Romania
Mª Helena F. Rodrigues, University of Basque Country, Spain
Tolga Cimilli, Bogazici University, Turkey
John E. Mottershead, The University of Liverpool, United Kingdom
David Thompson, ISVR - University of Southampton, United Kingdom
Alexander F. Vakakis, Uninersity of Illinois at Urbana-Champaig, USA
Sreenivas Alampalli, Transportation Research and Development Bureau, USA
Manuel Pagá, Intevep, S.A., Venezuela

AN OVERVIEW OF THE FUNDAMENTALS OF MODAL ANALYSIS

J. M. MONTALVÃO E SILVA
Instituto Superior Técnico, Dep. Eng^a. Mec.,
Av. Rovisco Pais, 1049-001 Lisboa, Portugal

1. Introduction

Modal analysis is an increasingly more important engineering tool that was first applied around 1940 in the search for a better understanding of aircraft dynamic behaviour. Till the end of the 60's developments were slow and experimental techniques were based on the use of expensive and cumbersome narrow band analogue spectrum analysers. The modern era of modal analysis can be taken as starting at the beginning of the 70's, based upon the commercial availability of Fast Fourier Transform (FFT) spectrum analysers, transfer function analysers (TFA) and discrete acquisition and analysis of data, together with the availability of increasingly smaller, less expensive and more powerful digital computers to process the data.

Modal analysis is primarily a tool for deriving reliable models to represent the dynamic behaviour of structures. In general, it can be said that the applications of modal analysis cover a broad range of objectives, such as identification and evaluation of vibration phenomena, validation, correction and updating of analytical dynamic models, development of experimentally based dynamic models, structural integrity assessment, structural modification and damage detection, model integration with other areas of dynamics such as acoustics and fatigue, etc..

Understanding modal analysis implies knowledge of a broad range of physical and mathematical laws and concepts. It is obviously not possible to cover every aspect in a short text and therefore, three basic assumptions will be established: i) the structure is a linear system whose dynamic behaviour can be described by a model represented by a set of second order differential equations; ii) the structure obeys Maxwell's reciprocity theorem; iii) the structure is time invariant.

2. Single degree-of-freedom systems. Harmonic vibrations

The dynamic properties of a mechanical system are described by mass, stiffness and damping, responsible respectively for inertia, elastic and dissipative forces. These properties are distributed in space and therefore, modelling a real system is a very complex or even impossible task. However, in most cases, satisfactory results may be achieved if the basic properties are considered as separated into simple discrete elements

1

J.M.M. Silva and N.M.M. Maia (eds.), Modal Analysis and Testing, 1–34.

which, properly combined, can represent the dynamic properties of the system to sufficient accuracy.

Consider a system (Figure 1) with a single degree-of-freedom (SDOF), for which inertia is represented by an infinitely rigid constant mass m, elasticity is represented by an ideal massless spring of constant stiffness k, and damping is represented by an ideal massless viscous damper with constant damping coefficient c).

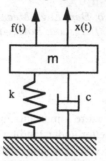

Figure 1 Discretised representation of a SDOF system.

Assuming that f(t) and x(t) are respectively a time dependent excitation force applied to the system and the corresponding displacement response, the spatial model of the system is described by

$$m\,\ddot{x}(t) + c\,\dot{x}(t) + k\,x(t) = f(t) \tag{1}$$

The solution of (1) is the sum of the solution of the corresponding homogeneous equation (where f(t)=0) with a particular integral of the nonhomogeneous equation and can be found in any basic text book. Setting $f(t) = 0$, it may easily be shown that

$$x(t) = e^{-\xi \omega_n t}\left(C_1\, e^{i\omega_n t\sqrt{1-\xi^2}} + C_2\, e^{-i\omega_n t\sqrt{1-\xi^2}}\right) \tag{2}$$

where $i = \sqrt{-1}$. Equation (2) represents the free vibration motion of the system for the particular case where $\xi < 1$ (underdamped system). In (2) we have

$$\omega_n = \sqrt{\frac{k}{m}} \qquad \text{undamped natural frequency} \tag{3}$$

$$\xi = \frac{c}{c_c} \qquad \text{damping ratio} \tag{4}$$

$$c_c = 2\sqrt{k\,m} = 2\,m\sqrt{\frac{k}{m}} = 2\,m\,\omega_n \qquad \text{critical damping coefficient} \tag{5}$$

C_1 and C_2 are constants determined by the initial conditions at $t = 0$. If $c \geq c_c$ ($\xi \geq 1$) there is no oscillating motion and the solution of (1) is given by

$$x(t) = e^{-\omega_n t}\left(C_1 + C_2\, t\right) \qquad \text{for } \xi = 1 \text{ (critically damped system)} \tag{6}$$

$$x(t) = e^{-\xi \omega_n t}\left(C_1\, e^{\omega_n t\sqrt{\xi^2-1}} + C_2\, e^{-\omega_n t\sqrt{\xi^2-1}}\right) \quad \text{for } \xi > 1 \text{ (overdamped system)} \quad (7)$$

While the undamped solution ($\xi = 0$) corresponds to a constant amplitude harmonic motion of frequency ω_n, the oscillating damped solution ($0 < \xi < 1$) describes a behaviour closer to what we observe in real life, decaying exponentially to zero (Figure 2). In this case the frequency of oscillation, known as the damped natural frequency, is given by:

$$\omega_d = \omega_n \sqrt{1-\xi^2} \qquad (8)$$

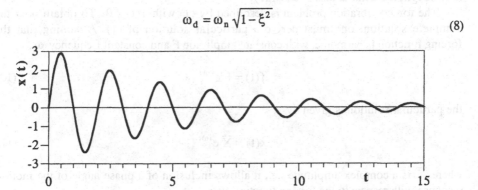

Figure 2 Example of free decaying viscously damped oscillation.

This decaying behaviour is very useful as it may be used as a means to evaluate the dynamic characteristics of a given system. From the plot of a simple free vibration test one may take the value of the peak amplitude X_i at a certain instant of time and the value of the peak amplitude X_{i+n} after n complete cycles of vibration. From these data and from (2) it is possible to derive a quantity known as the logarithmic decrement:

$$\delta_n = \ln \frac{X_i}{X_{i+n}} = \frac{2n\pi\xi}{\sqrt{1-\xi^2}} \qquad (9)$$

There are many practical situations where the dynamic behaviour of real systems may be represented by a SDOF model. The basic problem is the derivation of the spatial properties of the model. This is possible, for example, through the realisation of a simple test yielding a plot of a decaying free vibration. From this plot and from (9) one extracts the value of ξ. Counting n complete cycles of vibration and reading, from the plot, the corresponding time interval, enables us to derive the value of the damped natural frequency ω_d. The next step is the calculation of the undamped natural frequency ω_n from (8). The identification of the spatial properties m, k and c will then be an easy task provided one knows the value of one of these quantities. In most cases it is easy to know (or measure) m and therefore the problem is solved. The value of k will subsequently be obtained from $\omega_n = \sqrt{k/m}$ and finally, the value of c will be derived from (4) and (5).

Extracting dynamic characteristics from experimental data is the aim of the so-called identification procedures. This subject will be dealt with, in more depth, in a different chapter. For the moment it will suffice to emphasise that an experimental plot of the free vibration response of a system is not sufficient for extracting all the dynamic properties.

Another important aspect to emphasise is the fact that, for most real structures, the damping ratio is small (typically below 10%). Thus, it is possible to conclude that the free vibration (also known as transient) component of the complete response solution of equation (1) tends to disappear very rapidly. The consequences of the above reasoning are that, for most of the situations encountered in practice, it is only of interest to consider the particular integral of the forced response solution i.e., the initial transient component may be ignored without loss of accuracy.

The forced vibration problem is described by (1) with $f(t) \neq 0$. To obtain now the complete solution, one must derive a particular solution of (1). Assuming that the forcing function is harmonic, with constant amplitude F and constant frequency ω,

$$f(t) = F e^{i\omega t} \tag{10}$$

the particular solution is given by

$$x(t) = \overline{X} e^{i\omega t} \tag{11}$$

where $\overline{X}$ is a complex amplitude i.e., it allows inclusion of a phase angle of the motion response with respect to the forcing function f(t):

$$\overline{X} = X e^{i\theta} \tag{12}$$

Substituting (11) into (1), it follows that

$$\overline{X} = \frac{F}{\left(k - \omega^2 m\right) + i\omega c} \tag{13}$$

Taking into consideration that any complex number of the form $x + iy$ can be written as $R e^{i\theta}$, with $R = (x^2 + y^2)^{1/2}$ and $\tan\theta = y/x$, we obtain:

$$\overline{X} = \frac{F}{\sqrt{(k - \omega^2 m)^2 + (\omega c)^2}} e^{i\theta} \tag{14}$$

with

$$\tan\theta = \frac{-\omega c}{k - \omega^2 m} \tag{15}$$

The particular solution of (1), for the harmonic forcing function defined by (10), is therefore given by

$$x(t) = \frac{F}{\sqrt{(k - \omega^2 m)^2 + (\omega c)^2}} e^{i(\omega t + \theta)} \tag{16}$$

which is a harmonic function with constant amplitude (as is the exciting force). The response x(t) is delayed with respect to the forcing function f(t), this delay being described, in angular terms, by θ. This solution and the vibration it represents, in this case, are called steady-state solution and steady-state vibration.

The complete solution is therefore given given by:

$$x(t) = e^{-\xi \omega_n t} \left(C_1 \, e^{i \omega_n t \sqrt{1-\xi^2}} + C_2 \, e^{-i \omega_n t \sqrt{1-\xi^2}} \right) + \frac{F}{\sqrt{(k - \omega^2 m)^2 + (\omega c)^2}} \, e^{i(\omega t + \theta)} \quad (17)$$

The previous equation shows that the motions corresponding to the particular solution of equation (1) and to the solution of the homogeneous equation are superposed i.e., they are added algebraically. As stated before, the transient solution is important only for a limited initial period of time and therefore, after enough time has elapsed, practically only the steady-state solution described by (16) remains.

Taking only the steady-state part of the solution of the forced vibration problem, we may represent it graphically as shown in Figure 3. It can be seen that when $\xi = 0$ (i.e., c=0) and $\omega = \omega_n$, the denominator of (16) is zero, meaning that the steady-state vibration has infinite amplitude X no matter how small the exciting force amplitude F is (it is important to note that the infinite amplitude can be shown to take some time to be reached. It is not an instantaneous result). This particular situation is called resonance.

Fortunately, in practice, ξ is never zero because there is always some degree of energy dissipation in real systems. This means that any dynamic model should include a damping mechanism and therefore a non-zero value of c. In this case, the amplitude at resonance is not infinite, though for low damping it can have very large values. It can be proven that the maximum value of the amplitude of the steady-state vibration occurs for $\omega = \omega_n \sqrt{1 - 2\xi^2}$. This particular feature is observable in Figure 3a) by the peak amplitudes occurring to the left of the $\omega = \omega_n$ line, the shift being larger for larger damping values. As stated previously, most real structures have low damping and therefore resonance is usually taken as occurring for $\omega = \omega_n$. The error is below 1% for $\xi = 0.1$ and below 10% for $\xi = 0.5$, thus justifying the assumption. Away from resonance the response is hardly influenced by damping and all the response curves coincide.

In what concerns the phase angle θ, (Figure 3b)) one may note that the response has a phase shift from an initial $0°$ value to a final $-180°$ value when it passes through resonance (where $\theta = -90°$). The meaning of this phase shift is that the response is delayed in time relative to the forcing function. In the ideal theoretical case where ξ is zero, the phase shift is instantaneous whereas it becomes increasingly gradual for larger values of ξ.

An alternative way of looking at the equations derived for the steady-state case is to consider the dynamic properties of our system which are contained in the mathematical expression relating the output x(t) to the input f(t)

$$\frac{x(t)}{f(t)} = H(\omega) = \frac{X}{F} = \frac{1}{(k - \omega^2 m) + i \omega c} \quad (18)$$

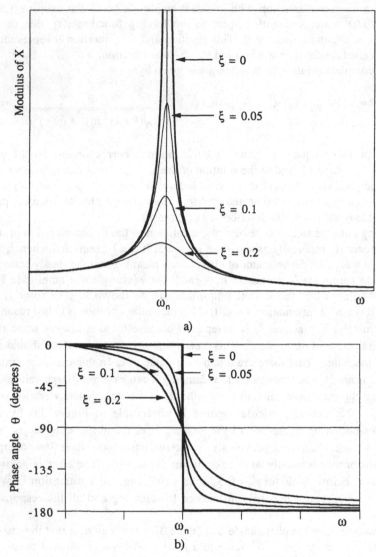

Figure 3 Graphical display of the modulus and phase of X.

The complex function of the frequency denoted by $H(\omega)$ is called the system's Frequency Response Function (FRF). This particular FRF where the response is described in terms of the displacement is known as Receptance.

3. Hysteretic damping as an alternative dissipation model

The dissipative process, in real vibrating systems, is the simultaneous result of several different mechanisms and is difficult to identify and to model accurately. By including a viscous damper in our SDOF model we are simply trying to represent the dissipative mechanism through the use of an equivalent linear element. The viscous damper is the simplest damping element from a theoretical point of view. It is the only strictly linear damper, in the sense that the equations of motion of a system incorporating this damping mechanism may be solved for any type of input. This characteristic makes viscous damping very simple to deal with mathematically and the consequence is its widespread use in most textbooks and papers. In actual fact, it often bears little resemblance to damping mechanisms encountered in practice, except if it represents an actual source of viscous damping such as an oil-filled dashpot.

The viscous damper opposes the relative velocity between its ends with a force which is proportional to that velocity ($f = c\dot{x}$). Considering the system represented in Figure 4, the corresponding harmonic load/deflection curve exhibits an elliptic loop denoting the energy dissipation phenomenon. The energy ΔE dissipated per cycle of oscillation is given by the area enclosed in the oscillation loop, i.e.,

$$\Delta E = \int_0^{2\pi/\omega} f(x)\,dx = \pi\,X^2\,c\,\omega \qquad (19)$$

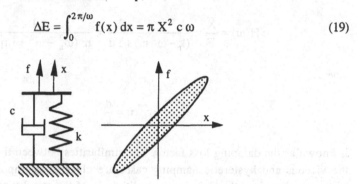

Figure 4 Load/deflection curve for a system with viscous damping.

Therefore, the energy dissipated per cycle of oscillation is frequency dependent. This frequency dependence differs from that observed in most common materials and real structures whose behaviour is found to be closer to a frequency independent (or weakly dependent) dissipation mechanism which is roughly proportional to the square of the displacement amplitude. Though rubber and other viscoelastic materials do exhibit frequency dependence, it is far less pronounced than that associated with a viscous damper and, over a limited range of frequencies, the properties may often be assumed constant. Thus, what is required is a damper model which opposes the relative motion between its ends with a force that is proportional to displacement and not to velocity (though still in phase with velocity). This is equivalent to using a viscous damper but making the viscous damping rate vary inversely with frequency, i.e., $c = d/\omega$. This is known as a hysteretic, solid or structural damper and the parameter d is called the hysteretic damping coefficient. This designation results from the fact that such a mechanism closely describes the load/deflection hysteresis behaviour of most materials.

The hysteretic damper is only a linear device in the sense that it is described by a linear frequency response relationship. Its use is limited to steady-state vibrations as it presents difficulties to rigorous free vibration or shock response analyses. In such cases there is little option but to return to the viscous damper, however inadequate this may be.

This alternative damping model has the advantage of not only describing more closely the energy dissipation mechanism exhibited by most real structures, but also of providing a much simpler analysis for multiple degree-of-freedom systems. In this case, the energy dissipated per cycle of oscillation is given by

$$\Delta E = \int_0^{2\pi/\omega} f(x) \, dx = \pi X^2 \, d \tag{20}$$

and the steady-state equation of motion of a hysteretically damped SDOF system, harmonically excited, may be written as

$$(-\omega^2 m + k + i \, d) \, \overline{X} \, e^{i\omega t} = F \, e^{i\omega t} \tag{21}$$

from which we obtain

$$H(\omega) = \frac{\overline{X}}{F} = \frac{1}{(k - \omega^2 m) + i \, d} = \frac{1}{m \, (\omega_n^2 - \omega^2 + i \, \eta \, \omega_n^2)} \tag{22}$$

where

$$\eta = \frac{d}{k} \tag{23}$$

is known as the damping loss factor. The similarities between the FRF expressions for the viscous and hysteretic damping cases are clear. It is important to note that, for hysteretic damping, the maximum amplitude X of the steady-state response is always obtained for $\omega = \omega_n$ while for viscous damping the maximum occurs at $\omega = \omega_n \sqrt{1 - 2\xi^2}$.

However, and as stated before for low damping values, a system with viscous damping may be assumed with sufficient accuracy as exhibiting its maximum steady-state response amplitude at $\omega = \omega_n$ and therefore its response amplitude behaviour may be assumed as equivalent to the behaviour of a hysteretically damped system. Thus, it may be concluded that the viscous and hysteretic models are approximately equivalent with $\eta = 2 \, \xi$.

4. Nonharmonic excitation. Fourier analysis

Excitations may be of many different types other than harmonic. In fact if one considers real excitation sources such as, for example, earthquakes, wind, road surfaces and all types of machinery, it is easy to understand that the forcing functions may only be harmonic in very particular cases.

Dynamic signals may be generally classified as deterministic or random. The former can be described by an analytical expression of their magnitude, as a function of time, while the latter cannot. Random signals are characterised by analysing their statistical properties and may be classified as stationary (i.e., they have constant statistical properties along their length) and non-stationary. Deterministic signals may be periodic or transient. A periodic signal is one that repeats itself after a period T of time, i.e., $f(t) = f(t + T)$. A transient signal is one that occurs only during a short period of time. If a function is periodic and satisfies certain conditions then it can be represented by a summation of harmonic functions known as the Fourier series. Hence, Fourier analysis of periodic functions yields discrete frequency spectra representing the amplitudes (and phases) of the discrete harmonic components plotted against frequency.

As we are considering linear behaviour, the mathematical statement of the principle of superposition applies. The validity of the application of the principle of superposition means that the response of a linear system to a periodic forcing function f(t) can be obtained by adding the responses to the separate harmonic forcing functions obtained from the decomposition of f(t) into its harmonic components. Each distinct steady-state response is obtained from the application of (16). Though a Fourier series have an infinite number of harmonic terms, in practice, consideration of just a few (typically less than 10) initial terms of the series yields results that are sufficiently accurate for most applications.

When the forcing functions are nonperiodic (transient), they cannot be handled directly through the use of Fourier series. However, it is not difficult to accept that a transient signal may be viewed as a periodic signal with period $T = \infty$. By considering the limit which is approached by a Fourier series as the period becomes infinite, it will be found that, under certain conditions, an arbitrary function f(t) can be described by an integral $F(\omega)$ given by:

$$F(\omega) = \mathcal{F}\left[f(t)\right] = \int_{-\infty}^{+\infty} f(t)\, e^{-i\omega t}\, dt \qquad (24)$$

where $F(\omega)$ is known as the Fourier transform of f(t). Conversely,

$$f(t) = \mathcal{F}^{-1}\left[F(\omega)\right] = \frac{1}{2\pi} \int_{-\infty}^{+\infty} F(\omega)\, e^{i\omega t}\, d\omega \qquad (25)$$

Equations (24) and (25) constitute what is known as a Fourier transform pair. Applying this same reasoning to equation (18), f(t) representing an arbitrary nonperiodic excitation, will yield

$$X(\omega) = H(\omega)\, F(\omega) \qquad (26)$$

i.e., the Fourier transform of the response is simply the product of the complex frequency response function $H(\omega)$ and the Fourier transform of the excitation. The response x(t) is then obtained from $X(\omega)$:

$$x(t) = \mathcal{F}^{-1}\left[X(\omega)\right] = \frac{1}{2\pi} \int_{-\infty}^{+\infty} X(\omega)\, e^{i\omega t}\, d\omega \qquad (27)$$

The frequency spectrum is now a continuous function of ω, in contrast to the frequency spectrum obtained for periodic time functions which consists of discrete components only. As a consequence, the $F(\omega)$ values represent an amplitude which is continuously distributed along the frequency and therefore represent units of amplitude per unit frequency, i.e., what is known as a spectral density.

To obtain $x(t)$ it is then necessary to evaluate the integral in (27) which often leads to difficulties from a mathematical point of view. On the other hand, there are a number of situations where the Fourier transform analysis is inadequate, yielding completely meaningless solutions of the integral (as in the case where $f(t)$ is a step function, for example). Attempting to solve these situations, through adequate mathematical manipulations, results in the use of a 'modified' Fourier transform, known as the Laplace transform, which can be easily found in the appropriate litterature.

In practice, the forcing function may be quite irregular, even if it is periodic, and may be determined only experimentally. Such cases correspond to having a graphical representation of the signal and no analytical expression to describe it. These situations can still be handled by means of adequate discretisation and numerical procedures applied to the signal.

5. Random excitation

Random signals cannot be treated in the same way as deterministic signals (they are nonperiodic and do not obey the Dirichlet condition). Given their properties, the analysis of random signals entails the use of probabilistic concepts. The way to avoid some mathematical difficulties is to assume that our random signals are stationary and ergodic, i.e., both averaging across many time history records at a given instant in time and averaging over time using just one time history give the same mean properties (mean, mean square and statistical distributions). This assumption is made in most practical cases and is one that we are going to consider.

Let us then take a signal $f(t)$ as our random forcing function and compute, along the time axis, the average or 'expected' value of the product $f(t)f(t+\tau)$

$$R_{ff}(\tau) = \lim_{T \to \infty} \frac{1}{T} \int_{-T/2}^{+T/2} f(t)f(t+\tau).dt \qquad (28)$$

where $f(t+\tau)$ designates the magnitude of the function $f(t)$ observed after a time delay τ has elapsed. Equation (28) is known as the auto-correlation function. In physical terms, it describes how a particular instantaneous amplitude value of our random time signal depends upon previously occurring instantaneous amplitude values. Thus, the original stationary random time history has been transformed into a new function of time, $R_{ff}(\tau)$, which is even, real valued, goes to zero as τ becomes large and obeys the Dirichlet condition. As a consequence, $R_{ff}(\tau)$ can be Fourier transformed:

$$S_{ff}(\omega) = \mathcal{F}\left[R_{ff}(\tau)\right] = \int_{-\infty}^{+\infty} R_{ff}(\tau) e^{-i\omega\tau} d\tau \qquad (29)$$

$S_{ff}(\omega)$ is known as the auto-spectral density (ASD) or power spectral density (PSD), which is also a real and even function of frequency. Therefore, we can describe a stationary random process through the following Fourier transform pair:

$$S_{ff}(\omega) = \int_{-\infty}^{+\infty} R_{ff}(\tau)\, e^{-i\omega\tau}\, d\tau$$

$$R_{ff}(\tau) = \frac{1}{2\pi} \int_{-\infty}^{+\infty} S_{ff}(\omega)\, e^{i\omega\tau}\, d\omega \tag{30}$$

The ASD provides a frequency description of the original signal f(t). However, its physical meaning is not exactly the same as the one provided by the Fourier transform of a transient signal. In the case of a random process, the physical meaning of the ASD can be better understood if we set $\tau = 0$ in (28) and combine the result with (30):

$$R_{ff}(0) = \lim_{T\to\infty} \frac{1}{T} \int_{-T/2}^{+T/2} f^2(t)\, dt = \frac{1}{2\pi} \int_{-\infty}^{+\infty} S_{ff}(\omega)\, d\omega \tag{31}$$

Thus, the ASD has units of amplitude mean square per unit frequency and, for that reason, it is sometimes referred to as the mean square spectral density. What is now described, continuously along the frequency range, is not the signal amplitude but rather a quantity squared that can be taken as an energy content indicator. This is the reason why this function is also called the power spectral density. The previous concepts can be extended in order to consider simultaneously the random force and the random response functions. We can therefore define

$$R_{fx}(\tau) = \lim_{T\to\infty} \frac{1}{T} \int_{-T/2}^{+T/2} f(t)\, x(t+\tau)\, dt = \frac{1}{2\pi} \int_{-\infty}^{+\infty} S_{fx}(\omega)\, e^{i\omega\tau}\, d\omega$$

$$S_{fx}(\omega) = \int_{-\infty}^{+\infty} R_{fx}(\tau)\, e^{-i\omega\tau}\, d\tau \tag{32}$$

and

$$R_{xf}(\tau) = \lim_{T\to\infty} \frac{1}{T} \int_{-T/2}^{+T/2} x(t)\, f(t+\tau)\, dt = \frac{1}{2\pi} \int_{-\infty}^{+\infty} S_{xf}(\omega)\, e^{i\omega\tau}\, d\omega$$

$$S_{xf}(\omega) = \int_{-\infty}^{+\infty} R_{xf}(\tau)\, e^{-i\omega\tau}\, d\tau \tag{33}$$

as the cross-correlation and the cross-spectral density functions, respectively. It is important to note that the cross-spectral densities are complex frequency spectra, containing real and imaginary parts (and therefore magnitude and phase information) whereas the PSD is a real function containing only magnitude (squared) information. Also, it can be seen that the cross-correlation functions obey the relationship

$$R_{xf}(\tau) = R_{fx}(-\tau) \tag{34}$$

and that the cross-spectral density functions are complex conjugates:

$$S_{xf}(\omega) = S_{fx}^*(\omega) \tag{35}$$

Having established the previous concepts, it is possible to go through a series of mathematical manipulations and demonstrate that:

$$S_{xx}(\omega) = |H(\omega)|^2 \, S_{ff}(\omega) \qquad\qquad (36)$$

$$S_{fx}(\omega) = H(\omega) \, S_{ff}(\omega) \qquad\qquad (37)$$

$$S_{xx}(\omega) = H(\omega) \, S_{xf}(\omega) \qquad\qquad (38)$$

Equations (37) and (38) are alternative adequate relationships that enable the frequency response function of a system to be determined from knowledge of the input force and output response characteristics.

6. FRFs representation and properties (SDOF)

The previously defined frequency response function $H(\omega)$ is just one of the possible forms of an FRF. It is called Receptance and is usually represented by $\alpha(\omega)$ or $\alpha(i\omega)$. This complex quantity fully describes the relationship between the displacement response and the excitation force applied to a system and thus fully characterises its dynamic properties.

As $\alpha(\omega)$ is a complex function of the frequency, there are three quantities (real part, imaginary part and frequency) to be taken into account whenever plotting FRF data. Thus, a full representation of an FRF in a single plot should be done using a three-dimensional display. It is obvious that this is not a convenient way of graphically representing the FRF. As an alternative, we may display the FRF data in two separate plots - real and imaginary parts against frequency - as shown in Figure 5. It is interesting to note that the real part of the receptance $\alpha(\omega)$ crosses the frequency axis at resonance while, in the same frequency region, the imaginary part reaches a minimum (this is only exactly true for hysteretic damping. In the case of viscous damping, the statement may only be taken as true for low damping values).

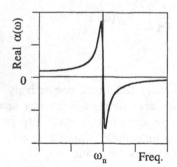 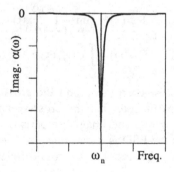

Figure 5 Real and imaginary parts of the receptance plotted against frequency.

The most common representation of a frequency response function is the Bode diagram, where the FRF magnitude and phase (instead of real and imaginary parts) are plotted as functions of the frequency (Figure 6). This representation allows an easy visual interpretation of the information contained in $\alpha(\omega)$.

Finally, if one plots $\alpha(\omega)$ in the Real/Imag. plane (complex plane) the result is a circular loop that contains all the information. However, in such a plot, one is normally unable to identify the frequency value corresponding to any point on the curve unless each data point (as shown in Figure 7) is accompanied by a caption indicating the corresponding frequency value. This representation is known as a Nyquist plot and it has the particular advantage of enhancing the resonant region as the circular loop occurs only close to resonance (180° phase shift of the FRF).

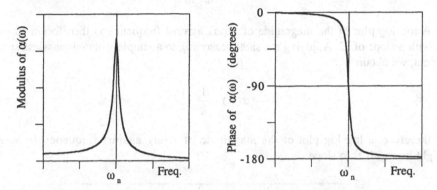

Figure 6 Modulus and phase of the receptance plotted against frequency.

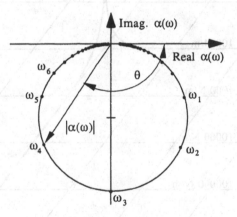

Figure 7 Nyquist plot of the receptance.

The Bode plots are the most commonly used in modal analysis though the magnitude-frequency graph is not presented as in Figure 6 but rather using logarithmic scales or, at least, a magnitude (vertical) logarithmic scale. This is due to the fact that the dynamic range of the responses may very easily cover a relatively wide range of values and, as a consequence, linear scales tend to completely lose detail at lower levels of the response.

If logarithmic scales are used both for the magnitude and frequency axes, the data that are displayed as curves on linear scales become asymptotic to straight lines. This feature provides a simple means of checking the validity of a plot and also allows for easily

14

identifying the mass and stiffness characteristics of the system under study. For example, it can easily be seen that, if we consider a rigid mass m, free in space, upon which a harmonic force f is applied, the receptance of this simple system is given by

$$\alpha(\omega) = -\frac{1}{\omega^2 m} \qquad (39)$$

A log-log plot of the magnitude of $\alpha(\omega)$ against frequency is therefore a straight line with a slope of -2. Applying the same reasoning to a simple isolated massless spring element, we obtain

$$\alpha(\omega) = \frac{1}{k} \qquad (40)$$

and therefore, a log-log plot of the magnitude of $\alpha(\omega)$ against frequency is now a straight line with zero slope.

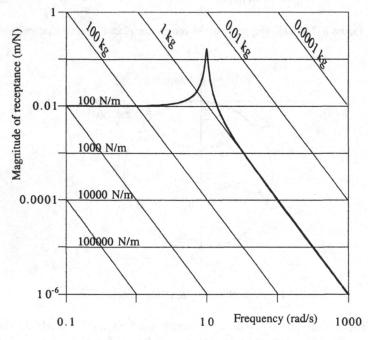

Figure 8   Example of an annotated log-log plot of the magnitude of the receptance *versus* frequency .

In order to illustrate these properties of FRF log-log plots, Figure 8 displays the magnitude of the receptance of a SDOF system for which m=1 kg, k=100 N/m and c=0.6 Ns/m. Constant mass and constant stiffness lines have been added to the plot. It is obvious that, outside the resonance region, the receptance is asymptotic to straight mass and stiffness lines. It is very easy to understand the physical meaning of this: while at very low frequencies (below resonance), the system approaches a static load/deflection behaviour which is dominated by the spring stiffness, at frequencies above resonance it is the inertia of the mass that dominates the system response.

It is therefore obvious that one can extract the mass and stiffness characteristics of a SDOF system from a log-log plot of experimental data. Damping characteristics can be obtained as well, as will be explained further on. Though at a very simple level, this reasoning can be viewed as a first step towards the so-called system identification techniques which aim at deriving the dynamic characteristics from experimental data.

Recalling that stiffness and inertia effects cancel each other ($k - \omega_n^2 m = 0$) when the system is oscillating at its natural frequency, it is only natural that the stiffness and mass lines intersect at a point corresponding to the natural frequency of the system. Under these conditions, the only dominant force left in the system to counterbalance the applied force is the damping force.

It is important to point out that magnitude log scales are not normally expressed in linear units (as in Figure 8) but rather in dB (decibel). The dB log scale used in vibration analysis is defined in analogy to the dB log scale used in acoustics. Taking a given variable such as, say, the receptance amplitude α, its value in dB is defined as

$$\alpha(dB) = 20 \log_{10}\left(\frac{\alpha}{\alpha_{ref}}\right) \tag{41}$$

where α_{ref} is a reference value that has to be known beforehand. When there is no indication of the reference value to be used it is usual to assume it as unity.

Usually vibration is measured in terms of motion and therefore the corresponding FRF may also be presented in terms of velocity or acceleration and not necessarily in terms of displacement as we have done so far. The terminology used in this text is:

$$\alpha(\omega) = \frac{\text{displacement response}}{\text{force excitation}} = \text{Receptance}$$

$$Y(\omega) = \frac{\text{velocity response}}{\text{force excitation}} = \text{Mobility}$$

$$A(\omega) = \frac{\text{acceleration response}}{\text{force excitation}} = \text{Accelerance}$$

The designation Mobility is also widely accepted as a general designation for any of the motion/force FRF forms.

Displacement, velocity and acceleration are mathematically interrelated response quantities. Therefore, knowledge of an FRF in terms of any one of the motion parameters will allow immediate derivation of any of the other FRF forms. Considering harmonic vibration:

$$Y(\omega) = \frac{\dot{x}(t)}{f(t)} = \frac{i \omega \overline{X} e^{i\omega t}}{F e^{i\omega t}} = i \omega \frac{\overline{X}}{F} = i \omega \, \alpha(\omega) \tag{42}$$

$$A(\omega) = \frac{\ddot{x}(t)}{f(t)} = \frac{-\omega^2 \, \overline{X} \, e^{i\omega t}}{F \, e^{i\omega t}} = -\omega^2 \, \alpha(\omega) \tag{43}$$

It follows that log-log plots of mobility or acceleration will show some differences with respect to the receptance plot, resulting from the fact that mass and stiffness, though still displayed as straight lines, have different slopes. This is shown in Figure 9 where the mobility and accelerance magnitudes of the system described by Figure 8 are plotted against frequency (a dB magnitude scale is now used).

One must not forget that the magnitude plot does not contain all the information. There is still the need to consider the phase or argument of the complex FRF. Phase plots may use logarithmic scales only for the frequency axis. All present a phase shift of $180°$ (viscous damping model) near resonance and they differ only on the vertical axis range.

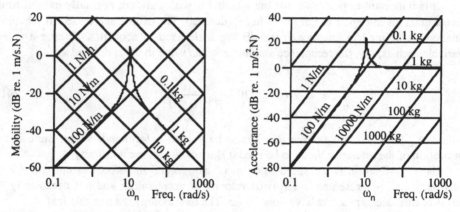

Figure 9   Annotated log-log plots of the magnitudes of mobility and accelerance *versus* frequency (same system as in Figure 8).

Recalling the Nyquist plot of receptance (Figure 7), similar plots can be obtained for the other FRF forms (Figure 10). An interesting feature of FRF Nyquist plots results from the shape of the path traced out by the data. It is clear from Figures 7 and 10 that, in each plot, the data describes a loop that looks like a circle. This particular behaviour is shown by both the hysteretic and viscous damped models that look very similar at a first glance. In fact, as will be shown later, hysteretically damped systems yield FRF data that plot exactly as a circle when receptance is considered. Mobility and Acceleration data plots are distorted circles. For viscously damped systems, it is the mobility which traces out an exact circle while receptance and acceleration trace out as distorted circles. In both cases, the amount of distortion depends on the damping values.

Evaluation of the damping value from an FRF plot may be based on the use of the so-called half-power points. Taking a SDOF system with hysteretic damping under steady-state harmonic vibration, the energy dissipated per cycle of oscillation at resonance is given by

$$\Delta E_{max} = \pi \, X_{max}^2 \, d = \pi \, |\alpha(\omega)|_{max}^2 \, F^2 \, k \, \eta \tag{44}$$

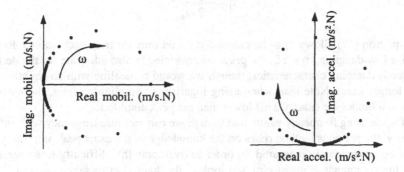

Figure 10 Nyquist plots of mobility and accelerance.

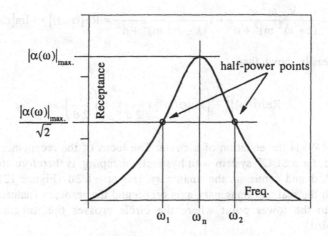

Figure 11 Half-power points definition.

Now, considering measurements on the flanks of the response peak (where damping is still effective) for which the energy dissipated per cycle of vibration is half of ΔE_{max}, we have two such points (Figure 11). For these data points, denoted by subscripts 1 and 2,

$$\Delta E_{1,2} = \frac{\Delta E_{max}}{2} \tag{45}$$

and therefore

$$|\alpha(\omega)|_{1,2} = \frac{|\alpha(\omega)|_{max}}{\sqrt{2}} \tag{46}$$

These points are called the half-power points and it can be easily found that

$$\eta = \frac{\omega_2^2 - \omega_1^2}{2\,\omega_n^2} \tag{47}$$

Equation (47) allows η to be calculated based only on frequency values. Recalling that, for low damping, $\eta \approx 2\xi$, the previous reasoning is also adequate for the derivation of viscous damping characteristics, though we would be dealing with an equation which is no longer exact. Note that, when using logarithmic amplitude scales, the half-power points amplitudes are exactly 3 dB lower than the peak amplitude.

Despite using frequency values and though we can measure frequency with sufficient accuracy, the previous method relies on the knowledge of the exact peak amplitude which cannot be so accurately measured. In order to overcome this difficulty it is necessary to recall the receptance Nyquist plot and look at its shape. Let us take equation (22) and rewrite it as:

$$\alpha(\omega) = \frac{k - \omega^2\,m}{(k - \omega^2\,m)^2 + d^2} - i\,\frac{d}{(k - \omega^2\,m)^2 + d^2} = \mathrm{Re}\left[\alpha(\omega)\right] + i\,\mathrm{Im}\left[\alpha(\omega)\right] \tag{48}$$

It can be easily shown that

$$\mathrm{Re}\left[\alpha(\omega)\right]^2 + \left\{\mathrm{Im}\left[\alpha(\omega)\right] + \frac{1}{2\,d}\right\}^2 = \left(\frac{1}{2\,d}\right)^2 \tag{49}$$

Equation (49) is the equation of a circle. The locus of the receptance data on the complex plane, for a SDOF system with hysteretic damping, is therefore an exact circle with radius $1/2\,d$ and centre on the imaginary axis at $-1/2\,d$ (Figure 12). The circle passes through the real and imaginary axis origin and the resonant (natural) frequency corresponds to the lower point where the circle crosses the imaginary axis (at $\mathrm{Im}\left[\alpha(\omega)\right] = -1/d$).

This means that, even without having measured it, one may derive both the resonant frequency and the resonant amplitude directly from the Nyquist plot. Due to the geometric properties of this plot, the half-power points (ω_1 and ω_2) are also easy to locate as they correspond to the points where the circle is intersected by its diameter parallel to the real axis.

Mobility and accelerance data for hysteretically damped systems do not plot as exact circles on the complex plane. The same is true if one considers receptance and accelerance for viscously damped systems. However, the mobility FRF of a SDOF system with viscous damping will plot as an exact circle in the Argand plane. In fact, for such a system, it can be shown that the the mobility satisfies the relationship

$$\left\{\mathrm{Re}[Y(\omega)] - \frac{1}{2\,c}\right\}^2 + \left\{\mathrm{Im}[Y(\omega)]\right\}^2 = \left(\frac{1}{2\,c}\right)^2 \tag{50}$$

Thus, a plot of the real part against the imaginary part of the mobility $Y(\omega)$ traces out a circle, on the Argand plane. This circle is of radius $1/2\,c$ and its centre is on the real axis at $\mathrm{Re}[Y(\omega)] = 1/2\,c$. The resonant frequency ω_n is derived from the point where the circle crosses the real axis.

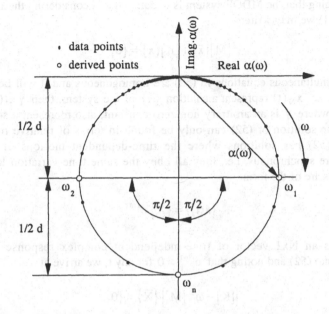

Figure 12 Nyquist plot of the receptance (hysteretically damped system) superimposed to a circle.

7. Multiple degree-of-freedom (MDOF) systems

Real structures are continuous and nonhomogeneous elastic systems which have an infinite number of degrees of freedom. Therefore, their analysis always entails an approximation which consists of describing their behaviour through the use of a finite number of degrees of freedom, as many as necessary to ensure enough accuracy. Usually, continuous structures are described as discretised (also referred to as lumped-mass) multiple degree-of-freedom (MDOF) systems.

Any basic text book on mechanical vibrations will show that, assuming that each mass may be forced to move by an external force $f_i(t)$ $(i = 1, 2, ..., N)$ and establishing the equilibrium of the forces acting on them, the motion of the system is governed by N second order differential equations, each of which requires two initial conditions in order to resolve the complete response at the N motion coordinates. Neither equation can be solved by itself because they are coupled, i.e., the motion response at a single coordinate depends on the motion at the other coordinates. A convenient method for solving such a system of equations is to use matrices:

$$[M]\{\ddot{x}\} + [C]\{\dot{x}\} + [K]\{x\} = \{f\} \tag{51}$$

where [M], [C] and [K] are NxN mass, damping and stiffness symmetric matrices, respectively, describing the spatial properties of the system. The column matrices $\{\ddot{x}\}$,

$\{\ddot{x}\}$ and $\{x\}$ are Nx1 vectors of time-varying acceleration, velocity and displacement responses, respectively, and $\{f\}$ is an Nx1 vector of the time-varying external excitation forces. Assuming that the MDOF system is undamped and considering the free vibration solution of (51) we may write:

$$[M]\{\ddot{x}\}+[K]\{x\} = \{0\} \tag{52}$$

The N simultaneous equations in (52) are homogeneous and it will be seen that if $x_1(t)$, $x_2(t)$, ..., $x_N(t)$ represent a solution $\{x\}$ of the system, then $\gamma x_1(t)$, $\gamma x_2(t)$, ..., $\gamma x_N(t)$, where γ is an arbitrary non-zero constant, also represent a solution. This means that the solution of (52) can only be found in terms of relative motions. It is known that (52) has solutions where the time-dependent motions of the system coordinates are synchronous, i.e., they all obey the same time-variation law, and that those solutions are of the form

$$\{x(t)\} = \{\overline{X}\}\, e^{i\omega t} \tag{53}$$

where $\{\overline{X}\}$ is an Nx1 vector of time-independent complex response amplitudes. Substituting into (52) and noting that $e^{i\omega t} \neq 0$ for any t, we arrive at

$$\left[[K] - \omega^2\,[M]\right]\{\overline{X}\} = \{0\} \tag{54}$$

For (54) to have a non-trivial solution, it is necessary to satisfy

$$\det\left[[K] - \omega^2\,[M]\right] = 0 \tag{55}$$

where 'det' stands for determinant. This algebraic equation, known as the characteristic equation, yields N possible positive real solutions ω_1^2, ω_2^2, ..., ω_N^2, known as the eigenvalues of (55) which are the undamped natural frequencies of the system. Substituting each natural frequency value in (54) and solving each of the resulting sets of equations for $\{\overline{X}\}$, we obtain N possible vector solutions $\{\psi_r\}$ (r = 1, 2, ..., N), known as the mode shapes of the system under analysis, which are the eigenvectors of our problem. Each $\{\psi_r\}$ contains N elements that are real quantities (positive or negative) and are only known in relative terms. Therefore we know the direction of the vectors but not their absolute magnitude.

What was found is that the system can vibrate freely, with synchronous motion, for N particular frequency values ω_r, each of which implies a particular configuration or 'shape' of the free motion, described by $\{\psi_r\}$. Each pair ω_r and $\{\psi_r\}$ is known as a mode of vibration of the system and $\{\psi_r\}$ is usually called the mode shape. The subscript r denotes the mode number and varies from 1 to N.

A graphical representation of a model in its static equilibrium position, superimposed on the same model with the coordinates displaced by values proportional to the values of the elements of $\{\psi_r\}$, is often used as it gives a clear view of how the system moves at that particular mode. This representation is very easy to perform given the fact that the elements in $\{\psi_r\}$ are real (positive or negative), the change in their sign indicating a phase shift of $180°$, i.e., that the motion is in opposite directions.

The complete free vibration solution is very often expressed in two NxN matrices

$$\begin{bmatrix} \ddots \omega_r^2 \ddots \end{bmatrix} = \begin{bmatrix} \omega_1^2 & 0 & \cdots & 0 \\ 0 & \omega_2^2 & \cdots & 0 \\ \vdots & \vdots & \ddots & \vdots \\ 0 & 0 & \cdots & \omega_N^2 \end{bmatrix} \tag{56}$$

and

$$[\Psi] = \begin{bmatrix} \{\psi_1\} & \{\psi_2\} & \cdots & \{\psi_N\} \end{bmatrix} \tag{57}$$

which contain a full description of the dynamic characteristics of the system. Equations (56) and (57) constitute what is known as the Modal Model, i.e., they describe the system through its modal properties (natural frequencies and mode shapes), as opposed to the Spatial Model where the system was described by its spatial properties ([M], [C] and [K]). [Ψ] is commonly known as the modal matrix.

The mode shape vectors possess very important properties known as the orthogonality properties which are described by

$$\{\psi_s\}^T [M] \{\psi_r\} = 0; \qquad (r \neq s) \tag{58}$$

$$\{\psi_s\}^T [K] \{\psi_r\} = 0; \qquad (r \neq s) \tag{59}$$

and

$$\{\psi_r\}^T [K] \{\psi_r\} = \omega_r^2 \{\psi_r\}^T [M] \{\psi_r\} \tag{60}$$

or

$$\omega_r^2 = \frac{\{\psi_r\}^T [K] \{\psi_r\}}{\{\psi_r\}^T [M] \{\psi_r\}} = \frac{k_r}{m_r} \tag{61}$$

where k_r and m_r are commonly known as the modal or generalised stiffness and mass, respectively, of mode r.

Thus, considering all the possible combinations of r and s we may state the modal model orthogonality properties as follows:

$$[\Psi]^T [M] [\Psi] = \begin{bmatrix} \ddots & m_r & \ddots \end{bmatrix}$$

$$[\Psi]^T [K] [\Psi] = \begin{bmatrix} \ddots & k_r & \ddots \end{bmatrix} \tag{62}$$

The mode shape vectors, due to their orthogonality properties, are linearly independent and therefore they form a basis in the N-dimensional space. As a consequence, any other vector in the same space can be expressed as a linear combination of the N linearly independent mode shape vectors.

It has been shown that the mode shapes are known within an indeterminate scaling factor. Thus, k_r and m_r cannot be taken separately as their values are also known within a scaling factor. It is the ratio $k_r/m_r = \omega_r^2$ that has a well defined value. Presentation of the mode shape vectors is therefore always subjected to a previous scaling or normalisation procedure. This normalisation is often based on making the largest element in each vector equal to unity. However, in modal analysis, it is common to scale the mode shape vectors so that

$$[\Phi]^T [M][\Phi] = [I] \tag{63}$$

where $[I]$ is the identity matrix and $[\Phi]$ is the mass-normalised modal matrix built up from mode shape vectors $\{\phi_r\} = \gamma_r \{\psi_r\}$ each of which obey the relationship

$$\{\phi_r\}^T [M]\{\phi_r\} = \{\gamma_r \psi_r\}^T [M]\{\gamma_r \psi_r\} = \gamma_r^2 \{\psi_r\}^T [M]\{\psi_r\} = 1 \tag{64}$$

for each mode r. Therefore,

$$\gamma_r = \frac{1}{\sqrt{\{\psi_r\}^T [M]\{\psi_r\}}} = \frac{1}{\sqrt{m_r}} \tag{65}$$

Therefore, the mass-normalised modal matrix orthogonality properties may be described by:

$$[\Phi]^T [M][\Phi] = [I]$$
$$[\Phi]^T [K][\Phi] = \left[\,\ddots\, \omega_r^2 \,\ddots\, \right] \tag{66}$$

These particular properties of the modal matrix may be used to our advantage in order to find the free vibration solution of (52). Defining the coordinate transformation

$$\{x(t)\} = [\Phi]\{q(t)\} \tag{67}$$

substituting into (52) and performing some mathematical manipulations, it can be shown that one arrives at

$$\{\ddot{q}(t)\} + \left[\,\ddots\, \omega_r^2 \,\ddots\, \right]\{q(t)\} = \{0\} \tag{68}$$

which represents a set of N uncoupled SDOF equations of motion. Thus, through a simple coordinate transformation, our MDOF system has been transformed into N independent SDOF systems, that can be solved separately. After solving for $q_i(t)$, the final free vibration solution, in terms of $x_i(t)$, is easily obtained through the coordinate transformation (67). The response coordinates $\{q(t)\}$ are known as the modal or principal coordinates and the mode shape vectors $\{\phi_r\}$ are said to represent the normal modes of the system.

Recalling now the equations of motion (51) of our general MDOF system with viscous damping, assuming $\{f\} = \{0\}$ and applying the same techniques as before, based on the modal matrix for the undamped system, we obtain

$$[\Phi]^T[M][\Phi]\{\ddot{q}(t)\} + [\Phi]^T[C][\Phi]\{\dot{q}(t)\} + [\Phi]^T[K][\Phi]\{q(t)\} = \{0\} \qquad (69)$$

or

$$\{\ddot{q}(t)\} + [\mathcal{C}]\{\dot{q}(t)\} + \left[\,{}^{\backprime}\omega_r^2\,{}_{\backprime}\right]\{q(t)\} = \{0\} \qquad (70)$$

where $[\mathcal{C}]$ is, in general, a non-diagonal NxN matrix. This characteristic is explainable by the fact that the modal matrix $[\Phi]$ was derived using only mass and stiffness information. It might be said that the mode shape vectors $\{\phi_r\}$ 'knew' nothing about $[C]$ when they were calculated, so there is no reason for them to diagonalise the damping matrix. We are, therefore, confronted with a difficult problem due to the fact that damping is providing additional coupling between the equations of motion which cannot be decoupled by the above modal transformation.

However, if damping is proportional i.e., the viscous damping matrix $[C]$ is directly proportional to the stiffness matrix, to the mass matrix or to a linear combination of both, we may write

$$[C] = \varepsilon[K] + \nu[M] \qquad (71)$$

where ε and ν are constants. It is immediately obvious that, for this case, the undamped modal matrix orthogonality properties will lead to

$$\{\ddot{q}(t)\} + [\Phi]^T[\varepsilon[K] + \nu[M]][\Phi]\{\dot{q}(t)\} + \left[\,{}^{\backprime}\omega_r^2\,{}_{\backprime}\right]\{q(t)\} = \{0\} \qquad (72)$$

or

$$\{\ddot{q}(t)\} + \left[\,{}^{\backprime}\nu + \varepsilon\omega_r^2\,{}_{\backprime}\right]\{\dot{q}(t)\} + \left[\,{}^{\backprime}\omega_r^2\,{}_{\backprime}\right]\{q(t)\} = \{0\} \qquad (73)$$

and therefore, taking the SDOF system for analogy, we may write

$$\{\ddot{q}(t)\} + \left[\,{}^{\backprime}2\xi_r\omega_r\,{}_{\backprime}\right]\{\dot{q}(t)\} + \left[\,{}^{\backprime}\omega_r^2\,{}_{\backprime}\right]\{q(t)\} = \{0\} \qquad (74)$$

where

$$\xi_r = \frac{\nu}{2\omega_r} + \frac{\varepsilon\omega_r}{2} \qquad r = 1, 2, ..., N \qquad (75)$$

is defined as a modal damping ratio for mode r. As in the case of the undamped system, we have now a set of N uncoupled damped SDOF equations each of which can be solved separately. Again, the final free vibration response will be derived using the coordinate transformation relationship and knowledge of the initial conditions.

In general, damping is not proportional. In many situations, when damping is small, it is acceptable to neglect the off-diagonal elements in $[\mathcal{C}]$ without a great loss in accuracy and an approximate solution may be reached. However, if damping is large, such an approximation cannot be done. The solution for this more general situation entails the use of complex state vectors and can be found in most text books. It suffices to say that we end up by obtaining a modal solution with complex modes.

We have seen that undamped systems exhibit mode shapes with real amplitudes that are only known within a multiplicative constant. Now, in the general case of damped systems, we have complex mode shapes, this meaning that we are dealing both with amplitudes and with phase angles. Thus, the mode shape vectors are not only known within a multiplicative constant, as far as the amplitudes are concerned, but also within a constant angular shift, as far as the phase angles are concerned.

Coming back to our MDOF system and assuming that the dissipative mechanism is hysteretic, we may write

$$[M]\{\ddot{x}(t)\} + i\,[D]\{x(t)\} + [K]\{x(t)\} = \{f(t)\} \tag{76}$$

where $[D]$ is the NxN hysteretic damping matrix. The problem now is the fact that this type of damping has been defined only for the particular case of forced harmonic vibrations and, as stated in 1.2.8, presents some difficulties to rigorous free vibration or shock response analysis. Resolution of (128) in the case where $\{f(t)\} = \{0\}$ may be thus considered a questionable decision.

Following a reasoning similar to the one used when dealing with the viscous damped model, we may start by assuming that damping is proportional, i.e.

$$[D] = \varepsilon[K] + \nu[M] \tag{77}$$

where ε and ν are constants. Assuming that a solution exists which is of the form

$$\{x(t)\} = \{\overline{X}\}\,e^{i\lambda t} \tag{78}$$

where $\{\overline{X}\}$ is an Nx1 vector of time-independent response amplitudes, and substituting into the homogeneous form of (76), we arrive at

$$\left[\left[[K] - \lambda^2[M]\right] + i\left[\varepsilon[K] + \nu[M]\right]\right]\{\overline{X}\} = \{0\} \tag{79}$$

which represents a complex eigenvalue problem leading to a solution in terms of N complex eigenvalues λ_r^2 and N real eigenvectors $\{\psi_r\}$ that are the same as for the undamped case. As the eigenvectors are the same mode shape vectors as for the undamped case, they may be taken as having the same physical meaning. Also, for the case under analysis, it is obvious that their properties of orthogonality will decouple the equations of motion. Thus, we may take λ_r^2 as containing information on the natural frequencies of the system and write

$$\lambda_r^2 = \omega_r^2 (1 + i \, \eta_r) \tag{80}$$

where ω_r^2 and η_r are the natural frequency and damping loss factor, respectively, for mode r, defined as being given by

$$\omega_r^2 = \frac{k_r}{m_r} \tag{81}$$

and

$$\eta_r = \varepsilon + \frac{\nu}{\omega_r^2} \tag{82}$$

The generalised modal stiffness and modal mass, k_r and m_r respectively, have been already defined in (61).

If non-proportional damping is assumed, the corresponding complex eigenproblem yields not only N complex eigenvalues λ_r^2 but also N complex eigenvectors $\{\psi_r\}$.

Let us now turn our attention to the forced response solution of MDOF systems and of the corresponding set of equations. As in the case of SDOF systems, we are going to neglect the transient part of the complete response and consider solely the steady-state situation. From (51) and taking the excitation force vector $\{f(t)\} = \{F\} e^{i \omega t}$ and the response vector $\{x(t)\} = \{\overline{X}\} e^{i \omega t}$, we arrive easily at

$$\{\overline{X}\} = \left[[K] - \omega^2 [M] + i \, \omega [C]\right]^{-1} \{F\} = [\alpha(\omega)] \{F\} \tag{83}$$

and similarly, for the hysteretic model,

$$\{\overline{X}\} = \left[[K] - \omega^2 [M] + i [D]\right]^{-1} \{F\} = [\alpha(\omega)] \{F\} \tag{84}$$

where $[\alpha(\omega)]$ is the NxN receptance matrix containing all the information on the system dynamic characteristics. Each element α_{jk} of the matrix corresponds to an individual FRF describing the relation between the response at a particular coordinate j and a single force excitation applied at coordinate k. The receptance matrix $[\alpha(\omega)]$ constitutes another form of modelling our system and is known as the Response Model as opposed to the Spatial Model and the Modal Model previously mentioned.

Despite their apparent simplicity, equations (83) and (84) tend to be very inefficient for numerical applications. In fact, although it is possible to calculate the values of $[\alpha(\omega)]$ at any frequency of interest, this operation requires the inversion of an NxN matrix for the chosen frequency value. When dealing with systems with a large number of degrees of freedom this may be a highly time-consuming operation. However, if we consider the modal properties, it is possible to derive more useful expressions for $[\alpha(\omega)]$. We shall consider solely the hysteretic model, which is far easier to manipulate from the mathematical point of view. The interested reader may find the viscous model case in most modal analysis text books.

Let us recall that, for the hysteretic model, there is a set of N complex eigenvalues λ_r^2 and associated eigenvectors $\{\psi_r\}$ which satisfy the homogeneous equation

$$\left[[K] + i[D] - \lambda_r^2[M]\right]\{\psi_r\} = \{0\} \tag{85}$$

As the N eigenvectors form a linearly independent set of vectors in N-space, possessing orthogonality properties, any vector in N-space, such as $\{\overline{X}\}$, can be expressed as a linear combination of the eigenvectors, i.e.,

$$\{\overline{X}\} = \sum_{r=1}^{N} \gamma_r \{\psi_r\} \tag{86}$$

Substituting into (85) and pre-multiplying by $\{\psi_s\}^T$,

$$\{\psi_s\}^T [[K] + i[D]] \sum_{r=1}^{N} \gamma_r \{\psi_r\} - \omega^2 \{\psi_s\}^T [M] \sum_{r=1}^{N} \gamma_r \{\psi_r\} = \{\psi_s\}^T \{F\} \tag{87}$$

Taking into consideration the orthogonality properties of the eigenvectors,

$$\gamma_r \{\psi_r\}^T [[K] + i[D]]\{\psi_r\} - \omega^2 \gamma_r \{\psi_r\}^T [M]\{\psi_r\} = \{\psi_r\}^T \{F\} \tag{88}$$

or

$$\gamma_r k_r - \omega^2 \gamma_r m_r = \{\psi_r\}^T \{F\} \tag{89}$$

Thus,

$$\gamma_r = \frac{\{\psi_r\}^T \{F\}}{k_r - \omega^2 m_r} \tag{90}$$

Substituting into (86) we end up with the definition of the steady-state response for hysteretic damping, in terms of the complex modes

$$\{x(t)\} = \{\overline{X}\} e^{i\omega t} = \sum_{r=1}^{N} \frac{\{\psi_r\}^T \{F\}\{\psi_r\}}{k_r - \omega^2 m_r} e^{i\omega t} \tag{91}$$

and therefore

$$\{\overline{X}\} = \sum_{r=1}^{N} \frac{\{\psi_r\}^T \{F\}\{\psi_r\}}{m_r (\omega_r^2 - \omega^2 + i\,\eta_r\,\omega_r^2)} \tag{92}$$

If we are interested in extracting a single receptance element, for example, the response at coordinate j due to a single harmonic force excitation applied at coordinate k, this means that vector $\{F\}$ will have just one non-zero element and therefore we may easily arrive at

$$\alpha_{jk}(\omega) = \frac{\overline{X}_j}{F_k} = \sum_{r=1}^{N} \frac{\psi_{jr}\,\psi_{kr}}{m_r\,(\omega_r^2 - \omega^2 + i\,\eta_r\,\omega_r^2)} \tag{93}$$

where ψ_{jr} and ψ_{kr} are elements j and k, respectively, of the mode shape vector $\{\psi_r\}$. Thus, we have arrived at a general expression for the elements of the receptance matrix, in terms of the modal properties. Equation (93) may be interpreted as stating that the total response is the result of a summation of contributions from N separated SDOF system responses (note the similarity of each element in the summation with equation (22) obtained for a SDOF system). In the general case of non-proportional damping, the numerator of (93) is complex whereas in the undamped or proportionally damped cases it is real. Taking into consideration the mass-normalised mode shape vectors

$$\alpha_{jk}(\omega) = \frac{\overline{X}_j}{F_k} = \sum_{r=1}^{N} \frac{\phi_{jr}\,\phi_{kr}}{\omega_r^2 - \omega^2 + i\,\eta_r\,\omega_r^2} \tag{94}$$

or

$$\alpha_{jk}(\omega) = \frac{\overline{X}_j}{F_k} = \sum_{r=1}^{N} \frac{{}_r\overline{A}_{jk}}{\omega_r^2 - \omega^2 + i\,\eta_r\,\omega_r^2} \tag{95}$$

where ${}_r\overline{A}_{jk} = {}_rA_{jk}\,e^{i\,{}_r\varphi_{jk}}$ is a complex quantity known as the Modal Constant, for which

$$ {}_rA_{jk} = \left| \frac{\psi_{jr}\,\psi_{kr}}{m_r} \right| = \left| \phi_{jr}\,\phi_{kr} \right| $$

$$ {}_r\varphi_{jk} = \arg\left(\frac{\psi_{jr}\,\psi_{kr}}{m_r} \right) = \arg\,(\phi_{jr}\,\phi_{kr}) \tag{96} $$

are constants for a given r, j and k. Two important conclusions may be extracted from the above derivations. First, it is clear that the receptance matrix is symmetric and therefore

$$\alpha_{jk} = \frac{\overline{X}_j}{F_k} = \alpha_{kj} = \frac{\overline{X}_k}{F_j} \tag{97}$$

this property being known as the principle of reciprocity and second, the modal constants obey a relationship described by

$$ {}_r\overline{A}_{jk} = \phi_{jr}\,\phi_{kr} $$

$$ {}_r\overline{A}_{jj} = \phi_{jr}^2 \qquad \text{or} \qquad {}_r\overline{A}_{kk} = \phi_{kr}^2 \tag{98} $$

known as the modal constants consistency equations. According to (97) and (98), if one knows a full line (or column) of the matrix $[\alpha(\omega)]$, then the whole matrix can be evaluated.

8. FRFs representation and properties (MDOF)

We have seen that the response model of a MDOF system consists of a set of different FRF functions and it has been shown that a system with N degrees of freedom is described by a modal model with N natural frequencies and N mode shapes. Also, it was shown that each FRF may be written under the form of a series of terms, each of which refers to the contribution to the total response of each mode of vibration.

Taking, for example, an undamped system with five degrees of freedom, the Bode plot of a direct point receptance FRF (i.e., response and excitation respectively measured and applied at the same coordinate) may be as shown in Figure 13 where the magnitude is displayed against frequency, in linear scale.

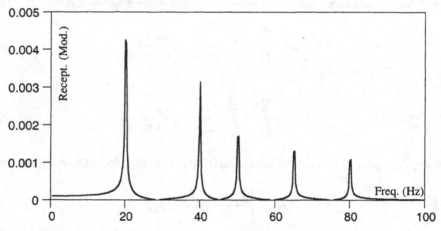

Figure 13 Plot of the magnitude (linear scale) of a direct point receptance for a 5 DOF system.

As can be seen, there are five peak amplitudes, corresponding to the five natural frequencies of the system. Accordingly, one is now confronted with five different resonances. In analogy with what we saw for SDOF systems, it is to be expected that, for each resonance, there will be a $180°$ phase shift. However, Figure 14 shows clearly that there are more than five such phase shifts. They not only occur at each resonance but also for intermediate frequency values that have no apparent special behaviour as far as the magnitude plot is concerned.

If, instead of linear, we use a logarithmic amplitude scale, we obtain what is shown in Figure 15. Now, we can also see detail at the lower levels of the response and the FRF shows that, in those regions, there are some 'inverted' peaks, each of which occurs in between the resonance peaks. These are called antiresonances and they have an important feature which is a phase change just like the phase change associated with resonances.

For an undamped system, the antiresonance corresponds to no motion at all at the coordinate where the response is being considered. This situation can be explained if one recalls equation (95) and takes it for zero damping

$$\alpha_{jk} = \sum_{r=1}^{N} \frac{{}_r A_{jk}}{\omega_r^2 - \omega^2} \qquad (99)$$

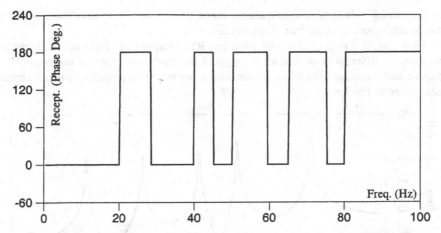

Figure 14 Plot of the phase angle of a direct point receptance for a 5 DOF system.

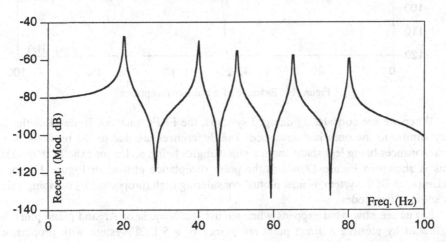

Figure 15 Plot of the magnitude (dB scale) of the receptance displayed in Figure 13.

The modal constant $_rA_{jk}$ is now a real quantity. Considering a direct point measurement, say α_{kk}, the modal constant $_rA_{kk}$ is always positive due to it being the product of element k of the eigenvector for mode r, by itself. Thus, in the lower frequency region, all the terms in the summation are positive and the receptance value is positive and dominated by the first mode ($r = 1$), for which the denominator $\omega_r^2 - \omega^2$ is smaller than for the other terms in the summation. After the first resonance, $\omega_1^2 - \omega^2$ becomes negative and therefore the first term in the series becomes negative, while still dominating the response, and therefore α_{kk} becomes negative. This change of sign corresponds to a phase shift from $0°$ to $-180°$. As we approach ω_2, there will be a frequency value for which the first term of the series is cancelled out by the sum of all the other terms and, subsequently, there is a new change of sign of α_{kk} which becomes positive. Accordingly, the phase angle suffers a new shift and becomes zero. The frequency for which the cancelling out occurs is the antiresonance. The same reasoning, for increasing values of the frequency, leads one to conclude that there will be an antiresonance between each pair of resonances.

However, if one is considering transfer FRFs (response and escitation respectively measured at different coordinates), the sign of the modal constants is no longer always positive and the occurrence of an antiresonance between two resonances is not certain as exemplified in Figure 16.

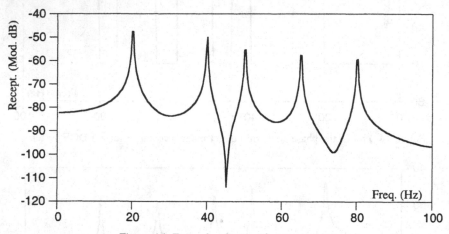

Figure 16 Example of a transfer receptance.

When we are considering damped systems, the FRFs trace out Bode plots that are very similar to the ones just described. The differences are due to the resonances and antiresonances being less sharp and the phase angles being no longer exactly 0° or -180° This is shown in Figure 17 where the point receptance plotted in Figure 15, for an undamped 5 DOF system, is now plotted considering high (proportional) damping values for some of the modes.

Let us see now what happens when we use the Nyquist or Argand plane plots. We shall start by plotting a direct point receptance of a 5 DOF system with proportional damping (Figure 18). Though the natural frequency regions plot as circular loops, it can be seen that the loops are not exactly centred with respect to the imaginary axis as in the case of a SDOF system. This can be easily explained if we recall equation (95). The modal constants $_rA_{kk}$ are real quantities due to the fact that damping was assumed to be proportional. Considering, for example, the first loop in our plot and recalling that each loop occurs for a frequency region close to the corresponding natural frequency, then it may be assumed that, for that particular frequency range (95) can be approximated by

$$\alpha_{kk}(\omega) \approx \frac{_1A_{kk}}{\omega_1^2 - \omega^2 + i\,\eta_1\,\omega_1^2} + B_{kk} \tag{100}$$

where B_{kk} is a constant complex quantity accounting for the contribution of the remaining modes to the total receptance value, which is dominated by the first mode. The first term of the summation plots as a circle with its centre on the imaginary axis, just like the receptance of a SDOF system. The only difference from a SDOF system is the fact that there is a real scaling factor (which alters the circle diameter), due to the existence of a modal constant $_1A_{kk}$ in the numerator. Summing a complex quantity B_{kk} will then produce a translation of the circle, displacing it from the original position.

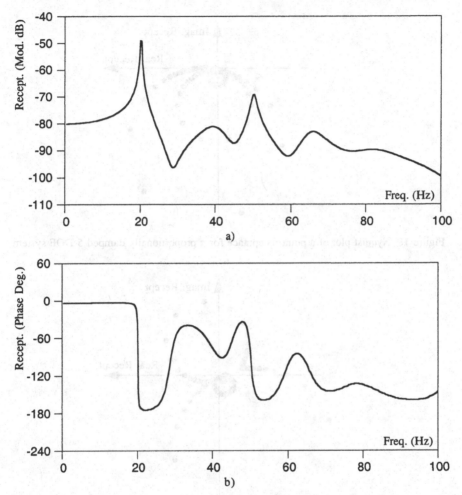

Figure 17 Point receptance of a 5 DOF system with hysteretic damping.

It can also be seen in Figure 18 that all circular loops are in the lower half of the complex plane. As we are considering a direct point receptance, the modal constants are all positive and, therefore, the loops remain in the lower half of the complex plane. A different situation occurs if we are plotting a transfer receptance. In this case, the modal constants may be positive or negative and the opposing signs of these quantities may cause one or more loops to be in the upper half of the complex plane.

If we consider the situation where damping is non-proportional, the modal constants become complex quantities, i.e., they have a magnitude and a phase. Thus, the circular loop displacement and scaling effect remain and are due to the contribution of the off-resonant modes and to the magnitude of the modal constant, respectively. In addition to the previous effects, the phases of the modal constants produce rotations of the modal loops which are no longer in the 'upright' position, as illustrated in Figure 19.

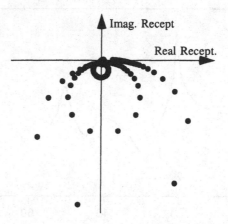

Figure 18 Nyquist plot of a point receptance for a proportionally damped 5 DOF system.

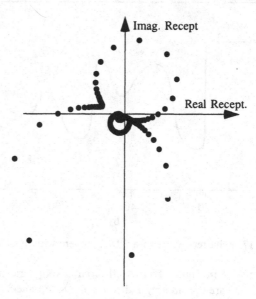

Figure 19 Nyquist plot of a point receptance for a non-proportionally damped 5 DOF system.

9. Incomplete models

In the previous sections we have always assumed that our systems are described by complete models, i.e., that all their mass, stiffness and damping properties are known, or that all the eigenvalues and all the elements in the eigenvectors are known, or that all the elements in the FRF matrix are known. Though this is a valid assumption from a theoretical point of view, in practice it corresponds to an impossibility. The knowledge that any real system has an infinite number of degrees of freedom is sufficient for understanding of the problem.

When we refer to incomplete models we are considering seceral forms of incompleteness. In fact, even if we assume that when reducing a real system to a model with N DOFs we have a complete model, in practice and from an experimental point of view, it is not usually possible to measure all the coordinate motion responses, or to apply all force excitations, or even to cover and analyse all modes.

The limitations introduced by the fact that, in practice, we shall be using reduced or incomplete models, and the techniques to overcome the inherent problems, are dealt with in the appropriate literature. Here, we shall only talk about the problem in a superficial way.

Let us consider, as an example, an N DOFs response model and assume that, if completely known, it gives us a thorough description of the dynamic behaviour of our system. Let us assume that we have to consider S rotation coordinates. We know that experimental measurement of rotational responses is a very difficult task. We may therefore have to decide to limit our model to include only the translation coordinates. The resulting reduced FRF model will therefore be of order (N-S)x(N-S). What is important to note is that, despite the fact that we limited our system description to a reduced number of coordinates, we have not altered the basic system which still has 5 degrees of freedom. We simply ceased to be able to describe all of the initial DOFs. When using a response model, reducing the coordinates of interest from N to N-S is a simple operation of elimination of the relevant S rows and columns of the frequency response matrix. However, performing such a row and column elimination would obviously be wrong if we were considering a spatial model.

Another important reduction that occurs in practical situations is related to the number of modes of vibration one can include in the analysis. The frequency range of an experimental analysis is limited, and therefore at least the high frequency modes will be omitted. Thus, despite the fact that the total number of coordinates N used in the model may remain without being reduced, one may have to consider a smaller number $m \leq N$ of modes, that is, equation (95) would be of the form

$$\alpha_{jk}(\omega) = \sum_{r=1}^{m \leq N} \frac{{}_r \overline{A}_{jk}}{\omega_r^2 - \omega^2 + i\, \eta_r\, \omega_r^2} \tag{101}$$

and, taking advantage of the orthogonality properties of the eigenvectors, the corresponding NxN receptance matrix containing a reduced modal information, can be written as

$$[\alpha(\omega)]_{NxN} = [\Phi]_{Nxm} \left[\,{}^{\backprime}(\lambda_r^2 - \omega^2)\,{}_{\backprime} \right]_{mxm}^{-1} [\Phi]_{mxN}^T \tag{102}$$

Any of the previous equations lacks information that may be of great importance. The fact that $m < N$ in (102) will lead to an NxN matrix $[\alpha(\omega)]$ that cannot be inverted

34

due to its rows and columns being not linearly independent, i.e., the order of the matrix is N but the rank is m. If $m \geq N$, this inversion problem does not exist. Finally, the consequence of considering only a limited number of modes m is the fact that we end up with a square eigenvalue matrix of order mxm and the eigenvector matrix will be rectangular (Nxm).

Let us take now the response model and consider a system which is to be modelled with a finite number N of degrees of freedom. These DOFs will correspond to the coordinates of interest for the analysis. Let us assume as well that our modal model is to be derived from the experimental data (through modal identification). The analysis will have to be based on a necessarily limited experimental frequency range and, therefore, on a limited number of modes m. Thus, the measured response model will consist of an NxN matrix of elements α_{jk} expressed in terms of experimental data.

However, our response model described now by the identified NxN matrix $[\alpha(\omega)]$ will contain errors due to omitting all the out-of-range modes. These errors are usually visible when comparing the measured FRF data with the corresponding identified FRFs if represented only by (101). One way of minimising the consequences of using such a model is to introduce corrections on the identified FRFs, so that they approximate the measured data in the frequency range of interest, by including an extra term in the response equation

$$\alpha_{jk}(\omega) = \sum_{r=1}^{m} \frac{r \overline{A}_{jk}}{\omega_r^2 - \omega^2 + i \, \eta_r \, \omega_r^2} + \overline{R}_{jk}(\omega) \tag{103}$$

where $\overline{R}_{jk}(\omega)$ is a complex residual term accounting for the contribution of the out-of-range modes.

Reduced models can also be obtained directly from the original NxN spatial model. In this case, reducing the number of coordinates entails reducing the mass, stiffness and damping matrices. Such a reduction cannot be performed by simply eliminating rows and columns as we stated before. The consequence would be to obtain a system with completely different properties. In this case one must use specific techniques (such as condensation techniques) for redistribution of the system mass, stiffness and damping properties amongst the retained coordinates.

10.   Bibliography

1.   Ewins, D. J. (1984), Modal Testing: Theory And Practice, *Research Studies Press Ltd.*, England.
2.   Maia, N. M. M., Silva, J. M. M. et al (1997), "Theoretical and Experimental Modal Analysis", *Research Studies Press Ltd.*, England.
3.   McConnell, K. G. (1995), Vibration Testing: Theory And Practice, *Wiley Interscience*.
4.   Inman, D. J. (1994), Engineering Vibration, *Prentice Hall*, U.S.A.
5.   Heylen, W., Lammens, S. and Sas, P. (1998), Modal Analysis Theory and Testing, *KU Leuven*, Belgium.

INTRODUCTION TO SIGNAL PROCESSING

Part I: Fundamentals of Signal Processing

J. K. HAMMOND
Institute of Sound and Vibration Research
University of Southampton

1. Introduction

Signal processing is the science of applying transformations to measurements, to facilitate their use by an observer or a computer and the analysis of data involves the three phases of data acquisition, processing and interpretation. The objective of data analysis is to highlight/extract information contained in a signal that direct observation may not reveal.

Figure 1 illustrates a broad categorization of time histories (Bendat and Piersol, 1971). Very often, processes are mixed and the demarcations of Figure 1 are not easily applied. Additionally, it is often convenient to regard a time history as arising from some 'data generation model' as in Figure 2.

It is important to be aware that procedures based on prior assumptions may bias the outcome. One should keep in mind the First Principle of Data Reduction (Ables 1974). "The result of any transformation imposed on experimental data shall incorporate and be consistent with all relevant data and be maximally noncommital with regard to unavailable data."

2. Fourier analysis of continuous time signals

If a signal repeats itself exactly every T_p seconds, then $x(t)$ may be represented as a sum of sinusoids and cosinusoids

$$x(t) = \sum_{n=-\infty}^{\infty} C_n \, e^{j2\pi nt / T_p} \tag{1}$$

The amplitude C_n is complex, representing both amplitude and phase of each component. If $x(t)$ is a transient then the discrete set of sinusoids and cosinusoids becomes a continuum and we write

$$x(t) = \int_{-\infty}^{\infty} X(f) \, e^{j2\pi ft} \, df \tag{2}$$

35

J.M.M. Silva and N.M.M. Maia (eds.), Modal Analysis and Testing, 35–52.
© *1999 Kluwer Academic Publishers. Printed in the Netherlands.*

where X(f) is given by

$$X(f) = \int_{-\infty}^{\infty} x(t) e^{-j2\pi ft} dt \tag{3}$$

The frequency f is in Hz and $\omega = 2\pi f$ is the frequency in radians/s.

It is logical to ask if the representations (1) and (2) may be generalised to allow the representation of random signals. In summary (Priestley, 1967), a stationary random signal may be expressed as

$$x(t) = \int_{-\infty}^{\infty} e^{j\omega t} dZ_x(\omega) \tag{4}$$

which is a generalisation of (2) (this is a Fourier-Stieltjes form - only reducing to the Riemann form (2) when $Z_x(\omega)$ is differentiable as for 'transients' above).

3. Some results in signal and system analysis

3.1.   GROUP DELAY

If a signal x(t) has Fourier transform X(f), then a delayed signal x(t - t_0) has Fourier transform $e^{-j2\pi ft_0} X(f)$, i.e., only the phase is affected. The derivative of the phase with respect to f is $d\phi(f)/df = -2\pi t_0$. Written in terms of ω we can write $-d\phi/d\omega = t_0$. The quantity $-d\phi/d\omega$ is called the group delay of the signal. In this case, the delay is the same for all frequencies (there is no dispersion). If $-d\phi/d\omega$ is a *nonlinear* function of ω then dispersion is present and phase distortion of the signal occurs.

3.2.   CONVOLUTION

The *convolution* of two time histories is often introduced via the concept of linear filtering. If x(t) is the excitation of a linear system with impulse response function h(t) producing response, y(t), then

$$y(t) = \int_{-\infty}^{t} h(t - t_1) x(t_1) dt_1 \tag{5}$$

The upper limit is t if the system is causal. Using the substitution t - $t_1 = \tau$, this may be expressed as

$$y(t) = \int_{0}^{\infty} h(\tau) x(t - \tau) d\tau \tag{6}$$

when h(τ) now has the role of a weighting (or memory) function.

This expression is considerably simplified by taking the Fourier transform, which yields

$$Y(f) = H(f) X(f) \qquad (7)$$

i.e., convolutions transform to products. H(f) is the frequency response of the linear filter and is the Fourier transform of the impulse response function.

3.3. WINDOWING

Suppose x(t) is known only for $-T/2 < t < T/2$, or we choose to truncate the signal to that length. It is as though we are looking at the data through a 'window' w(t) where, in this case,

$$\begin{aligned} w(t) &= 1 \qquad & |t| < T/2 \\ &= 0 \qquad & |t| > T/2 \end{aligned}$$

giving the truncated signal $x_T(t) = w(t)\, x(t)$.

Fourier transforming $x_T(t)$ we obtain (for a general w(t))

$$X_T(f) = \int_{-\infty}^{\infty} X(g)\, W(f-g)\, dg \qquad (8)$$

i.e., the Fourier transform of the *product* of two time functions is the convolution of their transforms. For the particular case of the 'rectangular' data window above,

$$W(f) = \frac{T \sin(\pi f T)}{(\pi f T)}$$

and through the convolution operation this widens and distorts the 'true' Fourier transform X(f). This distortion is sometimes called smearing (due to the main lobe) and 'leakage' since components of X(f) at values other than $g = f$ 'leak' through the side lobes to contribute to the value of X_T at f. For example, suppose $x(t) = 2\pi p t$, then

$$X(f) = \frac{1}{2}\, [\delta(f+p) + \delta(f-p)]$$

and so

$$X_T(f) = \frac{1}{2}\, [W(f+p) + W(f-p)]$$

showing that the shape of the spectral window (in the frequency domain) replaces the delta function. Comparison of theoretical and achieved spectra are depicted in Figure 3.

If two (or more) closely spaced harmonic components occur in a signal, then the distortion may mean that the components cannot (easily) be resolved. To separate two peaks at frequencies f_1, f_2, it is necessary to use a record length T of order $T \geq 2/(f_2 - f_1)$ for a rectangular window.

We note the rectangular window is a particularly 'poor' window as far as the side lobes are concerned (the highest side lobe is 13 dB below the main peak with asymptotic roll off of 6 dB/octave). The sharp corners of the data window result in this rippling effect. However, the main lobe is narrower than for other windows. Numerous alternatives have been suggested to reduce leakage in Fourier transform calculations. By tapering the windows to zero, the side lobes are reduced but at the expense of widening the main lobe (increasing smearing). The paper by Harris (1978) contains a comprehensive treatment of windows and their effects.

3.4. THE UNCERTAINTY PRINCIPLE

A property of the Fourier transform of a signal is that the narrower a signal characterisation is in one domain the wider is its characterisation in the other. An extreme example is a delta function $\delta(t)$ whose Fourier transform is a constant. This fundamental property of signals is summarised in the so-called 'Uncertainty Principle' as follows (Hsu, 1970).

If $x(t)$ is a signal with finite energy, i.e., $\|x\|^2 = \int_{-\infty}^{\infty} x^2(t)dt < \infty$, and has Fourier transform $X(\omega)$, then defining the following

$$\bar{t} = \frac{1}{\|x\|^2} \int_{-\infty}^{\infty} t\, x^2(t)dt \qquad \Delta t^2 = \frac{1}{\|x\|^2} \int_{-\infty}^{\infty} (t - \bar{t})^2 x^2(t)dt$$

and

$$\|X\|^2 = \int_{-\infty}^{\infty} |X(\omega)|^2 d\omega \qquad \bar{\omega} = \frac{1}{\|X\|^2} \int_{-\infty}^{\infty} \omega |X(\omega)|^2 d\omega$$

$$(\Delta\omega)^2 = \frac{1}{\|X\|^2} \int_{-\infty}^{\infty} (\omega - \bar{\omega})^2 |X(\omega)|^2 d\omega$$

($\bar{t}$, $\bar{\omega}$ are measures of location and Δt, $\Delta\omega$ are measures of spread), then it can be shown that

$$\Delta\omega\, \Delta t \geq \frac{1}{2} \qquad\qquad (9)$$

The Bandwidth-Time (BT) product of a signal has a lower bound of $1/2$. It is this inequality that is at the heart of 'difficulties' in time-frequency analysis methods which are Fourier based, i.e., if one wants a 'local' Fourier transform then increasing the 'localisation' leads to poorer resolution in frequency and *vice-versa*, i.e., one cannot simultaneously obtain arbitrarily fine 'resolution' in *both* frequency and time domains.

3.5. THE HILBERT TRANSFORM

The Hilbert transform of a signal $x(t)$ is defined as the convolution of $x(t)$ with $1/\pi t$ and is denoted $\hat{x}(t)$, i.e.,

$$\hat{x}(t) = \frac{1}{\pi t} * x(t) \qquad\qquad (10)$$

Despite the singular nature of the non-causal filter $1/\pi t$, many useful analytic results can be obtained. For example, the Hilbert transform of $\cos\Omega t$ is $\sin\Omega t$ (and that of $\sin\Omega t$ is $-\cos\Omega t$) and hence a Hilbert transform is sometimes referred to as a 90° phase shifter. Transformation to the frequency domain is helpful as the Fourier transform of the convolution operation can be expressed

$$\hat{X}(\omega) = H(\omega) X(\omega) \tag{11}$$

where
$$\begin{aligned} H(\omega) &= j & \omega &< 0 \\ &= -j & \omega &> 0 \\ &= 0 & \omega &= 0 \end{aligned} \tag{12}$$

This result may be used to obtain simple analytic results as above (rather than the time domain convolution).

The Hilbert transform has considerable significance in signal processing analysis and processing. We note that it is used in the definition of the 'instantaneous frequency' of a signal in the following way.

Combining the original signal $x(t)$ with its Hilbert transform $\hat{x}(t)$ in the form

$$\sigma_x(t) = x(t) + j\hat{x}(t)$$

creates what is referred to as the 'analytic' signal or 'pre-envelope' signal $\sigma_x(t)$, which is complex and hence expressible as

$$\sigma_x(t) = A_x(t) e^{j\phi_x(t)} \tag{13}$$

$A_x(t) = \sqrt{x(t)^2 + \hat{x}(t)^2}$ is called the instantaneous amplitude,

$\phi_x(t) = \tan^{-1}(\hat{x}(t)/x(t))$ is the instantaneous phase, and

$\dot{\phi}_x$ is the instantaneous frequency.

For the trivial case $x(t) = \cos\Omega t$, the analytic signal is $e^{j\Omega t}$ and the instantaneous frequency is Ω, as one would expect. These definitions of instantaneous amplitude, phase and frequency have come to be accepted for a wide class of signals, and are particularly relevant to amplitude-modulated and frequency-modulated (monocomponent) signals, e.g., for a chirp, when tracking a 'variable' frequency may be significant.

4. The effects of sampling

This section introduces the concepts associated with converting a continuous time signal $x(t)$ into a sequence of numbers obtained by observing the signal every Δ seconds.

4.1. ANALOGUE-TO-DIGITAL CONVERSION

An analogue-to-digital converter (ADC) is a device that operates on a continuous time history (input) and produces a sequence of numbers (output) that are sample values of the input. It is convenient to (conceptually)regard the process as consisting of two stages, namely 'sampling' and 'quantization'. $x(n\Delta)$ is the exact value attained by time history $x(t)$ at time $t = n\Delta$ (Δ is the sample interval for uniform rate sampling). $\tilde{x}(n\Delta)$ is the representation of $x(n\Delta)$ in a computer and differs from $x(n\Delta)$ since a finite number of bits are used to represent each number.

The problem of quantization error is treated by Otnes and Enochson (1978). The output $\tilde{x}(n\Delta)$ can be written as $\tilde{x}(n\Delta) = x(n\Delta) + e(n\Delta)$, and finite word length effects are described by treating the error $e(n\Delta)$ as random 'noise'. The probability distributions describing the error depend on the particular way in which the quantization is done. Often the error is ascribed a uniform distribution (with zero mean) over one quantization step and furthermore is assumed to be stationary and 'white'. For a b-bit word length (excluding sign), and if the full range of the ADC corresponds to X volts, say, then a measure of signal-to-noise ratio is $S/N = 10\log_{10}(\sigma_x^2/\sigma_e^2)$. Choosing $\sigma_x = X/4$ (to 'minimise' clipping) then $S/N \approx 6b$ dB. This is reduced further by practical considerations relating to the quality of the acquisition system.

It is vital that the rate at which $x(t)$ is sampled should be high enough to avoid aliasing, see Section 4.3. This usually means that $x(t)$ should be low-pass filtered using analogue filters preceding the ADC. The cut-off frequency and cut-off rate of these 'anti-alias' filters should be chosen with particular applications in mind, but very roughly speaking if the 3 dB point is a quarter of the sampling frequency and the cut-off rate better than 48 dB/octave, then this at least ensures a 40-50 dB reduction at the folding frequency (half the sampling frequency; see Section 4.3), and so probably an acceptable level of aliasing (though we must emphasize that this may not be adequate for some applications). Selection of sample rate is very important. Whilst aliasing is to be avoided, unnecessarily high sample rates are wasteful. The specific application (the bandwidth of interest) and the characteristics of the antialias filter should dictate the selection of an 'optimal' rate.

4.2. THE FOURIER TRANSFORM OF A SEQUENCE

4.2.1. *Impulse train modulation*

One way of introducing the Fourier transform of a sequence involves the use of the mathematical notion of 'ideal sampling' of a continuous wave. If an analogue signal $x(t)$ is to be sampled every Δ seconds, it is convenient to model the sampled signal as the product of the continuous signal with a 'train' of delta functions $i(t)$ where

$$i(t) = \sum_{n=-\infty}^{\infty} \delta(t - n\Delta) \tag{14}$$

so that the sampled signal, $x_s(t)$, is modelled as $x_s(t) = x(t).i(t)$. The Fourier transform of $x_s(t)$ is $X_s(f)$

$$X_s(f) = \sum_{n=-\infty}^{\infty} x(n\Delta)e^{-j2\pi f n\Delta} \tag{15}$$

from which

$$x(n\Delta) = \Delta \int_{-1/2\Delta}^{1/2\Delta} X_s(f)e^{j2\pi fn\Delta}df \qquad (16)$$

Equations (15) and (16) relate the *sequence of numbers* $x(n\Delta)$ to the quantity $X_s(f)$ which is termed the Fourier transform of the sequence.

A natural question to ask at this stage is 'How is $X_s(f)$ (or as it is often written $X(e^{j2\pi f\Delta})$) related to $X(f)$?' The answer to this can be obtained as follows. Since i(t) is periodic, we represent it as a Fourier series, i.e.,

$$i(t) = \frac{1}{\Delta} \sum_{n=-\infty}^{\infty} e^{(j2\pi nt)/\Delta} \qquad (17)$$

Now substituting this into $x_s(t) = x(t)\ i(t)$ and Fourier transforming gives an alternative right-hand side to (15), namely

$$X(e^{j2\pi/\Delta}) = \frac{1}{\Delta} \sum_{n=-\infty}^{\infty} X(f-(n/\Delta)) \qquad (18)$$

This important equation (depicted graphically in Figure 4) relates the Fourier transform of a continuous signal and the Fourier transform of the sequence formed by sampling the signal at equispaced intervals.

4.3. ALIASING

Equation (18) describes how the frequency components of the sampled waveform relate to that of the continuous waveform, and Figure 4 explains this pictorially. In Figure 4(a) it is assumed that $X(f) = 0$ for $|f| > f_0$ (say), and Figure 4(b) is a plot of $\Delta X(e^{j2\pi/\Delta})$ assuming that the sampling rate $f_s = 1/\Delta$ is such that $f_0 < 1/2\Delta$. Some commonly used terms are defined on the diagram. If $f_0 > f_s/2$, there is an overlapping of the shifted versions of $X(f)$ resulting in a distortion of the frequency description for $|f| < 1/2\Delta$ as in Figure 4(c). This 'distortion' is due to the fact that high frequencies in the data are indistinguishable from lower frequencies owing to the sampling rate not being fast enough.

To avoid aliasing, the sample rate must be chosen to be greater than twice the highest frequency contained in the signal (see the discussion in Section 4.2 regarding the 'anti-alias' filters).

4.4. THE DISCRETE FOURIER TRANSFORM

The equation (15) is still not a practical proposition, and a finite sum would be more appropriate which, for an N-point sequence, could be written

$$\sum_{n=0}^{N-1} x(n\Delta)e^{-j2\pi fn\Delta} \qquad (19)$$

Obviously, if this involved truncating the data there would be some distortion explained earlier by the discussion on windowing. Assuming for the moment that we can ignore this, we can think of simplifying our task by only evaluating f at specific points, say at $k/N\Delta$ Hz, $k = 0, ..., N - 1$. Then

$$X(e^{j2\pi k/N}) = \sum_{n=0}^{N-1} x(n\Delta)e^{-j(2\pi/N)nk} \tag{20}$$

Or, adopting the more usual notation (and setting Δ to unity for convenience),

$$X(k) = \sum_{n=0}^{N-1} x(n)e^{-j(2\pi/N)nk} \tag{21}$$

leading to

$$x(n) = \frac{1}{N}\sum_{k=0}^{N-1} X(k)e^{j(2\pi/N)nk} \tag{22}$$

The pair (21) and (22) constitute the Discrete Fourier Transform (DFT) and are the form that are suitable for machine computation.

It is important to realise that even though the original sequence $x(n)$ may be zero for n outside the range $0 \rightarrow N - 1$, the act of 'sampling in frequency' imposes a *periodic structure* to the sequence, i.e., it follows from (22) that $x(n + N) = x(n)$.

The evaluation of the DFT can be accomplished efficiently by a set of algorithms known as the fast Fourier transform (FFT). These algorithm exploit the periodicity and symmetry properties of the complex exponentials and so reduce the number of multiply and add operations needed to calculate $X(k)$ from about N^2 to approximately $N\log_2 N$ which for $N = 1024$ is a reduction by a factor of about 100.

5. Random processes

This section is devoted to the treatment of non-deterministic signals. We shall restrict ourselves to stationary random processes characterised by their first and second moments (mean and variance) and auto- and cross-correlation functions and auto- and cross-spectra.

5.1. COVARIANCE (CORRELATION) FUNCTIONS

For a random process $\{x(t)\}$ a widely used average is the *autocovariance function* which is defined as

$$E\left[\left(x(t_1) - \mu_x(t_1)\right)\left(x(t_2) - \mu_x(t_2)\right)\right] = R_{xx}(t_1, t_2) \tag{23}$$

This is a measure of the 'degree of association of the signal at time t_1 with *itself* at time t_2 ($t_2 \geq t_1$). If the mean value is not subtracted, $E[x(t_1) x(t_2)]$ is often referred to in the engineering literature as the *autocorrelation function*. For stationary processes the statistical properties are unchanged under a time shift, so

$$E\big[(x(t_1) - \mu_x)(x(t_2) - \mu_x)\big] = R_{xx}(t_2 - t_1) \tag{24}$$

i.e., a function of the difference $(t_2 - t_1)$ only. We usually write $t_1 = t$ and $t_2 = t + \tau$, so

$$E\big[(x(t) - \mu_x)(x(t+\tau) - \mu_x)\big] = R_{xx}(\tau) \tag{25}$$

τ is called the lag. $R_{xx}(\tau)$ is an even function of τ.

5.1.1.   *The cross covariance (cross correlation)*
When two random processes are involved $\{x(t),y(t)\}$, e.g., input/output processes, then we generalise the above and define the cross covariance function as

$$R_{xy}(\tau) = E\big[x(t)\,y(t+\tau)\big] \tag{26}$$

Note: $R_{xy}(\tau)$ is *not* an even function. In fact,

$$R_{xy}(\tau) = R_{yx}(-\tau)$$

5.2.   SPECTRA

This section is concerned with frequency domain analysis of random signals. We shall side-step the formally correct Fourier-Stieltjes form referred to in Section 2 as follows.

Consider a sample function of a random process $x(t)$ and let us look at a truncated version $x_T(t)$

$$x_T(t) \;=\; x(t), \qquad |t| < \frac{T}{2}$$
$$= 0 \qquad \text{elsewhere}$$

Since $x_T(t)$ is 'pulse-like', it has a Fourier representation

$$x_T(t) = \int_{-\infty}^{\infty} X_T(f) e^{j2\pi ft} df$$

The total energy of the signal is $\int_{-\infty}^{\infty} x_T^2(t)dt$ which tends to infinity as T gets large and so we consider the average $\dfrac{1}{T}\displaystyle\int_{-\infty}^{\infty} x_T^2(t)dt$, which is the average power. By Parseval's theorem

$$\frac{1}{T}\int_{-\infty}^{\infty} x_T^2(t)dt = \frac{1}{T}\int_{-T/2}^{T/2} x^2(t)dt = \frac{1}{T}\int_{-\infty}^{\infty} |X_T(f)|^2 df \tag{27}$$

The quantity $|X_T(f)|^2/T$ is called the sample (raw) spectral density and is given the notation $\hat{S}_{xx}(f)$ and since the left hand side of the above equation is the average power of the sample function it is tempting to define $\lim_{T\to\infty}\left(|X_T(f)|^2/T\right)$ as the power spectral density. Unfortunately this is not useful since this quantity does not converge in a statistical sense, i.e., if this quantity is evaluated from a data length $T' > T$ the results are just as erratic as for the shorter data length. However, a very useful theoretical function may be introduced by averaging this quantity to 'remove' the erratic behaviour, i.e., consider the average

$$E\left\{\lim_{T\to\infty}\frac{1}{T}\int_{-T/2}^{T/2}x^2(t)dt\right\} = E\left\{\int_{-\infty}^{\infty}\lim_{T\to\infty}\frac{|X_T(f)|^2}{T}df\right\} \tag{28}$$

This means that the variance of the process is

$$\sigma_x^2 = \int_{-\infty}^{\infty}S_{xx}(f)df \tag{29}$$

where

$$S_{xx}(f) = \lim_{T\to\infty}\frac{E|X_T(f)|^2}{T} \tag{30}$$

This states that the average power of the process (the variance) is decomposed over frequency in terms of the function $S_{xx}(f)$. This function is the *power spectral density function* of the process. This is a useful function which has a direct physical interpretation. It can also be related directly to the autocovariance function in the following way.

$$S_{xx}(f) = \int_{-\infty}^{\infty}R_{xx}(\tau)e^{-j2\pi f\tau}\,d\tau \tag{31}$$

and

$$R_{xx}(\tau) = \int_{-\infty}^{\infty}S_{xx}(f)e^{j2\pi f\tau}\,df \tag{32}$$

5.2.1 *The cross spectral density function*
Generalising the above, the cross-spectral density is

$$S_{xy}(f) = \int_{-\infty}^{\infty}R_{xy}(\tau)e^{-j2\pi f\tau}\,d\tau \tag{33}$$

with the inverse

$$R_{xy}(\tau) = \int_{-\infty}^{\infty} S_{xy}(f)e^{j2\pi f\tau}\,df \tag{34}$$

Alternatively, $S_{xy}(f)$ may be written

$$S_{xy}(f) = \lim_{T\to\infty} E\frac{\left\{X_T^*(f)\,Y_T(f)\right\}}{T} \tag{35}$$

where $X_T(f)$, $Y_T(f)$ are Fourier transforms of $x(t)$, $y(t)$ for $-\dfrac{T}{2} \le t \le \dfrac{T}{2}$.

The cross spectral density is in general complex, i.e., $S_{xy}(f) = \left|S_{xy}(f)\right|e^{j\arg S_{xy}(f)}$. $S_{xy}(f)|$ is the *cross amplitude spectrum* and it shows whether frequency components in one time series are associated with large or small amplitudes at the same frequency in the other series. Arg $S_{xy}(f)$ is the *phase spectrum* and this shows whether frequency components in one series lag or lead the components at the same frequency in the other series.

6. Input-output relationships and system identification

We shall now develop input-output relationships for time invariant linear systems operating with stationary random inputs.

6.1. SINGLE INPUT SINGLE OUTPUT SYSTEMS

Starting with the convolution form of the input-output relation (equation (6)) it is possible to show that

$$S_{yy}(f) = |H(f)|^2 S_{xx}(f) \tag{36}$$

and

$$S_{xy}(f) = H(f)\,S_{xx}(f) \tag{37}$$

The expression (37) is often the basis of system identification schemes based on measurements of input and output.

6.1.1. *The ordinary coherence function*
The coherence between input $x(t)$ and output $y(t)$ is defined

$$\gamma_{xy}^2(f) = \frac{\left|S_{xy}(f)\right|^2}{S_{xx}(f)S_{yy}(f)} \tag{38}$$

and from the inequality $|S_{xy}(f)|^2 \le S_{xx}(f)\,S_{yy}(f)$ it follows that

$$0 \leq \gamma_{xy}^2(f) \leq 1 \qquad (39)$$

Applying (38) to a single input-single output system, $\gamma_{xy}^2(f) = 1$ for all f. So, if two processes are *linearly* related, the coherence is unity, and if x(t) and y(t) are completely unrelated (linearly) the coherence function is zero. If the coherence function is greater than zero but less than one, x and y are in part linearly related but, in addition,

(i)   the measurements are 'noisy'

(ii)   x and y are not *only* linearly related

(iii)   y is an output due to input x *and other inputs*.

The quantity $\gamma_{xy}^2(f) S_{yy}(f)$ is termed coherent output power, i.e., that part of the power of y accounted for by linear operations on x. and $\left(1 - \gamma_{xy}^2(f)\right) S_{yy}(f)$ is noise (uncoherent) output power.

6.2.   SYSTEM IDENTIFICATION

Least squares (optimal) methods may be applied to 'identifying' the linear relation between two time histories. Two common estimators are

$$H_1(f) = \frac{S_{xy}(f)}{S_{xx}(f)} \qquad \text{(unbiased with respect to noise on the output)} \qquad (40)$$

$$H_2(f) = \frac{S_{yy}(f)}{S_{yx}(f)} \qquad \text{(unbiased with respect to noise on the input)} \qquad (41)$$

If there is noise on both, then an approach based on Total Least Squares leads to results that are somewhat more complicated and lead to the so-called H_v estimator (Leuridan *et al*, 1986) and sometimes called the H_s estimator (Wicks and Vold, 1986).

It is often the case that feedback is present in a dynamical system (see Figure 5). r is the (controlled) input and n is an extraneous input. H is the system to be determined.

Forming $H_1 = S_{xy}/S_{xx}$ leads to $(HS_{rr} + G*S_{nn})/(S_{rr} + |G|^2 S_{nn})$ which is certainly not the required H. Indeed, if S_{nn}/S_{rr} is large then the ratio S_{xy}/S_{xx} is 1/G. It can be easily verified, however, that the use of the three signals r, x and y and forming the ratio of two cross spectral densities S_{ry}/S_{rx} (sometimes referred to as estimator H_3) yields the required H(f).

6.3.   MULTI-INPUT MULTI-OUTPUT SYSTEMS

For a multi-input multi-output system (mimo) system with, say, m inputs written in vector form $\mathbf{x}(t)$ and n outputs written in vector form $\mathbf{y}(t)$, then the Fourier transforms of input and output are related by

$$\mathbf{Y}(f) = H(f)\,\mathbf{X}(f) \qquad (42)$$

where $H(f)$ is an $n \times m$ matrix of frequency response functions.

If all the signals involved are stationary random processes and defining spectral density matrices, then

$$S_{yy}(f) = H(f) \, S_{xx}(f) \, H^T(f) \tag{43}$$

and

$$S_{xy}(f) = S_{xx}(f) \, H^T(f) \tag{44}$$

which are generalisations of equations (36) and (37) for the single-input single-output cases.

Obviously these matrix equations provide a starting point for analyses of mimo systems, e.g., the system might be identified from input-output measurements by forming

$$H^T(f) = S_{xx}^{-1}(f) \, S_{xy}(f) \tag{45}$$

assuming invertibility of $S_{xx}(f)$. However, other approaches have been developed which reveal more insight. These approaches attempt to interpret the matrix relationships using the concepts of residual random variables, residual spectra, partial and multiple coherence and principal component analysis.

6.4. RESIDUAL RANDOM VARIABLES

As a starting point for this, consider two processes $x(t)$ and $y(t)$ and regard $y(t)$ as made up of two components, one of which is linearly related to x and the other not. Figure 6 depicts the situation (Bendat and Piersol, 1980).

The least squares optimal linear filter linking x and y is $L(f) = S_{xy}(f)/S_{xx}(f)$. y_1 is fully coherent with x and $S_{y_1 y_1}(f) = \gamma_{xy}^2(f) S_{yy}$ is the coherent output power, whilst y_2 is uncoherent with x and the uncoherent (or noise) output power is $\left(1 - \gamma_{xy}^2(f)\right) S_{yy}(f)$ The process y_2 is what is left of y after the 'linear effects; of x have been removed and is a residual random variable. We speak of y as having been conditioned with respect to x.

This idea can be extended to multi-variate processes. To explain this, consider three random variables written as x_1, x_2, x_3. The concepts above can be applied sequentially as indicated in Figure 7.

Referring in particular to 'stage 1' in the figure, it is noted that x_2 and x_3 have been conditioned with respect to x_1 (this choice is arbitrary), and from the earlier work one can write down the following:

(i)   The 'coherent output power', i.e., that proportion of the power of x_2 accounted for by linear operations on x_1 is $\gamma_{12}^2 S_{22}$ [N.B. $S_{22} = S_{x_2 x_2}$, etc.].

(ii) The 'noise' power $S_{y_3 y_3}$) is a residual power spectral density written $S_{22.1}$ and is $S_{22.1} = (1 - \gamma_{12}^2) S_{22}$.

(iii) The optimal linear filter L_1 is S_{12}/S_{11}.

Identical expressions hold for processes x_1, x_3 with 3 replacing 2 above.

It is important to realise that L_1 and L_2 are not the elements of the H matrix but can be related to them. It is possible to show that the residual cross-spectral density is

$$S_{23.1} = S_{23} - \frac{S_{21} S_{13}}{S_{11}}$$

i.e., the residual spectral density is evaluated in terms of ordinary spectra.

These concepts introduced at 'stage 1' can be extended in similar fashion to 'stage 2' where x_3 can be further conditioned, now with respect to x_2. y_6 is that part of $x_{3.1}$ coherent with $x_{2.1}$, and y_7 is that part of x_3 that is uncoherent with both x_1 and x_2 or in other words that part of x_3 not accounted for by linear operations on x_1 and x_2. This means y_7 is the residual random variable formed by conditioning x_3 with respect to both x_1 and x_2 and is written $x_{3.12}$.

At stage 2 (by analogy with stage 1) one can write down the following:

(i)   The output noise power spectral density is $S_{33.12} = (1 - \gamma_{23.1}^2) \, S_{33.1}$ where

(ii)   the partial coherence function $\gamma_{23.1}^2$ is

$$\gamma_{23.1}^2 = \frac{S_{23.1} \, S_{32.1}}{S_{22.1} \, S_{33.1}}$$

and

(iii)   the optimal filter L_3 is $\dfrac{S_{23.1}}{S_{22.1}}$.

The total output power spectral density can be decomposed as the sum of three terms (and this can be done directly from the diagram):

$$S_{33} = \qquad \gamma_{13}^2 \, S_{33} \qquad + \qquad \gamma_{23.1}^2 \, S_{33.1} \qquad + \qquad S_{33.12}$$

part fully coherent with x_1	part fully coherent with x_2 after x_1 has been removed from x_2 and x_3	uncoherent with both x_1 and x_2

The multiple coherence function (written $\gamma_{y.x}^2$) is defined by analogy with the ordinary coherence function as for a single-input system, i.e., it is that fraction of power accounted for in the output via linear relationships between input and output, i.e.,

$$\gamma_{y.x}^2 = \frac{S_{33} - S_{33.12}}{S_{33}}$$

and by the expressions above can be written

$$\gamma_{y.x}^2 = 1 - (1 - \gamma_{13}^2)(1 - \gamma_{23.1}^2)$$

We emphasise that the multiple coherence function is a measure of how well the two inputs x_1 and x_2 account for the measured response of the system, whilst the partial coherence is a measure of how well an additional signal (in this case x_2) improves the predicted output.

It is interesting to note that this approach to conditioning is equivalent to solution of the equation (44) by Gaussian elimination.

6.5.   PRINCIPAL COMPONENT ANALYSIS

A shortcoming of residual spectral analysis is that some prior ranking of the input signal would be helpful, i.e., some *a priori* knowledge. Principal component analysis is a general approach that explores correlation patterns in multiple signals and is used in source identification (Otte *et al*, 1988)

Suppose we have three processes x_1, x_2 and x_3; then we can begin by forming the cross-spectral density matrix $S_{xx}(f)$ (which is Hermitian). If there is a linear relationship between the x_i's then the determinant of this matrix is zero, i.e., its rank is < 3. If there is no linear relationship then its rank is 3.

For the full rank case the matrix $S_{xx}(f)$ can be expressed, i.e., $U \Lambda U^+$ where Λ is a diagonal matrix of eigenvalues of S_{xx} (which are real and positive because S_{xx} is Hermitian) and U is the matrix of eigenvectors. From this it can be argued that the elements of Λ are the power spectra of fictitious random processes which are mutually uncorrelated which pass through fictitious filters characerised by elements of U. These fictitious (or virtual) processes have power spectra whose magnitudes Λ define the principal components referred to. From these considerations one can deduce the number and relative magnitude of these principal components and introduce concepts such as virtual coherence.

If the rank of S_{xx} is not full, then the eigenvalue decomposition is replaced by the singular value decomposition of S_{xx} and the above considerations are generalised appropriately.

7. References

1.   Ables, J. G. (1974), Maximum Entropy Spectral Analysis,*Astronomy and Astrophysics Supplement Series*.
2.   Bendat, J.S. and Piersol, A.G. (1971), Random Data: Analysis and Measurement Procedures, *Wiley Interscience*.
3.   Bendat, J.S. and Piersol, A.G. (1980), Engineering Applications of Correlation and Spectral Analysis, *Wiley and Sons*.
4.   Harris, F.J. (1978), On the use of windows for harmonic analysis with the discrete Fourier transform, *Proc. IEEE 66(1)*.
5.   Hsu, H.P. (1970), Fourier Analysis, *New York: Simon and Schuster*.
6.   Leuridan, J., De Vis, D., Van der Auweraer, H. and Lembregts, F., (1986), A comparison of some frequency response measurement techniques *Proc. 4th International Modal Analysis Conference, Los Angeles*, 908-918.
7.   Otnes, R.K. and Enochson, L. (1978), Applied Time Series Analysis, Vol. 1, Basic techniques, *Wiley and Sons*.
8.   Otte, D., Sas, P. and Van de Ponseele, P. (1988), Principal component analysis for noise source identification, *Proc. 6th IMAC, Florida*, 1207-1214.

9. Priestley, M.B. (1967), Power spectral analysis of nonstationary processes, *Journal of Sound and Vibration* , 6, 86-97.
10. Wicks, A.L. and Vold, H. (1986), The H_s frequency response function estimator, *Proc. 4th International Modal Analysis Conference, Los Angeles*, 897-899.

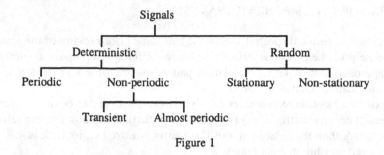

Figure 1

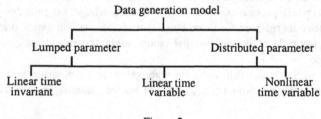

Figure 2

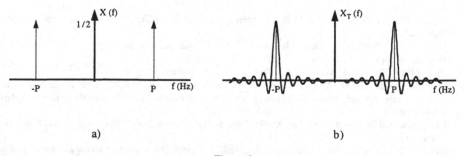

a)   b)

Figure 3

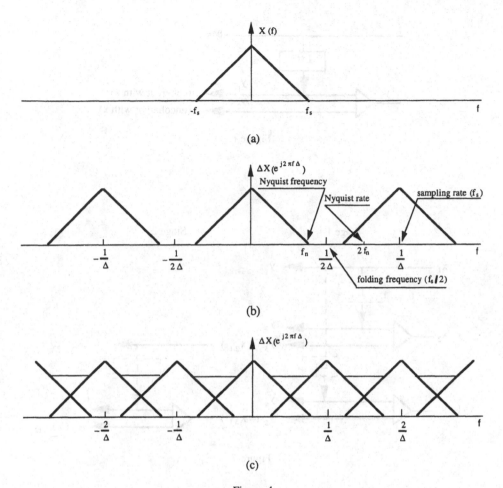

(a)

(b)

(c)

Figure 4

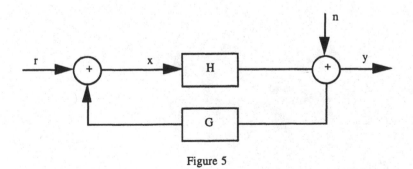

Figure 5

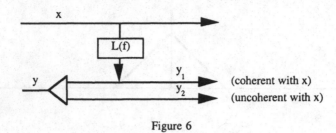

Figure 6

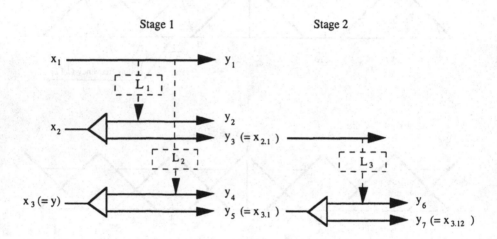

Figure 7

INTRODUCTION TO SIGNAL PROCESSING

Part 2: Advanced Signal Processing

J. K. HAMMOND
Institute of Sound and Vibration Research
University of Southampton

1. Introduction

This chapter is intended to provide an introduction to some topics in signal processing that are finding wide application. These are optimal (and adaptive) filtering using least squares criteria, deconvolution (including cepstral analysis), non-stationary signals with particular reference to time-frequency methods and, finally, higher-order spectra (for non-Gaussian and nonlinear processes).

2. Optimal filters

A classic problem in signal processing is that of designing an optimal linear filter. Suppose we have a signal x(t) and we wish to operate on it to make it look like another signal d(t). Figure 1 depicts the situation. In mathematical terms, we seek filter impulse response f(t) so that some function of e is minimised. We shall restrict this to be an average of $e^2(t)$. Treating x and d as random signals, then the average is $E[e^2(t)]$. The necessary condition satisfied by the optimal filter is

$$E[e(t)\,x(t')] = 0 \tag{1}$$

i.e., the error e(t) is uncorrelated with the observation x(t). This is sometimes called the orthogonal projection lemma or the Wiener-Hopf condition. If t' is restricted to t' < t, then this imposes further conditions on the optimal filter.

Substituting for e(t) = d(t) - y(t) and noting y(t) is the convolution of x(t) with filter impulse response f(t) gives the equation from which f(t) is to be found (for stationary random processes and a time invariant filter)

$$R_{dy}(\tau) = \int_a^b R_{yy}(\tau - \tau_1) f(\tau_1)\,d\tau_1 \tag{2}$$

If f(t) is not restricted to be causal then the limits are $a = -\infty$, $b = \infty$ and the solution may be found by Fourier transformation.

If f(t) is causal then the limits are $a = 0$ and $b = \infty$ and the equation is true for $\tau > 0$. A direct Fourier transform cannot now be taken. The solution of this follows the method of spectral factorisation (Laning and Battin, 1956).

J.M.M. Silva and N.M.M. Maia (eds.), Modal Analysis and Testing, 53–64.
© *1999 Kluwer Academic Publishers. Printed in the Netherlands.*

2.1. OPTIMUM DISCRETE TIME FILTERS

We now reformulate the above when the time histories have been sampled and the filter f is a digital filter. If the signals are non-random one may use $1/N\left(\sum e^2(n)\right)$ rather than $E[e^2(n)]$. The procedures above are applied to yield the equation satisfied by the optimal filter, i.e., $R_{xx}(n) * f(n) = R_{xd}(n)$. If $f(n)$ is restricted to being an L-point finite impulse response (FIR) filter, this equation may be written

$$\sum_{k=0}^{L-1} f(k) \sum_{n} x(n-k)x(n-i) = \sum_{n} d(n)x(n-i) \qquad i = 0, 1,..., L-1$$

or, in matrix form

$$Rf = g \qquad (3)$$

where R is the correlation matrix formed from $x(n)$, f is the vector of filter weights and g is a cross-correlation vector.

These equations are known as the *normal equations*. The specification of the range of summation of n leads to two distinct formulations (Makhoul, 1975) known as the autocorrelation and the covariance methods, respectively. There are several variations on this basic theme ultimately concerned with the specification of the optimal filter coefficients as the solution of a set of linear equations, often employing the so-called 'Levinson procedure' or the Durbin modification (Durbin, 1959).

2.2. ITERATIVE SOLUTIONS

Rather than attempting to solve for f as $R^{-1}g$ directly, an iterative solution may be employed. If $J = \dfrac{1}{N}\sum_{n=0}^{N-1} e^2(n)$ and if f_k is the set of filter weights at iteration k, then a gradient based approach to give f_{k+1} is

$$f_{k+1} = f_k - \frac{\alpha}{2}\nabla J_k$$

$\nabla J = 2(Rf_k - g)$ so the iterative algorithm is

$$f_{k+1} = [I - \alpha R]f_k - \alpha g \qquad (4)$$

This is the steepest descent algorithm and it can be shown that it converges to f_{opt} subject to the restriction

$$0 < \alpha < 2/\lambda_{max}$$

where λ_{max} is the largest eigenvalue of R (Widrow and Stearns, 1985).

2.3. THE LMS ADAPTIVE FILTER

To implement the above, we need to use *all* the data at each step. This is not helpful if the data is non-stationary. Using the instantaneous squared error $e^2(n)$ instead of $\sum e^2(n)$ leads to the celebrated LMS update equation (Widrow and Stearns, 1985)

$$f_{k+1} = f_k + \alpha e(k) x_k \qquad (5)$$

where $e(k)$ is the current (instantaneous) error and x_k denotes the vector of the last L data points. The range of application of this formulation and its many derivatives and extensions is quite enormous. Despite its complexity (nonlinearity, non-stationarity) it has proved effective, resilient and robust.

3. Inverse problems and deconvolution

Deconvolution is a term which embraces a wide range of signal processing theory. The aim is to unravel the effects of a convolution of two signals (or a system and signal) so as to restore one of the signals. For example, if a signal x is operated upon by a system with impulse response h to produce an output y, we seek to design and apply some operator f which will restore x.

It is assumed that the operation on x is linear and that h and y are known. The operator f is the 'inverse filter' corresponding to h and may be denoted h^{-1}. However, it may not always be possible to realise an exact inverse, in which case it is necessary to make do with an approximation to h^{-1} and a corresponding approximation to the signal x. We therefore write the output of f as $\hat{x}$ (the 'hat' notation denotes an estimate of x) and define an error $e = x - \hat{x}$. We now try and find the filter f that minimises some function of the error. The problem posed relates to the deconvolution problem in its barest form. More usually the measurement y is noise contaminated and/or the system dynamics h are not known exactly (indeed the system may be nonlinear). It is these embellishments to the basic theme which account for the large number of techniques that have been developed with the objective of deconvolution.

3.1. 'DIRECT' INVERSION

We shall look at the simplest case here, namely, when measurement y is noise-free and h is known exactly. The treatment will be in discrete time.

3.1.1. *Inverse digital filtering*
Suppose that h is a discrete time system with known transfer function H(z) which may be expressed as

$$H(z) = \frac{b_0 + b_1 z^{-1} + \dots b_M z^{-M}}{a_0 + a_1 z^{-1} + \dots a_N z^{-N}} = \frac{b_0 \prod\limits_{i=0}^{M}\left(1 + \alpha_i z^{-i}\right)}{a_0 \prod\limits_{j=0}^{N}\left(1 + \beta_j z^{-j}\right)} \qquad (6)$$

with zeros at $-\alpha_i$ and poles at $-\beta_j$ (the poles β_j satisfy $|\beta_j| < 1$ for stability). We seek f such that $h(n) * f(n) = \delta(n)$ or equivalently $H(z)F(z) = 1$, and we write, formally, $F(z) = 1/H(z)$. Unfortunately, of course, such a solution is not generally possible since the zeros of $H(z)$ need not lie within the unit disc in the z-plane (i.e., $H(z)$ may not be minimum phase). Thus $1/H(z)$ will have poles outside the unit disc and will therefore be unstable if realised as a difference equation in forward time'.

Three possibilities now present themselves:

(a) We form the inverse of the minimum phase version of $H(z)$. More specifically, any general transfer function $H(z)$ may be written $H(z) = H_{eq}(z)H_{ap}(z)$. $H_{eq}(z)$ denotes a transfer function having identical modulus to $H(z)$ but with zero locations adjusted to be within $|z| = 1$, and $H_{ap}(z)$ is an 'all pass' transfer function having unit modulus but with zeros outside the unit disc.

The all pass component has zeros and poles at conjugate reciprocal locations. The portion $H_{eq}(z)$ will have a stable, realisable inverse in forward time resulting in a signal $\hat{x}$ that is distorted by the all pass component. The *modulus* of the Fourier transforms of x and $\hat{x}$ are indistinguishable but the phase structures will differ.

(b) If we do not restrict ourselves to processing only in forward time then a factorisation of $H(z)$ into the form $H(z) = H_{min}(z)H_{max}(z)$ where $H_{min}(z)$ denotes the minimum phase part of $H(z)$ and $H_{max}(z)$ the maximum phase part of $H(z)$, permits complete deconvolution. This is demonstrated for a single zero lying outside the unit disc. Suppose we have $H(z) = 1 + kz^{-1}$, $|k| > 1$. Clearly, $1/H(z)$ is unstable in forward time. However, it can be argued that the inverse is stable in *reverse* time.

(c) The third possibility is an approximate version of (b). If it is imperative that processing is done in forward time, then the scheme (b) can be modified by calculating the impulse response function of the 'reverse time inverse' which is non-causal as far as the original data is concerned, truncating it at some suitable point, delaying it by its duration (to make it causal) and implementing it as a finite impulse response filter. The delay imposes a linear phase structure to the output but also the truncation introduces distortion to amplitude and phase.

As a final discussion on 'direct' methods, we consider the 'Fourier' procedure of undoing $h(n)$ by dividing the Fourier transform of y by the Fourier transform of h and inverse transforming the result. If $x(n)$ is an N_1 point sequence, $h(n)$ an N_2 point sequence, then $y(n) = h(n) * x(n)$ is an $L = N_1 + N_2 - 1$ point sequence. To obtain this convolution using the discrete Fourier transform (DFT) and to avoid distortions in the result due to 'circular convolution' zeros must be appended to $x(n)$ and $h(n)$ to make both, say, P point sequences $(P \geq L)$. Then the product of the P point DFT of $h(n)$ and $x(n)$ gives the DFT of a periodic sequence, one period of which is the required sequence $y(n)$. So we can write $X(k) = Y(k)/H(k)$. It is necessary, however, to resist the temptation to say that the sequence $h_1(n)$ whose DFT is $1/H(k)$ is the impulse response function of the inverse filter of $h(n)$ (Oppenheim and Schafer, 1975).

We note in conclusion that these discussions, relating as they do to ideal inversion, take no account of additive noise. It is worth noting that a compromise between inverse filter bandwidth and output signal-to-noise ratio may be considered by relaxing the requirement that $h * f = \delta$ and instead of the delta function use a more 'spread' function.

3.1.2. *Optimal deconvolution*

A fundamental feature of inversion problems is that many are inherently ill-conditioned, i.e., small perturbations to the output result in large changes under the inversion process. 'Optimal methods' may be employed to assist, resulting in so-called regularisation of the inversion. For example, the concepts of Section 2 may be recast as follows.

Suppose we wish to find an operator f(n) to invert an impulse response sequence h(n), then seek f(n) such that h(n)*f(n) δ(n). This can be posed as depicted in Figure 1 when x(t) = h(n), d(t) = δ(n), in which case equation (3) still gives the optimal filter where now R is the correlation matrix formed from h(n) and g is the cross-correlation of h(n) with δ(n), i.e., a vector with just one non-zero entry.

From the preceding discussion, clearly if h(n) is non-minimum phase then the output of the filter y cannot be δ(n) exactly. In fact, since the autocorrelation function is 'phase blind', then R is interpreted as the equivalent minimum phase version of h(n).

In order to 'properly' account for any non-minimum phase form of h(n) the 'desired signal' in Figure 1 is changed from δ(n) to δ(n - k), i.e., we try to approximate a delayed delta function. This changes the vector g and provides inversion of h(n) for an 'optimal' k (chosen by successive calculation of f and monitoring of the mean squared error). Furthermore, the inversion of R may be difficult if this problem is ill-conditioned and this may be 'improved' by writing $R + \sigma^2 I$ for R when σ^2 is a regularising parameter.

We note also that matrix methods may be employed in a somewhat different way by writing the input-output convolution relationships between an input x to a system with output y as

$$y = Ax.$$

The optimal recovery of x is written A^+y, where A^+ is a pseudo-inverse matrix. Methods for obtaining this are based on singular value decomposition. See Press *et al* (1992) for an excellent treatment of this and related matters.

4. Cepstral analysis

Cepstral analysis, sometimes referred to as homomorphic deconvolution, offers another approach to separating convolved signals. We consider the so-called complex cepstrum only and use discrete time.

4.1. DISCRETE TIME FORMULATION

The discrete time formulation is normally written using the z-transform even though the implementation is achieved using the FFT (see Oppenheim and Schafer, 1975).If $x_1(n)$ and $x_2(n)$ are two real sequences and if x(n) is formed as the convolution of the two, i.e., $x(n) = x_1(n) * x_2(n)$, then the complex cepstrum of x(n) defined in Figure 2 is $x^c = x_1^c + x_2^c$, i.e., the complex cepstrum splits convolution into addition. *N.B.* 'Complex' refers to the complex logarithm. $x^c(n)$ is real. Evaluation is normally done on the unit circle |z| = 1. Some comments are now in order regarding the definition of the complex logarithm. From Figure 2 we see

$$X^c(z) = ln[X_1(z)X_2(z)] = ln|X_1(z)| + ln|X_2(z)| + j(\arg X_1(z) + \arg X_2(z)) \qquad (7)$$

Whilst there is no ambiguity in defining the real part of the complex logarithm (when it exists), the imaginary part is multivalued (and so non-unique). The use of the principal value of the argument is inadmissible since the resulting discontinuities in the imaginary component are not permitted for a real $x^c(n)$. These problems are circumvented in the following way. If $x^c(n)$ is assumed to be a stable, real sequence then $X^c(z)$ is analytic on $|z| = 1$ so that the real and imaginary parts $X_R^c(e^{j\omega})$ and $X_I^c(e^{j\omega})$ are continuous (and periodic) functions of ω, and since $\hat{X}_R(e^{j\omega}) = ln|X(e^{j\omega})|$ and $\hat{X}_I(e^{j\omega}) = \arg(X(e^{j\omega}))$ then these functions, too, are continuous (and periodic). So it is the following conditions that dictate how the imaginary part of the complex logarithm should be defined, i.e., $\arg X(e^{j\omega})$ should be (i) uniquely defined; (ii) continuous; (iii) periodic with period 2π; (iv) the value at $\omega = \pi$ is zero.

Once the phase is satisfactorily defined then, the complex cepstrum separates convolved signals into additive form, where they *may* be separated and then the inverse operations performed to recover one or other of the signals.

5. Non-stationarity and time-frequency representations

A non-stationary signal is one whose 'internal structure' changes as a function of time. Examples include speech, noise and vibration signals from accelerating vehicles, Doppler shifted sound, chirp signals, the impulse response of a damped non-linear Duffing oscillator and trajectories of a Lorenz system. In each case the phenomena exhibit "non-stationarity" though we might not be specific about the form this takes.

Methods addressing non-stationarity by decomposing signals over two dimensions have been developed. Such decompositions may lead to "double-frequency" spectra or the more physically meaningful time-frequency spectra. The latter allow time histories to be examined in a way that the individual time and frequency domains do not. The advantages are that the time-frequency decompositions permit projections of the time history on to a space that allows separation of components of the signal facilitating enhancement, detection, filtering, classification and resynthesis (Hammond and White, 1996).

5.1. BASIC CONSIDERATIONS

5.1.1. *The Uncertainty Principle*
Fourier analysis provides a decomposition of a time history in the frequency domain through the linear representation and defining a suitable bandwidth and duration, one can show that the bandwidth-time product is never less than 1/2 (see Part I, Section 3.4). In essence, this Uncertainty Principle says that the more precisely one specifies "when" for a signal (i.e. T reducing) the less specifically can one specify what frequencies are involved.

The group delay (see Part I, Section 3.1) of a signal x(t) is obtained from the Fourier transform $X(\omega)$ by

$$\tau_g(\omega) = -\frac{d}{d\omega} \arg X(\omega) \tag{8}$$

This may be interpreted as the time at which frequency component ω makes its maximum contribution to a signal. This is, therefore, an attempt to link the concepts of time and frequency based on the Fourier transform. An alternative approach uses the concept of the Hilbert transform of a signal (see Part I, Section 3.5).

The *instantaneous frequency* is defined for $x(t)$ by first forming the analytic signal, $\sigma_x(t)$, where.

$$\sigma_x(t) = x(t) + j\hat{x}(t) = A(t)e^{j\phi(t)}$$

The instantaneous frequency is

$$w_i(t) = \dot{\phi}(t) \tag{9}$$

The two apparently different relationships, equations (8) and (9), link time and frequency for the signal $x(t)$ and one might ask if they are essentially the same - specifically whether they are inverse functions of each other. In general they are not - and the papers referred to in Hammond and White (1996) describe conditions that must be met for these two time-frequency laws to be equivalent. These are that the signal should be "asymptotic", i.e., that the BT product should be "large" and that the signal should be "monocomponent", i.e., $w_i = \dot{\phi}$ should be invertible.

5.2.   SPECTRAL REPRESENTATIONS AND NON-STATIONARITY

In this section we shall give a brief overview of the various approaches to the definition of time-frequency spectra.

A common starting point is the concept of the time-varying periodogram, i.e., obtained by sliding a window across a time record, then performing a Fourier transform to obtain the Short Time Fourier Transform (STFT)

$$S(t,\omega) = \int x(t') w(t-t') e^{-j\omega t'} dt' \tag{10}$$

A method of displaying this is to form $|S(t,\omega)|^2$ and plotting this as a function of ω and t. This squared magnitude of this quantity is called the periodogram or spectrogram and this three-dimensional plot shows how the energy of the process is decomposed over time and frequency.

5.2.1.   *Linear representations*
This section is based on the idea of representing a signal in terms of a set of other (simpler) components which when added up yield the original signal.

The evolutionary spectral density. Priestley (1967) proposed a modification of the Fourier representation for random signals alluded to in Part I, Section 2. The essence of Priestley's idea is that the basic building blocks in the representation are *amplitude modulated* sines and cosines so that

$$x(t) = \int A(t,\omega)e^{j\omega t}dZ_x(\omega) \tag{11}$$

For the class of processes $x(t)$ admitting this representation (called oscillatory processes) this leads to the definition of the evolutionary spectral density,

$$S_e(t,\omega) = |A(t,\omega)|^2 \, S(\omega) \tag{12}$$

The Gabor representation (see Hammond and White, 1996) considers a representation of signals in terms of basic building blocks that are Gaussian pulse modulated sines and cosines of the form

$$w(t - n\Delta)e^{j\omega_k t} \tag{13}$$

These basic elements have the lowest BT product achievable. The signal is considered to be a summation of these: i.e.,

$$x(t) = \sum \sum c_{nk} w(t - n\Delta)e^{j\omega_k t} \tag{14}$$

It is interesting to note that the short-time Fourier transform is closely related to the above. If one considers the STFT of $x(t)$, as defined by (10), then it can be shown that for any finite energy function $v(t)$ where

$$\int v(\tau)w(\tau)d\tau = C_s \neq 0$$

then

$$x(t) = \frac{1}{C_s} \int\int S(t',\omega)\, v(t - t')\, e^{j\omega t'} d\omega dt' \tag{15}$$

In this context the Gabor representation can be considered as a sampled version of the STFT, where the sampling is two dimensional and occurs in the time-frequency plane. It is important to note that it is not possible to create a Gabor representation in which the basis functions, defined in (13), are both orthogonal and have good localising properties in the time and frequency domains. However, wavelet transforms, discussed in the next section, can be constructed which do possess both of these desirable properties .

Wavelets. Put in a simplified way, wavelet analysis may be considered as a generalisation of using the Gabor decomposition (14) in which both the modulating function and the oscillatory component are scaled. This allows high 'frequencies' to be resolved closely in time and low frequencies to be well resolved in frequency, akin to a constant 'Q' frequency analysis.

Like the STFT, the wavelet transform creates a function of two variables from a time history. In the case of wavelets the time history is decomposed as a function of time and scale, a. The wavelet transform is calculated by forming inner products of the time history with versions of a second function w(t), (the *mother* wavelet). One only has to ensure that the mother wavelet has zero mean, ensuring that it must oscillate and decay in some manner. The continuous wavelet transform is defined as

$$W(t,a) = \int x(\tau)w((t-\tau)/a)^* d\tau \qquad (16)$$

Often, only the squared modulus of the wavelet transform is considered; this is called the scalogram. This relationship mimics that of the STFT to the spectrogram.

From the wavelet transform one can recover the original time history by using

$$x(t) = \frac{1}{C_g} \int\int \frac{1}{\sqrt{a}} W(t,a) w((t-\tau)/a) \frac{da\, d\tau}{a^2} \qquad (17)$$

where C_g is a constant.

5.2.2. *Quadratic forms*
This section is concerned with what is sometimes referred to as quadratic representations or the "energetic" point of view on time-frequency methods.

The autocorrelation form. The spectral density of a stationary signal may be obtained from the autocorrelation function, i.e.,

$$S(\omega) = \frac{1}{2\pi} \int R(\tau)e^{-j\omega\tau} d\tau \qquad (18)$$

For a non-stationary signal the concept of a "local" autocorrelation function is logical, i.e., one that quantifies the behaviour of the average of products $x(t_1)x(t_2)$ for t_1 and t_2 "in the vicinity" of time t. Then a corresponding time-dependent spectral density might be defined as

$$S_Q(t,\omega) = \int R(t,\tau)e^{-j\omega\tau} d\tau \qquad (19)$$

(the subscript referring to quadratic). This will have the property of an "energetic" description if $S_Q(t,\omega)$ describes the energy (or, in the stochastic case, power) of the process over the t-ω plane such that

$$E_x = \int\int S_Q(t,\omega)\, dt\, d\omega \qquad (20)$$

A variety of approaches to the definition of the local autocorrelation may be conceived. A candidate is

$$R(t,\tau) = \int g(u-t,\tau)x^*(u-\tau/2)x(u+\tau/2)du \qquad (21)$$

One can envisage a variety of functions g that lead to greater or lesser concentration "near" time t.

The Wigner distribution. Let us take the very special case of the "ultimate" concentrating function g above, namely a delta function, i.e., $g(u - t) = \delta(u - t)$, then

$$R(t,\tau) = x^*(t - \tau/2) \, x(t + \tau/2) \tag{22}$$

and

$$S_w(t,\omega) = \int x^*(t - \tau/2) x(t + \tau/2) e^{-j\omega\tau} d\tau \tag{23}$$

This is the Wigner-Ville distribution. It picks out the instantaneous frequency law perfectly for a linear chirp signal and it is tempting to think that this will be mirrored for other signals - but it turns out that it is only for a single (monocomponent) linear chirp that it "works" so well. Its more disconcerting characteristics, however, relate to two other properties. Despite being an energy density distribution it is *negative* for some values. The second disconcerting attribute arises from the existence of "cross-terms". This is a consequence of the bilinear nature of its definition resulting in unwelcome non-zero values in the t-ω plane.

The Cohen class of distributions. Cohen generalised the definition of time-frequency distributions in such a way as to include a wide variety of different distributions (Cohen, 1989). One can define Cohen's class as the Fourier transform, with respect to τ, of the generalised local correlation function in equation (21), where the function $g(t,\tau)$ is referred to as the kernel function, the choice of which defines the different distributions.

5.3. INTERLINKING OF DISTRIBUTIONS

We briefly summarise some results of the recent past in which the Wigner distribution plays a pivotal role. Specifically, if x(t) has a Wigner distribution $S_w(t,\omega)$ and if $S_g(t,\omega)$ denotes a (general) distribution belonging to the Cohen class, then

$$S_g(t,\omega) = \int S_w(\tau,v) \Pi(\tau - t, v - \omega) d\tau \, dv \tag{24}$$

where $\Pi(t, \omega)$ is an arbitrary function of time and frequency.

Smoothing of this nature is referred to as "regular" smoothing. An associated remarkable result uses the concept of affine smoothing. This result states that the scalogram is related to the Wigner-Ville distribution by affine smoothing: i.e.,

$$|W(t,a)|^2 = \int\int S_w(\tau,v) H_w\left(\frac{\tau - t}{a}, av\right) d\tau \, dv \tag{25}$$

where $H_w(t,\omega)$ now represents the Wigner distribution of the mother wavelet function.

Thus, with the Wigner distribution as the core distribution along with the concepts of regular and affine smoothing, we obtain a unified perspective on a class of distributions. However, the relationship between the members of the Cohen class and the evolutionary spectral density (for stochastic processes) does not conform to this, but is amenable to an alternative analysis. It has been shown (Hammond and White, 1996) that if $S_w(t,\omega)$ is the Wigner distribution, and $S_e(t,\omega)$ is the evolutionary spectral density of a process with modulating function $A(t,\omega)$, then

$$W(t, v) = \frac{1}{2\pi} \int V(t, \tau, \omega) S_e(t, \omega) \, d\omega \qquad (26)$$

where V is a normalised Wigner distribution of the modulating function.

6. Higher order spectra

Traditionally, a zero mean random process is described via its second order statistics, i.e., its autocorrelation function or, equivalently, its power spectrum. These measures are in general only a partial description of an arbitrary random process. To see this, one need only consider two discrete random processes commonly found in software packages: Gaussian white and uniform white noises. These two processes have the same spectral characteristics (and hence the same correlation structure) but their time series look very different; the uniform noise process has strict upper and lower bounds unlike its Gaussian counterpart. In this case, second order characterisations are unable to distinguish between these two processes and one may naturally ask what descriptors of a random process can one consider which are capable of characterising the differences between these two signals? One natural solution is to extend the principles of power spectra and correlation functions to orders greater than two; such methods are collectively known as Higher Order Spectra (HOS) (some authors prefer the terminology Higher Order Statistics). See Collis (1996) for details.

Gaussian processes are intimately connected with the study of second order methods since it is only Gaussian processes which can be completely described by their second order properties. Hence if a process is Gaussian then the HOS contain no new information, i.e., no information which cannot be obtained from the second order statistics. This leads us to one desirable property of HOS, namely, that we would like them to be identically zero for a Gaussian process.

HOS yield information about a signal's non-Gaussianity. The mechanism by which this non-Gaussianity arises is application dependent and is often the subject of *a priori* assumptions. Considering the signal as an output of some system, then it is common to assume either that the input is Gaussian and the system is nonlinear or that the input is non-Gaussian and the system is linear. The case of a Gaussian input to a linear system leads to a Gaussian output and so HOS yield no information, whilst the case of non-Gaussian inputs to nonlinear systems leads to problems whose complexity is often too great to deal with.

The first of the HOS is the third order spectrum, given the name bispectrum. This quantity has received most attention in the literature since it is the simplest of the HOS. However, the bispectrum only yields information in cases where the random process has a skewed distribution. In a significant number of physical problems, systems are

64

symmetrical and yield unskewed output signals; in these circumstances the bispectrum is an uninformative measure. See Collis (1996) which shows the concepts associated with the bispectrum carry over to the fourth order spectrum, referred to as the trispectrum, and to discuss how the trispectrum can be used to analyse symmetric nonlinearities.

7. References

1. Cohen, L. (1988), Time-frequency distributions - A review, *Proceedings of the IEEE 77(7)*, 941-981.
2. Collis, W. B. (1996), Higher order spectra and their application to nonlinear mechanical systems, *University of Southampton Ph.D. Thesis*.
3. Durbin, J. (1959), Efficient estimation of parameters in moving-average models, *Biometrika 46, Parts I and 2*.
4. Hammond, J. K. and White, P. R. (1996), The analysis of non-stationary signals using time-frequency methods, *Journal of Sound and Vibration 190(3)*, 419-447.
5. Laning, J. H., and Battin, R. H. (1956), Random Processes in Automatic Control, *New York: McGraw Hill*.
6. Makhoul, J. (1975), Linear prediction: a tutorial review, *Proceedings of the IEEE, 63*, 561-581.
7. Oppenheim, A. V., and Schafer, R. W. (1975), Digital Signal Processing. *Prentice Hall*.
8. Press, W. H., Teukolsky, S. A., Vetterling, W. T. and Flannery, B.P. (1992), Numerical Recipes, *Cambridge University Press*.
9. Priestley, M. B. (1967), Power spectral analysis of non-stationary processes, *Journal of Sound and Vibration 6*, 86-97.
10. Widrow, B. and Stearns (1985), Adaptive Signal Processing, *Prentice Hall*.

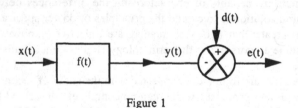

Figure 1

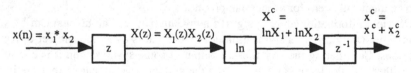

Figure 2

RULES FOR THE EXCHANGE AND ANALYSIS OF DYNAMIC INFORMATION

Part I: Basic Definitions and Test Scenarios

P. S. VAROTO
Dept de Engenharia Mecanica
Escola De Engenharia De São Carlos,
USP
São Carlos – SP – 13560-250, Brasil

K. G. McCONNELL
Dept. Aerospace Engineering and
Engineering Mechanics
Iowa State University
Ames, IA 50011 USA

Abstract

Three major structures are involved in laboratory simulations of field vibration environments: the test item, the vehicle, and the vibration exciter. The test item is the structure under study and is attached to the vehicle in the field environment. The word vehicle is used here to refer to any structure used in the field to transport, or simply to support the test item. In laboratory simulations, the test item is attached to one or more vibration exciters. The ultimate goal of a laboratory simulation is to define appropriate test item inputs such that its field dynamic behavior can be reasonably simulated, particularly stress levels, natural frequencies, and mode shapes. This paper describes a general theoretical model for laboratory simulations of field dynamic environments. Frequency domain input-output relationships are used to model the structures involved in the simulation process. Matrix partitioned frequency domain equations of motion are written for each structure. Manipulation of these equations with suitable boundary conditions leads to general expressions for the test item interface forces and motions in both field and laboratory environments. These expressions are used to describe four laboratory test scenarios.

Nomenclature

$[E]$	Exciter FRF matrix	$[H]$	FRF matrix
$[E_{cc}]$	Exciter connector FRF matrix	$[H_{cc}]$	Connector FRF matrix
$[E_{ce}]$	Exciter connector-external FRF matrix	$[H_{ce}]$	Connector-external FRF matrix
		$[H_{ee}]$	External FRF matrix
$[E_{ee}]$	Exciter external FRF matrix	[LED]	Laboratory Environment Dynamic matrix
$\{F\}$	Test item field-input force		
$\{F_c\}$	Test item field connector force	$\{P\}$	Vehicle input force
$\{F_e\}$	Test item external force	$\{P_c\}$	Vehicle interface force
[FED] Field Environment Dynamic matrix		$\{P_e\}$	Vehicle external force
		$\{Q_c\}$	Exciter connector force

J.M.M. Silva and N.M.M. Maia (eds.), Modal Analysis and Testing, 65–82.
© *1999 Kluwer Academic Publishers. Printed in the Netherlands.*

66

$\{Q_e\}$	Exciter external force	$[V_{ee}]$	Vehicle external points FRF matrix
$\{R\}$	Test item laboratory-input force		
$\{R_c\}$	Test item laboratory connector force	$\{X\}$	Test item motion
$\{Q\}$	Exciter input force	$\{X_c\}$	Test item connector motion
$\{R_e\}$	Test item laboratory external force	$\{X_e\}$	Test item external motion
		$\{X_c\}_c$	Test item connector motion (connector forces)
$[T]$	Test item FRF matrix	$\{X_c\}_e$	Test item connector motion (external forces)
$[T_{cc}]$	Test item connector FRF matrix		
$[T_{ce}]$	Test item connector-external FRF matrix	$\{X_e\}_c$	Test item external motion (connector forces)
$[T_{ec}]$	Test item external-connector FRF matrix	$\{X_e\}_e$	Test item external motion (external forces)
$[T_{ee}]$	Test item external points FRF matrix	$\{Y\}$	Vehicle output motion
		$\{Y_c\}$	Vehicle connector output motion
$[TV]$	Test item-Vehicle combined matrix	$\{Y_e\}$	Vehicle external output motion
		$\{Y_c\}_c$	Vehicle connector motion (connector forces)
$[TE]$	Test item-Exciter combined matrix	$\{Y_c\}_e$	Bare vehicle interface motions
$\{U\}$	Test item laboratory motion	$\{Y_e\}_c$	Vehicle external motion (connector forces)
$\{U_c\}$	Test item interface motion		
$\{U_e\}$	Test item external motion	$\{Y_e\}_e$	Vehicle external motion (external forces)
$\{U_c\}_c$	Test item connector motion (Connector forces)	$\{Z\}$	Exciter output motion vector
$\{U_c\}_e$	Test item connector motion (external forces)	$\{Z_c\}$	Exciter connector output motion
		$\{Z_e\}$	Exciter external output motion
$\{U_e\}_c$	Test item external motion (connector forces)	$\{Z_c\}_c$	Exciter connector motion (connector forces)
$[V]$	Vehicle FRF matrix	$\{Z_c\}_e$	Bare exciter interface motion
$[V_{cc}]$	Vehicle connector FRF matrix	$\{Z_e\}_c$	Exciter external motion (connector forces)
$[V_{ce}]$	Vehicle connector-external FRF matrix		
$[V_{ec}]$	Vehicle external-connector FRF matrix	$\{Z_e\}_e$	Exciter external motion (external forces)

1. Introduction

This paper describes a theoretical framework to perform laboratory simulations of field vibration environments and is a subset of the more general subject of Vibration Testing. Vibration testing can be defined as [1]"...*the art and science of measuring a structure's response while exposed to its dynamic environment and simulating this environment in a satisfactory manner to ensure that the structure will either only survive or function properly while exposed to this dynamic environment*'. Clearly, this definition consists of two parts: the first part is concerned with measuring the structure's dynamic response while it is exposed to its field vibration environment; and the second part is concerned with simulating the structure's response in the laboratory environment. Some important reasons for conducting vibration tests are [2]: (1) Verify a theoretical model of a structure; (2) Determine the structure's modes of failure; (3) Develop adequate assurance test methods; (4) Qualify a structure to meet a set of specifications; (5) Develop dynamic inputs from field data for use in either laboratory or finite element simulations or other analysis methods.

In the field environment, the test item is attached to a vehicle, and forms a combined structure. This test item can be subjected to three different forms of excitations that can be classified as: internal, boundary, and external excitations. Test item motions, as measured in the field, are due to these three types of excitation sources. Knowledge of the forces that act on a test item while in the field environment is one of the most important issues in vibration testing since these forces are needed in realistic laboratory simulations.

In the laboratory environment, the test item is attached to one or more vibration exciters to create a laboratory structure. The choice of appropriate inputs to the test item such that field measurements are reasonably simulated constitutes an important issue in laboratory testing where field interface forces (and possibly motions) are natural candidates as test item inputs. However, interface force measurements are usually difficult since they require force transducers to be placed in the force paths at the interface between the test item and vehicle, a requirement that may not be satisfied in some field test configurations due to design and/or space limitations.

This paper describes a theoretical model that can be used to generate suitable inputs from field vibration data. The structures involved in the simulation process are modeled by frequency domain input-output relationships that are used to guide us in selecting appropriate laboratory test configurations and inputs.

2. Frequency domain modeling

The frequency domain equations of motion for a *linear structure* having N degrees of freedom can be written in matrix form as

$$\{X\} = [H]\,\{F\} \tag{1}$$

where $\{X\} = \{X(\omega)\}$ is the output motion vector, $\{F\} = \{F(\omega)\}$ is the input vector, $[H] = [H(\omega)]$ is the structure's FRF matrix, and is the forcing frequency. Input and output variables are described in terms of forces and linear motions, respectively, but they could be used to represent moments and angular motions as well. The output variable $\{X\}$ can represent either displacements, velocities, or accelerations that result from the application of the input vector $\{F\}$. The FRF matrix $[H]$ may be expressed in terms of receptance, mobility, or accelerance FRFs dependent on the output variable used.

When Eq. 1 is partitioned as [3]

$$\begin{Bmatrix} \{X_c\} \\ \{X_e\} \end{Bmatrix} = \begin{bmatrix} [H_{cc}] & [H_{ce}] \\ [H_{ec}] & [H_{ee}] \end{bmatrix} \begin{Bmatrix} \{F_c\} \\ \{F_e\} \end{Bmatrix} \tag{2}$$

it can be used to describe not only the input-output relationship for a single structure, but it also can be applied to any number of independent structures that are coupled at a finite number of locations. Subscripts c and e in Eq. 2 refer to connection (or interface) points and external points, respectively. Connection points are points on the structure that are directly connected to another structure so that connection (or interface) points are points where coupling occurs between the test item and either the vehicle or vibration exciter. External points are points on the structure that are not directly involved in the coupling process. Interface motions $\{X_c\}$ occur at connecting points while external motions $\{X_e\}$ occur at the remaining points on the structure. Interface forces $\{F_c\}$ occur at interface points and are due to coupling effects only. The external force vector $\{F_e\}$ contains all remaining forces applied to the structure that can include acoustic, aerodynamic, electro-mechanical, etc. type of loads.

We shall see that it is important to distinguish between motions caused by interface forces from those caused by external forces. Hence, we partition the FRF matrix [H] into four sub-matrices, as seen from Eq. 2. In this case, $[H_{cc}]$ defines input-output FRFs for connection points, $[H_{ee}]$ defines input-output FRFs for external points, and $[H_{ce}] = [H_{ec}]^T$ define the FRFs between connectors and external points, respectively.

An expansion of Eq. 2 gives

$$\{X_c\} = [H_{cc}]\{F_c\} + [H_{ce}]\{F_e\} \tag{3}$$

$$\{X_e\} = [H_{ec}]\{F_c\} + [H_{ee}]\{F_e\} \tag{4}$$

The connector motions vector in Eq. 3 is composed of two parts, the first due to interface forces and the second due to external forces. Similarly, Eq. 4 shows that the external motion vector is formed by summing the motion due to external forces and those due to connector forces. Generally speaking, the motion of any point of the structure is composed of the sum of two motions, one part due to the connector forces and the other due to the external forces. Thus, the structure's motion can be written as

$$\{X_c\} = \{X_c\}_c + \{X_c\}_e \tag{5}$$

$$\{X_e\} = \{X_e\}_c + \{X_e\}_e \tag{6}$$

The vectors on the right hand side of Eqs. 5 and 6 now carry a double subscript. The first subscript refers to the location of the point on the structure (connector motions and external motions) while the second subscript refers to the location of excitation application to the structure (connector forces and external forces). Thus, terms $\{X_c\}_c$ and $\{X_e\}_c$ correspond to motions at connections and external points, respectively, that are caused by connector forces. Similarly, terms $\{X_c\}_e$ and $\{X_e\}_e$ correspond to motions at connector and external points, respectively, that are caused by external forces. Hence $\{X_c\}_e$ and $\{X_e\}_e$ correspond to motions that occur when only external forces are active.

3. Field dynamic environment

Figure 1a illustrates the field dynamic environment where the test item and vehicle are connected at N_c locations. The resulting combined structure is then subjected to field

69

external forces, $\{F_e\}$ and $\{P_e\}$ which in turn, cause forces $\{F_c\}$ and motions $\{X_c\}$ to occur at the interface points. Knowledge of field interface forces and test item motions is vital for successful laboratory simulations. Determination of these forces and motions requires definition of appropriate boundary conditions for the coupling points between test item and vehicle when these structures are connected in the field.

Two approaches can be used to define boundary conditions at interface points. In the first approach Eq. 1 can be used to define input-output relationships for the connectors. In this case, connectors are independent coupling structures. In the second approach, a simpler interface boundary condition can be used, where connectors are assumed to be part of either one structure or the other. This last approach is used here to define boundary conditions at the interface points between test item and vehicle. Connectors are considered to be part of the test item. Since test item and vehicle are connected through a finite number of discrete points N_c, compatibility of motions at interface connecting points require

$$\{X_c\} - \{Y_c\} = \{0\} \qquad (7)$$

where $\{X_c\}$ and $\{Y_c\}$ are N_c x 1 vectors that define test item and vehicle motions at the interface points, respectively. Both $\{X_c\}$ and $\{Y_c\}$ are assumed to be positive in the same direction.

The interface forces must satisfy

$$\{F_c\} + \{P_c\} = \{0\} \qquad (8)$$

where $\{F_c\}$ and $\{P_c\}$ are N_c x 1 vectors that represent the test item and vehicle interface forces, respectively. All matching forces are positive in the same direction.

When Eq. 3 is applied to the test item, we can write the test item interface motions as

$$\{X_c\} = [T_{cc}]\{F_c\} + [T_{ce}]\{F_e\} \qquad (9)$$

where $\{F_e\}$ is the test item's external force vector, $\{F_c\}$ is the test item interface force. $[T_{cc}]$ is the test item interface FRF matrix, and $[T_{ce}]$ is the test item's FRF matrix relating connection and external points, respectively.

Similarly, we can express the interface motion on the vehicle's side as

$$\{Y_c\} = [V_{cc}]\{P_c\} + [V_{ce}]\{P_e\} \qquad (10)$$

where $\{P_e\}$ is the external force applied to the vehicle, $[V_{cc}]$ is the vehicle interface FRF matrix, and $[V_{ce}]$ is the vehicle FRF matrix, relating interface and external points respectively.

Substitution of Eqs. 9 and 10 into Eq. 7 and noting the interface force requirement of Eq. 8, we obtain

$$([T_{cc}] + [V_{cc}])\{F_c\} = [V_{ce}]\{P_e\} - [T_{ce}]\{F_e\} \qquad (11)$$

The right hand side of Eq. 11 corresponds to the relative motion at interface connecting points. The relative interface motion is caused by external forces $\{F_e\}$ and $\{P_e\}$ being applied to test item and vehicle, respectively. Thus, using Eqs. 5 and 6, Eq. 11 can be rewritten as

$$([T_{cc}]+[V_{cc}])\{F_c\}=\{Y_c\}_e-\{X_c\}_e \tag{12}$$

Matrix $([T_{cc}] + [V_{cc}])$ in Eq. 12 is a square N_c x N_c matrix whose entries are the sum of test item and vehicle FRFs at the interface points. The main diagonal entries of this matrix are the sum of test item's and vehicle's driving point FRFs while the off-diagonal entries are the sum of test item's and vehicle's transfer point FRFs between connecting points. Solution of Eq. 12 for interface forces $\{F_c\}$ requires inversion of this matrix at each frequency value so that Eq. 12 becomes

$$\{F_c\}=[TV]\left(\{Y_c\}_e-\{X_c\}_e\right) \tag{13}$$

where $[TV] = ([T_{cc} + V_{cc}])^{-1}$ is called the test item-vehicle combined matrix. A particular case of Eq. 13 happens when there is no external forces acting on the test item in the field environment. In this case, $\{X_c\}_e = 0$, and Eq. 13 reduces to

$$\{F_c\}=[TV]\{Y_c\}_e \tag{14}$$

In this case, the interface forcer vector depends on the vehicle's connecting point motions only ($\{Y_c\}_e$). These motions are called the *bare vehicle interface motions*.

Once interface forces are determined, they can be used to calculate the test item field motions. Substitution of Eq. 13 into Eqs. 3 and 4, and using Eqs. 5 and 6, we have the following result for test item motions

$$\{X_c\}=[T_{cc}][TV]\left(\{Y_c\}_e-\{X_c\}_e\right)+\{X_c\}_e \tag{15}$$

$$\{X_e\}=[T_{ec}][TV]\left(\{Y_c\}_e-\{X_c\}_e\right)+\{X_e\}_e \tag{16}$$

Equations 15 and 16 show that test item field motions depend on the relative interface motion vector and on the motions caused by external forces acting on the test item. The test item motions in Eqs. 15 and 16 are also dependent on the test item and vehicle FRFs, since the relative interface motion ($\{Y_c\}_e - \{X_c\}_e$) is pre-multiplied by the product of matrices containing test item FRFs $[T_{cc}]$ and $[T_{ec}]$ as well as the combined system matrix $[TV]$.

Equations 15 and 16 can be rewritten in matrix form as

$$\begin{Bmatrix}\{X_c\}\\\{X_e\}\end{Bmatrix}=\begin{bmatrix}[T_{cc}]\,[TV]&-[T_{cc}]\,[TV]\\{}[T_{ec}]\,[TV]&-[T_{ec}]\,[TV]\end{bmatrix}\begin{Bmatrix}\{Y_c\}_e\\\{X_c\}_e\end{Bmatrix}+\begin{Bmatrix}\{X_c\}_e\\\{X_e\}_e\end{Bmatrix} \tag{17}$$

or simply

$$\{X\} = [FED]\{\hat{X}_c\}_e + \{X_e\} \tag{18}$$

where matrix [FED], N_t x $2N_c$, is a slightly different form of the Field Environment Dynamic matrix [1].

Equations 17 or 18 express the test item motion as a function of motions caused by external forces only. Vector $\{\hat{X}_c\}_e = \{\{Y_c\}_e \ \{X_c\}_e\}^T$, $2N_c$ x 1, contains test item and vehicle interface motions due to external forces; $\{\hat{X}_e\} = \{\{X_c\}_e \ \{X_e\}_e\}^T$ is a N_t x 1 vector containing the test item's motions due to external forces for all points. The first N_c elements of $\{\hat{X}_e\}$ are connector motions and the remaining $N_t - N_c$ elements are external motions.

Eq. 17 and 18 assume the following form when there are no external forces acting on the test item in the field environment.

$$\begin{Bmatrix} \{X_c\} \\ \{X_e\} \end{Bmatrix} = \begin{bmatrix} [T_{cc}][TV] & -[T_{cc}][TV] \\ [T_{ec}][TV] & -[T_{ec}][TV] \end{bmatrix} \begin{Bmatrix} \{Y_c\}_e \\ \{0\} \end{Bmatrix} \tag{19}$$

and

$$\{X\} = [FED]\{\hat{X}_c\}_e \tag{20}$$

where $\{\hat{X}_c\}_e = \{\{Y_c\}_e \ \{0\}\}^T$.

From Eqs. 19 and 20, it is obvious that when the test item is not subjected to external forces, the test item's field motions depend exclusively on the bare vehicle connecting motion $\{Y_c\}_e$. Recall that this vector corresponds to the vehicle's motion at the connecting points when the test item is absent. Despite being a particular case of the field dynamic environment, this situation constitutes a important case, since it occurs often and offers a good chance for a successful laboratory simulation [1].

4. The Laboratory dynamic environment

In the laboratory environment, the test item is attached to the vibration exciter, as shown in Fig. 1b. In this case, the test item has N_t input-output points and the exciter has N_e input-output points. It is assumed that test item and vibration exciter are connected at E_c points. The test item dynamic characteristics are described by Eq. 1, written as

$$\{U\} = [\hat{T}]\{R\} \tag{21}$$

where $\{U\}$ is the test item output motion, $\{R\}$ is the test item input force in the laboratory environment, and $[\hat{T}]$ is the test item FRF matrix. Note that the laboratory test item may have different dynamic characteristics from the field test item due to manufacturing variations. Thus, the symbol $\left[\hat{T}\right]$ is used to distinguish laboratory test item from field test item.

Similarly, the vibration exciter input-output characteristics are given as

$$\{Z\} = [E]\{Q\} \tag{22}$$

where $\{Q\}$ and $\{Z\}$ are the exciter input and output vectors, and $[E]$ is the exciter FRF matrix, respectively.

The same frequency domain modeling technique employed in the field dynamic environment will be used in the laboratory simulation. Motions will be grouped in two distinct sets: connector motions, associated with coupling points between test item and vibration exciter, and external motions that do not have direct participation in the coupling process. The same differentiation will be used with forces: interface or connector forces apply for interface forces due to coupling, while external forces apply to forces applied at external points.

As before, we are interested in obtaining general expressions for interface forces between test item and vibration exciter as well as for test item's motions in the laboratory dynamic environment. This can be achieved by defining boundary conditions between test item and vibration exciter in terms of interface forces and motions. The same interface boundary condition used in the field environment will be used in laboratory simulations. We will consider the connectors as being part of the test item. In this case, compatibility of motions at the E_c connecting interface points requires that

$$\{U_c\} - \{Z_c\} = \{0\} \tag{23}$$

where $\{U_c\}$ and $\{Z_c\}$ are E_c x 1 vectors defining test items and exciter motions at the interface points respectively. They are positive in the same direction.

As in the field environment, interface forces satisfy the relationship

$$\{R_c\} + \{Q_c\} = \{0\} \tag{24}$$

where $\{R_c\}$ and $\{Q_c\}$ are E_c x 1 vectors containing the test item and exciter interface forces with the same direction being positive.

Interface motions for test item and vibration exciter in the laboratory are obtained from Eq. 3

$$\{U_c\} = \left[\hat{T}_{cc}\right]\{R_c\} + \left[\hat{T}_{ce}\right]\{R_e\} \tag{25}$$

$$\{Z_c\} = \left[E_{cc}\right]\{Q_c\} + \left[E_{ce}\right]\{Q_e\} \tag{26}$$

Substitution of Eqs. 25 and 26 into Eq. 23 while using the interface force relationship from Eq. 24 gives the following system of equations for the unknown laboratory interface forces

$$\left(\left[\hat{T}_{cc}\right]+\left[E_{cc}\right]\right)\{R_c\}=\{Z_c\}_e-\{U_c\}_e \tag{27}$$

where $\{Z_c\}_e$ and $\{U_c\}_e$ correspond to the vibration exciter and test item interface motions, respectively. These interface motions are due to external forces only, as previously defined in Eq. 5. The matrix $\left(\left[\hat{T}_{cc}\right]+\left[E_{cc}\right]\right)$ is the sum of test item and vibration exciter FRF matrices at the interface points. Diagonal terms are the sum of driving point FRFs and off-diagonal terms are the sum of transfer point FRFs for interface points. Solution of Eq. 27 for the laboratory interface forces can be expressed as

$$\{R_c\}=[TE]\left(\{Z_c\}_e-\{U_c\}_e\right) \tag{28}$$

where $[TE]=\left(\left[\hat{T}_{cc}\right]+\left[E_{cc}\right]\right)^{-1}$ corresponds to the test item-exciter combined matrix.

The laboratory interface force vector $\{R_c\}$ as given by Eq. 28 depends on the relative interface motion vector $\{Z_c\}_e-\{U_c\}_e$ between vibration exciter and test item. This relative interface motion is due to external forces that act on the vibration exciter ($\{Z_c\}_e$) and test item ($\{U_c\}_e$). In addition, calculation of laboratory interface forces requires the inversion of a square and symmetric $E_c \times E_c$ matrix.

When there are no external forces acting on the test item, the expression for the interface force becomes

$$\{R_c\}=[TE]\{Z_c\}_e \tag{29}$$

since $\{U_c\}_e=0$. In this case, the interface force vector depends on the <u>vibration exciter</u> connecting points <u>only</u> ($\{Z_c\}_e$).

The laboratory interface forces can be used to calculate laboratory test item motions. Using the same procedure as for the field environment, we get the following expression for the laboratory test item motion

$$\left\{\begin{matrix}\{U_c\}\\\{U_e\}\end{matrix}\right\}=\left[\begin{matrix}\left[\hat{T}_{cc}\right][TE] & -\left[\hat{T}_{cc}\right][TE]\\\left[\hat{T}_{ec}\right][TE] & -\left[\hat{T}_{ec}\right][TE]\end{matrix}\right]\left\{\begin{matrix}\{Z_c\}_e\\\{U_c\}_e\end{matrix}\right\}+\left\{\begin{matrix}\{U_c\}_e\\\{U_e\}_e\end{matrix}\right\} \tag{30}$$

or simply

$$\{U\}=[LED]\left\{\hat{U}_c\right\}_e+\left\{\hat{U}_e\right\} \tag{31}$$

where $\{U\}$ is a $N_t \times 1$ vector that contains all test item motions. Vector $\left\{\hat{U}_c\right\}_e=\left\{\{Z_c\}_e \ \ \{U_c\}_e\right\}^T$ is $2E_c \times 1$ and contains the vibration exciter and test item interface motions due to external forces. Vector $\left\{\hat{U}_e\right\}=\left\{\{U_c\}_e \ \ \{U_e\}_e\right\}^T$ is $N_t \times 1$ and contains test item motions due to external forces for all of the test item input-output points. Matrix $[LED]$ is the $Nt \times 2E_c$ <u>L</u>aboratory <u>E</u>nvironment <u>D</u>ynamic matrix [1].

When the test item is subjected to no external forces in the laboratory environment, Eqs. 30 and 31 reduce to

$$\begin{Bmatrix} \{U_c\} \\ \{U_e\} \end{Bmatrix} = \begin{bmatrix} \left[\hat{T}_{cc}\right][TE] & -\left[\hat{T}_{cc}\right][TE] \\ \left[\hat{T}_{ec}\right][TE] & -\left[\hat{T}_{ec}\right][TE] \end{bmatrix} \begin{Bmatrix} \{Z_c\}_e \\ \{0\} \end{Bmatrix} \tag{32}$$

and

$$\{U\} = [LED]\left\{\hat{U}_c\right\}_e \tag{33}$$

where $\left\{\hat{U}_c\right\}_e = \left\{\{Z_c\}_e \quad \{0\}\right\}^T$.

Thus, while in the laboratory environment, the test item undergoes motions that, according to Eqs.30 and 31 depend on the test item and the exciter motions that are caused by external forces only. For the particular case where test item is subjected to no external forces, the test item motion vector is dependent on the Laboratory Environment Matrix [LED] and on the vibration exciter interface motions $\{Z_c\}_e$.

5. Test scenarios for laboratory simulations

This section describes four different test scenarios that can be employed in vibration testing. In each scenario, a different control strategy is used to match field data in the laboratory environment. It is *assumed* that *the number of connectors used to attach the test item to the vehicle in the field and to the vibration exciter in the laboratory is the same*, i.e., $N_c = E_c$. This is a valid assumption when defining different test scenarios but may not be the case in real situations.

In each test scenario, one of the following strategies is considered:

- The bare vehicle interface motions $\{Y_c\}_e$ is used to define the test item inputs in the laboratory environment. This corresponds to the situation where the test item laboratory interface motion $\{U_c\}$ is controlled such that $\{Y_c\}_e$ is matched in the laboratory. This laboratory test scenario is frequently employed in real situations [4,5] when a single vibration exciter is used. In this case, the test item input is generally defined by enveloping the bare vehicle interface motion $\{Y_c\}_e$.

- The exciters are controlled such that laboratory and field interface forces are matched so that

$$\{R_c\} = \{F_c\} \tag{34}$$

- The exciters are controlled such that laboratory and field interface motions are matched so that

$$\{U_c\} = \{X_c\} \tag{35}$$

- The exciters are controlled such that the motions are external points on the test item are the same in the field and laboratory environments so that

$$\{U_e\} = \{X_e\} \tag{36}$$

In practice, these control strategies may require multiple input controlled vibration tests, since multiple connectors are used in both the field and the laboratory environments. This, in principle requires one vibration exciter for each of the test item interface points in the laboratory simulation (as long as external forces are negligible compared with interface forces). We will now discuss some issues involved in these laboratory test scenarios.

5.1. TEST SCENARIO 1

In this case, the bare vehicle interface motion $\{Y_c\}_e$ is used to define the test item inputs in the laboratory. Recall that the bare vehicle data corresponds to field measurements obtained when the test item is not attached to the vehicle. In addition, *it is assumed that no information involving the vehicle FRF characteristics is available.*

The set of input forces that is required to drive the test item at its N_c interface points is obtained by solving the following system of equations for the unknown laboratory interface forces

$$[T_{cc}]\{R_c\} = \{Y_c\}_e \tag{37}$$

or

$$\{R_c\} = [T_{cc}]^{-1}\{Y_c\}_e \tag{38}$$

Comparison of Eq. 38 with Eq. 13 that gives the true interface forces when the test item and vehicle are connected in the field reveals some important issues. First, matrix [TV] in Eq. 13 accounts for test item and vehicle interface FRF characteristics while Eq. 38 accounts only for the test item interface FRFs. Second, field external effects ($\{X_c\}_e$) are accounted for in Eq. 13 while they are non-existent in Eq. 37. It is clear that the laboratory interface forces obtained by employing Eq. 37 will not match the field interface forces. Thus, the bare vehicle interface motions do not represent suitable field data to define the test item inputs in the laboratory environment.

However, suppose that we have enough information about the field environment prior to attaching the test item to the vehicle such that external force effects can be neglected. In this case, $\{X_c\}_e = 0$ and, as long as the vehicle field interface FRF characteristics $[V_{cc}]$ are known, a correct transformation matrix can be employed on the right hand side of Eq. 38 such that $\{R_c\} = \{F_c\}$.

Thus, if no significant external force effects exist in the field and if the vehicle interface driving and transfer point FRFs are available, then the bare vehicle interface motion can be used to define suitable test item inputs in the laboratory environment. Notice that Eq. 37 assumes that the test item has the same FRF characteristics in both the field and laboratory environments.

5.2.   TEST SCENARIO 2

In this case, interface forces are matched in the laboratory environment, according to Eq. 34. By comparing the equations for interface forces in both the field and laboratory environments, we gain additional insights to the requirements that must be satisfied so that this test scenario gives reasonable simulation results. Field and laboratory interface forces are given by Eqs. 13 and 28, that are conveniently rewritten here

$$\{F_c\} = [TV](\{Y_c\}_e - \{X_c\}_e) \tag{39}$$

$$\{R_c\} = [TE](\{Z_c\}_e - \{U_c\}_e) \tag{40}$$

First, we notice that the field and laboratory interface forces depend on the test item-vehicle FRF matrix [TV] and on the test item-exciter FRF matrix [TE], respectively. These matrices are obtained by inverting the FRF matrix given by the sum of the test item interface FRF matrix and either the vehicle interface FRF matrix in the field or the vibration exciter interface FRF matrix in the laboratory. Assuming that the test item has the same FRF characteristics in both environments, the problem is then matching the vehicle interface FRF matrix in the laboratory.

When a single vibration exciter is used in the laboratory simulation, the test item is usually attached to the exciter's table through a test fixture. The implications of Eqs. 39 and 40 are obvious. The test fixture must match the vehicle interface FRF characteristics so that [TV] = [TE] and the exciter must be driven so $\{Z_c\}_e = \{Y_c\}_e$. The matching of interface FRF characteristics is nearly impossible to satisfy except in the simplest situations.

Second, we compare the relative motion vector that appears on the right hand side of Eqs. 39 and 40. In the field environment, the relative interface motion is given by subtracting the test item interface motion that is due to external forces $\{X_c\}_e$ from the bare vehicle interface motion $\{Y_c\}_e$. Similarly, the laboratory relative motion is obtained by subtracting the test item interface motion due to laboratory external forces $\{U_c\}_e$ from the bare exciter interface motion $\{Z_c\}_e$. The first important observation to be made in this case is that, if existing field external forces are not accounted for in the laboratory, motion $\{U_c\}_e$ is identically zero, and thus, it is clear that this simulation fails, even if [TE] = [TV]. Furthermore, $\{Y_c\}_e$ and $\{Z_c\}_e$ represent vehicle and exciter interface motions obtained when the test item is absent from the field and laboratory, respectively, and the chances that these motions will be the same is minimal. Thus, field external forces effects must be properly accounted for in the laboratory such that the difference $(\{Z_c\}_e - \{U_c\}_e)$ be as close as possible to $(\{Y_c\}_e - \{X_c\}_e)$. In the special case where there are no external forces acting on the test item, the interface motions due to these forces are zero. Then, we can solve for the required exciter interface motions from Eqs. 39 and 40 to obtain

$$\{Z_c\}_e = [TE]^{-1}[TV]\{Y_c\}_e \tag{41}$$

It is nearly impossible to satisfy Eq. 41 through the use of a *single vibration exciter* with a *test fixture* unless all interface points have essentially the same motion. Thus, we see that the use of multiple exciters is the only way to overcome this limitation for a single exciter and test fixture.

5.3.   TEST SCENARIO 3

In this test scenario, the N_c field interface motions are matched in the laboratory environment, according to Eq. 35. This condition can be examined by comparison of the test item field and laboratory interface motions that can be obtained from the first row of Eqs. 17 and 30, respectively

$$\{X_c\} = [T_{cc}][TV](\{Y_c\}_e - \{X_c\}_e) + \{X_c\}_e \tag{42}$$

$$\{U_c\} = [T_{cc}][TE](\{Z_c\}_e - \{U_c\}_e) + \{U_c\}_e \tag{43}$$

where it is assumed that $[\hat{T}_{cc}] = [T_{cc}]$.

Similar to the previous test scenario, Eq. 42 contains the test item-vehicle combined matrix [TV] while Eq. 43 contains the test item-exciter combined matrix so that they are different when a single exciter and test fixture are used. The effects of external forces on interface points must be properly accounted for in this case as well. Hence, it appears that multiple exciters are required to satisfy all requirements.

5.4.   TEST SCENARIO 4

In this case, the test item field external motions are matched in the laboratory as stated by Eq. 36. Field and laboratory external motions are obtained from the second row of Eqs. 17 and 30, respectively, and they are given by

$$\{X_e\} = [T_{ec}][TV](\{Y_c\}_e - \{X_c\}_e) + \{X_e\}_e \tag{44}$$

$$\{U_e\} = [T_{ec}][TE](\{Z_c\}_e - \{U_c\}_e) + \{U_e\}_e \tag{45}$$

where it is assumed that the test item has the same FRFs in both environments, as was the case in the previous test scenarios. In this case, proper simulation of the external force effects on external motions is required as well as exciter interface motions $\{Z_c\}_e$. It should be clear that this scenario has the same single test fixture problem as the other scenarios.

Test scenarios 2, 3, and 4 contain terms that involve [TE] and $\{Z_c\}_e$. It is clear that it is not economical to create a single test fixture mounted on a single vibration exciter and achieve the behavior demanded by these terms, i.e, that

$$[TV](\{Y_c\}_e - \{X_c\}_e) = [TE](\{Z_c\}_e - \{U_c\}_e) \tag{46}$$

In addition, we see that external forces must be applied in the laboratory that simulate those experienced in the field. The only way to side step this single exciter-test fixture dilemma is to use multiple exciters at the interface points.

6. Summary and Conclusions

In this paper, a general vibration testing model is developed that describes the variables and processes that are involved in laboratory simulations of field dynamic environments. Three structures are involved in this process: the test item, the vehicle, and the vibration exciter. The structural interactions that occur when the test item is attached to the vehicle in the field or to the vibration exciters in the laboratory are modeled by frequency domain linear input-output relationships. A substructuring technique is used to distinguish interface properties (forces and motions) from external properties. Suitable interface boundary conditions lead to general expressions in terms of field and laboratory interface forces and motions and for the test item external motions.

Figure 2 shows that there are two field environments that can be considered. Figure 2a corresponds to the bare vehicle case where the test item is absent. In this case, the vehicle is subject of a set of loads P_r, $r = 1, ..., (N_v - N_c)$ applied to the vehicle's external DOFs. The interface forces are zero in this case since the test item is absent. The field data obtained in this simulation corresponds to the N_c bare vehicle motions $\{Y_c\}_e$ that are measured at interface points C_{np}, where $p = 1, ..., N_c$.

The second field environment considered here is shown in Fig. 2b. In this case, the test item is attached to the vehicle through the N_c interface points, forming the combined structure. The test item is subjected to the interface forces F_p, $p = 1, ..., N_c$ and to the external forces F_q, $q = N_c + 1, ..., N_t$. Similarly, the vehicle is subjected to interface and external forces. The resulting data from this field environment correspond to the interface forces $\{F_c\}$, the test item interface motions $\{X_c\}$ and the test item external motions $\{X_e\}$.

Multiple exciters are required in the laboratory simulations of the field environments shown in Fig. 2. Figure 3 illustrates the laboratory test scenarios that can be used to simulate field data, as well as some requirements to properly simulate field data. Figure 3a shows the laboratory simulation corresponding to the bare vehicle field environment of Fig. 2a. Each test item interface point is attached to a vibration exciter, and each exciter must be controlled such that the correct motion is reproduced at the control interface point. A procedure commonly employed in single exciter tests is to attach the test item to a *rigid* test fixture at N_c interface points. The exciter input is generated by enveloping the bare vehicle field data [4,5]. Inappropriate enveloping may lead to extremely conservative tests, where the test item is over-tested at many frequencies. Thus, care should be taken when using enveloping techniques to define test item inputs in the laboratory.

The required laboratory test arrangement that corresponds to the combined test item-vehicle structural systems is shown in Fig. 3b. In this case, separate exciters are attached to each interface connector point as well as external loading points. It is clear that we need either interface motions or forces in order to control the connector points. Generally, the external loads are not measured in the field since they often come from sources such as acoustic or aerodynamic loadings, which are difficult to measure directly. We need to be able to estimate these external forces from measured test item responses. This leads into the inverse problem, which is addressed in the following papers for deterministic and random loads.

All laboratory simulations discussed in this paper require multiple exciters operating in closed loop in order to match the corresponding field data in the laboratory. This represents a real challenge in the vibration testing context since not only the exciters must be able to generate and control a given frequency spectrum, but also cross correlation between excitation sources must be maintained [6,7].

The results presented in this paper were obtained by using an approach base on frequency spectra. This means that the several variables used to express interface and external properties are complex variables that carry both magnitude and phase information. This approach is suitable for deterministic (periodic and transient) input and output signals but is inadequate in the case of dealing with random signals. The expressions for interface forces and motions as well as for the test item external motions can be derived in terms of random signals and in this case, spectral density matrices containing auto and cross spectral densities will be used with random signals [8].

Acknowledgments

The authors would like to thank the Iowa State University Engineering College and the Aerospace Engineering and Engineering Mechanics Department for supporting this research. Mr. Paulo S. Varoto, from Universidade de São Paulo – São Carlos – Brasil was financially sponsored by CNPq – Brazil during his PhD program.

7. References

1. McConnell, K.G. (1995), Vibration Testing: Theory and Practice *John Wiley & Sons*.
2. McConnell, K.G. (1995), From field vibration data to laboratory simulations, *Experimental Mechanics, Vol. 34, N. 3*, 181-193.
3. Gordis, J.H., Bielawa, R.L., Flannelly, W.G. (1991), A general theory for frequency domain structural synthesis, *Journal of Sound and Vibration, vol. 150(1)*, 139-158.
4. Scharton, T. (1990), Motion and force controlled vibration testing, *Proceedings of the IES*, 77-85.
5. Scharton, T. (1995), Vibration-test force limits derived from frequency-shift method, *Journal of Spacecrafts and Rockets, Vol. 32, No. 2*, 312-316.
6. Smallwood, D. (1982), A random vibration control system for testing a single test item with multiple inputs, *SAE paper No. 821482*, 4571-4577.
7. Smallwood, D. (1978), Multiple shaker random vibration testing with cross coupling, *Proceedings of the IES*, 341-347.
8. Varoto, P.S., McConnell, K.G. (1999), Rules for the Exchange and Analysis of Dynamic Information: Part IV – Numerically simulated and experimental results for a random excitation, in J.M.M. Silva and N. M. M. Maia (eds.), *Modal Analysis & Testing, Kluwer Academic Publishers, NATO Series*, Dordrecht, 137-177.

80

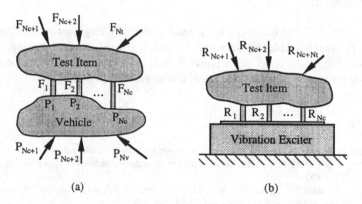

(a)   (b)

Figure 1   Test item in the field and laboratory environments: (a) Attached to the vehicle in the field; (b) attached to the vibration exciter in the laboratory.

81

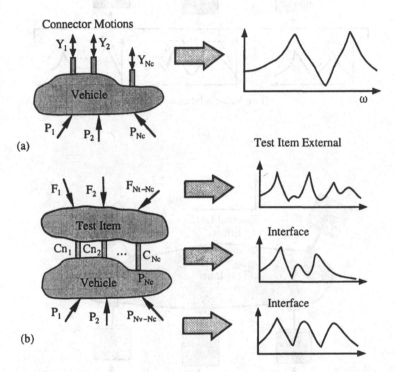

(a)

(b)

Figure 2   Field environments: (a) Test item is absent and field data corresponds to the bare vehicle interface motions $\{Y_c\}_e$; (b) Test item attached to the vehicle forming the combined structure and field data correspond to the interface force $\{F_c\}$, interface motion $\{X_c\}$, and test item external motion $\{X_e\}$.

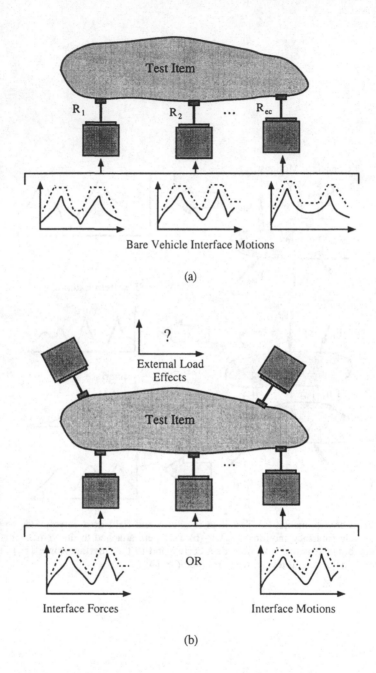

(a)

(b)

Figure 3   Laboratory test scenarios using field data: (a) test item imputs defined from the bare vehicle interface motion $\{Y_c\}_e$; (b) test item inputs defined from interface forces $\{F_c\}$ or external forces $\{F_e\}$ or interface motions $\{X_c\}$.

RULES FOR THE EXCHANGE AND ANALYSIS OF DYNAMIC INFORMATION

Part II: Numerically Simulated Results for a Deterministic Excitation with no External Loads

P. S. VAROTO
Dept de Engenharia Mecanica
Escola De Engenharia De São Carlos,
USP
São Carlos – SP – 13560-250, Brasil

K. G. McCONNELL
Dept. Aerospace Engineering and
Engineering Mechanics
Iowa State University
Ames, IA 50011 USA

Abstract

This paper presents a numerical example of the theory presented in Part I [1] where a model is developed to describe the requirements for laboratory simulations of field vibration environments. This example illustrates the application of such a model to simulated data in order to evaluate its accuracy in predicting field interface forces and motions, and in using field data to define appropriate test item inputs in laboratory simulations. It is assumed that no external forces act on the test item in either the field or laboratory environments. All Frequency response functions (FRFs) are calculated for all test structures from a multi degree of freedom (MDOF) discrete linear system. The equations developed in Ref. [1] are employed to estimate field interface forces and test item motions as well as to define test item inputs in the laboratory simulation. The test item motions obtained in all laboratory simulations are compared with the corresponding field motions.

Nomenclature

A_{pq}	Accelerance FRF	$\{R_c\}$	Test item laboratory interface forces
$\{F_c\}$	Test item field interface forces	$[T]$	Test item FRF matrix
$j = \sqrt{-1}$	Imaginary number	$[T_{cc}]$	Test item interface FRF matrix
N_t	Degrees of freedom for test item	$[T_{ce}]$	Test item interface-external FRF matrix
N_v	Degrees of freedom for vehicle	$[T_{ec}]$	Test item external-interface FRF matrix
N_c	Number of interface points	$[T_{ee}]$	Test item external FRF matrix
$[M]$	Mass matrix		
$\{R\}$	Test item laboratory input forces	T_{pq}	Test item FRF

83

[TV]	Combined Test Item Vehicle matrix	Y_2	Vehicle base input acceleration
{U}	Test item laboratory accelerations	$\{Y_c\}_e$	Bare vehicle interface accelerations
$\{U_c\}$	Test item laboratory interface accelerations	Greek	
V_{pq}	Vehicle FRF	Δf	Frequency resolution in Hz
VT_{pq}	Combined structure FRF	$\{\phi\}_r$	r^{th} mode shape
{X}	Test item field accelerations	$\Gamma(\omega)$	Acceleration transmissibility
$\{X_c\}$	Test item interface field accelerations	FRF	
		ω	Frequency in rad/s
$\{X_e\}$	Test item external field accelerations	ω_r	r^{th} natural frequency in rad/s
		ζ_r	r^{th} modal damping ratio
$\{Y_c\}$	Vehicle interface accelerations		

1. Introduction

The choice of appropriate test item inputs in the laboratory environment is a most important issue in simulating field vibration data. The ultimate goal of a laboratory simulation is to ensure that the test item will function properly when exposed to its field dynamic environment [1]. Forces and motions occurring at the N_c interface points between the test item and the vehicle are natural candidates for test item inputs in laboratory simulations. When the test item is subjected to no external forces in the field environment, interface forces and test item accelerations are calculated according to Eqs. 14 and 19 from Ref. [1]

$$\{F_c\} = [TV]\{Y_c\}_e \tag{1}$$

$$\begin{Bmatrix} \{X_c\} \\ \{X_e\} \end{Bmatrix} = \begin{bmatrix} [T_{cc}][TV] & -[T_{cc}][TV] \\ [T_{ec}][TV] & -[T_{ec}][TV] \end{bmatrix} \begin{Bmatrix} \{Y_c\}_e \\ \{0\} \end{Bmatrix} \tag{2}$$

In Eq. 1 $\{F_c\}$ is the N_c x 1 interface force vector, $\{Y_c\}_e$ is the *bare vehicle* interface acceleration, representing the motion measured at the vehicle interface points in the absence of the test item. Matrix [TV] is the combined test item and vehicle matrix that corresponds to the inverse of the N_c x N_c symmetric matrix whose entries are the sum of test item and vehicle interface point FRFs ([TV] = $[T_{cc} + V_{cc}]^{-1}$). The diagonal entries of $[T_{cc} + V_{cc}]$ are the sum of test item and vehicle driving point FRFs while the off diagonal entries are the sum of test item and vehicle transfer point FRFs between interface points.

The test item acceleration vector in Eq. 2, $\{\{X_c\}\{X_e\}\}T$ has N_t terms and is composed of a N_c x 1 vector $\{X_c\}$ that contains interface accelerations, and $(N_t - N_c)$ x 1 vector $\{X_e\}$ representing external point accelerations. In the absence of field external forces acting on the test item, interface forces and test item accelerations depend only on the test item and vehicle FRF characteristics, and on the bare vehicle interface accelerations $\{Y_c\}_e$ as seen in Eqs. 1 and 2.

The first subscript in the vectors defined in Eqs. 1 and 2 refers to the type of degree of freedom (DOF) where the motion occurs while the second subscript refers to the type of excitation that caused the corresponding acceleration. Thus, external forces occur at non-connector points and connector forces occur at points that are directly related to coupling, i.e., interface points. Similarly, external accelerations occur at non-connector points and connector accelerations occur at points that are directly related to coupling.

In the laboratory environment, the test item is attached to the vibration exciter and forces and motions occur at the interface points between test item and exciter as in the field. Expressions similar to Eqs. 1 and 2 can be obtained in the laboratory environment. Reference [1] shows that these expressions for laboratory interface forces and motions are dependent on motions caused by interface and external forces applied to both, the test item and vibration exciter, as well as on the FRFs of each structure. The appropriate choice of control strategy to be used with a particular laboratory test depends on the field data available, which in turn, is the basis for defining the test item inputs in the laboratory.

This paper illustrates the process of translating field data into suitable laboratory test item inputs by performing simulations on MDOF discrete linear systems. The test item is modeled as a free structure in space while the vehicle is grounded. Modal properties (natural frequencies, mode shapes, and modal damping ratios) are obtained for each structure and these properties are used to generate driving and transfer point FRFs for the test structures. Field interface forces and accelerations are calculated by using Eqs. 1 and 2, respectively. These forces and/or accelerations are used in the laboratory simulation to define the test item inputs in each test scenario considered. *No external forces are applied to the test item in either the field or laboratory environments in this paper.*

2. Theoretical response models of test structures

The response model of the structures involved in the simulation process requires the test structure's driving point and transfer point FRFs. The linear accelerance FRF $A_{pq}(\omega)$ which relates the output acceleration response at coordinate p due to a unit input force at coordinate q is calculated by [2]

$$A_{pq}(\omega) = -\omega^2 \sum_{r=1}^{N} \frac{\phi_{pr}\,\phi_{qr}}{m_r\left(\omega_r^2 - \omega^2 + j\,2\varsigma_r\omega_r\omega\right)} \tag{3}$$

where ϕ_{pr} and ϕ_{qr} are the pth and the qth elements of the rth mode shape, ω_r is the rth natural frequency (rad/s), ζ_r is the rth modal damping ratio, m_r is the modal mass of the rth mode shape, and ω (rad/s) is the excitation frequency.

When the input to the structure is motion at a single coordinate, the response model is given in terms of the transmissibility FRF vector, $\{\Gamma(\omega)\}$, that is defined as the ratio of the system's p^{th} coordinate response due to the rth coordinate input motion so that [6]

$$\{\Gamma(\omega)\} = \omega^2 \sum_{r=1}^{N} \frac{\{\phi\}_r\{\phi_r^T\}[M]\{1\}}{m_r\left(\omega_r^2 - \omega^2 + j\,2\varsigma_r\omega_r\omega\right)} + \{1\} \tag{4}$$

where [M] is the structure's mass matrix. Equations 3 and 4 define response models for viscously damped structures.

A four DOF lumped parameter model is used to define both the test item and vehicle structures as shown in Fig. 1. The physical parameters for the test item and the vehicle models are given in Table 1. When the test item and the vehicle are connected in the field at points 1 and 3, they form the combined structure shown in Fig. 1. The test item is attached to the vehicle through $N_c = 2$ connectors that are labeled C_{n1} and C_{n3}, respectively. Coordinate acceleration variables are assigned such that interface points carry the same coordinate number in both the test item and vehicle models. Thus, according to Fig. 1, Y_1 and Y_3 denote interface accelerations on the vehicle side and X_1 and X_3 denote interface accelerations on the test item side. The remaining coordinate variables (Y_2, Y_4, X_2, X_4) represent external accelerations.

Table 1: Physical Parameters

Mass	Test Item (Kg)	Vehicle (Kg)
M_1	0.20	0.50
M_2	0.20	0.50
M_3	0.25	0.30
M_4	0.15	0.50
Stiffness	Test Item N/m x 10^4	Vehicle N/m x 10^4
K_1	5.00	6.00
K_2	4.00	14.00
K_3	2.00	10.00

Compatibility of interface accelerations is defined for the combined structure according to [1]

$$\{X_c\} = \{Y_c\} \tag{5}$$

where $\{X_c\}$ and $\{Y_c\}$ are the N_c x 1 test item and vehicle interface acceleration vectors, respectively. From Eq. 5, we find that $X_1 = Y_1$ and $X_3 = Y_3$ so that we have $N_r = 6$ DOFs for the combined structure in Fig. 1.

Using standard modal analysis techniques, the natural frequencies, mode shapes, and modal damping ratios are calculated for the test item, the vehicle, and the combined structure. The test item is modeled as a free structure in space, while the vehicle and the combined structure are grounded to the vehicle excitation source through the mass M_2. The modal properties obtained for the test structures are listed in Table 2, where the modal damping ratios were obtained by assuming a proportional damping distribution. By using Eqs. 3 and 4, FRFs are calculated for all structures involved in the simulations. The following nomenclature is adopted for accelerance FRFs: symbols T_{pq} or V_{pq} are used for FRFs related to the test item and the vehicle when they are separated. The symbol VT_{pq} is used to represent the driving point accelerance FRFs for the combined structure.

Table 2: Modal Properties

Natural Frequency (Hz)	Test Item	Vehicle	Combined
f_1	0	30.00	23.86
f_2	52.60	86.17	58.94
f_3	85.70	164.72	85.05
f_4	131.00	–	112.00
f_5	–	–	160.21
Damping Ratio (%)	Test Item	Vehicle	Combined
ζ_1	0	0.28	0.24
ζ_2	0.50	0.81	0.56
ζ_3	0.81	1.55	0.91
ζ_4	1.23	–	1.06
ζ_5	---	–	1.54
Modal Masses	Test Item	Vehicle	Combined
m_1	0.20	0.43	0.38
m_2	0.17	0.48	0.17
m_3	0.21	0.37	0.55
m_4	0.20	–	0.21
m_5	---	–	0.34

Figure 2a shows the test item driving point accelerance FRF at interface points 1 and 3. These FRFs present a zero natural frequency corresponding to the rigid body mode, and three non-zero natural frequencies in the 0 – 200 Hz range as shown in Table 2. Figure 2b shows the vehicle driving point accelerance FRFs at its interface points 2 and 3. There are three resonance peaks in the 0 – 200 Hz bandwidth, corresponding to the values in Table 2. The driving point accelerance FRFs for the combined structure at the interface points are shown in Fig. 2c. In this case, the FRFs show five resonant peaks in the 0 – 200 Hz frequency range. Thus, test item and vehicle form a new structure when they are connected in the field, and this structure has unique dynamic characteristics. Interface transfer point FRFs are required in the calculation of field interface forces and accelerations as seen from Eq. 1 and 2. Figure 2d shows the connector transfer accelerance FRFs for the test item, vehicle, and combined structure.

Figures 3a and 3b show the acceleration transmissibility FRFs for the vehicle and combined structure at interface points 1 and 3, respectively. These transmissibility FRFs were obtained by employing Eq. 4 with the modal parameters of Table 2.

3. Field simulation

Two field simulations are considered. *First*, the bare vehicle interface accelerations $\{Y_c\}_e$ is obtained. The bare vehicle interface accelerations represent the only field data that is available for this case. *Second*, the test item is attached to the vehicle forming the combined structure. Interface forces and accelerations as well as test item external accelerations constitute the field data in this case. We need to emphasize that *no external forces are applied to test item in either of these field or laboratory environments.*

The vehicle excitation source:

In both field simulations, the vehicle excitation is given by an acceleration frequency spectrum applied to the vehicle base mass M_2, see Fig. 1. The magnitude of the base input acceleration frequency spectrum Y_2 is shown in Fig. 4a. It increases linearly with a slope of $1.28 \cdot 10^{-3}$ (m/s^2)/Hz in the $0 - 10$ Hz frequency range and then becomes constant with a magnitude of $1.28 \cdot 10^{-2}$ m/s^2 in the remaining $10 - 200$ Hz frequency range so that the total input vibration level is approximately 1.0 g_{RMS} (9.81 $(m/s^2)_{RMS}$). A total of 800 spectral lines are used in all simulations so that the frequency resolution is $\Delta f = 0.25$ Hz.

3.1. FIELD SIMULATION 1 – BARE VEHICLE DATA

The bare vehicle interface accelerations are obtained from the vehicle acceleration transmissibility FRFs and the vehicle input frequency spectrum Y_2 shown in Fig 4a by using Eq. 4 with $Y_p = Y_2$. The bare vehicle interface accelerations Y_1 and Y_3 are obtained and are shown in Fig 4b. Since the test item is absent from this field simulation, the bare vehicle interface accelerations constitute the field data that will be used later to define one of the test item inputs in the laboratory environment.

3.2. FIELD SIMULATION 2 – COMBINED STRUCTURE DATA

The test item is attached to the vehicle so that we have the combined structure shown in Fig. 1. The combined structure is subjected to the input base acceleration Y_2, and as a consequence, forces and motions occur at interface points between test item and vehicle. Interface forces are calculated for the combined structure by writing Eq. 1 as

$$\begin{Bmatrix} F_1 \\ F_3 \end{Bmatrix} = \begin{bmatrix} T_{11} + V_{11} & T_{13} + V_{13} \\ T_{13} + V_{31} & T_{33} + V_{33} \end{bmatrix}^{-1} \begin{Bmatrix} Y_1 \\ Y_3 \end{Bmatrix} \qquad (6)$$

Now, using Eq. 6 along with the bare vehicle interface accelerations shown in Fig. 4 and the test item and vehicle driving and transfer point FRFs shown in Fig. 2, we obtain the field interface forces F_1 and F_3 as shown in Fig. 5 where "exact" interface forces are plotted for comparison purposes. These "exact" interface forces are calculated by writing the equations of motions for the interface masses on either the test item or vehicle side and then solving these equations for the desired forces. A very good fit occurs between the interface forces predicted from Eq. 6 and those obtained through the "exact" method. Minor discrepancies are observed at frequencies around 85 Hz and 120 Hz in Fig. 5a and 84 Hz in Fig. 5b.

Similarly, all four test item accelerations X_1, X_2, X_3, and X_4 are calculated from Eq. 2. These accelerations are shown in Fig. 6 where "exact" test item accelerations are plotted for comparison purposes. These "exact" accelerations are obtained from

$$\{X\} = [T]\{F\} \qquad (7)$$

where the N_t x N_t matrix [T] contains the test item accelerance FRFs and {F} is the N_t x 1 vector of forces acting on the test item while in the field environment. This force vector contains only interface forces since no external forces are applied to the test item. The results of Eqs. 2 and 7 are indistinguishable in Fig. 6.

4. Laboratory simulation

The test item is attached to one or more vibration exciters and a suitable set of inputs must be chosen so that reasonable test item responses are obtained. Figure 7 shows the four DOFs test item in the laboratory environment, as well three field measurements that can be used to define test item inputs. These measurements are *the bare vehicle interface accelerations, the interface forces, and the test item interface accelerations*. Two exciters are employed in the laboratory simulation as shown in Fig. 7 since the test item and the vehicle are attached in the field at two locations. It is assumed that exciters 1 and 2 can be controlled in such a way that the correct inputs are generated to drive the test item according to any one of the three possibilities shown in Fig. 7.

Caution:

The results presented in this paper used frequency spectra that contain both magnitude and phase information for each time dependent variable such as accelerations and forces. This means that a specific cross correlation exists between each time variable pair even though no such function was explicitly calculated or used. These cross correlations are important for achieving reasonable laboratory simulations.

4.1. SIMULATION 1 – INPUT: BARE VEHICLE INTERFACE ACCELERATIONS

The *bare vehicle interface accelerations* shown in Fig. 4 are used to define the test item inputs. The vector containing the laboratory interface forces is obtained by solving the following system of equations for each frequency value

$$[T_{cc}] \{R_c\} = \{Y_c\}_e \tag{8}$$

where the N_c x 1 vector $\{R_c\}$ contains the laboratory interface forces that will be applied by exciters 1 and 2 at the test item interface points 1 and 3, respectively. The N_c x 1 vector $\{Y_c\}_e$ contains the bare vehicle interface accelerations, obtained from the field simulation, and the N_c x N_c matrix $[T_{cc}]$ contains the test item interface driving point and transfer accelerance FRFs. It is assumed in this paper that the test item has the same FRF characteristics in both the field and laboratory environments. This may not always be the case.

Now we solve Eq. 8 for the laboratory interface forces vector and expand the final equation for the system shown in Fig. 1 to obtain

$$\begin{Bmatrix} R_1 \\ R_3 \end{Bmatrix} = \begin{bmatrix} T_{11} & T_{13} \\ T_{31} & T_{33} \end{bmatrix}^{-1} \begin{Bmatrix} Y_1 \\ Y_3 \end{Bmatrix} \tag{9}$$

The N_t x 1 test item laboratory acceleration vector $\{U\}$ is obtained from

$$\{U\} = [T]\{R_c\} \tag{10}$$

where $[T]$ is the N_t x N_c test item FRF matrix. For the example shown in Fig. 2 $N_c = 2$ and $N_t = 4$ so that Eq. 10 expands to

$$\begin{Bmatrix} U_1 \\ U_3 \\ U_2 \\ U_4 \end{Bmatrix} = \begin{bmatrix} T_{11} & T_{13} \\ T_{31} & T_{33} \\ T_{21} & T_{23} \\ T_{41} & T_{43} \end{bmatrix} \begin{Bmatrix} R_1 \\ R_3 \end{Bmatrix} \tag{11}$$

The first two entries on the vector on the left hand side of Eq. 11 correspond to the interface accelerations while the remaining entries represent external accelerations.

Figures 8a and 8b show the laboratory interface forces calculated from Eq. 9 compared with the true field interface forces obtained from Eq. 1. It is seen that the laboratory interface forces and the corresponding field forces do not match in this case. Figure 9 shows the magnitude of the ratio of the laboratory interface forces divided by the field interface forces, i.e., R/F for both interface points. This ratio is seen to be greater than unity for most frequency components. In particular, there are frequencies for which R/F > 100, or, the force applied to the test item in the laboratory is at least 100 times larger than the corresponding field interface force. This suggests that at those frequencies the test item is being severely over-tested in the laboratory environment.

Figure 10 shows all four test item laboratory accelerations predicted from Eq. 11 and the corresponding field accelerations obtained from Eq. 2 when test item and vehicle are connected in the field. Clearly, there is a significant mismatch of acceleration in each case. The test item accelerations shown in Fig. 10 reveal that natural frequencies and vibration levels predicted in the laboratory are not the same as those predicted in the field environment. *Thus, a field data base composed of the bare vehicle interface accelerations only is not appropriate to define test item inputs in the laboratory.*

The incorrect laboratory accelerations obtained in this case can be explained by comparing the equations used to calculate interface forces and test item accelerations in both environments. The systems of equations expressed by Eq. 6 and Eq. 9 have different coefficient matrices. The entries in matrix $[TV]^{-1}$ in Eq. 1 or Eq. 6 contain the test item and vehicle driving and transfer point FRF information while the coefficient matrix $[T]$ in Eq. 9 contains only test item FRF information. Similarly, a comparison of Eqs. 2 and Eq. 10 shows that they do not represent the same linear transformation. Equation 2 has a coefficient matrix that contains FRF information from both the test item and the vehicle while Eqs. 10 and 11 have FRF information from test item only. Hence, incorrect results are obtained in the laboratory simulation when only bare vehicle interface accelerations are used without accounting for the vehicle driving and transfer point FRFs. However, if the field data contained both the bare vehicle interface accelerations and the vehicle interface FRFs, a successful laboratory simulation can be achieved since appropriate transformation matrices can be used in Eqs. 9 and 10. Thus, it is clear that driving and transfer point FRF measurements play an enormously important role in translating field data to a legitimate laboratory simulation. McConnell [3] arrived at similar conclusions both, analytically and experimentally by studying a system where test item and vehicle were attached at a single point.

4.2. SIMULATION 2 – INPUT: FIELD INTERFACE FORCES

In the second laboratory simulation, the field interface forces $\{F_c\}$ shown in Fig. 5 are used as test item inputs. Exciters 1 and 2 in Fig. 7 are controlled such that laboratory interface forces match the corresponding field measurements. Equation 7 is used to calculate the test item accelerations in this case, and it is rewritten as

$$\{U\} = [T]\{R_c\} \tag{12}$$

where $\{R_c\}$ is the N_c x 1 laboratory interface force vector that matches the field interface forces ($\{R_c\} = \{F_c\}$), $\{U\}$ is the N_t x 1 test item laboratory accelerations.

The resulting test item accelerations from Eq. 12 are shown in Fig. 11 and are compared with the corresponding field data. Field and laboratory test item accelerations are indistinguishable in this case. This result was expected since the test item inputs in the laboratory environment match the corresponding field interface forces and no field external forces were applied to the test item. Thus, controlling laboratory interface forces such that they match field interface forces leads to a successful laboratory simulation for the conditions of this example. *The trick is to get the exciters to be properly controlled since cross correlation must be maintained.*

4.3. SIMULATION 3 – INPUT: INTERFACE FORCES FROM FIELD COMBINED INTERFACE ACCELERATIONS

In the third test scenario, the field accelerations shown in Fig. 6 are used to generate the test item inputs in the laboratory environment. The field data in this case corresponds to the test item interface and external field accelerations. This laboratory simulation requires that exciters 1 and 2 in Fig. 7 be capable of generating and controlling the forces to drive the test item such that field motions be reproduced at the corresponding control points. Appropriate choice of the control points depend on the dynamic characteristics of the test item being tested as well as on the field data. Generally speaking, either interface or external motions can be used to define the test item inputs in the laboratory. In both cases, the problem is to find an appropriate set of test item inputs such that field motions can be reproduced at the control points in the laboratory environment. This corresponds to a force identification problem.

Force identification represents an inverse problem in vibration testing where a set of field accelerations are used to obtain a set of excitation forces that caused the measured motions. One method that is used for predicting forces is the so called pseudo-inverse technique [5]. The aim of this method is to identify a set of M excitation forces from a set of N measured motions. Generally N>M so that the solution for the unknown excitation forces is carried out in a least squares sense when using the following equation

$$\{F\} = [H]^+\{X\} \tag{13}$$

The superscript "+" denotes the pseudo-inverse of the FRF matrix, and is given by

$$[H]^+ = \left[[H]^H \, [H] \right]^{-1} [H]^H \tag{14}$$

where symbol "H" denotes the Hermitian transpose conjugate operator. Equations 13 and 14 employ a data containing magnitude and phase information, i.e., frequency spectra. In this present example, field interface accelerations will be used with the pseudo-inverse technique. In this case, Eq. 13 is written as

$$\begin{Bmatrix} R_1 \\ R_3 \end{Bmatrix} = \begin{bmatrix} T_{11} & T_{13} \\ T_{31} & T_{33} \end{bmatrix}^+ \begin{Bmatrix} X_1 \\ X_3 \end{Bmatrix} \tag{15}$$

The resulting forces from Eq. 15 match closely the field interface forces, as shown in Fig. 12, and this result is expected since there are no external forces applied to the test item while in the field environment. In addition, the pseudo-inversion reduces to the standard inversion in the case of Eq. 15 since the test item FRF matrix is square and it can be inverted exactly. Further investigation will reveal the required conditions as well as the major difficulties that arise while using the pseudo-inverses technique to define the test item laboratory inputs from field acceleration data [6,7].

The identified laboratory forces are then applied to the test item in the laboratory simulation producing the motions shown in Fig. 13 and calculated according to Eq. 10. The results present a nearly perfect match when compared to the corresponding field accelerations. Small discrepancies are observed in the laboratory accelerations for frequencies in the vicinity of acceleration valleys. These results for the test item laboratory responses are expected since the forces predicted using by the pseudo-inverse technique and the field interface forces are essentially the same.

5. Summary and Conclusions

There are a number of choices of how we take and interpret field data for the purpose of controlling a vibration test. This paper considers the simplest multi-connector vibration simulation problem where no external forces are involved in either the field or laboratory environments and the signals are all deterministic (periodic or transient) so that frequency spectra with magnitude and phase are used in all calculations. The test item and vehicle are each modeled by a discrete four DOF system. Several choices as to how we process the field data for controlling the test are evaluated.

Two field environments are considered as typical sources of field vibration data for use in laboratory simulations. In the first field environment, the test item is absent and bare vehicle accelerations are calculated. In the second field environment, the test item is attached to the vehicle at two locations, the vehicle has the same input as in the first simulation, an all interface forces and accelerations as well as external accelerations are calculated for this combined system. The interface accelerations (both bare vehicle and combined system) and forces are used as laboratory inputs in order to obtain typical test results.

First, the bare vehicle interface accelerations are applied to the test item using two vibration exciters in the laboratory. The resulting interface and external accelerations as well as interface forces showed different natural frequencies and magnitudes and resulted in many regions of either severe over-test or under-test. It is clear that blind application of bare vehicle data without accounting for the test item and vehicle interface FRF characteristics leads to significant test errors. If, however, these interface FRF characteristics are accounted for in transforming the bare vehicle interface accelerations, reasonable test results in terms of natural frequencies and magnitudes are obtained.

The second laboratory simulation choice is to use the combined interface forces as test item inputs. In this case, all interface and external accelerations matched the corresponding measured field accelerations.

The third laboratory simulation choice is to use the combined interface accelerations as test item inputs. The resulting external accelerations matched the field data in magnitude at nearly every frequency. The pseudo-inverse method is employed to estimate the interface forces in this case. It was found that the pseudo-inverse method worked well with this noise free data since the pseudo-inverse method gave a unique solution in this case.

Now, we move onto the case where external forces act on the test item in the field. These forces can cause significant motions to occur in the field environment and must be accounted for in any laboratory testing scheme.

Acknowledgements

The authors would like to thank the Iowa State University Engineering College and the Aerospace Engineering and Engineering Mechanics Department for supporting this research. Dr. Paulo S. Varoto, from Universidade de São Paulo – São Carlos – Brasil was financially sponsored by CNPq – Brazil during his PhD program.

6. References

1. Varoto, P.S., and McConnell, K.G. (1999), Rules for the Exchange and Analysis of Dynamic Information: Part I – Basic definitions and test scenarios, in J.M.M. Silva and N. M. M. Maia (eds.), *Modal Analysis & Testing, Kluwer Academic Publishers, NATO Series*, Dordrecht, 65-81.
2. McConnell, K.G. (1995), Vibration Testing: Theory and Practice, *John Wiley & Sons*, N.Y.
3. McConnell, K.G. (1995), From field vibration data to laboratory simulation, *Experimental Mechanics, Vol 34*, 181-193.
4. Gordis, J.H., Bielawa, R.L., Flannelly, W.G. (1991), A general theory for frequency domain structural synthesis, *Journal of Sound and Vibration, Vol. 150(1)*, 139-158.
5. Ewins, D. (1984), Modal Testing: Theory and Practice, *Research Studies Press*, London.
6. Varoto, P.S. and McConnell, K.G. (1999), Rules for the Exchange and Analysis of Dynamic Information: Part V – Q transmissibility matrix vs single point transmissibility in test environments, in J.M.M. Silva and N. M. M. Maia (eds.), *Modal Analysis & Testing, Kluwer Academic Publishers, NATO Series*, Dordrecht, 179-208.
7. Varoto, P.S., and McConnell, K.G. (1999), Rules for the Exchange and Analysis of Dynamic Information: Part IV – Numerically simulated and experimental results for a random excitation, in J.M.M. Silva and N. M. M. Maia (eds.), *Modal Analysis & Testing, Kluwer Academic Publishers, NATO Series*, Dordrecht, 137-177.

94

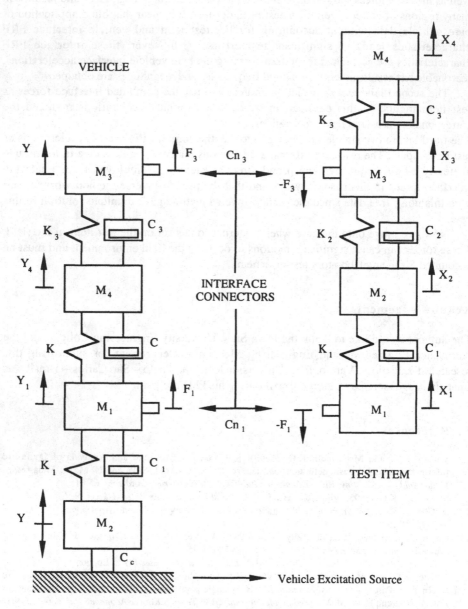

Figure 1 Test item attached to vehicle in the field environment.

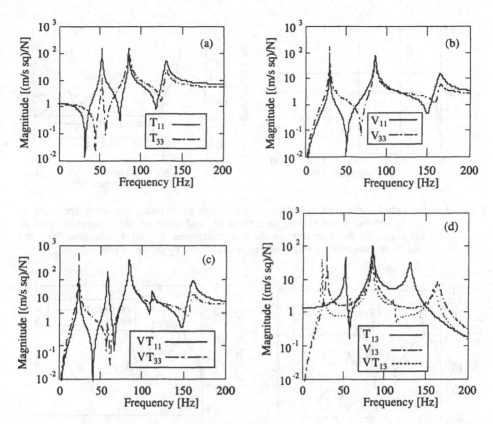

Figure 2   Interface driving and transfer accelerance FRF's for structures. Test item is modeled as a free structure in space; vehicle and combined structure are grounded through M2 (see Fig. 1): (a) Test item T_{11} and T_{33}; (b) Vehicle V_{11} and V_{33}; (c) Combined VT_{11} and VT_{33}; (d) Transfer FRF's T_{13} and VT_{13}.

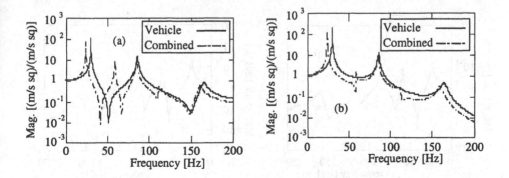

Figure 3   Vehicle and combined structure interface accelerance transmissibility FRF's: (a) Interface point 1; (b) Interface point 3.

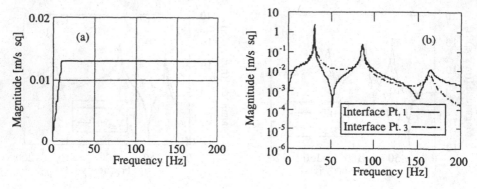

Figure 4   (a) Vehicle and combined structure input base acceleration frequency spectrum Y_2 that is applied to the vehicle base mass M_2 and gives an input vibration level of 1.0 g_{RMS}; (b) Bare vehicle interface accelerations Y_1 and Y_3 obtained from the vehicle transmissibility FRF's of Fig. 3 and from Y_2 shown in part (a).

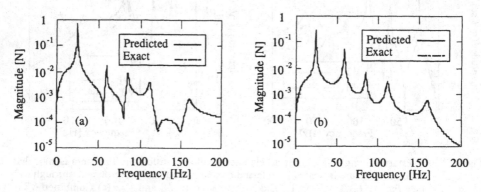

Figure 5   Field interface forces due to the input base frequency spectrum Y_2 shown in Fig. 4 and *no external forces are applied on test item*: (a) Interface point 1; (b) Interface point 3.

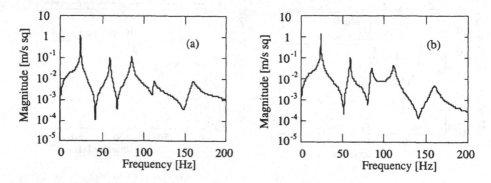

Figure 6   Test item field accelerations, *no external forces applied to the test item*: (a) Interface acceleration X_1 ; (b) External acceleration X_2 ;

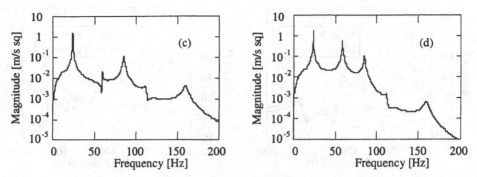

Figure 6   (*Cont.*) Test item field accelerations, *no external forces applied to the test item*: (c) Interface acceleration X_3 ; (d) External acceleration X_4.

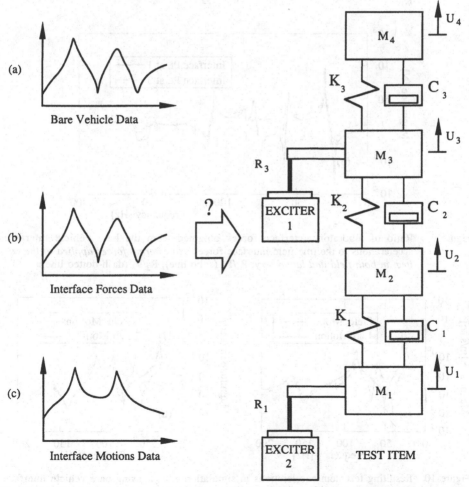

Figure 7   Definition of test item laboratory inputs from field data using: (a) Bare vehicle data; (b) Combined interface forces; (c) Combined interface motions.

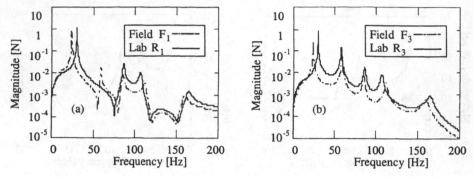

Figure 8   Comparison between field and laboratory interface forces for laboratory simulation using bare vehicle data: (a) Interface forces F_1, and R_1; (b) Interface forces F_3, and R_3.

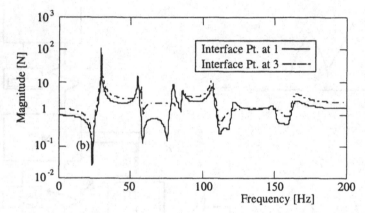

Figure 9   Ratio of laboratory interface forces obtained from the bare vehicle interface accelerations to the true field interface forces. *No external forces applied to the test item in both field and laboratory*: R_1/F_1 (solid line); R_3/F_3 (dash-dotted line).

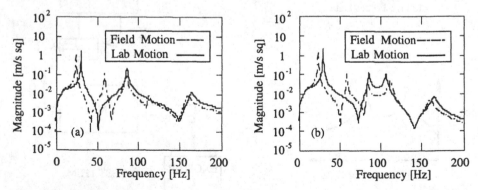

Figure 10   Resulting test item accelerations in simulation case 1, using bare vehicle interface data. *No external forces are applied to the test item in both environments*: (a) Interface acceleration U_1; (b) External acceleration U_2;

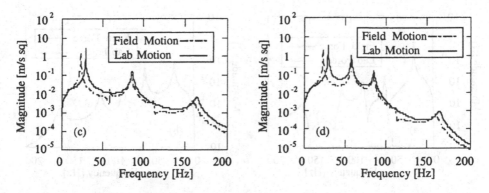

Figure 10   (*Cont.*) Resulting test item accelerations in simulation case 1, using bare vehicle interface data. *No external forces are applied to the test item in both environments*: (c) External acceleration U_3 ; (d) External acceleration U_4.

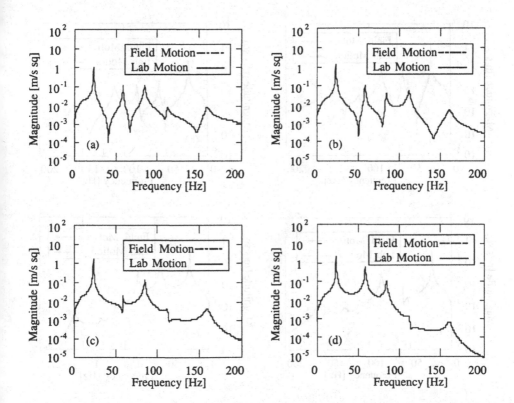

Figure 11   Test item accelerations from laboratory simulations case 2, using interface forces. *No external forces applied to the test item in both environments*: (a) and (c) Interface accelerations U_1 and U_3 ; (b) and (d) External accelerations U_2 and U_4.

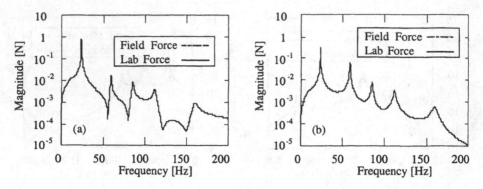

Figure 12   Predicted interface forces from field notions by the pseudo-inverse technique. *No field external forces applied to the test item*: (a) Interface force R_1 ; (b) Interface force R_3 .

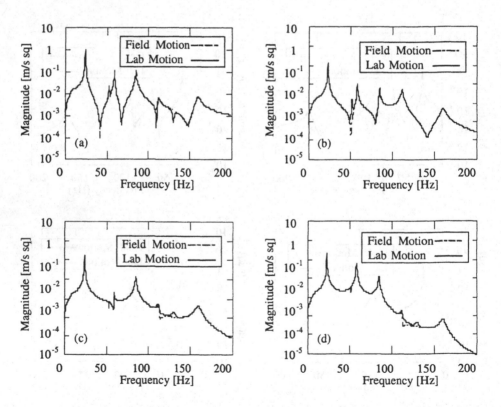

Figure 13   Test item accelerations from laboratory simulation case 3, using interface forces predicted from field motions. *No external forces applied to the test item in both environments*: (a) and (c) interface accelerations U_1 and U_3 ; (b) and (d) External accelerations U_2 and U_4 .

RULES FOR THE EXCHANGE AND ANALYSIS OF DYNAMIC INFORMATION

Part III: *Numerically Simulated and Experimental Results for a Deterministic Excitation with External Loads*

P. S. VAROTO
Dept de Engenharia Mecanica
Escola De Engenharia De São Carlos,
USP
São Carlos – SP – 13560-250, Brasil

K. G. McCONNELL
Dept. Aerospace Engineering and
Engineering Mechanics
Iowa State University
Ames, IA 50011 USA

Abstract

The ultimate goal of a laboratory simulation is to ensure that a given test item will survive when exposed to its field dynamic environment. Two previous papers [1,2] described the process of using field vibration data to define suitable test item inputs in laboratory simulations. The structural interactions that occur when the test item is attached to the vehicle in the field or to the vibration exciters in the laboratory were discussed in Ref. [1]. Numerical examples describing different test scenarios used in laboratory simulations were presented in Ref. [2]. In these test scenarios, the test item was subjected to *interface forces* only, i.e.; no *external forces* were applied to the test item while in the field environment. The first part of the present paper discusses laboratory simulations of field data when the test item is subjected to external forces in addition to interface forces in the field environment. Numerical examples of different laboratory test scenarios show that accounting for field external forces in the laboratory is of major importance. The second part of the paper deals with the force identification problem. The pseudo-inverse technique, commonly employed in modal testing is used to predict the test item input forces from field motions. Numerical examples show the feasibility of the pseudo-inverse method in predicting external loads. An experimental analysis is performed on a free beam to discuss the practical implications of this force identification technique when applied to realistic situations. *This paper deals only with transient and periodic forces and motions.*

Nomenclature

A_{pq}	Accelerance FRF	K_p	p^{th} test item and vehicle
Cn_1, Cn_2	Interface points		spring constants
C_p	p^{th} test item and vehicle damping coefficients	$\{F_c\}$	Test item field interface force
		F_1, F_3	Field interface forces
$j = \sqrt{-1}$	Imaginary number	F_4	Test item field external force

J.M.M. Silva and N.M.M. Maia (eds.), Modal Analysis and Testing, 101–135.

[M]	Mass matrix	[TV]	Test item Vehicle combined
M_p	p^{th} test item and vehicle		matrix
	masses	{U}	Test item laboratory
N_t	Degrees of freedom for test		accelerations
	item	$\{U_c\}$	Test item laboratory interface
N_v	Degrees of freedom for		accelerations
	vehicle	{X}	Test item output
N_c	Number of connectors		accelerations
T_{pq}	Test item accelerance FRF	$\{X_c\}$	Test item field interface
{R}	Test item laboratory input		accelerations
	force	$\{X_e\}$	Test item field external
$\{R_c\}$	Test item laboratory interface		accelerations
	force	$\{Y_c\}_e$	Bare vehicle interface
[T]	Test item accelerance FRF		accelerations
	matrix		
$[T_{cc}]$	Test item interface	Greek	
	accelerance FRF matrix	Δf	Frequency resolution in Hz
$[T_{ce}]$	Test item connector-external	$\{\phi\}_r$	r^{th} mode shape
	accelerance FRF matrix	$\Gamma(\omega)$	Acceleration transmissibility
$[T_{ec}]$	Test item external-connector		FRF
	accelerance FRF matrix	ω	Frequency in rad/s
$[T_{ee}]$	Test item external	ω_r	r^{th} natural frequency in rad/s
	accelerance FRF matrix	ζ_r	r^{th} modal damping ratio

1. Introduction

In the field environment, the test item is attached to the vehicle at several locations forming a combined field structure. During service, the test item is subjected to different types of loads, forces and moments that can be grouped as: (i) *Internal loads*; (ii) *External loads*; and (iii) *Interface loads*. Internal unbalance vibration sources are classified as internal loads. Loads that occur at the interface points between the test item and the vehicle due to coupling effects only are grouped as interface or connector loads. Other loads, such as aerodynamic and acoustic loads, as well as loads applied at the interface connections points that are not due to coupling effects as classified as external loads.

In the laboratory environment, the test item is attached to vibration exciters, and as previously shown [1], three different laboratory test scenarios can be used to match field data by defining suitable test item inputs. These test scenarios employ interface forces, interface motions and test item external motions as inputs to the test item. In each case, the vibration exciter is controlled such that a given field measurement is matched in the laboratory. Numerical simulations of these test scenarios [2] show that in the absence of field external forces, a successful laboratory simulation can be achieved, as long as interface forces and motions are used as test item inputs. However, in some situations, the test item is subjected to external forces in addition to interface forces while in the field environment [13]. The theoretical results presented in Ref. [1] show that unless external force effects are accounted for in the laboratory, the simulation results will be incorrect.

Thus, in addition to knowing field interface forces, knowledge of the external forces applied to the test item is required such that they can be accounted for in the laboratory.

This paper is divided in two major parts. The first part shows the effects of field external forces in laboratory simulations. This is achieved by performing numerical simulations on a multi degree of freedom (MDOF) discrete linear system. The equations developed in Ref. [1] are used to calculate field interface forces and accelerations when the test item is subjected to field external forces in addition to interface forces. The pseudo-inverse method is used in the laboratory simulation to calculate a set of forces from field accelerations that are applied to the test item in order to simulate field data.

The second part shows results of an experimental analysis performed on a free free beam in order to access the feasibility of the pseudo-inverse technique when dealing with experimental data. Deterministic excitation signals are used to drive the beam at several locations. The beam's output acceleration frequency spectra are measured at several locations and the pseudo-inverse technique is used to estimate the input forces.

2. Simulation calculation steps

Figures 1 and 2 show the sequence of calculation steps that are used in the field and laboratory simulations presented in the first part of this paper. Figure 1 contains two columns and represents the sequence of calculations performed in the simulation of the field environment. The first column shows the various situations analyzed while the second column shows the field data obtained in each case. We give now a brief description of each calculation step performed in the field simulation shown in Fig. 1:

- *Step 1*. This step is shown in Fig. 1a and corresponds to the test item acceleration response due to external forces *only*. In this case, the test item is modeled as a free structure in space and two acceleration vectors are obtained in this step as shown in Fig. 1a. These accelerations are used in obtaining the interface forces and test item accelerations in the field environment.

- *Step 2*. The results of this calculation are the bare vehicle interface accelerations as shown in Fig. 1b. In this case, the test item is not connected to the vehicle in the field and the resulting bare vehicle interface acceleration vector $\{Y_c\}_e$ is due to external forces $(P_1, P_2, ..., P_{N_v-N_c})$ acting on the vehicle.

- *Step 3*. This corresponds to the combined structure simulation where test item and vehicle are connected in the field, as shown in Fig. 1c. In this case, the interface forces $\{F_c\}$ as well as the test item interface and external accelerations $\{X_c\}$ and $\{X_e\}$, are obtained.

Similarly, the laboratory simulation calculation procedure shown in Fig. 2 is used. The first column in Fig. 2 represents the test setup employed in the simulation while the second column shows the corresponding field data used to define the test item inputs in each simulation. Three laboratory simulations are performed:

- *Step 1*. This laboratory simulation is shown in Fig. 2a and uses the bare vehicle data obtained in the field simulation of Fig. 1a to define the test item inputs. *No external force effects are accounted for in this simulation.*

- *Step 2.* In this case, interface forces from the field simulation shown in Fig. 1b are used as test item inputs as shown in Fig. 2b. *No external force effects are accounted for in this simulation.*

- *Step 3.* This laboratory simulation is shown in Fig. 3c and employs the test item interface and external motions obtained in the field simulation shown in Fig. 1c. Three cases are analyzed where the test item inputs are obtained using different motions in each case. *External force effects are accounted for in this simulation.*

Figures 1 and 2 will be referred throughout the numerical simulations presented in this paper.

3. Theoretical response models of test structures

The models used in the field and laboratory simulations are shown in Fig. 3. The test item and vehicle are modeled by a four DOF system and are attached to one another in the field environment through connectors Cn_1 and Cn_3 forming the combined field structure. Coordinate numbers are assigned for the test item and vehicle models such that interface points carry the same number in both structures. Physical parameters for both the test item and vehicle models are shown in Table 1.

Using standard modal analysis principles, modal properties are obtained for the test item, the vehicle, and the combined structure. *When test item and vehicle are not connected in the field environment, the test item is modeled as a free structure in space, while the vehicle is grounded through the mass M_2.* The results for modal parameters for all structures involved in the simulation are shown in Table 2.

Accelerance FRFs are obtained for the structures involved according to [4]

$$A_{pq}(\omega) = -\omega^2 \sum_{r=1}^{N} \frac{\phi_{pr}\,\phi_{pr}}{m_r\left(\omega_r^2 - \omega^2 + j\,2\zeta_r\,\omega_r\,\omega\right)} \tag{1}$$

where ϕ_{pr} and ϕ_{qr} are the pth and the qth elements of the rth mode shape, ω_r is the rth natural frequency (rad/s), ζ_r is the rth modal damping ratio, m_r is the modal mass of the rth mode shape, and ω (rad/s) is the excitation frequency.

Table 1: Physical Parameters

Mass	Test Item (Kg)	Vehicle (Kg)
M_1	0.20	0.50
M_2	0.20	0.50
M_3	0.25	0.30
M_4	0.15	0.50
Stiffness	Test Item N/m x 10^4	Vehicle N/m x 10^4
K_1	5.00	6.00
K_2	4.00	14.00
K_3	2.00	10.00

Table 2: Modal Properties

Natural Frequency	Test Item (H_z)	Vehicle (H_z)	Combined (H_z)
f_1	0	30.00	23.86
f_2	52.60	86.17	58.94
f_3	85.70	164.72	85.05
f_4	131.00	-	112.00
f_5	-	-	160.21
Damping Ratio	Test Item (%)	Vehicle (%)	Combined (%)
1	0	0.28	0.24
2	0.50	0.81	0.56
3	0.81	1.55	0.91
4	1.23	-	1.06
5	-	-	1.54
Modal Masses	Test Item (Kg)	Vehicle (Kg)	Combined (Kg)
m_1	0.20	0.43	0.38
m_2	0.17	0.48	0.17
m_3	0.21	0.37	0.55
m_4	0.20	-	0.21
m_5	-	-	0.34

Similarly, for input motion at a single coordinate, the acceleration transmissibility FRF is obtained by [2]

$$\{\Gamma(\omega)\} = \omega^2 \sum_{r=1}^{N} \frac{\{\phi\}_r \{\phi\}_r^T [M] \{\delta\}}{m_r \left(\omega_r^2 - \omega^2 + j 2\zeta_r \omega_r \omega\right)} + \{\delta\} \tag{2}$$

where $[M]$ is the structure's mass matrix and $\{\phi\}_r$ is the structure's r^{th} mode shape. Equations 1 and 2 define response models for viscously damped structures.

An external force F_4 is applied to test item mass M_4 as shown in Fig. 3. The magnitude and phase angle of this external force are shown in Fig. 4. The test item acceleration response to this excitation is obtained from reference [1] as

$$\begin{Bmatrix} \{X_c\} \\ \{X_e\} \end{Bmatrix} = \begin{bmatrix} [T_{cc}] & [T_{ce}] \\ [T_{ec}] & [T_{ee}] \end{bmatrix} \begin{Bmatrix} \{F_c\} \\ \{F_e\} \end{Bmatrix} \tag{3}$$

where $\{X_c\}$ is the N_c x 1 test item interface acceleration and $\{X_e\}$ is the $(N_t - N_c)$ x 1 test item external acceleration. Vectors $\{F_c\}$, N_c x 1 and $\{F_e\}$, $(N_t - N_c)$ x 1 contain external forces applied to interface and external points, respectively. For the model of Fig. 3, we have $\{F_c\} = \{0\}$ and $\{F_e\} = \{F_4\}$. The resulting test item accelerations obtained from Eq. 3 are shown in Fig. 5. Figures 5a and 5c represent the test item interface accelerations while Figs. 5b and 5d correspond to the test item external accelerations.

4. Field simulation

The field simulations are shown in Fig. 1b and 1c. *First*, the test item is absent from the field so that the field data consists of only by the bare vehicle interface accelerations as shown in Fig. 1b. *Second*, the test item and vehicle are connected in the field so that both interface forces and test item accelerations constitute the field data as depicted in Fig. 1c.

The interface forces are calculated according to [1] as

$$\{F_c\} = [TV]\left(\{Y_c\}_e - \{X_c\}_e\right) \tag{4}$$

where $\{F_c\}$ contains the N_c interface forces. Matrix $[TV] = [[T_{cc}] + [V_{cc}]]^{-1}$ is N_c x N_c and is the test item-vehicle combined matrix that is given by the inverse of the matrix containing the sum of test item and vehicle interface accelerance FRFs T_{pq} and V_{pq} for p, q = 1...N_c, respectively. Vectors $\{Y_c\}_e$ and $\{X_c\}_e$ are N_c x 1 and represent the vehicle and the test item connector accelerations that are due to external forces, respectively. Vector $\{Y_c\}_e$ is called the *bare* vehicle interface accelerations. Since these vectors carry the second subscript, they are caused by external inputs acting on the vehicle and test item, respectively, when they are not connected.

The test item accelerations that occur in the field environment are obtained from the following equation from reference [1] as

$$\begin{Bmatrix} \{X_c\} \\ \{X_e\} \end{Bmatrix} = \begin{bmatrix} [T_{cc}][TV] & -[T_{cc}][TV] \\ [T_{ec}][TV] & -[T_{ec}][TV] \end{bmatrix} \begin{Bmatrix} \{Y_c\}_e \\ \{X_c\}_e \end{Bmatrix} + \begin{Bmatrix} \{X_c\}_e \\ \{X_e\}_e \end{Bmatrix} \tag{5}$$

Equation 5 shows that the test item field accelerations depend on the test item and vehicle dynamic characteristics (FRFs) and on the interface and external accelerations that are caused by external forces only.

The vehicle excitation source:

The vehicle excitation source is assumed to be a base motion Y_2 that is applied to mass M_2 as shown in Fig. 6a. This input base acceleration frequency spectrum covers the 0 - 200 H_z frequency range, increasing linearly with a slope of 1.28×10^{-3} (m/s²)/H_z in the 0-10 H_z frequency range and becoming constant with a magnitude of 1.28×10^{-2} m/s² in the remaining 10-200 H_z frequency range. A total of 800 spectral lines are used in all simulations, giving a frequency resolution of f = 0.25 Hz so that the input vibration level is approximately 1.0 g_{RMS} (9.81 (m/s²) $_{RMS}$).

4.1. FIELD SIMULATION 1: BARE VEHICLE INTERFACE ACCELERATIONS

This simulation corresponds to the situation shown in Fig. 1b. In this case, the test item is not connected to the vehicle. For the system of Fig. 3, the vehicle input is composed only by the input acceleration frequency spectrum Y_2 applied at M_2. Thus, the bare vehicle accelerations can be obtained from the acceleration transmissibility FRF given by Eq. 2 by using the following relationship

$$Y_p = \Gamma_{pq} Y_q \qquad (6)$$

where $p = 1 \ldots N_v$ and Y_q denotes the input motion at the qth coordinate. By using Eq. 6 with $Y_q = Y_2$, the bare vehicle interface accelerations shown in Fig. 6b are obtained.

4.2. FIELD SIMULATION 2: COMBINED STRUCTURE

This field simulation corresponds to the third calculation step in Fig. 1c. In this case, the test item and the vehicle are connected in the field as shown in Fig. 3. The combined structure has two inputs, the vehicle excitation acceleration spectrum Y_2 that is applied to the vehicle base mass M_2, and the external force F_4 that is applied to the test item mass M_4. Equations 4 and 5 must be used in this case so that the effects of F_4 are properly accounted for. Since these expressions involve test item accelerations caused by external forces, knowledge of these accelerations is required to employ these expressions in this field simulation. Recall that the test item accelerations due to the external force F_4 were obtained by using Eq. 3 and the results are shown in Fig. 5. Thus, since the bare vehicle interface acceleration $\{Y_c\}_e$ were obtained in the previous field simulation, and the test item and vehicle FF characteristics are known, Eqs. 4 and 5 can be used to predict the field interface forces and accelerations in this case.

However, the field environment analysis presented in this paper assumed that the external force applied to the test item is known a priori, and thus, the external accelerations $\{X_c\}_e$ and $\{X_e\}_e$ could be obtained by using Eq. 3. This seldom happens in real situations, where possible field external forces effects are not readily available [13].

As will be shown later, knowledge of field external forces is a most important issue in laboratory simulations, and in order to show how external forces affect the interface forces and test item motions in the field, two field simulations will be considered for the combined structure of Fig. 3.

First, Eqs. 4 and 5 are used to correctly predict the interface forces and the test item accelerations. Second, it is assumed that force F_4 is unknown as frequently occurs in real situations where aerodynamic and acoustic forces occur [13]. Thus, although the force F_4 is contributing for the interface forces and accelerations, we will ignore its effects by writing Eqs. 4 and 5 as

$$\{F_c\} \cong [TV] \{Y_c\}_e \qquad (7)$$

$$\begin{Bmatrix} \{X_c\} \\ \{X_e\} \end{Bmatrix} \cong \begin{bmatrix} [T_{cc}][TV] & -[T_{cc}][TV] \\ [T_{ec}][TV] & -[T_{ec}][TV] \end{bmatrix} \begin{Bmatrix} \{Y_c\}_e \\ \{0\} \end{Bmatrix} \qquad (8)$$

where it is assumed that $\{X_c\}_e \approx 0$ and $\{X_e\}_e \approx 0$ in this case.

The important issue addressed here is that of neglecting external forces effects in Eqs. 4 and 5 when these forces contribute significantly for the test item response leads to incorrect predictions for interface forces and accelerations. The approximated expressions in Eqs. 7 and 8 may not be totally inappropriate in situations where the external forces have only minor effect on the interface forces and test item accelerations.

Interface force results using Eqs. 4 (correct forces) and 7 (incorrect forces) are shown in Fig. 7. As expected, it is seen that significantly different interface forces are obtained when external force effects are not negligible. These differences are highly frequency dependent and cover a significant dynamic range.

The resulting test item output accelerations from Eqs. 5 (correct motions) and 8 (incorrect motions) are shown in Fig. 8. It is clear that incorrect results are obtained when the external force effects are not negligible. Again, these errors are large and a large dynamic range occurs.

5. Laboratory simulation

In the field simulation, suitable test item inputs must be defined such that the field environment can be properly simulated. The bare vehicle interface accelerations lead to incorrect laboratory test results when used to define test item inputs [2,5]. This conclusion is valid independent of the test item being subjected to field external forces or not since the bare vehicle interface accelerations are obtained when the test item is absent from the field environment. Thus, the laboratory simulation employing only the bare vehicle interface accelerations will not be repeated here. The reader is referred to Refs. [3,5] for more information about this laboratory test scenario. Instead, the laboratory test scenarios shown in Fig. 2 will be discussed.

5.1. LABORATORY SIMULATION 1: INTERFACE FORCES

This simulation corresponds to the situation shown in Fig. 2a. In this test scenario, interface forces are used as test item inputs in the laboratory environment. It is assumed that the interface forces can be generated and controlled in the laboratory environment such that field and laboratory interface forces are matched.

The test item acceleration response in the laboratory environment is calculated according to Eq. 3, rewritten as

$$\begin{Bmatrix} \{U_c\} \\ \{U_e\} \end{Bmatrix} = \begin{bmatrix} [T_{cc}] & [T_{ce}] \\ [T_{ec}] & [T_{ee}] \end{bmatrix} \begin{Bmatrix} \{R_c\} \\ \{R_e\} \end{Bmatrix} \tag{9}$$

where $\{\{U_c\}\{U_e\}\}^T$ is the N_t x 1 test item response vector that is composed of the laboratory interface acceleration vector $\{U_c\}$, N_c x 1, and the laboratory external acceleration vector $\{U_e\}$, $(N_t - N_c)$ x 1. Vector $\{\{R_c\}\{R_e\}\}^T$ contains the test item laboratory input forces in terms of forces applied to interface points $\{R_c\}$ and forces applied to external points $\{R_e\}$. Equation 9 *assumes* that the test item has the same FRF matrix in both the field and laboratory environments.

Matching interface forces requires that the following condition be satisfied

$$\{R_c\} = \{F_c\} \tag{10}$$

where it is *assumed* that the number of interface points in the laboratory is the same as in the field so that $\{R_c\}$ and $\{F_c\}$ are N_c x 1. Only field interface forces are used in this laboratory test scenario so that $\{R_e\} = \{0\}$. Two interface forces are applied to the test item at points 1 and 3.

Then, by using Eq. 9 with Eq. 0, the test item laboratory accelerations are obtained and are shown in Fig. 9. The test item field accelerations are plotted for comparison purposes. Figures 9a and 9c show the test item interface accelerations while Fig. 9b and 9d show test item external accelerations. As seen in Fig. 9, this laboratory simulation gave incorrect test item accelerations, and this is easily understood since only interface forces were applied to the test item in the laboratory and the effects of the external force F_4 were not accounted for. Note that the interface forces used in this test scenario are those shown in Fig. 7 and calculated through Eq. 4, which in turn, account for F_4. Hence, *matching the correct interface forces in the laboratory leads to incorrect simulation results if external force effects are not accounted for in addition to interface forces.*

5.2. LABORATORY SIMULATION 2: INTERFACE AND EXTERNAL MOTIONS

This test scenario is shown in Figs. 2b and 2c. In this case, the field accelerations shown in Fig. 8 are used to define the test item inputs in the laboratory. This constitutes a force identification problem, since a set of laboratory input forces is required to drive the test item from knowledge of field motions.

Force identification represents an inverse problem in mechanics. The pseudo-inverse technique [6] is used to estimate unknown applied forces to a given structure from knowledge of motion records and the structure's FRF characteristics.

The solution for M unknown forces from knowledge of N motion records (generally N >M) through the pseudo-inverse technique is carried out in the frequency domain through the following relationship

$$\{F\} = [A]^+ \{X\} \tag{11}$$

where superscript $^+$ denotes the pseudo-inverse of the FRF matrix, and is given by

$$[A]^+ = \left[[A]^H [A] \right]^{-1} [A]^H \tag{12}$$

where the superscript ^{H} denotes the *hermitian conjugate* (complex conjugate transpose) of the accelerance FRF matrix since [A] is complex. Equations 11 and 12 are employed when the structure's output acceleration frequency spectra (magnitude and phase) $\{X\}$ are known so that these relationships are restricted to deterministic (transient and periodic) signals.

As in many inverse problems, the pseudo-inverse technique is numerically ill conditioned at some frequencies; particularly natural frequencies since the calculation procedure requires the inversion of the structure's FRF matrix for each frequency value. The FRF matrix is usually rectangular (N > M); and hence, the pseudo-inversion of the FRF matrix is required in order to solve for the unknown forces.

Some important characteristics of the pseudo-inverse technique are:

- The calculation procedure is very sensitive to the accuracy of the mode shape data used in the structure's modal model [4]. This affects the quality of the FRFs used in the identification process. Frequency response function curve fitting has been proposed [6] to reduce noise effects on the structure's FRF matrix, improving the numerical stability of the pseudo-inversion;

- The number of forces that can be predicted at a given frequency depends on the number of mode shapes participating in the structure's response at that frequency. In other words, the FRF matrix may be rank deficient at some frequencies [7].

- Experimental and numerical procedures have been proposed to improve the conditioning of the inversion problem. Hillary and Ewins [8] proposed to use strain gage data instead of acceleration data in the identification of unknown forces. Han and Wicks [9] proposed to use rotational FRF data to improve the stability of the FRF matrix inversion.

A numerical study was performed on a simply supported beam in order to assess the feasibility of the pseudo-inverse technique [10].

Sinusoidal excitation at a single frequency was used. This investigation reinforced the fact that the ill conditioning is related to the number of participating modes at each frequency, and this causes rank deficiency to the FRF matrix, which in turn, produces errors in the predicted forces. In addition, this paper points to the difference between using experimental FRFs and curve fitted FRFs in the solution of the inverse problem.

In the vibration testing context, the pseudo-inverse technique is used to calculate a set of laboratory test item inputs from knowledge of field acceleration records. Since the test item motions are divided in interface motions and external motions, a natural question that arises when using field motions to define test item inputs in this laboratory simulation is: *What motions should be measured in the field such that a suitable set of test item input forces are obtained from the pseudo-inverse technique in the laboratory environment?*

In this laboratory simulation, three cases are discussed. In each case, the pseudo-inverse technique is employed with field accelerations to predict a set of test item laboratory input forces. These cases are summarized below:

- *Case #1*: A set of N_c interface forces is calculated from N_c interface motions according to the following expression

$$\{R_c\} = [T_{cc}]^+ \{X_c\} \tag{13}$$

where $\{R_c\}$ is $N_c \times 1$ and corresponds to the predicted laboratory interface forces, $\{X_c\}$ is $N_c \times 1$ and contains field interface accelerations, and $[T_{cc}]$ is a square of $N_c \times N_c$ matrix in this case, its pseudo-inverse coincides with the standard inverse.

- *Case #2*: A set of N_c interface forces is calculated from all test item N_t accelerations, and in this case, Eq. 11 gives

$$\{R_c\} = \left[\left[[T_{cc}][T_{cc}] \right]^T \right]^+ \{X\} \tag{14}$$

where $\{X\}$ is $N_t \times 1$ and contains the test item laboratory interface and external accelerations and $[[T_{cc}][T_{cc}]]^T$ is the $N_c \times N_t$ test item FRF matrix.

- *Case #3*: A set of N_t forces is calculated from N_t accelerations according to

$$\begin{Bmatrix} \{R_c\} \\ \{R_e\} \end{Bmatrix} = \begin{bmatrix} [T_{cc}] & [T_{ce}] \\ [T_{ec}] & [T_{ee}] \end{bmatrix}^+ \begin{Bmatrix} \{X_c\} \\ \{X_e\} \end{Bmatrix} \tag{15}$$

where in this case, a set of interface and external forces is obtained.

The first two cases attempt to reproduce field data by predicting interface forces only while the third case attempts to obtain interface as well as external forces. The results of these three cases will be presented for the test item described in Fig. 3.

5.2.1. *Case #1: N_c interface forces from N_c interface accelerations*
For the test item model shown in Fig. 3, Eq. 13 is rewritten as

$$\begin{Bmatrix} R_1 \\ R_3 \end{Bmatrix} = \begin{bmatrix} T_{11} & T_{13} \\ T_{31} & T_{33} \end{bmatrix}^+ \begin{Bmatrix} X_1 \\ X_3 \end{Bmatrix} \tag{16}$$

In this case, the test item field accelerations calculated according to Eq. 5 and shown in Fig. 8a and c (solid line) are used. The results for the interface forces obtained from Eq. 16 are shown in Fig. 10, where it is seen that the pseudo-inverse technique yielded a correct result for the interface force R_1, but an incorrect result for R_3.

The test item laboratory accelerations due to the interface forces R_1 and R_3 are calculated according to

$$\begin{Bmatrix} U_1 \\ U_2 \\ U_3 \\ U_4 \end{Bmatrix} = \begin{bmatrix} T_{11} & T_{13} \\ T_{21} & T_{23} \\ T_{31} & T_{33} \\ T_{41} & T_{43} \end{bmatrix} \begin{Bmatrix} R_1 \\ R_3 \end{Bmatrix} \tag{17}$$

and they are shown in Fig. 11, where field accelerations are plotted for comparison purposes. Figures 11a and c show that the interface acceleration obtained in the laboratory simulation matched the corresponding field data regardless the fact that $R_3 \neq F_3$.

Similarly, field and laboratory external accelerations X_2 and U_2 matched closely in Fig. 11b. Since there are no external forces acting on M_2, the field acceleration X_2 is governed by spring and damper forces at elements K_1, K_2, C_1, and C_2, which in turn, depend on interface displacements and velocities at points 1 and 3, respectively. Thus, if correct interface accelerations are obtained in the laboratory simulation, accelerations U_2 and X_2 will be identical. On the other hand, an incorrect result is obtained for U_4 as seen in Fig. 11d. In this case, the external force F_4 is applied to M_4 in the field, but is not accounted for in the laboratory simulation. *The resulting laboratory motion U_4 is due to* *interface forces only while in the field X_4 is caused by the interface forces F_1 and F_3 as well as by the external force F_4.*

The results of this simulation can be summarized as:

- Interface forces:

$$R_1 = F_1$$
$$R_3 \neq R_3 \tag{18}$$

- Test item accelerations:

$$U_1 = X_1$$
$$U_2 = X_2 \tag{19}$$
$$U_3 = X_3$$
$$U_4 \neq X_4$$

Two additional simulations were performed in case 1. Both simulations employed Eqs. 16 and 17 but the point of application of the external force F_4 was different in each case. The results of these additional simulations are not shown here, but the following was observed:

(a) First, the test item external force F_4 was removed from M_4 and applied to M_2. Interface forces and test item accelerations were recalculated using Eqs. 4 and 5, respectively. Then, Eqs. 16 and 17 were used to calculate the laboratory interface forces and test item accelerations, respectively. The results of this simulation are:

- Interface forces:

$$R_1 \neq F_1$$
$$R_3 \neq R_3 \tag{20}$$

- Test item accelerations:

$$U_1 = X_1$$
$$U_2 \neq X_2 \tag{21}$$
$$U_3 = X_3$$
$$U_4 = X_4$$

(b) Second, the external force F_4 is simultaneously applied to the test item masses M_2 and M_4. The results obtained are

- Interface forces:

$$R_1 \neq F_1$$
$$R_3 \neq R_3 \tag{22}$$

- Test item accelerations:

$$U_1 = X_1$$

$$U_2 \neq X_2 \tag{23}$$

$$U_3 = X_3 \tag{24}$$

$$U_4 \neq X_4 \tag{25}$$

Thus, the first laboratory simulation case employing interface motions shows the following trends: (i) Interface accelerations are matched in the laboratory independent of the matching of interface forces; (ii) External accelerations corresponding to points that are subjected to external forces in the field environment are not matched in the laboratory; (iii) In two cases, the motion at an external point matched the corresponding field measurement, but this cannot be seen as a general trend, since the model used in these simulations contains only one external point free of forces.

The simulated results presented in this test scenario agree with the theoretical developments shown in Ref. [1]. Recall that interface and external accelerations can be broken in two parts, one due to connector forces and the other due to external forces, according to [1]

$$\{X_c\} = \{X_{cc}\} + \{X_{ce}\} \tag{26}$$

$$\{X_e\} = \{X_{ec}\} + \{X_{ee}\} \tag{27}$$

where the first subscript in Eqs. 26 and 27 belongs to the motion group (connector or external) and the second subscript belongs to the force causing the corresponding acceleration. For the model shown in Fig. 3, with the external force applied to M_4, Eqs. 26 and 27 are written as

$$X_1 = X_{11} + X_{13} + X_{14} \tag{28}$$

$$X_3 = X_{31} + X_{33} + X_{34} \tag{29}$$

$$X_2 = X_{21} + X_{23} + X_{24} \tag{30}$$

$$X_4 = X_{41} + X_{43} + X_{44} \tag{31}$$

Similarly, in the laboratory simulation

$$U_1 = U_{11} + U_{13} \tag{32}$$

$$U_3 = U_{31} + U_{33} \tag{33}$$

$$U_2 = U_{21} + U_{23} \tag{34}$$

$$U_4 = U_{41} + U_{43} \tag{35}$$

Thus, by comparing Eqs. 28 and 29 with Eqs. 32 and 33, we notice that if accelerations X_{14} and X_{34} are negligible, then the problem is reduced to matching interface accelerations X_{11}, X_{13}, X_{31}, and X_{33} that are caused by interface forces only. This condition can be achieved by applying the pseudo-inverse technique to field interface accelerations in order to get a set of forces such that application of these forces to the test item in the laboratory will give the same interface accelerations as measured in the field.

Another way of comparing field and laboratory interface characteristics is by expressing the laboratory forces as a function of field forces. By imposing $U_1 = X_1$ and $U_3 = X_3$ and using Eqs. 13 for interface points 1 and 3, we get the following expression for the laboratory interface forces required in this case.

$$\begin{Bmatrix} R_1 \\ R_3 \end{Bmatrix} = \begin{Bmatrix} F_1 \\ F_3 \end{Bmatrix} + \begin{bmatrix} T_{11} & T_{13} \\ T_{31} & T_{33} \end{bmatrix}^{-1} \begin{Bmatrix} T_{14}F_4 \\ T_{34}F_4 \end{Bmatrix} \tag{36}$$

or expanding the inverse of the test item interface FRF matrix we get

$$\begin{Bmatrix} R_1 \\ R_3 \end{Bmatrix} = \begin{Bmatrix} F_1 \\ F_3 \end{Bmatrix} + \frac{1}{T_{11} T_{33} - T_{31} T_{13}} \begin{Bmatrix} T_{33} T_{14} - T_{13} T_{34} \\ T_{11} T_{34} - T_{31} T_{14} \end{Bmatrix} F_4 \tag{37}$$

According to Eq. 36 or its expanded form, Eq. 37, field and laboratory interface forces will match if and only if there are no external forces acting on the test item while in the field, i.e., $F_4 = 0$, or if the second vector on the right hand side of Eq. 37 equals zero. The first condition frequently occurs in real situations, but the second represents a special case that occurred for R_1 in Fig. 10.

5.2.2. *Case #2: N_c interface forces from N_t motions*
In this laboratory test scenario, a set of N_c interface forces are calculated and applied to the test item according to Eq. 14. For the system shown in Fig. 3, Eq. 14 is written as

$$\begin{Bmatrix} R_1 \\ R_3 \end{Bmatrix} = \begin{bmatrix} T_{11} & T_{13} \\ T_{31} & T_{33} \\ T_{21} & T_{23} \\ T_{41} & T_{43} \end{bmatrix}^{+} \begin{Bmatrix} X_1 \\ X_3 \\ X_2 \\ X_4 \end{Bmatrix} \tag{38}$$

The interface forces calculated through Eq. 38 are shown in Fig. 12. These interface forces do not match the field forces in this case.

Test item accelerations are calculated according to Eq. 17, and they are shown in Fig. 13. All laboratory accelerations resulted incorrect in this simulation.

5.2.3. *Case #3: N_t forces from N_t motions*
In this case scenario, N_t laboratory forces are calculated from all N_t accelerations so that Eq. 15 reduces to

$$\begin{Bmatrix} R_1 \\ R_3 \\ R_2 \\ R_4 \end{Bmatrix} = \begin{bmatrix} T_{11} & T_{13} & T_{12} & T_{14} \\ T_{31} & T_{33} & T_{32} & T_{34} \\ T_{21} & T_{23} & T_{22} & T_{24} \\ T_{41} & T_{43} & T_{42} & T_{44} \end{bmatrix}^{+} \begin{Bmatrix} X_1 \\ X_3 \\ X_2 \\ X_4 \end{Bmatrix} \tag{39}$$

Three forces resulted from Eq. 39, R_1, R_3, and R_4, and they are shown in Fig. 14 where we see that both interface forces and the external force closely match the corresponding field forces. The result for R_2 were practically zero, presenting magnitudes varying from 10^{-16} to 10^{-19} for all frequency components since no force is applied to M_2 in the field.

The test item accelerations are obtained from

$$\begin{Bmatrix} U_1 \\ U_3 \\ U_2 \\ U_4 \end{Bmatrix} = \begin{bmatrix} T_{11} & T_{13} & T_{14} \\ T_{31} & T_{33} & T_{34} \\ T_{21} & T_{23} & T_{24} \\ T_{41} & T_{43} & T_{44} \end{bmatrix} \begin{Bmatrix} R_1 \\ R_3 \\ R_4 \end{Bmatrix} \tag{40}$$

and the calculated results closely match the field accelerations as shown in Fig. 15.

It is clear that the pseudo-inverse technique was able to accurately predict the field interface forces in this numerical example when more motions were measured than excitation forces existed and the time signals were either transient or periodic. In addition, it is clear that the test item experienced essentially the same accelerations in the laboratory as in the field if we are able to properly control the input forces in the laboratory.

6. Summary of numerical results

The results obtained in the numerical simulations can be summarized as follows:

- Field external force effects must be accounted for in the laboratory in addition to interface force effects in order to obtain reasonable laboratory simulation results.

- The pseudo-inverse technique was successfully applied to the test item field interface and external acceleration frequency spectra in order to define the laboratory inputs. Cross correlation requirements between variables are automatically satisfied since the field data contained both magnitude and phase information.

7. Experimental analysis

An experimental study was performed on a typical structure in order to investigate the feasibility of the pseudo-inverse technique when dealing with experimental data. The test item is a 92 x 1.25 x 1.0 in cold rolled steel beam that is supported by flexible cords in order to simulate the free free boundary condition. Two test setups were employed as shown in Fig. 16. In the first experimental setup shown in Fig. 16a, the beam is excited at two locations by two vibration exciters where the first is attached to the beam's mid point (MB model Modal 50) and the second is attached at the beam's left end (B&K model 4808). The second setup is shown in Fig. 16b and employs the same setup as in Fig. 16a except that in this case, two 0.22 K_g ($\approx$ 0.5 lb) dummy masses are used to introduce a third unknown excitation force to the beam (inertia force).

Deterministic (periodic and transient) excitation signals are used to drive the beam in both tests shown in Fig. 16. The excitation forces are measured by piezoelectric force transducers to provide a comparison basis for the identified forces from the pseudo-inverse technique. A total of four acceleration measurements are taken along the beam's length at points A_1, A_2, A_3, and A_4 by using piezoelectric accelerometers. The characteristics of the sensors used to measure all vibration signals are listed in Table 3.

Table 3: Characteristics of sensors used in tests

Sensor Type	Variable and Position on beam	Sensor Model	Charge or Voltage Sensitivity
Force	F3	PCB 208/A03	50.84 mv/N
Force	F4	PCB 208/A03	50.84 mv/N
Accel.	A1	PCB 302A	10.9 mv/g
Accel.	A2	Endveco 2222C	1.70 pcb/g
Accel.	A3	PCB 302A	10.18 mv/g
Accel.	A4	PCB 302A02	10.04 mv/g

Input and output vibration signals were gathered by a 486 PC equipped with the Data Physics Dp420© data acquisition and signal processing board. Frequency response function (FRF) measurements were performed on the bare beam in the 0 - 625 Hz frequency range that contains the first 7 natural frequencies. A total of 1000 spectral lines were used thus giving a frequency resolution of $\Delta f = 0.6357$ Hz and an analysis period of $\Delta t = 1.57$ s. Pseudo-random excitation was used in the FRF measurements. Table 4 contains the beam's first 7 bending natural frequencies in the 0 - 625 Hz frequency range that were obtained by curve fitting the experimental accelerance FRFs using ICATS© modal analysis software [11]. The experimental natural frequencies are compared with values obtained analytically [12].

Table 4: Natural frequencies for free free beam

fn	Theoretical Hz	Experimental Hz
1	24.30	25.30
2	67.60	67.98
3	132.40	131.22
4	218.00	216.60
5	327.00	322.53
6	457.00	450.60
7	608.00	597.62

In the first set of measurements, two harmonic signals having approximately the same magnitude but different frequencies are used to drive the beam in both setups shown in Fig. 16. A single frequency sinusoidal signal of 200 H_z is applied to the beam through the vibration exciter to drive the beam's mid point and its corresponding frequency spectrum is measured by the force transducer F3 mounted at point 3. Simultaneously, a second single frequency sinusoidal excitation signal of 350 Hz is applied by the vibration exciter at point 4 and its frequency spectrum is measured by the force transducer F4. Hanning windows are used with the input and output channels. The beam's output acceleration frequency spectra are measured at all four positions shown in Fig. 16.

Chirp excitation was employed in the second set of measurements. In this case, both vibration exciters were driven by the same chirp pulse that was generated by the Dp420 system. This chirp was designed to sweep the 0 - 625 Hz frequency range with a duration of $\Delta t_c = 100$ ms, thus occupying less than 10% of the total analysis period ($\Delta t = 1.57$ s). Rectangular windows are used with input and output channels in this case.

A potential problem when calculating the pseudo-inverse of the FRF matrix is instrumentation noise. Frequency response function curve fitting can be used in this case in order to remove noise from measurements [4]. All measured FRFs are curve fitted by a suitable algorithm, modal parameters are then extracted for the various modes present in the frequency range of interest. These modal parameters are subsequently used to regenerate the FRFs that will be used in Eqs. 41 or 42. Although this technique greatly reduces the effects of instrumentation noise on the pseudo-inversion of the FRF matrix, potential curve fitting problems may occur and this can lead to an inaccurate mathematical model of the structure, giving false results for the predicted loads. *All results presented in this section were obtained by using the beam's experimental accelerance FRFs.* Figure 17 shows driving and transfer point accelerance measurements for the free free beam of Fig. 16.

7.1. FORCE PREDICTION RESULTS

The acceleration signals obtained from these experiments were used along with the beam's FRFs in Eq. 11 in order to predict the beam's input loads when sinusoidal or chirp excitation are used. Table 5 summarizes the tests performed as well as the forces obtained in each test.

Table 5: Measured and predicted forces for all test cases

Case	Setup	Input	Measured	Predicted
a	1	Sinusoidal	F_3, F_4	F_3, F_4
b	1	Chirp	F_3, F_4	F_3, F_4
c	2	Sinusoidal	F_1, F_3, F_4	F_1, F_3, F_4
d	2	Chirp	F_1, F_3, F_4	F_1, F_3, F_4

7.1.1. *Case (a)*

In this case, Eq. 11 is written as

$$\begin{Bmatrix} F_1 \\ F_2 \\ F_3 \\ F_4 \end{Bmatrix} = \begin{bmatrix} A_{11} & A_{12} & A_{12} & A_{12} \\ A_{21} & A_{22} & A_{23} & A_{24} \\ A_{31} & A_{32} & A_{33} & A_{34} \\ A_{41} & A_{42} & A_{43} & A_{44} \end{bmatrix}^+ \begin{Bmatrix} X_1 \\ X_2 \\ X_3 \\ X_4 \end{Bmatrix} \tag{41}$$

118

where F_q, $q = 1...4$ are the unknown input force frequency spectra, X_p, $p = 1...4$ are the measured output acceleration frequency spectra, and A_{pq} are the beam's accelerance FRFs. Equation 41 corresponds to the situation where the number and location of inputs are unknown; and hence, as many input forces as output measurements are calculated. On the other hand, when the input forces locations are known, as it is the case for the forces applied by the exciters in Fig. 16a, then Eq. 41 can be simplified to

$$\begin{Bmatrix} F_3 \\ F_4 \end{Bmatrix} = \begin{bmatrix} A_{13} & A_{14} \\ A_{23} & A_{24} \\ A_{33} & A_{34} \\ A_{43} & A_{44} \end{bmatrix}^+ \begin{Bmatrix} X_1 \\ X_2 \\ X_3 \\ X_4 \end{Bmatrix} \tag{42}$$

The results for sinusoidal forces F_3 and F_4 predicted through Eq. 42 and corresponding to case (a) are shown in Fig. 18 where each excitation peak is isolated for clarity. Figures 18a and b contains the measured and predicted forces F_3 while Figs. 18c and d show the measured and predicted forces F_4. In both cases, the measured and predicted forces present two peaks, one at 200 Hz and the other at 350 Hz and this is due to the cross coupling between the excitation sources, i.e., the exciter's armature at 3 reacts to the force applied to the beam by the exciter at 4, and vice and versa. Consequently, these reaction forces are introduced to the beam as a secondary form of excitation. Thus, the force transducer mounted at 3 measures two excitation force components, as shown in Figs. 18a and b. The first component is at 200 Hz and corresponds to the force generated by the sinusoidal input applied by the mid point exciter while the second component is at 350 Hz and it is due to the excitation force applied by the exciter at point 4. Similarly, Figs. 18c and d show that the force transducer attached to the beam at point 4 measures the sinusoidal excitation force applied to the beam by the exciter at point 4 with a 350 Hz and a second force component at 200 Hz that is due to the vibration exciter attached at point 3. Since the output acceleration at any point on the beam is due to both force components, the forces predicted from the pseudo-inverse technique and shown in Fig. 18 also present two peaks.

Figures 18a and d correspond to the main components caused by the cross coupling effects. Figure 18b corresponds to the force component at 350 Hz caused by the exciter at 4. Values for the predicted force in this case are larger than the corresponding measured values. Figure 18c shows the force components at 200 Hz that is caused by the exciter at 3 and in this case, measured and predicted forces agree closely at frequencies close to the peak frequency, but disagree at frequencies away from the peak.

7.1.2. *Case (b)*

Chirp excitation was used in this case. The predicted forces were obtained through Eq. 42 and they are shown in Fig. 19. Figure 19a shows the results for the chirp applied to the beam's mid point. In this case, measured and predicted forces do not present a good agreement in the 0 – 50 Hz bandwidth. The measured force present a notch at $f \approx 20$ Hz while the predicted force presents a peak at this frequency. This behavior was observed before [6]. For frequencies above 50 Hz both F_3 and F_4 and the corresponding measured signals present a better agreement.

7.1.3. *Case (c)*
In this case the two 0.22 Kg steel lumped masses were attached to the beam's free end 1 as shown in Fig. 16b. This is done in order to introduce a third force (inertia force) to the beam. The forces are calculated by rewriting Eq. 42 as

$$\left\{ \begin{array}{c} F_1 \\ F_3 \\ F_4 \end{array} \right\} = \left[\begin{array}{ccc} A_{11} & A_{13} & A_{14} \\ A_{21} & A_{23} & A_{24} \\ A_{31} & A_{33} & A_{34} \\ A_{41} & A_{43} & A_{44} \end{array} \right]^+ \left\{ \begin{array}{c} X_1 \\ X_2 \\ X_3 \\ X_4 \end{array} \right\} \tag{43}$$

The results for the test case (c) are shown in Fig. 20. Figures 20a and b shows the results for the inertia load F_1. In this case, as occurred in case (a) measured and predicted forces present two peaks in the $0 - 625$ Hz frequency range. Figure 20a shows the measured and predicted forces in the range containing the 200 Hz excitation frequency while Fig. 20b shows the inertia force component in the range containing the 350 Hz excitation frequency. In both cases, the inertia loads used for comparison (dashed line) are calculated from the acceleration signal from accelerometer A_1 and the lumped masses attached to the beam through Newton's second law in order to avoid problems with direct force measurement at point 1.

Measured and predicted results for forces F_3 and F_4 are shown in Figs. 20c, d, e, and f, respectively. The main component of the predicted force F_3 in Fig. 20c presents a good agreement with the measured result for frequencies below 210 Hz while the amplitude values of the predicted F_4 in Fig. 20f gave larger results for most frequency components in the 300-400 Hz range. This same amplitude difference is observed for the forces caused by the cross coupling effects, Figs. 20d and e.

7.1.4. *Case (d)*
In this case, the experimental setup of Fig 16b is used with chirp excitation. As in case (b), the same chirp signal is used to feed both exciters at 3 and 4. The predicted forces are obtained through Eq. 43 and they are shown along with the corresponding measured signals in Fig. 21. Figure 21a shows the results for the inertia load. In this case, measured and predicted results agree closely for frequencies in the neighborhood of peaks and for frequencies away from peaks the measured and predicted inertia loads follow the same trend except at the beam's natural frequencies where the predicted force presents notches as seen in Fig. 21a. These notches are caused by the pseudo-inversion of the FRF matrix in Eq. 43 since this matrix generally is rank deficient at frequencies coinciding with the structure's natural frequencies; hence, this causes numerical difficulties in the calculation of the pseudo-inverse.

Figures 21b and c show the measured and predicted results for forces F_3 and F_4, respectively. In both cases, we notice that measured and predicted results follow basically the same trend, except in 0-50 Hz frequency range where they do not show agreement. The presence of notches coinciding with the beam's natural frequencies are noticed in Fig. 21c as it was the case for the predicted inertia force.

Finally, Fig. 21d shows the measured and predicted results for F_4 in case (d) when a different vibration exciter is employed at point 4 in the setup of Fig. 16b. In this case, the B&K 4808 vibration exciter was replaced by a Unholtz-Dickie T206 vibration exciter. The Unholtz-Dickie vibration exciter's armature is approximately 9 Kg while the MB Dynamics 50 and the B&K exciters have armature masses of approximately 0.18 Kg. Thus, a large discrepancy in armature masses occurs when the UD T206 is attached at point 4 and the MB Dynamics at point 3, since the first has an armature that is about 50 times heavier than the other. By comparing the results for force F_4 when using the Unholtz-Dickie at point 4 instead of the B&K 4808 with the previous result of Fig. 21c, *it becomes clear that vibration exciter armature mass can seriously affect the results in laboratory simulations.*

8. Summary of experimental results

The results obtained in the experimental analysis can be summarized as follows:

- The pseudo-inverse technique is feasible when working with experimental *deterministic* signals (transient and periodic). Acceleration *frequency spectra containing magnitude and phase information* was used with this technique to predict the external loads applied to the test item. This means that a *specific cross correlation exists between each time variable pair* even though no such function was explicitly calculated or measured,

- The solution for the external forces was obtained in a least squares sense since more accelerations were measured than forces were predicted. This lease squares formulation of the inverse problem helps in obtaining a unique solution for the unknown forces and helps to reduce measurement noise problems.

- The solution obtained from the pseudo-inverse technique appears to be sensitive to the acceleration measurement locations. In addition, the locations were the predicted forces were applied was known since the corresponding FRFs and accelerations at these locations were used. This is a requirement imposed by the nature of this inverse problem.

9. Summary and Conclusions

This paper presents numerically simulated results of a model for vibration testing tailoring when the test item is subjected to external forces while connected to the vehicle in the field vibration environment. Test item and vehicle are modeled as MDOF linear systems. Modal properties and response models are calculated for all structures by using standard modal analysis techniques. Equations for interface forces and test item motions are used to perform field and laboratory simulations. The pseudo-inverse technique is employed to calculate the required forces to drive the test item in laboratory simulations. Numerically simulated results show that accounting for external force effects in the laboratory is of major importance for a successful simulation. An experimental analysis is performed on a free free beam in order to investigate the feasibility of the pseudo-inverse technique when applied to experimental data. Experimental FRFs are used in the pseudo-inverse and the excitation forces applied to the beam are calculated for both sinusoidal and transient signals. Results from this experimental analysis show that the pseudo-inverse technique is an important tool in predicting test item interface and external forces when dealing with deterministic signals.

Acknowledgements

The authors would like to thank the Iowa State University Engineering College and the Aerospace Engineering and Engineering Mechanics Department for supporting this research. Dr. Paulo S. Varoto, from Universidade de São Paulo – São Carlos – Brasil was financially sponsored by CNPq – Brazil during his PhD program.

10. References

1. Varoto, P.S., and McConnell, K.G. (1999), Rules for the Exchange and Analysis of Dynamic Information: Part I – Basic definitions and test scenarios, in J.M.M. Silva and N.M.M. Maia (eds.), *Modal Analysis & Testing, Kluwer Academic Publishers, NATO Series*, Dordrecht, 65-81.
2. Varoto, P.S. and McConnell, K.G. (1999), Rules for the Exchange and Analysis of Dynamic Information, Part II: Numerically simulated results for a deterministic excitation with no external forces, in J.M.M. Silva and N.M.M. Maia (eds.), *Modal Analysis & Testing, Kluwer Academic Publishers, NATO Series*, Dordrecht, 83-100.
3. McConnell, K.G. (1995), Vibration Testing: Theory and Practice, *John Wiley & Sons*, N.Y.
4. Ewins, D. (1984), Modal Testing: Theory and Practice, *Research Studies Press*, London.
5. McConnell, K.G. (1995), From field vibration data to laboratory simulation, *Experimental Mechanics, Vol 34*, 181-193.
6. Hillary, B. (1983), Indirect measurement of vibration excitation forces, *PhD Thesis, Imperial College of Science Technology and Medicine*, London.
7. Fabumni, J. (1986), Effects of structural modes on vibratory force determination by the pseudoinverse technique, *AIAA Journal, V. 24, N. 3*, 504-509.
8. Hillary, B. and Ewins, D.J. (1984), The use of strain gages in force determination and frequency response measurements, *Proceedings of the II International Modal Analysis Conference*, 627-634, Orlando, FL.
9. Han, M-C and Wicks, A.L. (1990), Force determination with slope and strain response measurement', *Proceedings of the VIII International Modal Analysis Conference, Kissimmee, FL, V. 1*, 365-372.
10. Fregolent, A., and Sestieri, A. (1990), Assessment of procedures for force identification from experimental response, *Proceedings of the XV International Seminar on Modal Analysis, Part II*, 825-838, Leuven, Belgium.
11. Gregory, D.C., Priddy, T.G., Smallwood, D.O. (1987), Experimental determination of the dynamic forces acting on non-rigid bodies, *SAE paper No. 861791*.
12. Scharton, T.D. (1994), Force limited vibration testing "Top Ten" research problems, *Handouts from the tutorial session "Vibration Test Tailoring", XIII International Modal Analysis Conference*, Nashville, TN.
13. Szymkowiak, E., A. and Silver II W. (1990), A Captive Store Flight Vibration Simulation Project, *Proceedings of the 36th Annual meeting, IES*, 531-538, New Orleans.

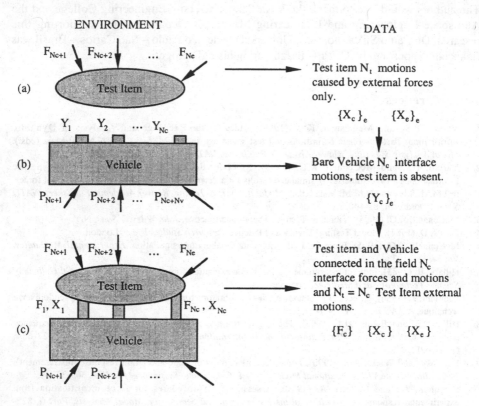

ENVIRONMENT   DATA

(a) Test item with F_{Nc+1} F_{Nc+2} ... F_{Ne} — Test item N_t motions caused by external forces only.

$$\{X_c\}_e \qquad \{X_e\}_e$$

(b) Vehicle with Y_1 Y_2 ... Y_{Nc} and P_{Nc+1} P_{Nc+2} ... P_{Nc+Nv} — Bare Vehicle N_c interface motions, test item is absent.

$$\{Y_c\}_e$$

(c) Test Item and Vehicle connected: F_{Nc+1} F_{Nc+2} ... F_{Ne}, F_1, X_1, F_{Nc}, X_{Nc}, P_{Nc+1} P_{Nc+2} ... — Test item and Vehicle connected in the field N_c interface forces and motions and $N_t = N_c$ Test Item external motions.

$$\{F_c\} \qquad \{X_c\} \qquad \{X_e\}$$

Figure 1   Sequence of field calculations: (a) Test item output response due to external forces, test item is a free structure in space; (b) Bare vehicle data; (c) Interface forces and test item motions form combined structure.

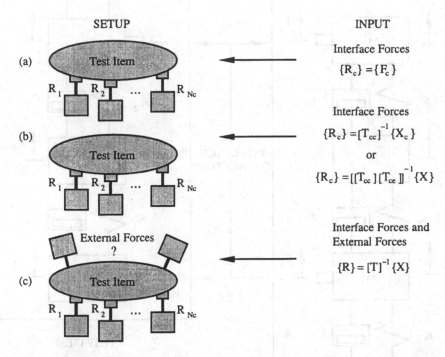

SETUP   INPUT

(a) Test Item — R_1 R_2 ... R_{Nc}

Interface Forces
$$\{R_c\} = \{F_c\}$$

(b) Test Item — R_1 R_2 ... R_{Nc}

Interface Forces
$$\{R_c\} = [T_{cc}]^{-1}\{X_c\}$$
or
$$\{R_c\} = [[T_{cc}][T_{ce}]]^{-1}\{X\}$$

(c) External Forces ? Test Item — R_1 R_2 ... R_{Nc}

Interface Forces and External Forces
$$\{R\} = [T]^{-1}\{X\}$$

Figure 2   Sequence of laboratory calculations defining test item inputs: (a) Using interface forces, *no external force effects accounted for*; (b) Using interface forces from interface motions, *no external force effects accounted for*; (c) Using interface and external forces.

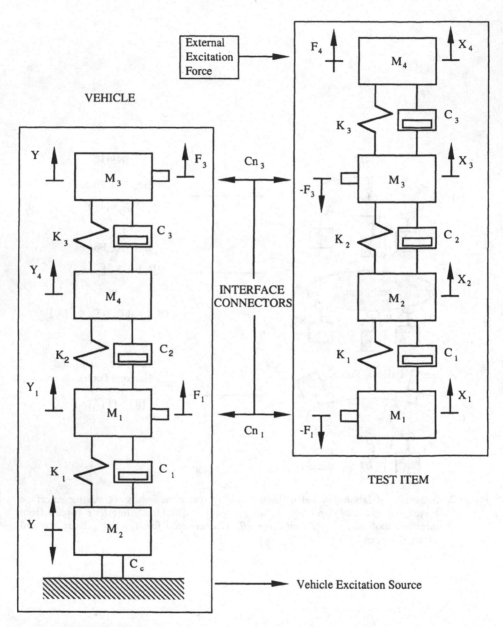

Figure 3   Test item attached to vehicle by rigid connectors Cn_1 and Cn_3 in the filed environment. External inputs to the combined structure given in terms of the vehicle base input acceleration Y_2 and the external force F_4 applied to the test item mass M_4.

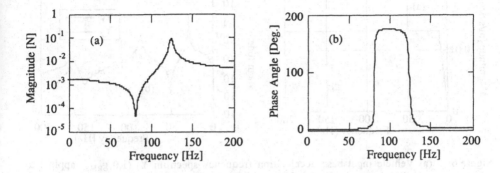

Figure 4 External force frequency spectrum F_4 applied to the test item mass M_4 in the field corresponding to Fig. 1a: (a) Magnitude [N]; (b) Phase angle [Deg].

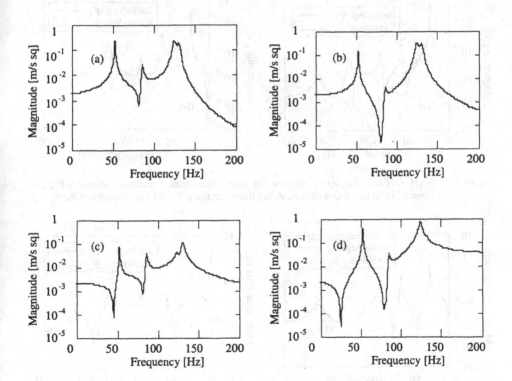

Figure 5 Test item field accelerations due to external load F_4 that is applied to the test item mass M_4 in the field. Test item is a free structure as shown in Fig. 1a: (a) and (c) Interface accelerations X_1 and X_3; (b) and (d) External accelerations X_2 and X_4, respectively.

126

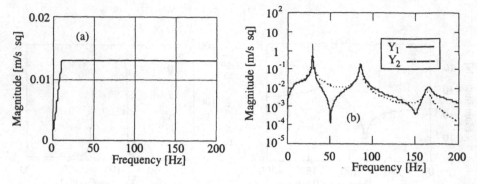

Figure 6   (a) Vehicle input base acceleration frequency spectrum Y_2 (1.0 g_{RMS}) applied to vehicle base mass M_2; (b) Bare vehicle interface acceleration frequency spectra $\{Y_c\}_e$ due to base input motion Y_2 shown in Fig. 6a, corresponding to the case shown in Fig. 1b.

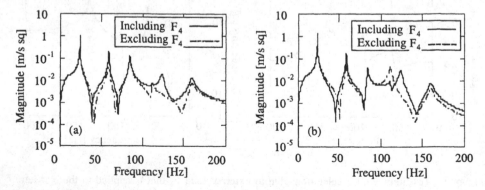

Figure 7   Interface force frequency spectra for field simulation 2 corresponding to Fig. 1c when F_4 is included and excluded: (a) Interface force F_1, (b) Interface force F_3.

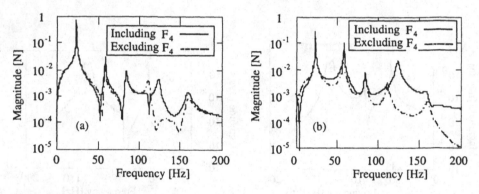

Figure 8   Test item acceleration responses for field simulation 2 corresponding to Fig. 1c when F_4 is included and excluded: (a) Interface X_1, (b) Interface X_3;

127

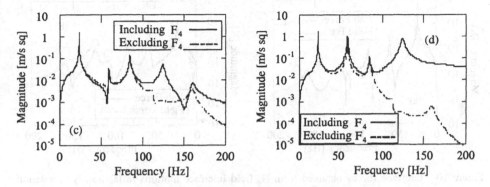

Figure 8   (*Cont.*) Test item acceleration responses for field simulation 2 corresponding to Fig. 1c when F_4 is included and excluded: (c) External X_2; and (d) External X_4.

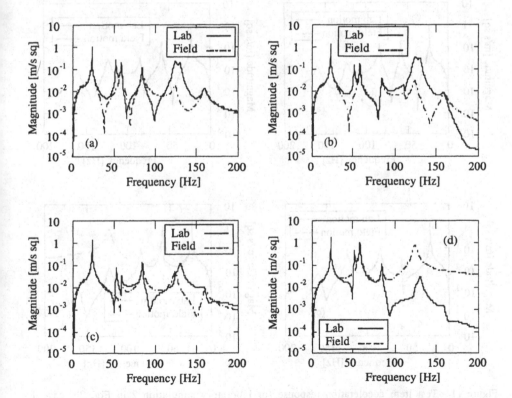

Figure 9   Test item acceleration response spectra in laboratory simulation 1, Fig. 2a, using interface forces as test item inputs. *No field external forces are accounted for in the laboratory*: (a) and (c) interface accelerations U_1 and U_3; (b) and (d) External accelerations U_2 and U_4, respectively.

128

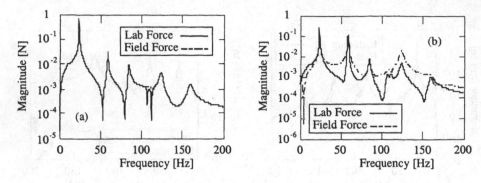

Figure 10   Interface forces obtained from N_c field interface motions in laboratory simulation
2, case 1 shown in Fig. 2b. *No external force effects accounted for in the
laboratory*: (a) Force R_1; (b) Force R_3.

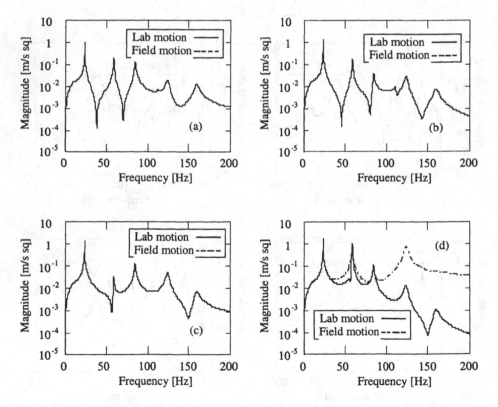

Figure 11   Test item acceleration response for laboratory simulation 2 in Fig. 2b, case 1
where N_t accelerations are obtained by driving the test item with N_c interface forces
obtained from N_c interface motions. *No external forces are accounted for*: (a) and
(c) Interface accelerations U_1 and U_3; (b) and (d) External accelerations U_2 and U_4,
respectively.

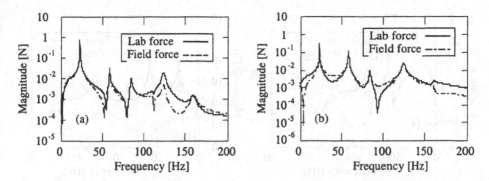

Figure 12   Interface forces obtained from N_t field motions in laboratory Simulation 2 in Fig. 2b, case 2. *No external forces are accounted for in the laboratory*: (a) Interface force R_1; (b) Interface force R_3.

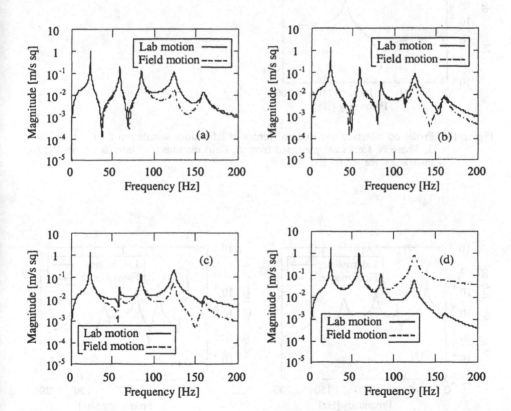

Figure 13   Test item acceleration response for laboratory simulation 2 in Fig. 2b, case 2 where N_t accelerations are obtained by driving the test item with N_c interface forces from N_t field accelerations: (a) and (c) Interface accelerations U_1 and U_3; (b) and (d) Interface accelerations U_2 and U_4.

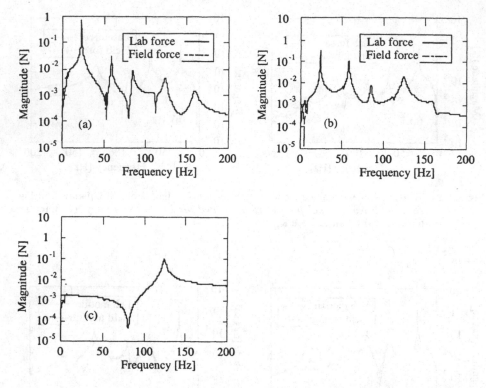

Figure 14   Predicted interface and external forces in laboratory simulation 2 in Fig. 2c, case
3, where N_t forces are predicted from N_t field motions: (a) Interface force R_1; (b)
Interface force R_3; (c) Ext. force R_4.

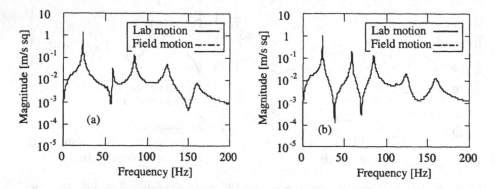

Figure 15   Test item acceleration response in laboratory simulation 2 in Fig. 2c, case 3 where
test item is driven by two interface forces and one external force obtained from N_t
field accelerations: (a) Interface acceleration U_1; (b) External acceleration U_2.

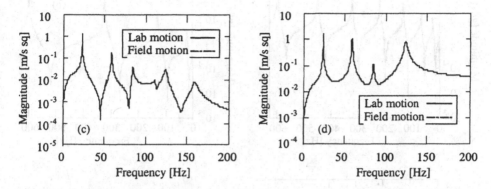

Figure 15 (*Cont.*) Test item acceleration response in laboratory simulation 2 in Fig. 2c, case 3 where test item is driven by two interface forces and one external force obtained from N_t field accelerations: (c) Interface acceleration U_3; (d) External acceleration U_4.

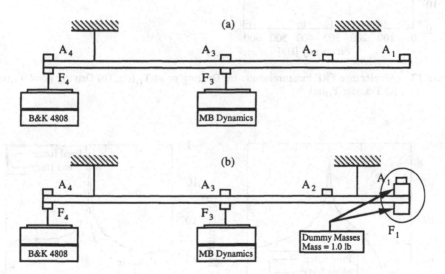

Fgure 16 Experimental setup for force predictions: (a) Two unknown forces; (b) Three unknown forces.

132

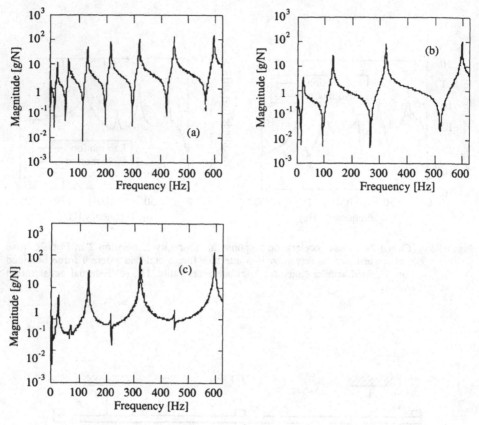

Figure 17   Accelerance FRF measurements: (a) Driving point $T_{11}(\omega)$; (b) Driving point $T_{33}(\omega)$; (c) Transfer $T_{13}(\omega)$.

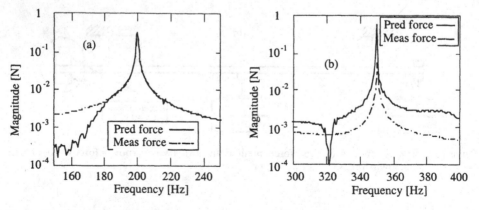

Figure 18   Forces predicted from pseudoinverse method, case a: (a) and (b) F_3, excitation at point 3.

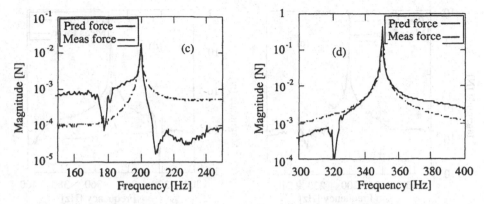

Figure 18 (*Cont.*) Forces predicted from pseudoinverse method, case a: (c) and (d) F_4, excitation at point 4.

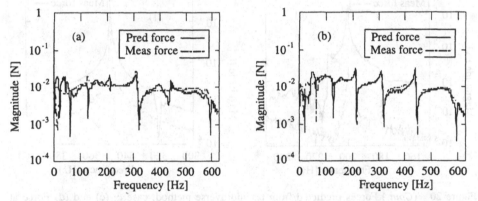

Figure 19 Forces predicted from pseudoinvers method, case b: (a) F_3, excitation at point 3; (b) F_4, excitation at point 4.

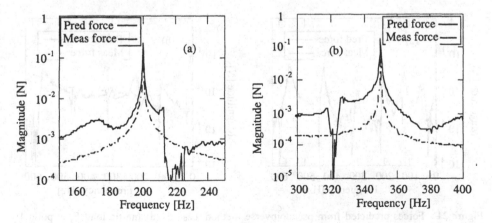

Figure 20 Forces predicted from pseudoinverse method, case c: (a) and (b) Inertia load F_1.

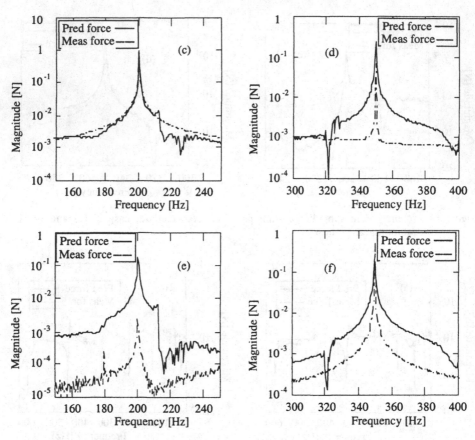

Figure 20   (*Cont.*) Forces predicted from pseudoinverse method, case c: (c) and (d) Force at point 3; (e) and (f) Force at point 4.

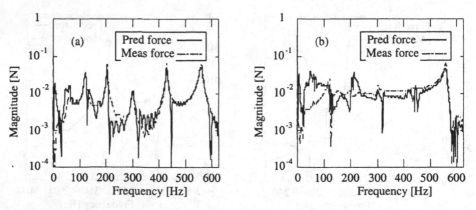

Figure 21   Forces predicted from pseudoinverse method, case d: (a) Inertia load F_1 at point 1; (b) F_3, excitation at point 3.

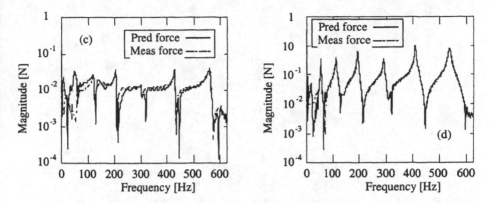

Figure 21   (*Cont.*) Forces predicted from pseudoinverse method, case d: (c) F_4, excitation at point 4; (d) F_4, excitation at point 4 using UD exciter.

RULES FOR THE EXCHANGE AND ANALYSIS OF DYNAMIC INFORMATION

Part IV: Numerically Simulated and Experimental Results for a Random Excitation

P. S. VAROTO
Dept de Engenharia Mecanica
Escola De Engenharia De São Carlos,
USP
São Carlos – SP – 13560-250, Brasil

K. G. McCONNELL
Dept. Aerospace Engineering and
Engineering Mechanics
Iowa State University
Ames, IA 50011 USA

Abstract

Random loads often occur in field dynamic environments. This paper illustrates how the results obtained in Ref. [1] can be used when we are dealing with random signals. The expressions for field interface forces and test item motions derived in terms of frequency spectra are used to obtain new expressions in terms of auto spectral densities and cross spectral densities for random signals. These new expressions are used to numerically simulate field and laboratory dynamic environments for a multi degree of freedom (MDOF) discrete system. First, field interface force and test item acceleration spectral density matrices are calculated when the test item and the vehicle are connected in the field environment through multiple connectors. Second, the auto and cross spectral densities obtained in the field are used to define suitable test item inputs in different laboratory test scenarios. As previously shown in Ref. [3], knowledge of external forces acting on the test item while in the field environment is a major issue in properly simulating the test item vibration characteristics in the laboratory. Thus, the force identification problem is addressed in the laboratory simulations of field vibration employing random loads. *Numerically simulated and experimental results show that a common assumption of the test item motions being <u>uncorrelated</u> is untrue and that cross-spectral densities must be accounted for in all situations regardless of forces being either <u>correlated</u> or <u>uncorrelated</u>.*

Nomenclature

A_{pq}	Accelerance FRF	$\{F_c\}$	Field interface force
Cn_1, Cn_3	Connectors	F_4	Test item field external force
C_q	q^{th} test item and vehicle damping coefficients	$[Gf_{ff}]$	Input force spectral density
		$[Gf_{cc}]$	Interface force spectral density
K_q	q^{th} test item and vehicle spring constants		
$\{F\}$	Generic input force	$[Gf_{ce}]$	Connector-external force spectral density

137

J.M.M. Silva and N.M.M. Maia (eds.), Modal Analysis and Testing, 137–177.

$[Gf_{ee}]$	External force spectral density
$[G_{xx}]$	Field acceleration spectral density
$[Gx_{cc}]$	Interface acceleration spectral density
$[Gx_{ce}]$	Connector-external acceleration spectral density
$[Gx_{ee}]$	External acceleration spectral density
$[Gu]$	Laboratory acceleration spectral density
$[Gr]$	Laboratory force spectral density
$[H_{cc}]$	Connector acceleration FRF matrix
$[H_{ce}]$	Connector-external acceleration FRF matrix
$[H_{ec}]$	External-connector acceleration FRF matrix
$[H_{ee}]$	External acceleration FRF matrix
$j = \sqrt{-1}$	Imaginary number
M_q	q^{th} test item and vehicle masses
$[M]$	Mass matrix
N	Degrees of freedom for generic system
N_t	Degrees of freedom for test item
N_v	Degrees of freedom for vehicle
N_c	Number of connectors
T_{pq}	Test item acceleration FRF
$\{R\}$	Test item laboratory force
$\{R_c\}$	Test item laboratory connector force
$\{R_e\}$	Test item laboratory external force
$[T]$	Test item acceleration FRF matrix
$[T_{cc}]$	Test item connector acceleration FRF matrix
$[T_{ce}]$	Test item connector-external acceleration FRF matrix
$[T_{ec}]$	Test item external-connector acceleration FRF matrix
$[T_{ee}]$	Test item external acceleration FRF matrix
$[TV]$	Test item-Vehicle combined matrix
$\{U\}$	Test item laboratory acceleration
$\{U_c\}$	Test item laboratory interface acceleration
$\{U_e\}$	Test item laboratory external acceleration
V_{pq}	Vehicle acceleration FRF
$\{X\}$	Test item field acceleration
$\{X_c\}$	Test item field interface acceleration
$\{X_e\}$	Test item field external acceleration
$\{Y_c\}_e$	Bare Vehicle interface acceleration

Greek

ε	Expected value
Δf	Frequency resolution in Hz
$\{\phi\}_r$	r^{th} mode shape
ϕ_{pr}	r^{th} element of pth mode shape
ϕ_{qr}	r^{th} element of qth mode shape
$\Gamma(\omega)$	Acceleration transmissibility FRF
ω	Frequency in rad/s
ω_r	r^{th} natural frequency in rad/s
τ	Analysis period in seconds
ζ_r	r^{th} modal damping ratio

1. Introduction

When connected in the field dynamic environment, the test item and the vehicle form an unique structure that is subjected to a variety of external loads. In order to properly simulate the field environment experienced by the test item, it is important to have knowledge of the interface foces and external forces acting on the test item as well as its interface and external motions. Random excitations that result from aerodynamic and acoustic loads occur frequently so that field data should include both auto spectral densities (ASDs) and cross spectral densities (CSDs) due to the random characteristics of the excitation and response signals.

In the laboratory environment, the test item is attached to one or more vibration exciters and random excitation is often employed in order to generate the test item inputs. A common control strategy is to generate and control a single input acceleration ASD which is applied to the test item through a test fixture. The shape of the input acceleration ASD may vary widely, depending on the type of structure being tested, the type of field environment being simulated, and the capabilities of the excitation system available. The blind application of a general test specification has led to situations where the exciter system was unable to generate the required input forces and motions since the test specification was based on bare vehicle vibration data [4]. The presence of the test item reduced the required inputs by factors on the order of ten to one. Thus, the direct use of bare vehicle motion data can lead to excessive test levels. In another case, multiple exciters operating in closed loop were required to simulate field external loads effects [5].

This paper employs a previously developed [1] vibration testing model that is extended to the case of random signals. Expressions for field interface forces and test item motions that were initially derived in terms of frequency spectra are used here in order to obtain a new set of equations that must be used with random signals. The new expressions for interface forces and test item motions are given in terms of spectral density matrices, where the diagonal contains the ASDs that are real valued functions of frequency while the off-diagonal entries are the corresponding cross spectral densities (CSDs) that contain magnitude and phase information as a function of frequency. These expressions are used in numerical simulations in order to illustrate the process of using field data in the laboratory environment when dealing with random signals.

Accounting for external force effects in the laboratory is of major importance for a successful simulation [3, 5]. Thus, part of this paper investigates the problem of predicting multiple random forces that are obtained from test item acceleration measurements. This prediction process represents an ill-conditioned inverse problem at some frequencies, but the pseudo-inverse technique [6] is commonly employed to predict a set of pseudo forces. This technique requires the inversion of the test item accelerance FRF matrix at each frequency component. Since the number of measured accelerations is generally larger than the number of unknown forces, the test item accelerance FRF matrix is not square, and the pseudo-inversion process is required. The input forces obtained from this technique may or may not coincide with the actual forces acting on the test item, depending primarily on the locations where the test item output accelerations are measured. *The pseudo-inversion of the FRF matrix may be numerically ill-conditioned at the test item natural frequencies since the number of forces that can be successfully identified depends on the number of structural modes participating on the test item response at those frequencies [7].*

When employing the pseudo-inverse technique to predict random forces, it is commonly accepted to assume that the unknown forces acting on the test item are statistically _uncorrelated_ [6]. An implicit consequence of this assumption is that the output accelerations are _uncorrelated_ as well so that only output acceleration ASDs are required in solving for the unknown force ASDs. *However, each measured motion exhibited by the test item is due to all forces applied to the test item, and thus, the measured motions are correlated, independently of whether the applied forces are correlated or uncorrelated.*

The results presented in this paper can be divided in two major parts. In the first part, expressions for interface force and test item motion spectral density matrices are derived for random signals. These expressions are employed in numerical simulations of field environments and the results of these simulations are used in order to define the test item dynamic environment in the laboratory.

The second part of this paper describes an experimental analysis performed on a free free beam in order to investigate the feasibility of the pseudo-inverse technique when dealing with experimental random signals. Two approaches are used to define multiple random inputs that are applied to the beam at several locations. The first approach employs independent random input signals to drive the amplifiers of the vibration exciters that are used to drive the beam. In the second approach, the same random excitation signal is used to drive all exciters. These tests are performed in order to check the assumptoin of _correlated_ and _uncorrelated_ inputs and outputs when using the pseudo-inverse technique to solve for the excitation forces. Two experimental setups are employed. In the first setup, two vibration exciters are used to drive the beam at two locations. The applied forces are either _correlated_ or _uncorrelated_ and two force ASDs and one CSD are predicted in this case. In the second setup, a lumped mass is added to the beam to simulate a third excitation source that is given by an inertia force. Three force ASDs and two CSD are predicted in this second experiment.

Both numerical and experimental results presented in this paper indicate that the assumption of uncorrelated inputs and outputs leads to inaccurate force predictions and that CSDs must be employed in addition to ASDs when solving the inverse problem.

2. Simulation calculation steps

Figures 1 and 2 show the sequence of calculation steps that will be used in the field and laboratory simulations presented in the first part of this paper. Figure 1 contains two columns and represents the sequence of calculations performed in the simulation of the field environment. The first column shows the various situations analyzed while the second column shows the field data obtained in each case. Now, we present a brief description of each calculation step performed in the field simulation shown in Fig. 1:

- *Step 1:* This step is shown in Fig. 1a and corresponds to the test item acceleration response due to external forces. In this case, the test item is modeled as a free structure in space and three acceleration spectral density matrices are obtained in this step as shown in Fig. 1a. As shown later, these acceleration ASDs and CSDs will be used to obtain the interface force and test item acceleration spectral density matrices.

- *Step 2:* The result of this calculation is the *bare* vehicle interface acceleration ASDs and CSDs, as shown in Fig. 1b. In this case, the test item is not connected to the vehicle in the field and the resulting bare vehicle interface acceleration spectral density matrix $[Gy_{cc}]_e$ is due to external forces (P_{Nv+1}, P_{Nv+2}, ..., P_{Nv}) acting on the vehicle.

- *Step 3:* This corresponds to the combined structure where the test item and the vehicle are connected in the field, as shown in Fig. 1c. In this case, the interface force spectral density matrix $[Gf_{cc}]$ as well as the test item acceleration spectral density matrices $[Gx_{cc}]$, $[Gx_{ee}]$, and $[Gx_{ce}]$ are obtained.

Similarly, the laboratory simulation calculation steps are shown in Fig. 2 where the first column shows the test setups employed in the simulation while the second column shows the corresponding field data used to define the test item inputs in each simulation. three laboratory simulations are performed:

- *Step 1:* This laboratory simulation is shown in Fig. 2a and uses the bare vehicle data obtained in the field simulation of Fig. 1a to define the test item inputs. *No external force effects are accounted for in this simulation.*

- *Step 2:* In this case, interface forces from the field simulation shown in Fig. 1b are used as test item inputs as shown in Fig. 2b. *No external force effects are accounted for in this simulation.*

- *Step 3:* This laboratory simulation is shown in Fig. 2c and employs the test item interface and external motions obtained in the field simulation shown in Fig. 1c. Three cases are analyzed where the test item inputs are obtained using different motions in each case. *External force effects are accounted for in this simulation.*

Figures 1 and 2 will be referred to throughout the numerical simulations in order to keep track of which simulation is being performed.

3. Theoretical response models for random excitation

Consider a NDOF linear structure with an acceleration FRF matrix that is denoted by [H] = [H()] and is subjected to a multiple random input forces. The structure input-output relationship can be expressed as [8].

$$[G_{xx}]=[H]^*[G_{ff}][H]^T \tag{1}$$

where symbols "*" and "T" denote the complex conjugate and non conjugate transpose of the structure's FRF matrix, respectively. Matrices $[G_{ff}]$ and $[G_{xx}]$ are the input and output spectral density matrices whose diagonal entries are the real valued input and output ASDs while the off-diagonal entries represent the complex valued input and output CSDs, respectively. These matrices are obtained from the input and output frequency spectra $\{F\} = \{F(\omega)\}$ and $\{X\} = \{X(\omega)\}$, respectively according to [8].

$$[G_{ff}] = \lim_{T\to\infty} \frac{2}{\tau}\varepsilon\left(\{F\}^*\{F\}^T\right) \tag{2}$$

$$[G_{xx}] = \lim_{T\to\infty} \frac{2}{\tau}\varepsilon\left(\{X\}^*\{X\}^T\right) \tag{3}$$

where τ is the analysis period and symbol "ε" denotes the "expected value" (in a statistical sense) of the product of vectors in Eqs. 2 and 3. *The limit terms in Eqs. 2 and 3 are subsequently dropped in all equations developed for simplicity of notation.*

When interface DOFs are differentiated from external DOFs, Eq. 1 assumes the following form

$$\begin{bmatrix} [Gx_{cc}] & [Gx_{ce}] \\ [Gx_{ec}] & [Gx_{ee}] \end{bmatrix} =$$
$$\begin{bmatrix} [H_{cc}] & [H_{ce}] \\ [H_{ec}] & [H_{ee}] \end{bmatrix}^* \begin{bmatrix} [Gf_{cc}] & [Gf_{ce}] \\ [Gf_{ec}] & [Gf_{ee}] \end{bmatrix} \begin{bmatrix} [H_{cc}] & [H_{ce}] \\ [H_{ec}] & [H_{ee}] \end{bmatrix}^T \tag{4}$$

where subscripts "c" and "e" denotes connector and external points, respectively. Note that $[H_{ce}] = [H_{ec}]^T$, but $[G_{ce}] = [G_{ec}]^H$, where symbol H denotes the Hermitian (complex conjugate transpose) operator.

When the test item and vehicle are connected in the field, the intereface forces frequency spectra are obtained by Eq. 13 of Ref. [1] and is given by

$$\{F_c\} = [TV]\left(\{Y_c\}_e - \{X_c\}_e\right) \tag{5}$$

where the N_c x 1 vector $\{Y_c\}_e$ contains the bare vehicle N_c interface accelerations that are caused by external loads acting on the vehicle in the absence of the test item, the N_c x 1 vector $\{X_c\}_e$, is the test item output interface accelerations that is caused by external loads acting on the test item, and the N_c x N_c matrix $[TV] = [T_{cc} + V_{cc}]^{-1}$ is the test item-vehicle combined matrix [1]. By using Eqs. 2 and 3, the following expression is obtained for the N_c x N_c interface forces spectral density matrix $[Gf_{cc}]$

$$[Gf_{cc}] = [TV]^*\left[[Gy_{cc}]_e + [Gx_{cc}]_e\right][TV] \tag{6}$$

The diagonal entries in $[Gf_{cc}]$ represent the interface force ASDs while the off-diagonal entries are the CSDs between interface points. The N_c x N_c matrices $[Gy_{cc}]_e$ and $[Gx_{cc}]_e$, contain acceleration ASDs and CSDs that originated from the bare vehicle and test item interface motions, respectively. These matrices carry an extra subscript "e" since they represent contributions of external forces acting on the vehicle and test item structures.

In obtaining Eq. 6, first the complex conjugate of both sides of Eq. 5 is post multiplied by the corresponding non conjugate transpose terms, as defined by Eqs. 2 and 3.

When expected values are taken and the term $2/\tau$ is canceled in both sides, the left hand side of the resulting equation is the interface force spectral density matrix $[Gf_{cc}]$ while four terms appear on the right hand side of Eq. 6. Terms $\{Y_c\}_e^* \{Y_c\}_e^T$ and $\{X_c\}_e^* \{X_c\}_e^T$ are the acceleration spectral density matrices shown on the right hand side of Eq. 6. The two remaining terms $\{Y_c\}_e^* \{X_c\}_e^T$ and $\{X_c\}_e^* \{Y_c\}_e^T$ are zero since *motions* $\{Yc\}_e$ and $\{Xc\}_e$ *occur when vehicle and test item are not connected and thus they are* <u>*uncorrelated*</u>.

The test item interface and external acceleration frequency spectra are obtained in the field according to Eqs. 15 and 16 from Ref. [1] and are given by

$$\{X_c\} = [T_{cc}][TV]\left(\{Y_c\}_e - \{X_c\}_e\right) + \{X_c\}_e \tag{7}$$

$$\{X_e\} = [T_{ec}][TV]\left(\{Y_c\}_e - \{X_c\}_e\right) + \{X_e\}_e \tag{8}$$

Equations 7 and 8 can be converted to random signals by following the same procedure as was used on Eq. 6. This procedure gives the following results for the test item interface and external acceleration spectral density matrices

$$
\begin{aligned}
[Gx_{cc}] = {}& [T_{cc}]^* [TV]^* \left[[Gy_{cc}]_e + [Gx_{cc}]_e \right][TV][T_{cc}] \\
& - [T_{cc}]^* [TV]^* [Gx_{cc}]_e - [Gx_{cc}]_e [TV][T_{cc}] + [Gx_{cc}]_e
\end{aligned}
\tag{9}
$$

$$
\begin{aligned}
[Gx_{ee}] = {}& [T_{ec}]^* [TV]^* \left[[Gy_{cc}]_e + [Gx_{cc}]_e \right][TV][T_{ec}]^T \\
& - [T_{ec}]^* [TV]^* [Gx_{ce}]_e - [Gx_{ec}]_e [TV][T_{ec}]^T + [Gx_{ee}]_e
\end{aligned}
\tag{10}
$$

Two new spectral density matrices, $[Gx_{ce}]_e = \left[Gx_{ec} \right]_e^H$ and $[Gx_{ee}]_e$ appear in Eq. 10. The N_c x N_e matrix $[Gx_{ce}]_e$ contains CSDs relating the N_c interface and N_e external points while the N_e x N_e matrix $[Gx_{ee}]_e$ contains ASDs and CSDs relating external points only. Both matrices are originated from test item motions that are caused by field external forces *only*. Symbol "т" was dropped from matrices [TV] and $[T_{cc}]$ in Eqs. 6, 9, and 10 since they are symmetric matrices.

The results obtained in Eqs. 9 and 10 correspond to the N_c x N_c and N_e x N_e matrices $[Gx_{cc}]$ and $[Gx_{ee}]$ that appear on the left hand side of Eq. 4, respectively. The remaining N_c x N_e acceleration spectral density matrix $[Gx_{ce}] = [Gx_{ec}]^H$ shown in Eqs. 4 is obtained by the same procedure used in Eqs. 9 and 10. Substitution of $\{X_{cc}\}^*$ obtained from Eq. 7 and of $\{X_{ee}\}^T$ given by Eq. 8 on the right hand side of Eq. 3 leads to the following result for the N_c x N_e CSD matrix $[Gx_{ce}]$

$$
\begin{aligned}
[Gx_{ce}] = {}& [T_{cc}]^* [TV]^* \left[[Gy_{cc}]_e + [Gx_{cc}]_e \right][TV][T_{ec}]^T \\
& - [T_{cc}]^* [TV]^* [Gx_{ce}]_e - [Gx_{cc}]_e [TV][T_{ec}]^T + [Gx_{ce}]_e
\end{aligned}
\tag{11}
$$

4. Field simulations

Two field simulations are performed. In the first simulation, the test item is not connected to the vehicle so that we are dealing with the bare vehicle interface accelerations only. In the second field simulation, the test item and vehicle are connected in the field at several points, forming a combined structure. In this case, the interface force and test item motion spectral density matrices are obtained by employing the equations derived in the previous section.

Theoretical MDOF model:

In order to illustrate the field and laboratory simulations for random signals, the model shown in Fig. 3 is used. The test item and vehicle structures are modeled by two lumped systems, each one having four DOF. They are attached in the field through rigid connectors Cn_1 and Cn_3 forming the combined structure. The masses in the vehicle and test item models are numbered such that interface points have the same number in each structure. The physical parameters for test item and vehicle models are shown in Table 1.

Table 1: Physical parameters of test structures

Mass	Test item (Kg)	Vehicle (Kg)
M_1	0.20	0.50
M_2	0.20	0.50
M_3	0.25	0.30
M_4	0.15	0.50
Stiffness	Test item N/m x 10^4	Vehicle N/m x 10^4
K_1	5.00	6.00
K_2	4.00	14.00
K_3	2.00	10.00

Using standard modal analysis procedures, the modal properties for test item, vehicle, and combined structure are calculated. These results are summarized in Table 2.

Accelerance FRFs are calculated for both structures according to

$$A_{pq}(\omega) = -\omega^2 \sum_{r=1}^{N} \frac{\phi_{pr}\,\phi_{qr}}{m_r\left(\omega_r^2 - \omega^2 + j\,2\,\zeta_r\,\omega_r\,\omega\right)} \tag{12}$$

where ϕ_{pr} and ϕ_{qr} are the pth and the qth elements of the rth mode shape, ω_r is the rth natural frequency (rad/s), ζ_r is the rth modal damping ratio corresponding to the rth mode shape and ω (rad/s) is the excitation frequency.

Similarly, for single base input motion at location 2, the transmissibility FRF is given by

$$\{\Gamma(\omega)\} = \omega^2 \sum_{r=1}^{N} \frac{\{\phi\}_r\,\{\phi\}_r^T\,[M]\,\{1\}}{m_r\left(\omega_r^2 - \omega^2 + j\,2\zeta_r\,\omega_r\,\omega\right)} + \{1\} \tag{13}$$

where [M] is the structure's mass matrix, Equations 12 and 13 define response models for viscously damped structures.

Table 2: Modal properties for test structures

Natural Frequency (Hz)	Test item	Vehicle	Combined
f_1	0	30.00	23.86
f_2	52.60	86.17	58.94
f_3	85.70	164.72	85.05
f_4	131.00	–	112.00
f_5	–	–	160.21
Damping Ratio (%)	Test Item	Vehicle	Combined
ζ_1	0	0.28	0.24
ζ_2	0.50	0.81	0.56
ζ_3	0.81	1.55	0.91
ζ_4	1.23	–	1.06
ζ_5	–	–	1.54
Modal Masses	Test Item	Vehicle	Combined
m_1	0.20	0.43	0.38
m_2	0.17	0.48	0.17
m_3	0.21	0.37	0.55
m_4	0.20	–	0.21
m_5	–	–	0.34

Vehicle excitation source:

Two excitation sources are simultaneously applied to the combined structure, as seen in Fig. 4. The first excitation source is a base acceleration ASD Gy_{22} as shown in Fig. 4a that is applied to the vehicle base mass M_2. This spectrum increases with a slope of 40 dB/decade from 0 to 10 Hz and becomes constant at 0.00517 g^2/Hz up to 200 Hz. This amplitude level is chosen such that an overall 1.0 g_{RMS} input acceleration vibration level is obtained for the 0-200 Hz frequency range. A total of 800 spectral lines is used in these simulations so that the frequency resolution is $\Delta f = 0.25$ Hz.

The test item external excitation source:

The second excitation source is the external force ASD Gf_{44} shown in Fig. 4b that is applied to the test item mass M_4. The N_t x N_t test item output response spectral density matrix $[Gx]_e$ due to this external random input is calculated by Eq. 4 for all of the test item N_t points. This corresponds to the case a shown in Fig. 1a. In this case, the input spectral density matrix $[Gf]_e$ is a matrix of size N_t x N_t with zeros at all entries, except at position (4, 4) where Gf_{44} is used. The test item accelerance FRFs used with Eq. 4 are those that correspond to the test item being modeled as a free structure in space since

[Gx]$_e$ corresponds to the test item response due to external loads only. The test item interface acceleration ASDs Gx$_{11}$ and Gx$_{33}$ due to Gf$_{44}$ are shown in Figs. 5a and b.

4.1.   FIELD SIMULATION 1: BARE VEHICLE INTERFACE MOTIONS

This field simulation corresponds to the second step shown in Fig. 1b. In some situations, laboratory tests involving the test item before it is attached to the vehicle in the field environment are required. This is done in order to predict what vibration levels the test item will be subjected to when in the actual field environment. In order to achieve this goal, suitable test item inputs must be defined so that when they are applied to the test item in the laboratory environment reasonable estimates of the test item actual dynamic behavior will occur. Since no field data involving the combined structure is available in this case, one alternative is to measure the vehicle interface motions and use these measurements to generate test item inputs in the laboratory simulation.

When the test item is absent from the field, the input base motion ASD Gy$_{22}$ represents the only external input to the vehicle in the model shown in Figure 3. Thus, the bare vehicle interface acceleration spectral density matrix [Gy$_{cc}$]$_e$ can be calculated by using the following expression

$$[Gy_{cc}]_e = [: \Gamma_{pq} :]^* [Gy_{qq}] [: \Gamma_{pq} :]^T \tag{14}$$

where index p = 1, ..., N$_c$ covers all interface points, and q = 2 in the case of Fig. 3 is fixed coordinate where the input motion Gy$_{22}$ is applied. Matrix [:Γ_{pq}:] is a N$_c$ x N$_c$ matrix whose qth column contains the acceleration transmissibility FRFs as defined by Eq. 13, and all remaining entries are zero. The N$_c$ x N$_c$ diagonal matrix [Gy$_{qq}$] contains zeros at all diagonal entries except at (2, 2) where the input base motion Gy$_{22}$ is applied. The diagonal elements of the resulting spectral density matrix [Gy$_{cc}$]$_e$ correspond to the bare vehicle interface acceleration ASDs while the off-diagonal entries are the CSDs among the vehicle interface points. The resulting bare vehicle interface ASDs Gy$_{11}$ and Gy$_{33}$ due to the input acceleration ASD Gy$_{22}$ are shown in Figs. 5a and 5b in comparison with the test item ASDs due to the external force ASD Gf$_{44}$.

4.2.   FIELD SIMULATION 2: COMBINED STRUCTURE

This field simulation corresponds to the third field calculation step shown in Fig. 1c. In this field simulation, the test item is connected to the vehicle in the field forming a combined structure, as shown in Fig. 3 and both inputs from Fig. 4 are active. The field interface force spectral density matrix is calculated from Eq. 6, and they are displayed in Fig. 6 along with the corresponding exact values for comparison purposes. These exact values can be obtained by writing the equations of motions for either the test item or vehicle in Fig. 3 and solving for the interface forces frequency spectra. The corresponding exact interface force spectral densities can then be obtained by employing Eq. 2. A good magnitude fit is observed for all frequencies with only small discrepancies occurring for frequencies around 160 Hz. An inversion of the predicted phase angle Gf$_{13}$ is seen to occur in the 160 - 200 Hz frequency range.

The test item motions are calculated according to Eqs. 9, 10, and 11. The test item acceleration ASDs are shown in Fig. 7, and they are seen to agree closely when compared with the exact values. Small discrepancies occur in the predicted Gx_{22} for frequencies in the vicinity of 142 Hz and 160 Hz, and in Gx_{11} and Gx_{33} for frequencies in the vicinity of 160 Hz.

5. Laboratory simulations

The goal of a laboratory simulation is to ensure that the test item will survive when exposed to its field dynamic environment. Field data is required in laboratory simulations in order to define realistic test item inputs. The data obtained from the previous field simulations will be used in this section to define test item inputs in the laboratory environment. The three laboratory simulations shown in Fig. 2 will be discussed for the test item shown in Fig. 3. In each test, one of the following field measurement results will be used to define the test item inputs: (i) bare vehicle interface motions, Fig. 2a; (ii) interface forces, Fig. 2b and; (iii) test item interface and external motions, Fig. 2c.

Two previous papers [2,3] showed that a set of forces must be determined when using field motions in order to define the test item inputs in the laboratory. The pseudo-inverse technique was successfully employed to calculate the forces acting on the test item in the field from the measured acceleration signals under certain conditions, i.e., the signals are deterministic [2,3]. This force identification technique presents problems at some frequencies since the procedure requires the pseudo-inversion of the test item accelerance FRF matrix which can be rank deficient at the test item natural frequencies.

When working with deterministic signals, the test item input forces in the laboratory are obtained through the following expression

$$\{R\} = \left[\hat{T}\right]^{+} \{X\} \tag{15}$$

where $\left[\hat{T}\right]$ is the test item FRF matrix in the laboratory environment, $\{X\}$ is the vector containing the test item field output frequency spectra, and $\{R\}$ contains the resulting laboratory input force frequency spectra. The symbol "+" denotes the pseudo-inverse of the FRF matrix $\left[\hat{T}\right]$. Equation 15 involves complex quantities carrying magnitude and phase information so that the cross correlation requirements among the variables are automatically satisfied when employing Eq. 15.

When dealing with random signals, Eq. 1 is used in order to solve for the unknown input spectral matrix $[G_{ff}]$ based on the measured output spectral density matrix $[G_{xx}]$. This equation is rewritten in terms of the test item laboratory acceleration and force spectral density matrices.

$$\left[G_{uu}\right] = \left[\hat{T}\right]^{*}\left[G_{rr}\right]\left[\hat{T}\right]^{T} \tag{16}$$

When we pre and post multiply both sides if Eq. 16 by $\left[\left[\hat{T}\right]^{*}\right]^{-1}$ and $\left[\left[\hat{T}\right]^{T}\right]^{-1}$, the following result is obtained for the unknown input spectral density matrix.

$$[G_{rr}] = \left[\left[\hat{T} \right]^+ \right]^* [G_{uu}] \left[\left[\hat{T} \right]^+ \right]^T \tag{17}$$

The resulting matrices on the left hand side of Eqs. 16 and 17 contain the test item acceleration and force ASDs and CSDs, respectively. These two expressions contain the correct phase relationship among the variables since the complex CSDs are accounted for in both cases. Thus, proper correlation between the corresponding time variables is accounted for when employing Eq. 16 to obtain the acceleration spectral matrix or Eq. 17 to solve the inverse problem for the unknown force auto spectral density matrix. *Consequently, Eqs. 16 and 17 are called the* **_correlated_** *expressions for acceleration and force spectral density matrices.*

$$\{G_{rr}\} = \left[\left| \hat{T} \right|^2 \right] \{G_{rr}\} \tag{18}$$

where $\{G_{rr}\}$ and $\{G_{uu}\}$ are real valued vectors containing the input and output ASDs, respectively. Thus, Eq. 18 shows that, under the assumption of _uncorrelated_ input forces, the resulting accelerations are equally _uncorrelated_ and does not provide any phase information since only output ASDs are available. Using the pseudo-inverse technique with Eq. 18, the following result is obtained for the unknown input force ASDs vector

$$\{G_{rr}\} = \left[\left| \hat{T} \right|^2 \right]^+ \{G_{xx}\} \tag{19}$$

A comparison of Eqs. 16 and 18 for the acceleration spectral density matrices as well as of Eqs. 17 and 19 for the solution of the inverse problem clearly shows the differences in neglecting the CSDs in both Eqs. 18 and 19. There is no phase information in Eqs. 18 and 19 so that no _correlation_ exists between the time variables. *Hence, these equations will be subsequently referred to as the* **_uncorrelated_** *expressions for acceleration and force spectral density matrices.*

5.1.   LABORATORY SIMULATION 1: BARE VEHICLE INTERFACE MOTIONS

This laboratory simulation corresponds to the first calculation step as shown in Fig. 2a. In this case, the bare vehicle interface acceleration ASDs and CSD shown in Fig. 5 are used with Eqs. 17 and 19 in order to calculate a set of test item inputs in the laboratory. First, the _correlated_ (Eq. 17) and _uncorrelated_ (Eq. 19) are rewritten as a function of the bare vehicle data and the resulting expressions are

$$[Gr_{cc}] = \left[[T]^+ \right]^* [Gy_{cc}]_e \left[[T]^+ \right]^T \tag{20}$$

$$\{Gr_{cc}\} = \left[|T|^2 \right]^+ \{Gy_{cc}\}_e \tag{21}$$

where the test item is assumed to have the same accelerance FRF characteristics in both the field and laboratory environments, i.e., $[T] = \left[\hat{T} \right]$. Equation 20 employs the bare vehicle interface acceleration ASDs and CSD while only ASDs are used with Eq. 21. Thus, Eq. 20 represents the case where bare vehicle motions are assumed to be _correlated_ while Eq. 21 uses the assumption of _uncorrelated_ motions.

Similarly, the test item acceleration spectral density matrix is obtained from Eq. 1 that is rewritten as

$$[Gu] = [T]^* [Gr_{cc}] [T]^T \qquad (22)$$

where the $N_t \times N_t$ matrix [Gu] is the test item acceleration spectral density matrix in the laboratory, and the $N_c \times N_c$ matrix [Gr_{cc}] fully populated wit interface force spectral densities when the _correlated_ form of Eq. 20 is used or is populated with elements of the $N_c \times 1$ vector {Gr_{cc}} on the main diagonal entries and zeros elsewhere when the _uncorrelated_ Eq. 21 is employed. Figures 8a and 8c show the resulting test item interface acceleration ASDs and Figs. 8b and 8d show the external acceleration ADSs obtained through Eq. 22 for [Gr_{cc}] obtained from either Eq. 20 and Eq. 21. The test item field acceleration ASDs are plotted for comparison purposes.

It is evident in Fig. 8 that there is more similarity between the field data and the results from the _correlated_ inputs situation than there is between the field data and the _uncorrelated_ input situation. The _uncorrelated_ ASD results are much larger in error and represent the poorest estimates. It is clear that neither method was really successful for two reasons.

First, we attempted to reproduce field data in the laboratory by applying forces at the test item interface points only when in reality the actual field environment is subjected to the test item interface forces F_1 and F_3 as well as the external force F_4 being applied to mass M_4. _Second_, the bare vehicle interface motions do not contain any information regarding coupling effects between test item and vehicle since the test item is absent when field data is obtained. Thus, it is concluded that the use of the bare vehicle interface acceleration data alone is inadequate for predicting how the test item will behave under field conditions unless corrections to take interface driving point and transfer FRFs of test item and vehicle into account.

5.2. LABORATORY SIMULATION 2: FIELD INTERFACE FORCES

This simulation corresponds to the second calculation step shown in Fig. 2b. Two cases are analyzed when using field interface force as test item inputs in the laboratory. _First_, the interface forces spectral density matrix obtained from Eq. 6 and shown in Fig. 6, is applied to the test item interface points in the laboratory. These interface forces were obtained when the external random input ASD shown in Fig. 4b is applied to the test item in the field.

Second, the test item external input ASD Gf_{44} is removed from the field environment and the interface force spectral density matrix is recalculated by Eq. 6. The new interface force spectral density matrix is used to drive the test item in the laboratory.

These two scenarios are used to check the accuracy of the predicted test item motions when external random forces are either present and absent from the field environment. In both cases, Eq. 1 is rewritten as

$$[Gu] = [T]^* [Gr_{cc}] [T]^T \qquad (23)$$

with $[Gr_{cc}] = [Gf_{cc}]$ where $[Gf_{cc}]$ represents the $N_c \times N_c$ interface forces spectral density matrix obtained through Eq. 6.

If *uncorrelated* field interface forces are assumed, Eq. 18 is written as

$$\{Gu\} = \left[|T|^2 \right] \{Gr_{cc}\} \qquad (24)$$

where $\{Gr_{cc}\}$ contains the interface forces ASDs from matrix $[Gf_{cc}]$.

The results of the first simulation are shown in Fig. 9. Figures 9a and c show the test item interface acceleration ASDs, and Figs. 9b and d show the external acceleration ASDs. Incorrect test item motions are obtained when either the *correlated* Eq. 23 or the *uncorrelated* Eq. 24 are used. *This is due to the fact that while in the field environment, the test item is subjected to an external force F_4 in addition to interface forces F_1 and F_3, and this external force is not accounted for in the laboratory simulation.*

The results of the second laboratory simulation employing interface forces as inputs to the test item are shown in Fig. 10. In this case, the external input ASD Gf_{44} is removed from the test item mass M_4 in the field environment. *Thus, the test item is subjected to interface forces only*. The new interface forces spectral density matrix is obtained by using Eq. 6, with $[Gx_{cc}]_e = 0$. Also the test item acceleration spectral density matrices given by Eqs. 9, 10, and 11 are simplified, since all matrices involving the test item accelerations ASDs and CSDs due to external forces are zero. Thus, field interface forces and test item acceleration spectral density matrices depend only on the bare vehicle interface acceleration data $[Gy_{cc}]_e$ in the absence of field external forces applied to the test item. Identical results were obtained in [1] for deterministic signals in terms of frequency spectra. *The test item acceleration ASDs that are obtained from the correlated Eq. 23 are shown in Fig. 10 and are seen to closely match the corresponding field ASDs. Small discrepancies are observed at some frequencies. The ASDs that are obtained from the uncorrelated Eq. 24, which do not include CSDs do not agree with the field data in this case.*

5.3. LABORATORY SIMULATION 3: TEST ITEM MOTIONS

This laboratory simulation refers to the third simulation calculation step in Fig. 2c. In this laboratory simulation, the test item field acceleration ASDs and CSDs are used to define test item inputs. Similarly to the case where the bare vehicle interface acceleration ASDs were used. This simulation requires that the test item inputs be obtained from field motions, and in this case, the pseudo-inverse technique is used again.

When using the test item field accelerations to define the laboratory input forces by the pseudo-inverse technique, a natural question that arises is: *"Which measured accelerations should I use, and how do the predicted input forces compare with the true forces occurring in the field?"* Since field data is divided in terms of interface and external forces and accelerations, we consider the following possibilities when using field motions to define the test item inputs in the laboratory [3]:

- (i) N_c interface forces are obtained from N_c interface motions.

- (ii) N_c interface forces are obtained from all N_t measured field motions.

- (iii) N_t forces are obtained from N_t motions.

These cases will be illustrated in the case of random signals for both _uncorrelated_ and _correlated_ inputs and outputs.

5.3.1. N_c interface forces using N_c field interface motions
In this case, only field interface motions are used to solve for interface forces in the laboratory environment. Thus, Eqs. 17 and 19 are rewritten as

$$[Gr_{cc}] = \left[[T_{cc}]^+\right]^* [Gx_{cc}] \left[[T_{cc}]^+\right]^T \tag{25}$$

$$\{Gr_{cc}\} = \left[|T_{cc}|^2\right]^+ \{Gx_{cc}\} \tag{26}$$

Since the test item item interface accelerance FRF N_c x N_c matrix $[T_{cc}]$ is square, the pseudo-inversion shown in Eqs. 25 and 26 reduces to the standard inverse. The test item output response to this set of forces is obtained through Eqs. 23 and 24, for the case of _correlated_ and _uncorrelated_ inputs, respectively. The results from this simulation are shown in Figs. 11 and 12. Figure 11 shows the interface forces ASDs calculated through Eqs. 25 and 26. _In this case, the correlated Eqs. 25 gave a good estimate for Gr_{11} when compared with the true interface force from the field but a poor estimate for Gr_{33}. The itnerface forces ASDs obtained from the uncorrelated Eq. 26 produced significantly incorrect estimates for both Gr_{11} and Gr_{33}._

The results for the test item acceleration ASDs are shown in Fig. 12. The interface ASDs shown in Figs. 12a and c match the corresponding field measurements for the case of _correlated_ inputs. The external ASD Gu_{22} is seen to match the field measurement as well. This result is expected since mass M_2 is located between M_1 and M_3, which have correct motions and there is not external force applied to M_2. Thus, if the interface accelerations motions Gu_{11} and Gu_{33} are correct in the laboratory environment, then Gu_{22} should match the field Gx_{22}. The external acceleration ASD estimate Gu_{44} that is based on _correlated_ inputs is seen to be seriously in error at higher frequencies. This error is due to the directly applied external force Gf_{44} that is directly applied to M_4 in the field but is unaccounted for in this simulation. _In all cases, the laboratory estimates based on ASD field information without using CSD for both interface forces (Fig. 11) and test item acceleration ASDs (Fig. 12) are seriously in error. It is noted that the use of correlated inputs caused the interface force inputs to be adjusted so that the corrected interface acceleration occurred at locations 1 and 3. Acceleration of mass M_2 is correct since the only forces acting on M_2 in this case are those due to the motions of masses M_1 and M_3. Similarly, the motion of M_4 is incorrect since an unaccounted for external force is acting on mass M_4. Thus, the motion of M_2 is isolated from the external force applied to M_4 by the distorted interface force acting on M_3._

Thus, this laboratory simulation employing only interface motions to obtain the test item input forces fails due to the fact that the external effects caused by F4 are not accounted for. In the absence of external forces, the procedure of generating N_c interface forces from N_c interface accelerations could result in correct laboratory predictions as long as the *correlated* relationship of Eq. 25 is used. However, this simulation case should be viewed with caution since the pseudo-inversion of the test item interface FRF matrix in Eq. 25 reduces to the standard inversion. This occurs since the number of motions is the same as the number of input forces that are predicted. In this case, the inversion of $[T_{cc}]$ can cause numerical problems due to rank deficiency at some frequencies. In these cases, the solution for the input forces must be obtained in theleast squares sense where more motions are used rather than forces predicted [6]. This procedure not only improves the numerical stability of the pseudo-inversion process but also helps to reduce experimental noise effects when dealing with actual data.

5.3.2. N_c interface forces using N_t field motions
In this simulation, N_c interface forces are calculated from all N_t field motions. Equations 25 and 26 are rewritten as

$$[Gr_{cc}] = \left[\left[\left[[T_{cc}]\ [T_{ce}]\right]^T\ \right]^+\ \right]^* [Gx] \left[\left[\left[[T_{cc}]\ [T_{ce}]\right]^T\ \right]^+\ \right]^T \tag{27}$$

$$\{Gr_{cc}\} = \left[\left|\left[[\ T_{cc}\ ]\ [\ T_{ce}\ ]\right]^T\right|^2\ \right]^+ \{Gx\} \tag{28}$$

where the $N_t \times N_t$ matrix $[Gx]$ contains all measured accelerations in the field. The results for the interface forces ASDs calculated from the *correlated* (Eq. 27) and *uncorrelated* (Eq. 28) relationships are shown in Fig. 13. Both the *correlated* and *uncorrelated* results do not match the corresponding field interface forces ASDs.

The test item output accelerations for *correlated* and *uncorrelated* inputs are obtained through Eqs. 23 and 24, respectively with the results being shown in Fig. 14. In this case, both the *correlated* and the *uncorrelated* estimates present surprisingly good agreement when compared one to another, but both results deviate from the true field measurement. The assumption of *uncorrelated* inputs does not appear to be totally inadequate in this case. However, the fact that the test item is subjected to interface forces only in the laboratory while in the field both interface and external forces exist is the cause for the incorrect laboratory simulation results.

5.3.3. N_t forces from N_t field motions
This laboratory simulation corresponds to determining N_t forces from N_t measurements. Thus, in addition to calculating interface forces, possible external forces acting on test item external points where motion measurements were taken are identified. In this case, the *correlated* (Eq. 17) and the *uncorrelated* (Eq. 19) relationships are used with the full $N_t \times N_t$ test item accelerance FRF matrix. The full $N_t \times N_t$ output spectral density matrix is used with Eq. 17 while only the output ASD functions are used in Eq. 19.

The results for the predicted forces are shown in Fig. 15. Figures 15a and b show the interface forces ASDs. In this case, the results when CSDs are accounted for (*correlated outputs*) match the corresponding field measurements. The results from the *uncorrelated* Eq. 19 are not in agreement with the field measurements in the 0 – 50 Hz frequency range but the agreement is considerably improved in the remaining 50 – 200 Hz frequency range. The results for the external force F4 applied to the test item in the field is shown in Fig. 15c and 15d. In this case, the result for *correlated* inputs match the corresponding Gf_{44} (see Fig. 15d for a plot of the estimated Gf_{44} in linear scale) while the result for *uncorrelated* inputs do not present a good agreement in the 0 – 120 Hz frequency range, but improves in the 120 – 200 Hz range.

The test item output spectral density N_t x N_t matrix [Gu] is obtained by using Eq. 23 with the corresponding ASDs results being shown in Fig. 16. In this case, all field acceleration ASDs are matched in the laboratory. This is due to the fact that the external force effects from Gf_{44} were accounted for in the laboratory. It is interesting to notice that, although the predicted *uncorrelated* input ASDs did not yield good results along the entire 0 – 200 Hz range in Fig. 15, the resulting test item accelerations ASDs matched closely with the field measurements. However, it should be recalled that when assuming *uncorrelated* inputs, Eqs. 19 and 22 are used to obtain the forces and resulting acceleration ASDs, respectively, and no information about the force and acceleration CSDs are obtained when the assumption of *uncorrelated* inputs is used. Although not shown here, when Eqs. 17 and 1 are used, the resulting force and acceleration CSDs match the corresponding field measurements.

This last simulation case represents a good example of the difficulties involved in obtaining the correct test item excitation forces from knowledge of accelerations when using the pseudo-inverse technique. The *uncorrelated* predicted interface and external force ASDs do not match the actual forces experienced by the test item in the field as shown in Fig. 15. However, when this set of predicted forces are applied to the test item, the exact field accelerations are reproduced in the laboratory. The solution for the laboratory forces was reduced to the standard inversion of the N_t x N_t test item FRF matrix since the number of motions and the number of forces are the same. Thus, it appears that FRF matrix rank deficiency problems occurred in employing Eq. 19. These numerical difficulties caused distortions on the predicted set of forces but, when these *pseudo-forces* were applied to the test item, the correct motions were obtained. *This is a classical example of a non-unique set of forces that produced the correct motions and it reinforces the idea that by over-determining Eqs. 17 and 19 (more motions than forces), a unique set of forces can be obtained in a lease squares sense.*

6. Summary of numerical results

The results obtained in the numerical simulations can be summarized as follows:

- Acceleration CSDs must be accounted for in addition to acceleration ASDs when obtaining the test item inputs from knowledge of field motions through the pseudo-inverse technique. This is due to the fact that motions are always *correlated* whether the input forces are *correlated* or *uncorrelated*. Numerical problems can occur when employing the pseudo-inverse technique due to rank deficiency of the test item FRF matrix at some frequencies. These difficulties can be overcome by seeking a set of pseudo-forces in a least squares sense. This set of pseudo-forces will, in some cases, resemble the actual forces acting on the test item depending primarily on the locations where the accelerations and FRFs were measured.

- The bare vehicle interface acceleration ASDs do not represent suitable test item inputs in laboratory simulations. These accelerations can give reasonable simulation results in the absence of field external forces and if the vehicle driving and transfer point FRFs are known. Otherwise, the bare vehicle data alone is meaningless.

- Field external forces must be accounted for in laboratory simulations so that the reasonable acceleration predictions can be obtained.

7. Experimental analysis

The results shown in the previous section revealed that knowledge of the external forces acting on the test item in the field environment is vital for a successful laboratory simulation. This section presents results from an experimental analysis in order to assess the feasibility of the pseudo-inverse technique when dealing with experimental random signals.

7.1. THE EXPERIMENTAL SETUP

A cold rolled steel beam (92 x 1.25 x 1.0 in) is chosen as the test item in this case. The beam is suspended by flexible cords in order to properly simulate the *free-free boundary conditions*. Two experimental setups are used in the tests and they are shown in Fig.17. Both setups employ two vibration exciters that are used to drive the beam at two locations. A MB (model Modal 50) exciter is used to drive the beam mid point at 3 and a B&K (model 4808) exciter is used to excite the beam end point at 4. The experimental setup shown in Fig. 17b employs two dummy masses attached to the beam free end at 1. Each dummy mass weighs 0.22 Kg ($\approx$ 0.5 lb) and they are used to simulate a third unknown load (inertia load) applied to the beam.

Random excitation is used with both experimental setups shown in Fig. 17. The excitation forces ASDs are measured by two force transducers mounted at points 3 and 4. The measured excitation forces will be used later as a comparison basis for the identified loads. The inertia load applied to the beam end by the dummy masses in Fig. 17b is not directly measured, but it is calculated from the beam acceleration signal at point 1. The beam output acceleration ASDs are measured by accelerometers at four locations as shown in Fig. 17. Table 3 shows the characteristics of all transducers employed in the tests.

Input and output ASDs and CSDs are obtained by a 486 PC equipped with the Data Physics Dp420© data acquisition board. The beam FRFs relating all input and output points are measured in the 0 – 625 Hz frequency range. A total of 1000 spectral lines are used so that the frequency resolution is of $\Delta f = 0.6357$ Hz. Hanning windows are used with all data channels. Table 4 contains the beam's first 7 bending natural frequencies in the 0 – 625 Hz frequency range that were obtained by curve fitting the experimental accelerance FRFs using ICATS© modal analysis software [11]. The experimental natural frequencies are compared with values obtained analytically [12].

Table 3: Characteristics of sensors used in tests

Sensor Type	Position on beam	Sensor Model	Charge or Voltage Sensitivity
Force	F3	PCB 208A03	50.84 mv/N
Force	F4	PCB 208A03	50.84 mv/N
Accel.	A1	PCB 302A	10.09 mv/g
Accel.	A2	Endveco 2222C	1.70 pc/g
Accel.	A3	PCB 302A	10.18 mv/g
Accel.	A4	PCB 302A02	10.04 mv/g

Table 4: Natural frequencies for free free beam

fn	Theoretical Hz	Experimental Hz
1	24.30	25.30
2	67.60	67.98
3	132.40	131.22
4	218.00	216.60
5	327.00	322.53
6	457.00	450.60
7	608.00	597.62

Two tests are performed with each experimental setup shown in Fig. 17. In the **first test**, the random signals that are used to drive the exciters at points 3 and 4 are generated by two independent random signal generators. *This is done in order to simulate the case where the input forces are statistically uncorrelated*. In the **second test**, the same random signal is used to feed both vibration exciters at points 3 and 4. *This corresponds to the case where the input signals to the vibration exciters are correlated*. Table 5 summarizes the tests performed as well as the forces identified in each test.

Table 5: Measured and predicted forces for all test cases

Case	Setup	Sources	Measured	Predicted
a	1	Two	F_3, F_4	F_3, F_4
b	1	One	F_3, F_4	F_3, F_4
c	2	Two	F_1, F_3, F_4	F_1, F_3, F_4
d	2	One	F_1, F_3, F_4	F_1, F_3, F_4

7.2. FORCE PREDICTION RESULTS

The *correlated* (Eq. 17) and *uncorrelated* (Eq. 19) relationships are used in each case shown in Table 5.

7.2.1. Case (a)

The experimental setup of Fig. 17a is used with two independent random signal generators feeding exciters at 3 and 4. The predicted random forces ASDs and CSDs from the _correlated_ (Eq 17) expression are shown in Fig. 18. The predicted ASDs from the _uncorrelated_ (Eq. 19) expression are shown in Fig. 19. By comparing predicted ASD results from Fig. 18 with those shown in Fig. 19, it is seen that even when two independent random sources are used with exciters at points 3 and 4, the _correlated_ (Eq. 17) expression yielded better results than the _uncorrelated_ (Eq. 19) expression. A reasonable prediction for Gf_{34} was obtained from Eq. 17 while Eq. 19 yields no CSD results.

7.2.2. Case (b)

Both vibration exciters in 3 and 4 are driven by the same input random excitation signal. The _correlated_ Eq. 17 and _uncorrelated_ Eq. 19 are used to obtain the predicted forces. Figures 20a and b show the results for Gf_{33} and Gf_{44} when Eq. 17 is used. By comparing these predicted ASD results with those obtained from Eq. 19 as shown in Figs. 21, it is clear that the assumption of _uncorrelated_ forces and motions is inadequate in this case. Predicted results for Gf_{34} in Figs. 20c and d show close agreement with the measured data.

7.2.3. Case (c)

This test employs the setup shown in Fig. 17b. Two independent random excitation signals are used to feed the exciters at positions 3 and 4. Three forces are predicted in this case, namely, the two measured forces at locations 3 and 4, and the inertia load at location 1 that is applied to the beam by the dummy masses attached to the beam end at 1. Force prediction estimates are shown in Figs. 22 and 23. Figures 22a, b, and c show the predicted ASDs from the _correlated_ Eq. 17. The measured inertia load ASD Gf_{11} in Fig. 22a was obtained from the acceleration ASD Gx_{11} and from the dummy masses according to Newton's second law. Figures 22d and e show the magnitude and phase angle of the predicted CSD Gf_{34}, along with the corresponding measured CSD. Predicted results for Gf_{13} and Gf_{14} are not shown but they present a reasonable good agreement with the corresponding measurements. The predicted ASDs from Eq. 19 are shown in Fig. 23a, 23b, and 23c. As in the two previous tests, the force estimates from the _correlated_ Eq. 17 gave better results than those obtained from the _uncorrelated_ Eq. 19.

7.2.4. Case (d)

The results of this last experiment are shown in Figs. 24 and 25 for forces predictions using Eqs. 17 and 19, respectively. This test employed a single signal that drive both exciters in the setup of Fig. 17b. The predicted force ASDs shown in Figs. 24 and 25 indicate that the results obtained from the _correlated_ (Eq. 17) expression present better agreement with measured forces than those obtained from the _uncorrelated_ (Eq. 19) expression.

Thus, the force prediction results for all cases shown in Table 5 demonstrate that CSDs should be accounted for in random force estimates from motion measurements when using the pseudo-inverse technique. This means that the _correlated_ Eq. 17 should be employed in solving this inverse problem for random signals, and that the assumption of *uncorrelated* test item motions is untrue even when the exciters are driven by statistically *uncorrelated* random signals. There are two reasons for this correlation requirement.

First, the motion at each point in a structure is due to all forces applied to the structure. When we examine the cross correlation between any two points, we find these motions are *correlated* because each force is perfectly *correlated* with itself even when it is completely *uncorrelated* with the other forces. In order to clarify this point, consider a structure being subjected to two random forces with frequency spectra respectively given by F_1 and F_2. The corresponding structure acceleration responses X_1 and X_2 can be written in terms of the structure driving and transfer point accelerances and they are respectively given by

$$X_1 = A_{11} F_1 + A_{12} F_2 \qquad (29)$$

$$X_2 = A_{21} F_1 + A_{22} F_2 \qquad (30)$$

Now, assume that the random forces F_1 and F_2 are statistically *uncorrelated*, i.e., $Gf_{12} = 0$. Substitution of X_1 and X_2 as given by Eqs. 29 and 30 into Eq. 3 and accounting for the fact that $Gf_{12} = 0$, gives the following result for the acceleration CSD between points 1 and 2

$$Gx_{12} = A_{11}^* A_{21} \, Gf_{11} + A_{12}^* \, A_{22} \, Gf_{22} \qquad (31)$$

where Gf_{11} and Gf_{22} are the input forces ASDs and A_{pq}, $p,q = 1...2$ represent the structure accelerance FRFs. Thus, since Gf_{11} and Gf_{22} *are not* zero, the resulting acceleration CSD Gx_{12} *is not* zero as well. ***Thus, there is correlation between motions at 1 and 2 independent of the input forces being either correlated or uncorrelated***.

Second, the force signals measured by force transducers at 3 and 4 will present some correlation even when independent input signals are used with both exciters due to the mechanical coupling of the exciters by the beam. This gives resulting the so called cross coupling effect among excitation sources that is due to exciter armature inertia [9,10].

8. Summary of experimental results

The results obtained in the experimental analysis can be summarized as follows:

- The pseudo-inverse technique is feasible when working with experimental random signals as long as the experimental acceleration CSDs are accounted for in addition to the acceleration ASDs when solving the inverse problem. This requirement is independent of forces being *correlated* or *uncorrelated* since motions are always *correlated*.

- The solution for the unknown input forces were obtained in least squares sense, since more motions were measured than forces predicted. The location of the input forces were assumed to be known since the accelerations as well as FRFs for the excitation locations were measured and accounted for in the solution of the inverse problem. This is a requirement imposed by the pseudo-inverse technique that if met along with the motion cross correlation requirement leads to reasonable estimates of the input forces.

9. Summary and Conclusions

This paper applies a vibration testing model for exchanging random vibration excitation and response signals. Field interface forces and motions spectral density matrices are derived and applied in field simulations of vibration data. The results of these field simulations are used in the laboratory environment to define suitable test item inputs. When using field motions to define test item inputs, the pseudo-inverse technique is employed to calculate the required input forces for the test item in the laboratory. In the case of random signals, it was numerically and experimentally shown that CSDs in addition to ASDs must be accounted for when predicting random loads from motion measurements. The assumption of _uncorrelated_ inputs and outputs was shown to be inadequate in most situations when trying to solve this difficult laboratory simulation problem.

Acknowledgements

The authors would like to thank the Iowa State University Engineering College and the Aerospace Engineering and Engineering Mechanics Department for supporting this research. Dr. Paulo S. Varoto, from Universidade de São Paulo – São Carlos – Brasil was financially sponsored by CNPq – Brazil during his PhD program.

10. References

1. Varoto, P.S., and McConnell, K.G. (1999), Rules for the Exchange and Analysis of Dynamic Information: Part I – Basic definitions and test scenarios, in J.M.M. Silva and N.M.M. Maia (eds.), *Modal Analysis & Testing, Kluwer Academic Publishers, NATO Series*, Dordrecht, 65-81.
2. Varoto, P.S. and McConnell, K.G. (1999), Rules for the Exchange and Analysis of Dynamic Information, Part II: Numerically simulated results for a deterministic excitation with no external forces, in J.M.M. Silva and N.M.M. Maia (eds.), *Modal Analysis & Testing, Kluwer Academic Publishers, NATO Series*, Dordrecht, 83-100.
3. Varoto, P.S. and McConnell, K.G. (1999), Rules for the Exchange and Analysis of Dynamic Information, Part III: Numerically simulated and experimental results for a deterministic excitation with external loads, in J.M.M. Silva and N.M.M. Maia (eds.), *Modal Analysis & Testing, Kluwer Academic Publishers, NATO Series*, Dordrecht, 101-135.
4. Rogers, J.D., Beightol, D.B., and Doggett, J.W. (1990), Helicopter flight vibration of large transportation containers – A case for test tailoring, *Proceedings of the IES*, 515-521.
5. Szymkowiak, E., A. and Silver II W. (1990), A captive store flight vibration simulation project, *Proceedings of the 36th Annual meeting, IES*, New Orleans, 531-538.
6. Hillary, B. (1983), Indirect measurement of vibration excitation forces, *PhD Thesis, Imperial College of Science Technology and Medicine*, London.
7. Fabummi, J. (1986), Effects of structural modes on vibratory force determination by the pseudoinverse technique, *AIAA Journal, Vol. 24, No. 3*, 504-509.
8. Bendat, J. and Piersol, A. (1971), Random Data, Analysis and Measurement Procedures, *2nd ed., John Wiley*.
9. Smallwood, D. (1982), A random vibration control system for testing a single test item with multiple inputs, *SAE Paper No. 821482*, 4571-4577.
10. Smallwood, D. (1978), Multiple shaker random vibration testing with cross coupling, Proceedings of the IES, 341-347.

11. "MODENT, Reference Manual ICATS" (1994), *Imperial College of Science, Technology and Medicine, Mech. Eng. Department*, London.
12. McConnell, K.G. (1995), Vibration Testing: Theory and Practice, *John Wiley & Sons*.

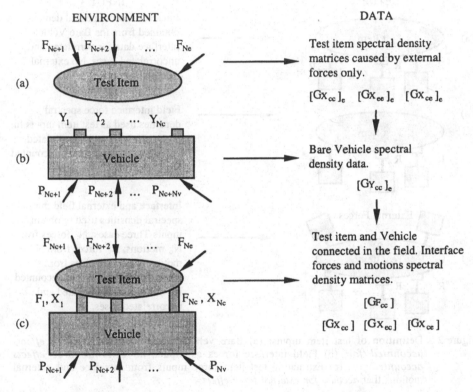

Figure 1   Sequence of field simulations with random external loads: (a) Test item output spectral densities due to external forces, test item is a free structure in space; (b) Bare vehicle data; (c) Interface forces and test item motions for the combined structure.

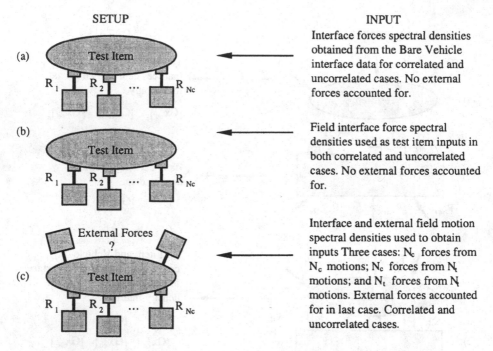

SETUP

INPUT

(a) Interface forces spectral densities obtained from the Bare Vehicle interface data for correlated and uncorrelated cases. No external forces accounted for.

(b) Field interface force spectral densities used as test item inputs in both correlated and uncorrelated cases. No external forces accounted for.

(c) Interface and external field motion spectral densities used to obtain inputs Three cases: N_c forces from N_c motions; N_c forces from N_t motions; and N_t forces from N_t motions. External forces accounted for in last case. Correlated and uncorrelated cases.

Figure 2   Definition of test item inputs: (a) Bare vehicle data, *no external force effects accounted for*; (b) Field interface forces as inputs, *no external force effects accounted for*; (c) Estimating test item force inputs from interface and external motions that *account for external force effects*.

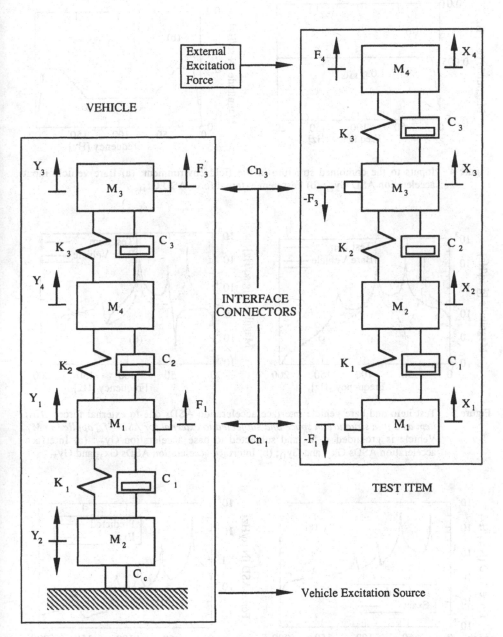

Figure 3   Test item attached to vehicle by rigid connectors Cn_1 and Cn_3 in the filed environment. External inputs to the combined structure given in terms of the vehicle base input acceleration Y_2 and the external force F_4 applied to the test item mass M_4.

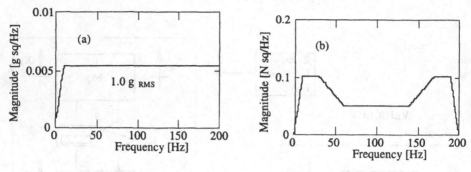

Figure 4   Inputs to the combined structure in the field environment: (a) Bare vehicle input acceleration ASD Gy_{22}; (b) Test item external force ASD Gf_{44}.

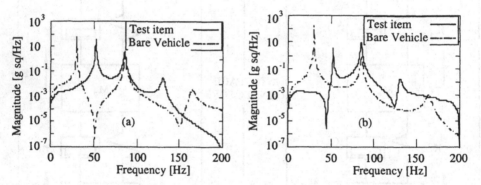

Figure 5   Test item and bare vehicle interface acceleration ASDs due to external forces. *Test item as a free structure in space and subjected to external for ASD Gf_{44} applied to M_4.* Vehicle is grounded at M_2 and subjected to base acceleration Gy_{22}: (a) Interface acceleration ASDs Gx_{11} and Gy_{11}; (b) Interface acceleration ASDs Gx_{33} and Gy_{33}

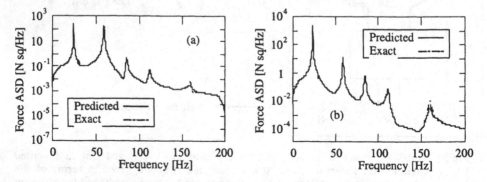

Figure 6   Field interface force ASDs and CSD *obtained due to input accelerations ASD Gy_{22} and the external force ASD Gf_{44} that are applied to the combined structure*: (a) Interface force ASD Gf_{11}; (b) Interface force ASD Gf_{33};

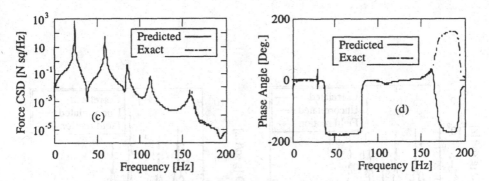

Figure 6   (*Cont.*) Field interface force ASDs and CSD *obtained due to input accelerations ASD Gy$_{22}$ and the external force ASD Gf$_{44}$ that are applied to the combined structure*: (c) Interface force CSD Gf$_{13}$; (d) Phase angle of CSD Gf$_{13}$.

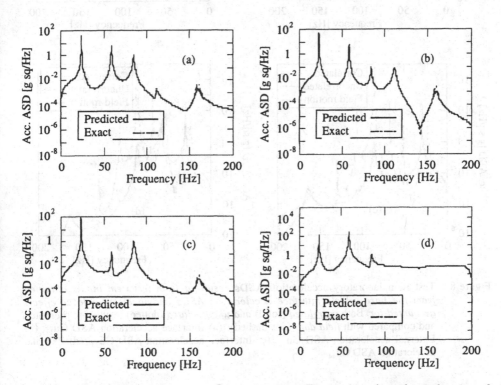

Figure 7   Test item interface and external acceleration ASDs *resulting from the application of input acceleration Gy$_{22}$ and external force ASD Gf$_{44}$ to the combined structure*: (a) Interface acceleration ASD Gx$_{11}$; (b) External acceleration ASD Gx$_{22}$; (c) Interface acceleration ASD Gx$_{33}$; (d) External acceleration ASD Gx$_{44}$.

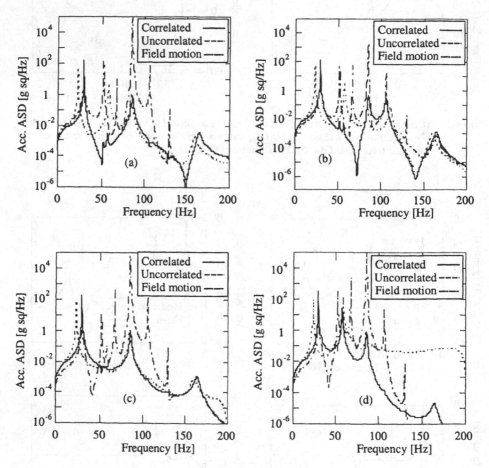

Figure 8   Test item laboratory acceleration ASDs *resulting from test item inputs obtained from the bare vehicle interface acceleration ASDs and no external force effects accounted for*. Both *correlated* (solid) *and uncorrelated* (dotted) results are presented and compared with *field data* (dot dashed): (a) Interface acceleration ASD Gu_{11}; (b) External acceleration ASD Gu_{22}; (c) Interface acceleration ASD Gu_{33}; (d) External acceleration ASD Gu_{44}.

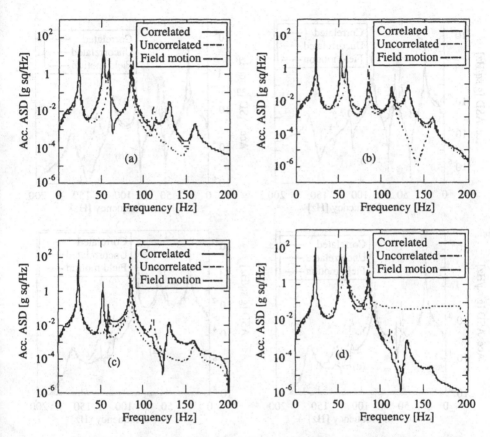

Figure 9   Test item laboratory acceleration ASDs *resulting from field interface forces as test item inputs when the external force ASD Gf₄₄ is applied in the field but not in the laboratory.* Both <u>*correlated*</u> (solid) *and* <u>*uncorrelated*</u> (dotted) results are presented and compared with *field data* (dot dashed): (a) Interface acceleration ASD Gu_{11}; (b) External acceleration ASD Gu_{22}; (c) Internal acceleration ASD Gu_{33}; (d) External acceleration ASD Gu_{44}.

166

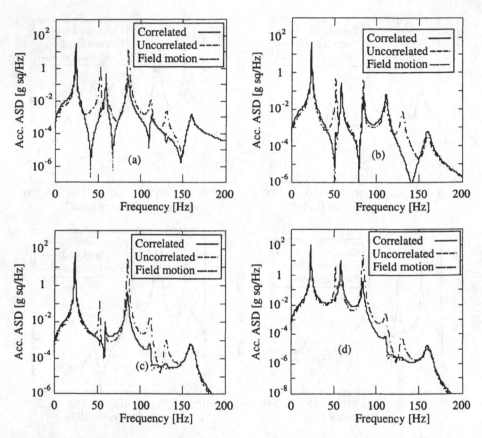

Figure 10   Test item laboratory acceleration ASDs resulting from using field interface forces as test item inputs when the external force ASD Gf$_{44}$ is not applied test item in the field. Both <u>correlated</u> (solid) and <u>uncorrelated</u> (dotted) results are presented and compared with field data (dot dashed). (a) Interface acceleration ASD Gu$_{11}$; (b) External acceleration ASD Gu$_{22}$; (c) Internal acceleration ASD Gu$_{33}$; (d) External acceleration ASD Gu$_{44}$.

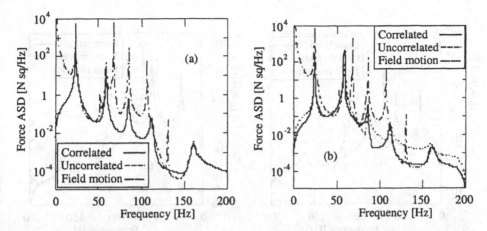

Figure 11   Laboratory Nc interface force ASDs obtained from Nc field interface accelerations by the pseudo-inverse technique. External force ASD Gf_{44} is applied in the field but is not predicted in this case. Both <u>correlated</u> (solid) and <u>uncorrelated</u> (dotted) are presented and compared with the field data (dash-dot): (a) Interface force ASD Gr_{11}; (b) Interface force ASD Gr_{33}.

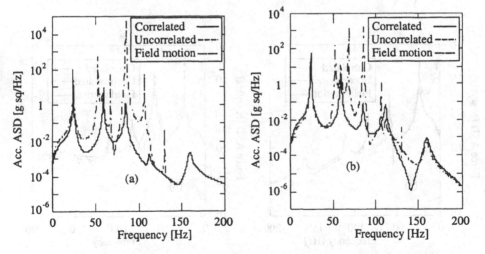

Figure 12   Test item laboratory acceleration ASDs resulting from N_c field interface forces predicted from N_c field interface accelerations. No external forces accounted for in the laboratory simulation. Both <u>correlated</u> (solid) and <u>uncorrelated</u> (dotted) are presented and compared with the field data (dash-dot): (a) Interface acceleration ASD Gu_{11}; (b) External acceleration ASD Gu_{22};

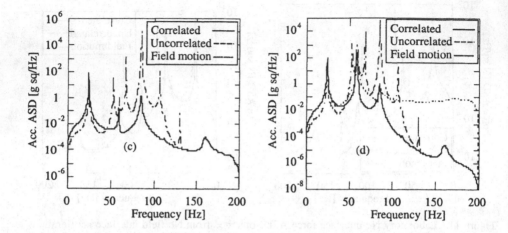

Figure 12   (*Cont.*) Test item laboratory acceleration ASDs resulting from N_c field interface forces predicted from N_c field interface accelerations. No external forces accounted for in the laboratory simulation. Both <u>correlated</u> (solid) and <u>uncorrelated</u> (dotted) are presented and compared with the field data (dash-dot): (c) Internal acceleration ASD Gu_{33}; (d) External acceleration ASD Gu_{44}.

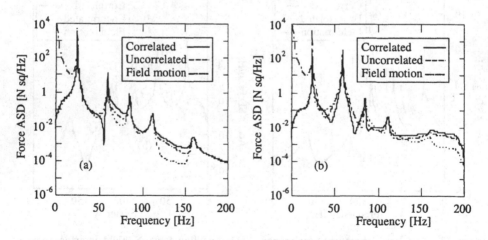

Figure 13   Laboratory N_c interface force ASDs obtained from N_t field interface accelerations by the pseudo-inverse technique. External force ASD Gf_{44} is applied in the field but is not predicted in this case. <u>Correlated</u> (solid), <u>uncorrelated</u> (dotted) and field data (dash-dot) are compared; (a) Interface force ASD Gr_{11}; (b) Interface force ASD Gr_{33}.

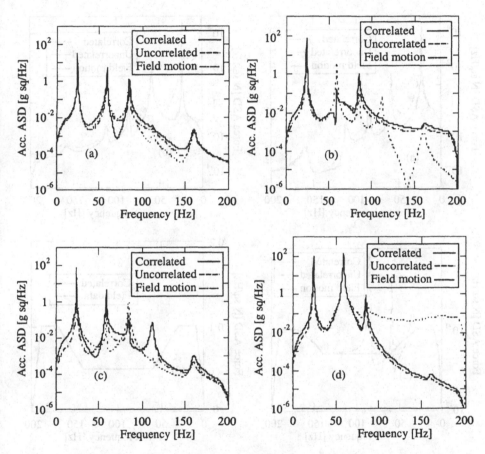

Figure 14 Test item laboratory acceleration ASDs resulting from Nc laboratory interface forces predicted from N_f field accelerations. No external forces accounted for in the laboratory simulation. Correlated (solid), uncorrelated (dotted), and field data (dash-dot) are compared: (a) Interface acceleration ASD Gu_{11}; (b) External acceleration ASD Gu_{22}; (c) Internal acceleration ASD Gu_{33}; (d) External acceleration ASD Gu_{44}.

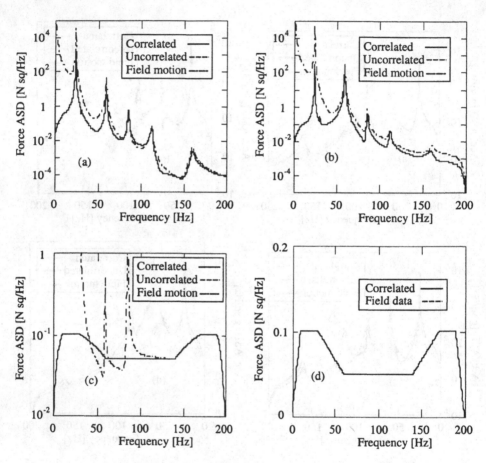

Figure 15   Laboratory interface and external force ASDs predicted from N_t field accelerations by pseudo-inverse technique. <u>Correlated</u> (solid), <u>uncorrelated</u> (dotted), and field data (dash dot) are compared: (a) Interface force ASD Gr_{11}; (b) Interface force ASD Gr_{33}; (c) External force ASD Gr_{44}; (d) External force ASD Gr_{44} in linear scale.

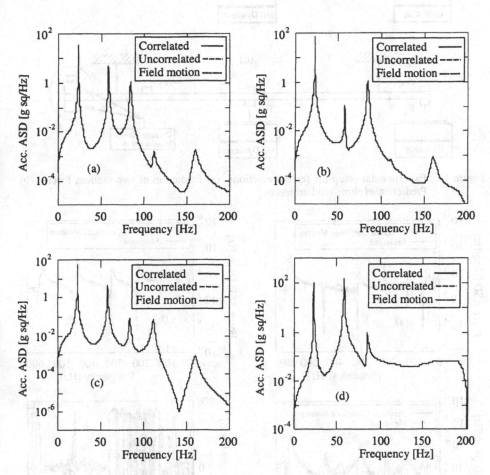

Figure 16   Test item laboratory acceleration ASDs resulting from N_t forces predicted from N_t field accelerations. External force effects are accounted for. <u>Correlated</u> (solid), <u>uncorrelated</u> (dotted) and field data (dash-dot) are compared: (a) Interface acceleration ASD Gu_{11}; (b) External acceleration ASD Gu_{22}; (c) Internal acceleration ASD Gu_{33}; (d) External acceleration ASD Gu_{44}.

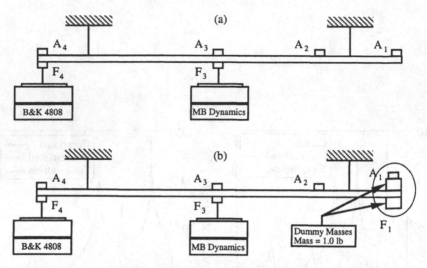

Figure 17 Experimental setup for force predictions: (a) Prediction of two random forces; (b) Prediction of three random forces.

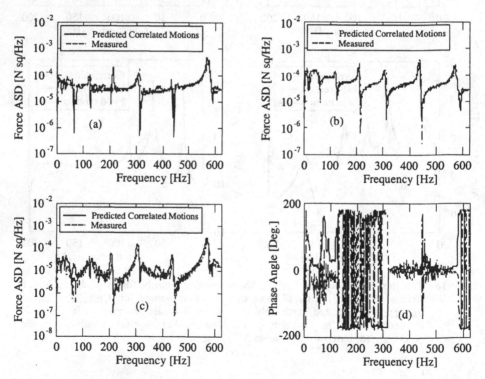

Figure 18 Predicted force ASDs and CSD for case (a) using CSDs (*correlated* motions). *Predicted forces from pseudo-inverse technique (solid line) are compared with measured forces (dash dot)*: (a) Force ASD Gf_{33}; (b) Force ASD Gf_{44}; (c) Magnitude of force CSD Gf_{34}; (d) Phase angle for force CSD Gf_{34}.

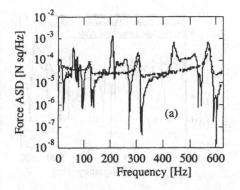
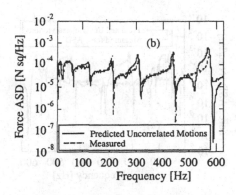

Figure 19 Predicted force ASDs for case (a) *neglecting* CSDs (***uncorrelated** motions*). *Predicted forces from pseudo-invers technique (solid) compared to measured forces (dash dot)*: (a) Force ASD Gf_{33}; (b) Force ASD Gf_{44}.

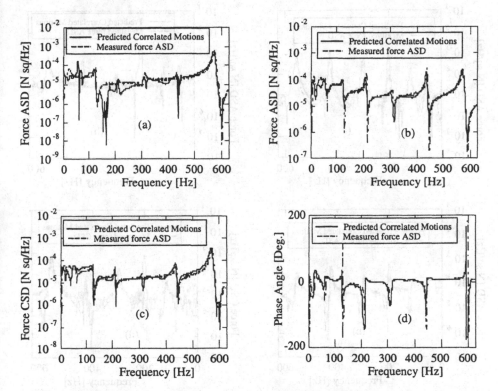

Figure 20 Predicted force ASDs and CSD for case (b) *including* CSDs (***correlated** motions*). *Predicted forces from pseudo-inverse technique (solid) compared to measured forces (dash dot)*: (a) Force ASD Gf_{33}; (b) Force ASD Gf_{44}; (c) Magnitude of force CSD Gf_{34}; (d) Phase angle of force CSD Gf_{34}.

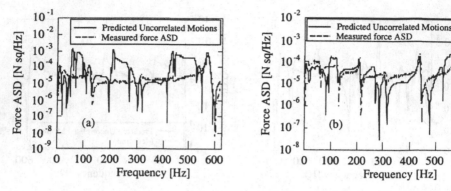

Figure 21   Predicted force ASDs and CSD for case (b) *neglecting* CSDs (**_uncorrelated_ _motions_**). *Predicted forces from pseudo-inverse technique (solid) compared to measured forces (dash dot)*: (a) Force ASD Gf_{33}; (b) Force ASD Gf_{44}.

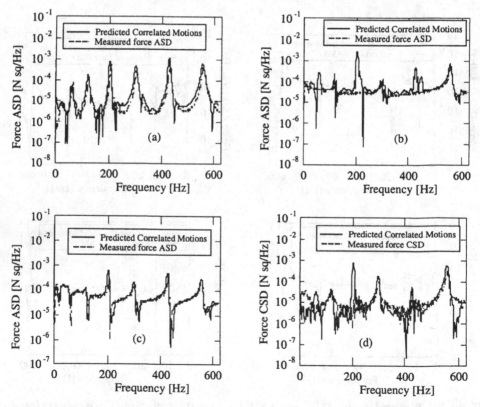

Figure 22   Predicted force ASDs and CSD for case (c) *including* CSD (**_correlated_ _motions_**). *Predicted forces from pseudo-inverse technique (solid) compared to measured forces (dash dot)*: (a) Inertia force ASD Gf_{11}; (b) Force ASD Gf_{33}; (c) Force ASD Gf_{44}; (d) Magitude of force CSD Gf_{34};

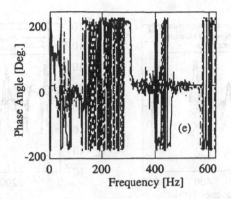

Figure 22 (*Cont.*) Predicted force ASDs and CSD for case (c): (e) Phase of force CSD Gf$_{34}$.

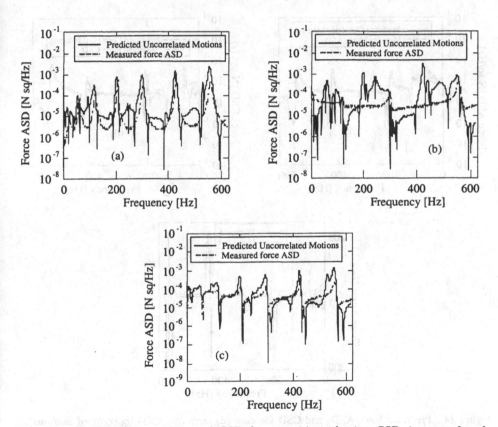

Figure 23 Predicted force ASDs and CSD for case (c) *neglecting* CSD (*<u>uncorrelated</u> motions*). *Predicted forces from pseudo-inverse technique (solid) compared to measured forces (dash dot)*: (a) Inertia force ASD Gf$_{11}$; (b) Force ASD Gf$_{33}$; (c) Force ASD Gf$_{44}$.

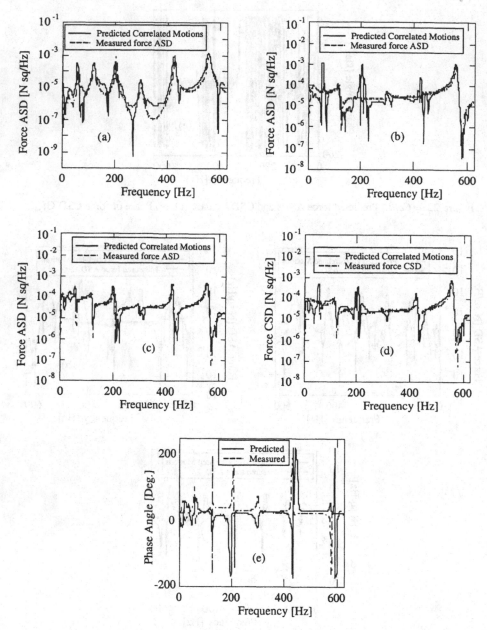

Figure 24   Predicted force ASDs and CSD for case (d) *including* CSD (***correlated*** *motions*). *Predicted forces from pseudo-inverse technique (solid) compared to measured forces (dash dot)*: (a) Inertia force ASD Gf_{11}; (b) Force ASD Gf_{33}; (c) Force ASD Gf_{44}; (d) Magnitude of force CSD Gf_{34}; (e) Phase of force CSD Gf_{34}.

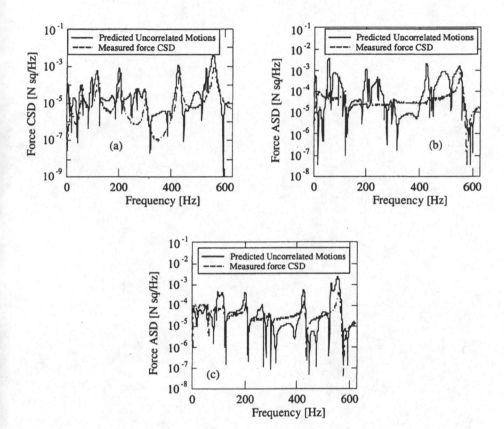

Figure 25   Predicted force ASDs and CSD for case (d) neglecting CSD (**uncorrelated motions**). Predicted forces from pseudo-inverse technique (solid) compared to measured forces (dash dot): (a) Inertia force ASD Gf_{11}; (b) Force ASD Gf_{33}; (c) Force ASD Gf_{44}.

RULES FOR THE EXCHANGE AND ANALYSIS OF DYNAMIC INFORMATION

Part V: Q-Transmissibility Matrix vs. Single Point Transmissibility in Test Environments

P. S. VAROTO
Dept de Engenharia Mecanica
Escola De Engenharia De São Carlos,
USP
São Carlos – SP – 13560-250, Brasil

K. G. McCONNELL
Dept. Aerospace Engineering and
Engineering Mechanics
Iowa State University
Ames, IA 50011   USA

Abstract

This paper discusses motion transmissibility concepts and their application to test environments. *When a **test item** is **attached** to a **vehicle** at a **single point** and field external force effects are negligible, test item accelerations can be predicted by using the acceleration transmissibility frequency response functions (FRF) and the test item's single point field interface motion is used as the input motion. When the **test item** has **multiple interface points** and field external force effects are negligible, the motion transmissibility concept is extended by defining a transformation such that the multiple field interface motions can be used as test item inputs in the laboratory.* This transformation is defined by the Q-transmissibility matrix that is obtained from the test item driving and transfer point acceleration FRFs and reduces to the standard single point transmissibility FRF for the case of a single interface point. The Q-transmissibility matrix approach is employed to numerically simulated data to predict the test item external motions and it is shown that the usual laboratory setup employing a single vibration exciter and a rigid test fixture leads to incorrect motion predictions and that multiple vibration exciters must be used to simulate field data. Experimental results indicate that: (i) the Q-transmissibility matrix transformation is feasible when dealing with actual data as long as the solution for the test item motions is carried out in a least squares sense since the test item interface FRF matrix may present rank deficiency problems in the solution process, and (ii) Curve fitted acceleration FRFs can be used to reduce experimental noise effects but the quality of the resulting motions is very sensitive to curve fitting errors, especially in the vicinity of natural frequency peaks and anti-resonance valleys.

Nomenclature

A_{pq}	Accelerance FRF	C_q	q^{th} test item and vehicle
Cn_1, Cn_3	Connectors		damping coefficients
$[C]$	Damping matrix		

J.M.M. Silva and N.M.M. Maia (eds.), Modal Analysis and Testing, 179–208.

K_q	q^{th} test item and vehicle spring constants	$\{U_e\}$	Laboratory external acceleration
$\{F\}$	Generic input force	V_{pq}	Vehicle accelerance FRF
$\{F_c\}$	Interface force	$x = x(t)$	Relative time displacement
$\{F_e\}$	External force	x_0	Displacement amplitude
$[Gf_{cc}]$	Interface force spectral density	$x^t = x^t(t)$	Absolute time displacement
		$x_b = x_b(t)$	Input base displacement
$[Gx_c]$	Interface Acceleration Spectral Density	$\{f\}_{eff}$	Effective excitation force
		$\{X\}$	Test item field acceleration
$[Gx_e]$	External Acceleration spectral density	$\{X_c\}$	Test item field interface acceleration
$j = \sqrt{-1}$	Imaginary number	$\{X_e\}$	Test item field external acceleration
$[K]$	Stiffness matrix		
$[M]$	Mass matrix	$\{Y_c\}_e$	Bare Vehicle interface acceleration
M_q	q^{th} test item and vehicle masses	Y_r	r^{th} modal amplitude
N_t	Degrees of freedom for test item	$y_r = y_r(t)$	r^{th} modal solution
N_v	Degrees of freedom for vehicle	**Greek**	
N_c	Number of connectors	$\{\delta\}$	Displacement influence coefficient
$[Q]$	Q-matrix		
Q_{pq}	Entry of Q-matrix	Δf	Frequency resolution in Hz
T_{pq}	Test item acceleration FRF	ω_r	r^{th} modal loading factor
$[T]$	Test item acceleration FRF matrix	$[\Phi]$	Undamped mode shape matrix
$[T_{cc}]$	Connector acceleration FRF matrix	$\{\phi\}$	r^{th} mode shape
$[T_{ce}]$	Connector-external accelerance FRF matrix	$\Gamma(\omega)$	Acceleration transmissibility FRF
$[TV]$	Test item-Vehicle combined matrix	ω	Frequency in rad/s
		ω_r	r^{th} natural frequency in rad/s
$\{U\}$	Test item laboratory acceleration	τ	Analysis period in seconds
$\{U_c\}$	Laboratory interface acceleration	ζ_r	r^{th} modal damping ratio

1. Introduction

This paper discusses motion transmissibility concepts and their application to predict test item external motions from knowledge of interface motions when the test item has multiple interface points. The single point transmissibility FRF [1] is a useful tool in predicting test item external motions when the test item input is given in terms of a base motion that is applied to a *single* point and no other external forces are applied to the test item. Most field and laboratory test scenarios involve the attachment of the test item to a second structure at several interface points. The use of single point transmissibility concepts is inappropriate for predicting test item motions in this case since multiple input motions are *simultaneously* applied to the test item interface points when in the field environment.

In this paper, the single point transmissibility FRF concept is extended to the case where the inputs to the test item are given in terms of a set of multiple interface motions that are *simultaneously* applied to the test item. An expression relating the test item unknown external accelerations to the input interface acceleration is obtained by writing the test item frequency domain equations of motion in matrix partition form where interface motions and forces are differentiated from external forces and motions. *It is assumed that no external forces are applied to the test item in both the field and laboratory environments*. The resulting expression for the external motions is given as the product of the interface motions vector and a transformation matrix that is called Q-transmissibility matrix (Q-T matrix). The Q-T matrix is obtained from the traditional test item driving point and transfer accelerance FRFs.

The Q-T matrix approach is applied to numerically simulated data where the test item has several interface points. Two laboratory test scenarios are considered. *First*, the test item is attached to a *single* vibration exciter by a *rigid* test fixture. *Second*, multiple vibration exciters are used to predict the test item motions. The results from these numerical simulations show that, in general, incorrect test item motion occurs when single exciter simulations are used and that multiple exciters must be employed in situations where the test item has multiple interface points.

An experimental study is performed in order to investigate the feasibility of the Q-T matrix approach when dealing with actual data. A free-free beam is the test item in this case. Multiple vibration exciters are used to excite the beam at two locations using random excitations. The output response auto spectral densities (ASDs) and cross spectral densities (CSDs) of the beam are measured at four locations. The Q-T matrix approach is used to estimate the beam external acceleration ASDs from knowledge of the measured interface data.

The experimental results indicate that the Q-T matrix approach can be successfully applied to real data as long as the solution is carried out in a least squares sense since the test test item interface FRF matrix that is used to form the Q-T matrix can be rank deficient, a situation that can seriously affect the resulting motions. If experimental noise is a major concern, curve fitted FRFs can be used in obtaining the Q-T matrix, but the curve fitting should be carefully performed since the solution process appears to be sensitive to curve fitting errors.

2. The single point transmissibility FRF

Figure 1a shows a N DOF test item attached to the vibration exciter's table and subjected to a base input motion $x_b = x_b(t)$. The test item equations of motion can be written using matrix notation as

$$[M]\left\{\ddot{x}^t\right\} + [C]\left\{\dot{x}\right\} + [K]\left\{x\right\} = \{0\} \tag{1}$$

where [M], [C], and [K] represent the test item N x N mass, damping, and stiffness matrices, respectively. The N x 1 vector $\{x^t\} = \{x^t(t)\}$ contains the test item absolute displacements with respect to a fixed reference system. The N x 1 vector $\{x\} = \{x(t)\}$ corresponds to the test item's relative displacement with respect to the exciter armature. The test item absolute and relative displacements are related to each other by the following equation [2]

$$\{x^t\} = \{x\} + \{\delta\} x_b \tag{2}$$

where $\{\delta\}$ is a constant N x 1 vector and is used to express the fact that a unit static displacement of the exciter's armature in the direction of x_b produces a unit static displacement of all test item DOFs in the same direction. Thus, in a sense, $\{\delta\}$ can be viewed as an influence coefficient vector that identifies all test item DOFs that are in the direction of the input base motion x_b. Thus, vector $\{\delta\}$ is composed of "ones" at the DOFs coinciding with the direction of the input motion and "zeroes" for the remaining orthogonal DOFs.

Substitution of Eq. 2 in Eq. 1 results in the following alternative form for the test item equations of motion in terms of the relative displacement $\{x\}$

$$[M] \{\ddot{x}\} + [C] \{\dot{x}\} + [K] \{x\} = \{f\}_{eff} \tag{3}$$

where $\{f\}_{eff}$ is the effective base excitation inertia force that is given by

$$\{f\}_{eff} = \{f(t)\}_{eff} = - [M] \{\delta\} \ddot{x}_b \tag{4}$$

where the minus sign on the right hand side of Eq. 4 expresses the fact that the inertia excitation force opposes the input base acceleration $\ddot{x}_b$.

Solution of Eq. 4 is carried out by employing basic modal analysis principles [2]. Assuming that the test item of Fig. 1a is proportionally damped, pre-multiplying both sides of Eq. 4 by the transpose of the undamped N x N mode shapes matrix [Φ], and using the following coordinate transformation relationship

$$\{x\} = [\Phi] \{y\} \tag{5}$$

we obtained a set of N uncoupled equations of motion in modal coordinate variable $\{y\}$.

The equation of motion for the rth coordinate is written as

$$\ddot{y}_r + 2\zeta_r\omega_r \dot{y}_r + \omega_r^2 y_r = - \eta_r \ddot{x}_b \tag{6}$$

where ω_r and ζ_r correspond to the rth natural frequency and the rth modal damping ratio, respectively. The ω_r's are obtained from the rth modal mass m_r and modal stiffness k_r parameters. The constant η_r on the right hand side of Eq. 6 is the *base motion modal participation factor* of the rth mode shape and is defined as

$$\eta_r = \frac{1}{m_r} \{\phi\}_r^T [M] \{\delta\} \tag{7}$$

Note that the base motion modal participation factor depends on $\{\delta\}$ so that $\{\delta\}$ dictates which test item mode shapes will be excited by the input base motion x_b.

Consider that the input base motion x_b and the solution for the motion y_r vary harmonically according to

$$x_b = x_o e^{j\omega t} \tag{8}$$

$$y_r = Y_r e^{j\omega t} \tag{9}$$

where x_0 and Y_r correspond to the amplitudes of the input motion and the solution of the rth modal equation, respectively. Substitution of Eqs. 8 and 9 into Eq. 6 leads to the standard result of

$$Y_r = \frac{\eta_r \omega^2 x_o}{\omega_r^2 - \omega^2 + j \, 2\zeta_r \omega_r \omega} \tag{10}$$

which, when substituted back into Eq. 9, gives the time solution for the rth modal coordinate. Once all equations of motion are solved in the modal domain, the solution of Eq. 3 in the physical domain is obtained by using Eq. 5, which can be conveniently rewritten as [2]

$$\{x\} = \sum_{r=1}^{N} \{\phi\}_r \, y_r \tag{11}$$

Thus, substitution of Eq. 9 along with Eq. 10 into Eq. 11 leads to the following time domain solution for the *relative displacement* $\{x\}$:

$$\{x\} = \omega^2 \sum_{r=1}^{N} \frac{\{\phi\}_r \{\phi\}_r^T [M] \{\delta\}}{m_r \left(\omega_r^2 - \omega^2 + j \, 2\zeta_r \omega_r \omega\right)} x_o e^{j\omega t} \tag{12}$$

The test item absolute motion $\{x^t\}$ is obtained from combining Eqs. 2, 8, and 12 as

$$\{x^t\} = \{\Gamma(\omega)\} \, x_o e^{j\omega t} \tag{13}$$

where $\{\Gamma(\omega)\}$ represents the N x 1 *absolute transmissibility FRF* vector for input motion at a *single* coordinate and is given by

$$\{\Gamma(\omega)\} = \omega^2 \sum_{r=1}^{N} \frac{\{\phi\}_r \{\phi\}_r^T [M] \{\delta\}}{m_r \left(\omega_r^2 - \omega^2 + j \, 2\zeta_r \omega_r \omega\right)} + \{\delta\} \tag{14}$$

The pth element of vector $\{\Gamma(\omega)\}$ gives the motion (displacement, velocity, or acceleration) of the pth coordinate point that is due to a single input motion (displacement, velocity, or acceleration) at the rth coordinate point with all remaining inputs being identically zero. Mathematically

184

$$\Gamma_{pr}(\omega) = \frac{X_p^t}{X_r}(\omega) \tag{15}$$

where $X_p^t = X_p^t(\omega)$ and $X_r = X_r(\omega)$ are the Fourier transforms of the test item's p^{th} absolute displacement x_p^t and the single *absolute* input motion x_r, respectively. In the case of Fig. 1a, $X_r = X_b$. Once the test item transmissibility FRF vector is known, it can be used to determine the motion at all test item DOFs that are due to the exciter input motion as long as X_b represents the *only* test item input that is applied at a single point.

3. Multiple input motions: the Q - T matrix approach

The transmissibility FRF concept discussed in the previous section represents a useful tool for determining external motions when the test item is subjected to an input motion at a single interface point. However, in most cases, the test item is attached to the vehicle in the field or to the vibration exciter in the laboratory at several interface points, as shown in Fig. 1b. In these cases, the transmissibility input-output relationship defined by Eq. 15 is not applicable.

This section describes a procedure that can be used to predict the test item responses at external points when the test item excitation consists of multiple interface motions that are simultaneously applied at its interface points. This procedure is called the Q-T matrix since we are dealing with multiple interface inputs and multiple external motion outputs. The Q-T matrix is defined by

$$\{X_e\} = [Q]\{X_c\} \tag{16}$$

where the N_e x 1 vector $\{X_e\}$ represents the external test item motions, the N_c x 1 vector $\{Xc\}$ contains the measured interface motions, and the N_e x N_c matrix [Q] is the Q-T matrix that defines the linear transformation between measured interface motions and the external motions. Equation 16 is similar to the well known force input-output motion linear transformation

$$\{X\} = [T]\{F\} \tag{17}$$

where the input forces vector $\{F\}$ is transformed into the motion vector $\{X\}$ by the test item FRF matrix [T]. In the following discussion, we shall *assume that [T] represents accelerance.*

3.1. RELATIONSHIP BETWEEN FRF'S AND Q-T MATRIX

The entries of the Q-T matrix in Eq. 16 are obtained from the test item driving point and transfer point accelerance FRFs by writing Eq. 17 as a partitioned matrix that separates *connector* and *external* points according to the scheme used in Ref. [3]

$$\begin{Bmatrix} \{X_c\} \\ \{X_e\} \end{Bmatrix} = \begin{bmatrix} [T_{cc}] & [T_{ce}] \\ [T_{ec}] & [T_{ec}] \end{bmatrix} \begin{Bmatrix} \{F_c\} \\ \{F_e\} \end{Bmatrix} \tag{18}$$

Equations 17 and 18 represent the same linear transformation except that Eq. 18 is written in a partitioned form where *interface forces and motions are differentiated* from *external forces and motions*. The N_c x 1 vector $\{X_c\}$ represents the test item interface (connector) accelerations while the N_e x 1 vector $\{X_e\}$ contains the test item external accelerations. Similarly, the excitation force vector $\{F\}$ on the right hand side of Eq. 17 is broken in two components, the N_c x 1 vector of interface forces $\{F_c\}$ that occur at the test item's N_c connector points, and the N_c x 1 vector of external forces $\{F_e\}$. *Interface forces are assumed to result from coupling effects only*, i.e., the force interactions that occur when the test item is attached to a second structure. The partition of input force and output acceleration vectors in terms of interface and external variable requires that the test item accelerance FRF N_t x N_t matrix [T] be broken in four submatrices. The N_c x N_c FRF matrix $[T_{cc}]$ contains the test item accelerance FRFs relating interface points only while the N_e x N_e matrix $[T_{ee}]$ contains the test item accelerance FRFs relating external points only. The N_c x N_e (N_e x N_c) matrix $[T_{ce}]$ ($[T_{ec}]$) contains the test item accelerance FRFs relating interface (external) to external (interface) points.

Expansion of Eq. 18 leads to the following expressions for interface and external motions

$$\{X_c\} = [T_{cc}]\{F_c\} + [T_{ce}]\{F_e\} \qquad (19)$$

$$\{X_e\} = [T_{ec}]\{F_c\} + [T_{ee}]\{F_e\} \qquad (20)$$

where it is clear that both the interface and external motions $\{X_c\}$ and $\{X_e\}$ are caused by both interface and external forces that are applied to the test item. Equations 19 and 20 can be simplified if we *assume that the test item is subjected to no external forces* so that $\{F_e\} = \{0\}$. In this case, we can obtain the interface force from Eq. 19 as

$$\{F_c\} = [T_{cc}]^{-1}\{X_c\} \qquad (21)$$

Substitution of Eq. 21 into Eq. 20 with the assumption that $\{F_e\} = \{0\}$ leads to the following equation relating the unknown external motions $\{X_e\}$ to the measured interface motions $\{X_c\}$

$$\{X_e\} = [T_{ec}][T_{cc}]^{-1}\{X_c\} \qquad (22)$$

which is the same as equation 16 with

$$[Q] = [T_{ec}][T_{cc}]^{-1} \qquad (23)$$

Thus, in principle, Eq. 22 allows us to obtain external test item accelerations by knowing the interface test item accelerations and the test item interface $[T_{cc}]$ and interface-external $[T_{ce}]$ accelerance FRF characteristics.

3.2. LIMITING CASE - SINGLE INPUT MOTION

If the test item has a single interface point ($N_c = 1$) that is subjected to an input motion X_r, then Eq. 22 reduces to

$$\{X_e\} = \frac{X_r}{T_{rr}} \{T_{er}\} \tag{24}$$

The external motion at the pth coordinate point is seen to be given by

$$X_p = \frac{T_{pr}}{T_{rr}} X_r = \Gamma_{pr} X_r \tag{25}$$

In this simple case, *the ratio of the test item external-connector FRF T_{pr} to the connector-connector FRF T_{rr} is the single point transmissibility FRF Γ_{pr}* between points p and r as given by Eqs. 14 or 15. This simple result is only valid for the case where the test item is subjected to a single input motion.

3.3. EFFECTS OF RIGID TEST FIXTURES

Now we consider the situation where $N_c > 1$ so that the test item is subjected to multiple input motions that are simultaneously applied to its interface points. In this case, the external motion at the pth coordinate point is given by

$$X_p = Q_{p1} X_1 + Q_{p2} X_2 + \cdots + Q_{pNc} X_{Nc} \tag{26}$$

where the Q_{pr}, $r = 1, ..., N_c$ are the entries of the pth row of the Q-T matrix defined by Eq. 23, and X_r, $r = 1, ..., N_c$, $r \neq p$ are the input interface accelerations. Standard laboratory testing procedures often recommend that the test item be attached to a *single* exciter through a *rigid test fixture*. In this case, the excitation motion that is generated by the exciter system is transmitted to the test item through each interface point causing all input motions to the test item to be the same, i.e., $X_1 = X_2 = ... = X_{Nc} = X_0$. Thus, Eq. 26 reduces to

$$X_p = \left(\sum_{r-1}^{Nc} Q_{pr} \right) X_o \tag{27}$$

Thus, when the same input motion is applied to all of the interface points, the test item acceleration response at the external point X_p is given by the product of the input motion X_0 and the sum of the pth row of the Q-T matrix. In this case, *the right hand side of Eq. 27 does not reduce to the transmissibility* FRF *that is experienced under field conditions*. The only exception occurs when multiple interface points all have the same field motions. The implications of using Eq. 27 in laboratory simulations will be investigated in the next section.

4. Numerical simulations - field environment

This section presents numerically simulated results using the single point transmissibility FRF and the Q-T matrix concepts.

The order of calculation for the field environment is shown in Fig. 2. *First*, the *bare* vehicle interface acceleration frequency spectra $\{Y_c\}_e$ are obtained as shown in Fig. 2a. In this case, the test item is not connected to the vehicle in the field and the vehicle interface accelerations $\{Y_c\}_e$ are due exclusively to external forces acting on the vehicle. *Second*, the combined structure is analyzed where the test item is attached to the vehicle at N_c interface points as shown in Fig. 2b. Interface forces $\{F_c\}$ and accelerations $\{X_c\}$ as well as the test item external accelerations $\{X_e\}$ are calculated as the field data in this case. *It is assumed in this paper that the test item is subjected to no external forces when connected to the vehicle in the field*. Thus, the only forces acting on the test item are the N_c interface forces that occur due to coupling effects between the structures. The effects of external forces applied to the test item were investigated in Refs. [4,5].

4.1. THEORETICAL MDOF MODELS

The model used in the simulation process are shown in Fig. 3. The test item and vehicle are each modeled by a four DOFs system that are attached to one another in the field environment through connectors Cn_1 and Cn_3 to form the combined field structure. Coordinate numbers are assigned to the test item and vehicle models such that interface points carry the same number in both structures. Physical parameters for the test item and the vehicle models are given in Table 1. The natural frequencies, modal damping ratios, modal masses, and mode shapes are obtained for the structures involved by standard modal analysis principles. The test item modal parameters are calculated by assuming that the test item is a free structure in space. The vehicle and the combined structure are grounded to the base excitation source (see Fig. 3) through the vehicle mass M_2. The resulting modal parameters for the test item, the vehicle, and the combined structure are shown in Table 2. The modal damping ratios were obtained by assuming a proportional damping distribution for all structures. Accelerance FRFs are obtained for the structures involved according to

$$A_{pq}(\omega) = -\omega^2 \sum_{r=1}^{N} \frac{\phi_{pr}\,\phi_{qr}}{m_r\left(\omega_r^2 - \omega^2 + j\, 2\zeta_r\omega_r\omega\right)} \tag{28}$$

Similarly, for input motion at a single coordinate, the acceleration transmissibility FRF given by Eq. 14 is used to calculate the vehicle motions.

Table 1: Physical Parameters For Lumped Models

Mass	Test Item (Kg)	Vehicle (Kg)
M_1	0.20	0.50
M_2	0.20	0.50
M_3	0.25	0.30
M_4	0.15	0.50
Stiffness	Test Item N/m x 10^4	Vehicle N/m x 10^4
K_1	5.00	6.00
K_2	4.00	14.00
K_3	2.00	10.00

Table 2: Modal Properties

Natural Frequency (Hz)	Test Item	Vehicle	Combined
f_1	0	30.00	23.86
f_2	52.60	86.17	58.94
f_3	85.70	164.72	85.05
f_4	131.00	–	112.00
f_5	–	–	160.21
Damping Ratio (%)	Test Item	Vehicle	Combined
ζ_1	0	0.28	0.24
ζ_2	0.50	0.81	0.56
ζ_3	0.81	1.55	0.91
ζ_4	1.23	–	1.06
ζ_5	–	–	1.54
Modal Masses	Test Item	Vehicle	Combined
m_1	0.20	0.43	0.38
m_2	0.17	0.48	0.17
m_3	0.21	0.37	0.55
m_4	0.20	–	0.21
m_5	–	–	0.34

4.2. VEHICLE EXCITATION SOURCE

The excitation source is shown in Fig. 4 and corresponds to a single acceleration frequency spectrum Y_2 that is applied to the vehicle base mass M_2. This input base acceleration frequency spectrum covers the 0 - 200 Hz frequency range, increasing linearly with a slope of 1.28×10^{-3} (m/s^2)/Hz in the 0-10 Hz frequency range and becoming constant with the magnitude of 1.28×10^{-2} m/s^2 in the remaining 10-200 Hz frequency range. A total of 800 spectral lines are used in all simulations, giving a frequency resolution of $\Delta f = 0.25$ Hz so that the input vibration level is approximately 1.0 g_{RMS} (9.81 (m/s^2) $_{RMS}$).

4.3. THE BARE VEHICLE INTERFACE ACCELERATIONS

This field simulation is shown in Fig. 2a. Since the vehicle input in the case shown in Fig. 3 is given by the single acceleration frequency spectra Y_2 that is applied to the single point of mass M_2, the bare vehicle interface acceleration $\{Y_c\}_e$ is obtained by employing the transmissibility concepts previously developed. Thus, rewriting Eq. 15 as

$$Y_p = \Gamma_{pr} Y_r \qquad (29)$$

for p = 1, 3 and $Y_r = Y_2$, and using Eq. 14 along with the vehicle modal parameters listed in Table 2 to calculate the vehicle transmissibility FRFs, we obtain the bare vehicle interface accelerations with magnitudes as shown in Fig. 5. These motions will be used as one source of inputs in the next field simulation as well as possible laboratory inputs.

4.4. THE COMBINED SYSTEM ACCELERATIONS

In this case, the test item is connected to the vehicle as shown in Fig. 2b. The test item accelerations are calculated according to Eq. 19 of Ref. [3] that is given by

$$
\left\{ \begin{array}{c} \{X_c\} \\ \{X_e\} \end{array} \right\} = \left[\begin{array}{cc} [T_{cc}][TV] & -[T_{cc}][TV] \\ [T_{ec}][TV] & -[T_{ec}][TV] \end{array} \right] \left\{ \begin{array}{c} \{Y_c\}_e \\ \{0\} \end{array} \right\}
\tag{30}
$$

where the N_c x 1 vector $\{X_c\}$ contains the interface test item accelerations, and the N_c x 1 vector $\{X_e\}$ contains the external test item accelerations. The N_c x N_c matrix $[TV] = [[T_{cc} + [V_{cc}]]^{-1}$ is the test item-vehicle combined matrix [3] and is given by the inverse of the sum of the test item and vehicle interface accelerance FRF matrices $[T_{cc}]$ and $[V_{cc}]$, respectively. Note, from Eq. 30, it is clear that *in the absence of external forces, the test item motions depend only on the bare vehicle interface accelerations and on the test item and vehicle accelerance FRF characteristics*. Then, using the bare vehicle interface accelerations shown in Fig. 5 and the test item and vehicle accelerances calculated according to Eq. 28, the test item acceleration frequency spectra magnitudes are obtained as shown in Fig. 6. Figures 6a shows the test item interface accelerations at connector points 1 and 3 while Fig. 6b shows the test item external accelerations at points 2 and 4. The interface accelerations of Fig. 6a will be used to illustrate the use of the Q-T matrix in the laboratory simulations that will be discussed in the next section while the external accelerations in Fig. 6b are used as references for success in the laboratory simulations.

5. Numerical simulations - the laboratory environment

Figure 7 shows two laboratory test scenarios that can be employed in order to predict field data. *First*, a test scenario usually recommended by the MIL-STD 810D [6] is shown in Figure 7a. In this case, the test item is attached to a single vibration exciter through a *rigid* test fixture. When available, MIL-STD 810D recommends that enveloped field data be used to define the test item input that is to be generated by the vibration exciter and transmitted to the multiple test item interface points by the *rigid* fixture. *Second*, a test scenario employing multiple vibration exciters is considered in Fig. 7b. In this test scenario, each test item interface point is attached to a single vibration exciter, and each exciter is independently controlled such that field data is matched at all interface points. *No external exciters are required in this case since no external forces occur in the field.*

The choice of suitable test item inputs is a major issue when attempting to predict field data by either using the single exciter test scenario of Fig. 7a or the multiple exciter scenario of Fig. 7b. A previous paper [7] showed that when the test item is not subjected to field external forces, interface forces and motions constitute suitable test item inputs in laboratory simulations. When interface motions are used to define the test item inputs, a set of interface forces is obtained by first solving the inverse problem through the pseudo-inverse technique [4,5], and then using this set of interface forces to drive the test item.

In this paper, the test item interface motions obtained in the field simulation are used as test item inputs by employing the Q-T matrix approach given in Eqs. 22 and 23. *In principle, the prediction of input forces is not required when employing Eq. 22 since the measured interface forces are used indirectly as test item inputs in the process of developing the Q-T matrix approach.*

5.1. TEST ITEM ATTACHED TO A SINGLE VIBRATION EXCITER THROUGH A *RIGID* TEST FIXTURE

This test scenario is shown in Fig. 7a. The test item is attached to the vibration exciter table through a rigid test fixture. The test item and vehicle models used in the field simulations and shown in Fig. 3 will be used in this laboratory simulation. The vehicle becomes the test fixture in this case when new spring constants are chosen so that a rigid test fixture is obtained as recommended by the MIL-STD 810D. The test fixture spring constants are $K_1 = 9.10^6$ N/m, $K_2 = 14.10^7$ N/m, and $K_3 = 10^8$ N/m. The test fixture masses are the same as the vehicle masses shown in Table 1. The resulting fixture natural frequencies are respectively $f_1 = 410$ Hz, $f_2 = 2,464$ Hz, and $f_3 = 5,161$ Hz. These natural frequencies are located well above the 0 - 200 Hz test frequency range. The test item is assumed to be the same in both environments. The test item and the test fixture are connected at interface points 1 and 3 as in the field. The excitation signal is generated by a single vibration exciter that is attached to the test fixture mass M_2.

Choice of the test item input in this case constitutes a problem since a single vibration exciter is employed in the simulation while there are two interface accelerations in the actual field environment as shown in Fig. 6a. Thus, a natural question that arises in simulations where a single vibration exciter is used is: **Which field motion should be used as the test item input**? Two cases are considered here:

- *Case 1:* The vibration exciter is controlled to reproduce either of the interface accelerations shown in Fig. 6a.

- *Case 2:* The vibration exciter is controlled to generate an input acceleration that is a combination of the measured field interface accelerations.

Note that the input accelerations applied at the test item interface points 1 and 3 will be the same due to the rigid test fixture. Thus, Eq. 27 must be employed in order to obtain the resulting test item external accelerations.

5.1.1. *Case 1: Matching field acceleration X_1*
In this case, the acceleration frequency spectrum at the interface point 1 X_1 is used to drive the test fixture and test item. Equation 27 is rewritten as

$$U_p = \left(\sum_{r=1}^{Nc} Q_{pr} \right) X_1 \tag{31}$$

where U_p (for $p = 2$ and 4) is p^{th} external test item acceleration. The Q-T matrix is obtained from Eq. 23. Figures 8a and 8b show the resulting test item external acceleration magnitudes U_2 and U_4 obtained in this test scenario as well as the field acceleration magnitudes X_2 and X_4 that are used for comparison purposes. Both external motions are incorrect since the *same* interface input motion X_1 is applied to *both* test item interface points 1 and 3 while in the field the actual interface motions are different due to the flexibility characteristics of the vehicle structure. It is clear that both over-testing and under-testing occur in this test arrangement.

5.1.2. *Case 2: Matching a combination of X_1 and X_3*

In this test scenario, a single test item input is obtained by using the average of the field interface accelerations X_1 and X_3 such that

$$X_o = \frac{X_1 + X_3}{2} \tag{32}$$

This acceleration is transmitted to the test item through the test fixture so that Eq. 27 is used again. The resulting external acceleration frequency spectra U_2 and U_4 are plotted in Figs. 9a and 9b along with the corresponding field responses. The laboratory external acceleration U_2 is shown in Fig. 9a where good agreement with field data occurs for frequency components in the 0 - 70 Hz test frequency range, a poor agreement between field and laboratory data occurs in the 70 - 105 Hz frequency range, and a nearly perfect agreement occurs between U_2 and X_2 for the remaining spectral lines. The external acceleration U_4 has good agreement with X_4 in the 0 - 30 Hz frequency range while there is poor agreement in the remaining 30 - 200 Hz frequency range. It is clear that both over-testing and under-testing occur in this testing arrangement as well.

5.2. TEST ITEM ATTACHED TO N_c VIBRATION EXCITERS

This test scenario is shown in Fig. 7b. In this case, each of the N_c vibration exciters that are connected to the test item interface points is controlled so that the test item is driven by a set of accelerations that match the field interface accelerations in magnitude and phase and are simultaneously applied to the test item interface points. This means that each of the N_c interface points will be subjected to a different input acceleration. Then, the test item external acceleration at the pth location is calculated by Eq. 26 where X_1, X_2, ..., X_{Nc} represent the N_c input accelerations that are generated by the vibration exciters and Q_{p1}, Q_{p2}, ..., Q_{pNc} is the pth row of the Q-T matrix that is obtained from Eq. 23.

For the test item in Fig. 3, the external motions are obtained by expanding Eq. 22 as

$$\begin{Bmatrix} U_2 \\ U_4 \end{Bmatrix} = \begin{bmatrix} T_{21} & T_{23} \\ T_{41} & T_{43} \end{bmatrix} \begin{bmatrix} T_{11} & T_{13} \\ T_{31} & T_{33} \end{bmatrix}^{-1} \begin{Bmatrix} X_1 \\ X_3 \end{Bmatrix} \tag{33}$$

In this simple case, Eq. 33 can be easily expanded to

$$U_2 = \frac{1}{\Delta_{cc}} \left[\left(T_{11} T_{23} - T_{33} T_{21} \right) X_1 - \left(T_{13} T_{21} - T_{31} T_{23} \right) X_3 \right] \tag{34}$$

$$U_4 = \frac{1}{\Delta_{cc}} \left[\left(T_{11} T_{43} - T_{33} T_{41} \right) X_1 - \left(T_{13} T_{41} - T_{31} T_{43} \right) X_3 \right] \tag{35}$$

where Δ_{cc} is the determinant of the test item interface accelerance matrix ($\Delta_{cc} = T_{11}T_{33} - T_{13}T_{31}$). From Eqs. 34 and 35, we can obtain the Q-T matrix for the test item

$$[Q] = \frac{1}{\Delta_{cc}} \begin{bmatrix} \left(T_{11}T_{23} - T_{33}T_{21}\right) & -\left(T_{13}T_{21} - T_{31}T_{23}\right) \\ \left(T_{11}T_{43} - T_{33}T_{41}\right) & -\left(T_{13}T_{41} - T_{31}T_{43}\right) \end{bmatrix} \tag{36}$$

and we notice that each entry on the Q-T matrix is obtained by multiplying an interface accelerance FRF by an external-interface accelerance FRF and dividing the result by Δ_{cc}.

The test item external accelerations obtained from Eqs. 34 and 35 are shown in Fig. 10. An excellent agreement is seen to occur between the accelerations obtained through the Q-T matrix approach (U_2 and U_4) and the corresponding field data (X_2 and X_4).

6. Summary of numerical results

From the numerical results presented in this section, we can summarize the following:

- *Both* simulations where the test item is attached to a *rigid* test fixture and a *single* vibration exciter is employed to excite the test item gave inaccurate results for the test item external accelerations. Equation 27 was used to predict the test item external motions U_2 and U_4 by using two different combinations of field interface accelerations as test item inputs.

- When the test item is attached to multiple vibration exciters, the Q-T matrix approach gave good estimates of external accelerations for both U_2 and U_4. In this case, Eq. 26 is used to calculate each external motion where each entry of the Q-T matrix is multiplied by a different interface acceleration. The resulting Q-T matrix in this example is given by Eq. 36 where we see that it is a function of Δ_{cc}, i.e., the *determinant of the test item interface accelerance FRF matrix*.

7. Experimental study

This section shows results of an experimental analysis performed on a free free beam in order to investigate the feasibility of the Q-T matrix approach while using experimental data.

Table 3: Characteristics of sensors used in tests

Sensor Type	Position on Beam	Sensor Model	Charge or Voltage Sensitivity	
Force	F3	PCB 208A03	50.84	mv/N
Force	F4	PCB 208A03	50.84	mv/N
Accel.	A1	PCB 302A	10.9	mv/g
Accel.	A2	Endveco 2222C	1.70	pc/g
Accel.	A3	PCB 302A	10.18	mv/g
Accel.	A4	PCB 302A02	10.04	mv/g

7.1.   THE EXPERIMENTAL SETUP

The experimental setup used in the experimental test is shown in Fig. 11. A cold rolled steel beam (92 x 1.25 x 1.0 in) represents the test item in this case. The beam is suspended by flexible cords in order to simulate free free boundary conditions. Two vibration exciters are used to drive the beam at two locations. A MB (Model Modal 50) exciter is used to drive the beam mid point at point 3 while a B&K (Model 4808) exciter is used to excite the beam end at point 4.

Random excitation is used with both vibration exciters to drive the beam. In this case, the beam output is given in terms of acceleration auto spectral densities (ASDs) [5] that are measured by piezoelectric accelerometers at four locations (A_1, A_2, A_3, and A_4) as shown. The corresponding cross spectral densities CSDs between all measurement locations are also measured since the resulting motions are *correlated*. This correlation must be accounted for in the calculations through the CSDs. The acceleration ASDs from the locations where the input excitation signals are applied (points A_3, and A_4) will be referred to as the interface accelerations while the acceleration ASDs at A_1, and A_2 will correspond to the external accelerations.

Our goal in performing this experiment is to predict the external acceleration ASDs and CSDs at points 1 and 2 from the measured case, the external acceleration ASDs are measured in order to provide a comparison basis for the resulting motions predicted through the results of the previous sections. Table 3 shows the characteristics of all transducers used in the experimental analysis.

The data acquisition is performed by a 486-PC equipped with the Data Physics Dp420© data acquisition and signal processing system. The beam FRFs relating all input and output points are measured in the 0 - 625 Hz frequency range. A total of 1000 spectral lines are used so that the frequency resolution is $\Delta f = 0.6357$ Hz. Hanning windows are used with input and output channels in all measurements. Table 4 contains the first 7 bending natural frequencies of the beam in the 0 - 625 Hz frequency range that were obtained by curve fitting the measured accelerance FRFs using ICATS© modal analysis software [11]. The natural frequencies obtained from the curve fitting process are compared in Table 4 with the corresponding natural frequencies predicted by a continuous model [1]. Figure 12 shows the measured and curve fitted interface accelerance FRFs. The driving point accelerances T_{33} and T_{44} are shown in Figs. 12a and 12b while the transfer accelerance T_{34} is shown in Fig. 12c.

Table 4: Natural frequencies for free free beam

fn	Theoretical Hz	Experimental Hz
1	24.30	25.30
2	67.60	67.98
3	132.40	131.22
4	218.00	216.60
5	327.00	322.53
6	457.00	450.60
7	608.00	597.62

194

7.2. THE Q-T MATRIX EXPRESSIONS OF RANDOM SIGNALS

In order to predict the beam external acceleration ASDs at locations 1 and 2 and the corresponding external acceleration CSD between 1 and 2, Eq. 22 must be reformulated for random signals since it was originally derived in terms of deterministic frequency spectra. In order to obtain the corresponding Q-T matrix expression for random signals, we use Eq. 3 of Ref. [5] that is given as

$$[Gx_e] = \lim_{r \to \infty} \frac{2}{\tau} \epsilon \left(\{X\}_e^* \{X\}_e^T \right) \tag{37}$$

where the N_e x N_e spectral density matrix $[Gx_e]$ contains the external acceleration ASDs in the main diagonal and the corresponding CSDs in the remaining off-diagonal entries where τ represents the analysis period. The symbol * denotes the conjugate of the complex external acceleration frequency spectrum $\{X_e\}$.

The application of this fundamental definition from Eq. 37 to both sides of Eq. 22 gives the following result for the external acceleration spectral density matrix

$$[Gx_e] = [T_{ec}]^* \left[[T_{cc}]^{-1} \right]^* [Gx_c] \left[[T_{cc}]^{-1} \right]^T [T_{ec}]^T \tag{38}$$

where the N_c x N_c spectral density matrix $[Gx_c]$ contains the interface acceleration ASDs and CSDs. The right hand side of Eq. 38 can be simplified by using the identities: $[T_{ce}] = [T_{ce}]^T$ and $[[T_{cc}]^{-1}]^T = [[T_{cc}]^T]^{-1} = [T_{cc}]^{-1}$, since $[T_{cc}]$ is symmetric. The resulting expression for the external acceleration spectral density matrix is

$$[Gx_e] = [T_{ec}]^* \left[[T_{cc}]^{-1} \right]^* [Gx_c] [T_{cc}]^{-1} [T_{ce}] \tag{39}$$

or simply

$$[Gx_e] = [Q]^* [Gx_c] [Q]^T \tag{40}$$

where in this case, the interface acceleration spectral density matrix $[Gx_c]$ is multiplied on the left by the complex conjugate of the Q-T matrix and on the right by the transpose of the Q-T matrix $[Q] = [T_{ec}][T_{cc}]^{-1}$, respectively.

7.3. PREDICTION OF EXTERNAL ACCELERATIONS

Equation 40 is used with the experimental beam data from the interface acceleration ASDs and CSD at locations 3 and 4 in order to predict the external acceleration ASDs and CSDs at points 1 and 2. In order to reduce the experimental noise in the calculation procedure, the curve fitted interface accelerances shown in Fig. 12 and the curve fitted external-interface accelerances (not shown) are used to construct the FRF matrices $[T_{cc}]$ and $[T_{ec}]$, respectively. These curve fitted accelerance FRFs were generated by using an accelerance FRF expression similar to Eq. 28. The modal parameters used in this expression are those that resulted from the curve fitting of the measured accelerance FRFs.

Figures 13a and 13b show the predicted external accelerations Gu_1 and Gu_2, respectively. The corresponding measured ASDs Gx_1 and Gx_2 are plotted for comparison purposes. In both cases, the predicted external acceleration ASD is in good agreement with the corresponding measured data for frequency components located in the vicinity of the acceleration peaks while a poor agreement is observed in the valleys located between two consecutive acceleration peaks. In these valleys, the estimated acceleration ASD is seen to present a fictitious acceleration peak followed by a notch or a notch followed by an acceleration peak. This fictitious behavior may cause the false impression that the beam has additional natural frequencies in the 0 - 625 Hz that were not detected in the FRF measurements. Since these uncertainties were not observed in the numerical simulation results employing the Q-T matrix approach presented in the previous section, we must ask ourselves the fundamental question: "What is going on?"

If we return to the original expressions that culminated in the Q-T matrix approach, i.e., Eqs. 22 and 23 or its reformulated version in terms of acceleration ASDs and CSDs in Eq. 39, we see that a matrix inversion is required for all frequency components in the frequency bandwidth covered by the tests. The matrix that must be *explicitly* inverted corresponds to the test item interface accelerance N_c x N_c FRF matrix $[T_{cc}]$ that is *square* and *symmetric*. Since our analysis is based on the assumption that the test item is subjected to interface forces *only*, the external motion $\{X_e\}$ in Eqs. 22 or 23 in terms of frequency spectra or $[Gx_e]$ in Eq. 39 in terms of spectral densities is due to interface forces *only*. Thus, Eq. 22 can be rewritten as

$$\{X_e\} = [T_{ec}]\{F_c\} \qquad (41)$$

while Eq. 39 can be rewritten as

$$[Gx_e] = [T_{ec}]^*[Gf_{cc}][T_{ce}] \qquad (42)$$

where the N_c x 1 vector $\{F_c\} = [T_{cc}]^{-1}\{X_c\}$ contains the interface forces frequency spectra and the N_c x N_c matrix $[Gf_{cc}] = [T_{cc}]^*[Gx_c][T_{cc}]^{-1}$ contains the interface force spectral densities, respectively. Thus, in a sense, although Eqs 22 and 39 express the test item input in terms of interface motions, the *inverse* solution for the interface forces, in terms of interface motions, is *implicitly* implied in these equations as seen in Eqs. 41 and 42.

The solution for the excitation forces from knowledge of measured motions corresponds to an *inverse* problem in mechanics, and is well known to offer numerical difficulties since the inversion of the structure's FRF matrix for each frequency spectral line is required in solving for the unknown forces [8,9,12]. These numerical difficulties occur due to the fact that the system FRF matrix tends to be rank deficient in the vicinity of the natural frequencies and this rank deficiency is caused by an insufficient number of modes participating at the structure's response at those frequencies [8]. This rank deficiency of the FRF matrix causes numerical problems and affects the uniqueness of the solution of frequencies close to the natural frequencies.

The pseudo-inverse technique is frequently employed to give a least squares solution for the N excitation forces from knowledge of M acceleration records when M > N, i.e., more motions are used than forces predicted. This over-determination of the system of equations that must be solved for each frequency component requires the pseudo-inversion process of the M x N structural FRF matrix in order to obtain a unique solution for the unknown forces in a least square sense. The predicted forces may or may not resemble the actual forces acting on the structure depending on the information contained in the measured motions and on the structure's FRF matrix. The least squares solution of the inverse problem also helps to reduce measurement noise effects at frequencies where the measured data has a poor signal to noise ratio. However, the numerical difficulties may still persist even when seeking an approximate least squares solution for the unknown forces since the solution process may be affected by a number of factors such as an unsuitable selection of measurement location, motions caused by unknown external forces, and moments that are not accounted for in the solution process. Several procedures have been proposed to reduce the ill conditioning of the inversion process [9,10,13,14].

The external acceleration ASDs predictions shown in Fig. 13 indicate that the free-free beam interface accelerance FRF matrix $[T_{cc}]$ is rank deficient at those frequencies where false peaks and valleys are observed in the predicted acceleration ASDs. Since the interface accelerance FRF matrix is 2x2 in this case, Eqs. 36 can be used to obtain the inverse of the beam interface FRF matrix. The determinant of $[T_{cc}]$ is $\Delta_{cc} = T_{33}T_{44} - T_{34}T_{43}$ and it is shown in Fig. 14 for both the experimental and curve fitted interface accelerance FRF matrices. These determinants present amplitude variations in terms of peaks and notches in the 0 - 625 Hz frequency range that vary from approximately 10^{-4} $(g/N)^2$ to 10^3 $(g/N)^2$. A close inspection at the notch frequencies of Δ_{cc} and the corresponding frequencies where the false peaks and valleys occur in Fig. 13, reveals that they are essentially the same. Thus, what is happening is that the pseudo-inverse of $[T_{cc}]$ in the present case coincides with standard inversion and the determinant Δ_{cc} that is in the denominator of $[T_{cc}]^{-1}$ in Eq. 39 becomes very small (in the order of $\approx 10^{-3}$ for the experimental FRFs or even $\approx 10^{-4}$ in the case of the curve fitted FRFs) at the frequencies where unwanted variations on the predicted motions are seen to occur.

In order to improve the conditioning of the inversion process shown in Eqs. 22 and 39, we propose to increase the order of the input acceleration vector $\{X_c\}$ in Eq. 22 and consequently in Eq. 39 such that the solution for the unknown external motions is obtained in a least squares sense. Since the input acceleration vector $\{X_c\}$ is formed by *all interface* motions, the only way of increasing the order of this input motion is if we introduce *external* motions as test item inputs in Eqs. 22 and 39. In this case, Eqs. 22 and 39 are rewritten as

$$\{X_e\} = [T_{ec}][\hat{T}_{cc}]^+ \{\hat{X}_c\} \tag{43}$$

$$[Gx_e] = [T_{ec}]^* \left[[\hat{T}_{cc}]^+ \right]^* [\hat{G}x_c][\hat{T}_{cc}]^+ [T_{ce}] \tag{44}$$

where the hat symbol is used to denote the fact the interface and external accelerations form the input motions frequency spectra and spectral densities in Eqs. 43 and 44. Although this new formulation of the Q-T matrix approach essentially violates the initial assumption that only interface motions would be used as test item inputs, it is expected that it can improve the conditioning of the problem since a least squares solution is not obtained from Eqs. 43 and 44. The pseudo-inverse of the FRF matrix $[\hat{T}_{cc}]$ is required in this case as denoted by symbol "+" in Eqs. 43 and 44. In this case, the singular value decomposition [16] is used to calculate the pseudo-inverse of the FRF matrix when it is required.

Equation 44 will be employed with the experimental data obtained in the free-free beam experiments. Since we need to use one of the two external measured acceleration ASDs at either points 1 or 2 in order to obtain the over-determined set of equations that are required by Eq. 44, the following two cases are considered:

- *Case 1:* We use the measured external acceleration ASD Gu_{22} and CSDs Gu_{23} and Gu_{24} on the right hand side of Eq. 44 and solve for Gu_{11}. This means that $[\hat{T}_{cc}]$ contains the additional T_{23} and T_{24} FRFs.

- *Case 2:* We use the measured external acceleration ASD Gu_{11} and CSDs Gu_{13} and Gu_{14} on the right hand side of Eq. 44 and solve for Gu_{22}. This means that $[\hat{T}_{cc}]$ contains the additional T_{13} and T_{14} FRFs.

The results for both cases 1 and 2 are shown in Figs. 15a and 15b, respectively. These results were obtained by using the beam accelerance FRFs obtained by the curve fitting process. We see that the results for both Gu_{11} and Gu_{22} in Figs. 15a and 15b are greatly improved when employing the least squares version of the Q-T matrix approach, Eq. 44. Small discrepancies are seen to occur in the predicted Gu_{11} and Gu_{22} for frequency components in the vicinity of the first and fifth amplitude peaks. A possible reason for these discrepancies is that the experimental and curve fitted FRFs are not in good agreement for frequencies that are close to the first and fifth natural frequencies as it is shown in Fig. 12b. In this case, the first and fifth natural frequencies peaks of the experimental T_{44} are sharper and have higher amplitude values than the corresponding T_{44} obtained by curve fitting.

In order to check how sensitive Eq. 44 is to curve fitting errors, we repeated the same two cases previously discussed by using the beam experimental FRFs without curve fitting to perform the calculations stated by Eq. 44. The results for cases 1 and 2 are shown in Figs. 16a and 16b, respectively. Figure 16a shows the results for the predicted Gu_{11} when the measured Gu_{22} is used as an input in Eq. 44. The results for this acceleration ASD show that the uncertainties that were obtained in the vicinity of the first and the fifth natural frequencies when the curve *fitted* FRFs were used are considerably reduced. The same trends are observed in the predicted acceleration ASD Gu_{22} of Fig. 16b for frequencies close to the first acceleration peak, but there is considerable uncertainty near the fifth acceleration peak where the experimental and predicted acceleration ASD appear to have serious noise contamination.

The point of this exercise is that reasonable estimate of test item external acceleration result when the correct interface input accelerations are applied to the test item. The real problem is to develop a multi-exciter control system that can handle the required ASD and CSD requirements. In this case, the inversion problem is automatically handled by the test item.

8. Summary of experimental results

The results obtained in this experimental analysis can be summarized as follows:

- When only *interface* acceleration ASDs and CSDs are used as test item inputs in the Q-T matrix procedure, the inversion of test item interface accelerance FRF tends to be rank deficient even though $[T_{cc}]$ is square and symmetric. *This rank deficiency seriously affects the results by causing false peaks and valleys to appear in the predicted external acceleration ASDs between acceleration peaks.*

- The rank deficiency problem may be overcome by using *external* accelerations in addition to *interface* accelerations as test item inputs. In this case, the resulting set of equations represents an over-determined problem where the solution for external motions is obtained in a least square sense and,

- The results obtained when curve fitted FRFs were used show that the inversion of the accelerance FRF matrix formed from interface and external FRFs is very sensitive to variations in the test item mode shapes since small errors present in the curve fitted FRF around natural frequencies tend to introduce large errors in the predicted external motions. Ewins [15] arrived at a similar conclusion by performing similar experiments on a cantilever beam. Thus, care should be taken when using curve fitted FRFs in the inversion process.

9. Summary and Conclusions

This paper discusses motion transmissibility FRF concepts in test environments. When the test item is connected to a vehicle by a *single* interface connector and *subjected to no field external forces*, the test item motions can be predicted from knowledge of the interface acceleration and the test item single point acceleration transmissibility FRFs. When several interface points exist between the test item and the vehicle, the Q-T matrix approach was developed to obtain the test item external motions from knowledge of the test item interface motions. Similar to single point transmissibility, the Q-T matrix transmissibility approach assumes that the test item is subjected to no external forces. It is shown that the Q-T matrix reduces to the single point transmissibility FRF for the case of a single interface point.

The Q-T matrix approach is used in numerical simulations of the laboratory environment to address the issue of using field interface motions as test item inputs. Two test scenarios are considered in these numerical simulations. In the first scenario, the standard procedure of attaching the test item to a single vibration exciter by a *rigid* test fixture is considered. In this case, two different test item inputs are used. *First*, one of the field interface accelerations is transmitted to the test item by the test fixture. *Second*, the test fixture and test item input is given in terms of an averaged interface acceleration. In both situations, the *same* input motion is transmitted to all of the test item interface points by the *rigid* test fixture. It is shown that the test item's p^{th} external motion is the product of the single motion that is applied the test item and the *sum* of the entries belonging to the pth row of the Q-T matrix. This expression *does not* correspond to the standard single point transmissibility concept. The numerical results showed that in general, incorrect external motions are obtained when employing the Q-T matrix approach with the rigid test fixture-test item combination.

The second numerical simulation employs the Q-T matrix approach when multiple vibration exciters are used in the laboratory environment. In this case, no test fixture is used and each test item interface point is connected to a different vibration exciter. Each interface point is driven with its corresponding field interface acceleration. The correct external accelerations are predicted in this test scenario.

The use of the Q-T matrix approach with actual experimental data revealed that the inversion of the test item interface accelerance FRF matrix can present rank deficiency problems and consequently cause serious distortions on the predicted motions. The order of the system of equations that describe the Q-T matrix transformation is increased by using external motions in addition to interface motions as test item inputs. The solution for the unknown external motions are carried out in a least squares sense thus ensuring that a unique solution in terms of external motions is obtained. If experimental noise represents a major problem when using real data, curve fitted FRFs can be used. In this case, the curve fitting of the experimental data should be carefully done since small fitting errors in the vicinity of natural frequency peaks can introduce large degrees of uncertainty in the predicted motions.

Finally, although the derivation of the Q-T matrix transformation was based on the assumption that *the test item is subjected to no external forces*, it appears that an adequate matrix partition of the test item frequency domain equations of motion could incorporate external forces effects depending on the nature of the external forces that are present in the test environment.

Acknowledgements

The authors would like to thank the Iowa State University Engineering College and the Aerospace Engineering and Engineering Mechanics Department for supporting this research. Dr. Paulo S. Varoto, from Universidade de São Paulo – São Carlos – Brasil was financially sponsored by CNPq – Brazil during his PhD program.

10. References

1. McConnell, K.G. (1995), Vibration Testing: Theory and Practice, *John Wiley & Sons*.
2. Clough, R. and Penizien, J. (1993), Dynamics of Structures, 2nd *ed., McGraw Hill*.
3. Varoto, P.S., and McConnell, K.G. (1999), Rules for the Exchange and Analysis of Dynamic Information: Part I – Basic definitions and test scenarios, in J.M.M. Silva and N.M.M. Maia (eds.), *Modal Analysis & Testing, Kluwer Academic Publishers, NATO Series*, Dordrecht, 65-81.
4. Varoto, P.S. and McConnell, K.G. (1999), Rules for the Exchange and Analysis of Dynamic Information, Part III: Numerically simulated and experimental results for a deterministic excitation with external loads, in J.M.M. Silva and N.M.M. Maia (eds.), *Modal Analysis & Testing, Kluwer Academic Publishers, NATO Series*, Dordrecht, 101-135.
5. Varoto, P.S. and McConnell, K.G. (1999), Rules for the Exchange and Analysis of Dynamic Information, Part IV – Numerically simulated and experimental results for a random excitation, in J.M.M. Silva and N.M.M. Maia (eds.), *Modal Analysis & Testing, Kluwer Academic Publishers, NATO Series*, Dordrecht, 137-177.
6. "Environmental test methods and engineering guidelines" (1983), *MIL-STD-810D*.

7. Varoto, P.S. and McConnell, K.G. (1999), Rules for the Exchange and Analysis of Dynamic Information, Part II: Numerically simulated results for a deterministic excitation with no external forces, in J.M.M. Silva and N.M.M. Maia (eds.), *Modal Analysis & Testing, Kluwer Academic Publishers, NATO Series*, Dordrecht, 83-100.

8. Fabumni, J. (1986), Effects of structural modes on vibratory force determination by the pseudoinverse technique, *AIAA Journal, Vol. 24, No. 3*, 504-509.

9. Hillary, B. and Ewins, D.J. (1984), The use of strain gages in force determination and frequency response measurements, *Proceedings of the II International Modal Analysis Conference*, Orlando, FL, 627-634.

10. Han, M-C and Wicks, A.L. (1990), Force determination with slope and strain response measurement, *Proceedings of the VIII International Modal Analysis Conference, Vol. 1*, 365-372, Kissimmee, FL.

11. "MODENT, Reference Manual ICATS" (1994), *Imperial College of Science, Technology and Medicine, Mech. Eng. Department*, London.

12. Starkey J.M. and Merrill, G.L. (1989), On the ill-conditioning nature of indirect force-measurement techniques, *IJAEMA, Vol. 4, No. 3*, 103-108.

13. Hansen, M. and Starkey J.M. (1990), On predicting and improving the condition of modal-model based indirect force measurement algorithms, *Proceedings of the VII International Modal Analysis Conference, Vol. 1*, 115-120, Orlando, FL.

14. Lee, J.K. and Park, Y. (1994), Response selection and dynamic damper application to improve the identification of multiple input forces of narrow frequency band, *Mechanical Systems and Signal Processing 8(6)*, 649-664.

15. Ewins, D.J. (1994), Modal Testing: Theory and Practice, *Research Studies Press*.

16. To, W.M. and Ewins, D.J. (1995), The role of the generalized inverse in structural dynamics, *Journal of Sound and Vibration, 186(2)*, 185-195.

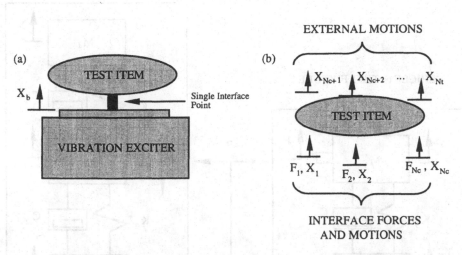

Figure 1   (a) Test item attached to a vibration exciter and subjected to a single interface input – *the single point transmissibility concept*; (b) Test item subjected to multiple interface inputs – *the* **Q-T** *matrix concept*.

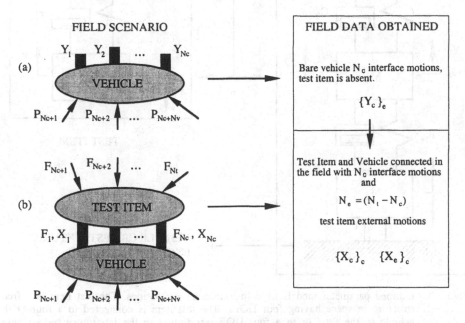

Figure 2   Field environment calculation procedure: (a) The *bare vehicle* interface motions, *test item is absent in the field in this case*; (b) The *combined* test item and vehicle structure.

202

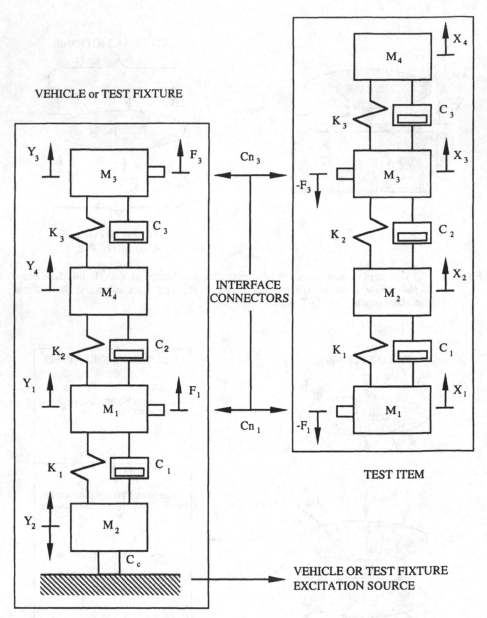

Figure 3   Lumped parameter models used in numerical simulations. The test item is a free structure in space having four DOFs. The test item is connected to a four DOF vehicle in the field or to a four DOF test fixture in the laboratory by a single interface rigid connector Cn_1. The excitation source is applied to the vehicle or test fixture base mass. *No external forces are applied to the test item in either the field or the laboratory environments.*

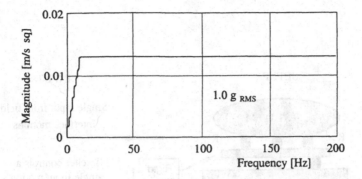

Figure 4   Input acceleration frequency spectrum Y_2 applied to the vehicle base M_2 to produce the *bare vehicle interface acceleration and the interface and external accelerations for the combined structure*.

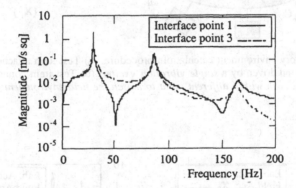

Figure 5   Bare vehicle interface acceleration frequency spectra for system shown in Fig. 3 due to base input acceleration Y_2.

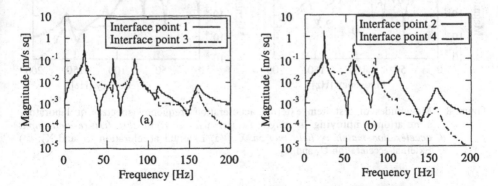

Figure 6   Test item acceleration frequency spectra in the field environment due to base input motion Y_2. *No external forces applied to the test item*: (a) Interface accelerations X_1 and X_3; (b) External accelerations X_2 and X_4.

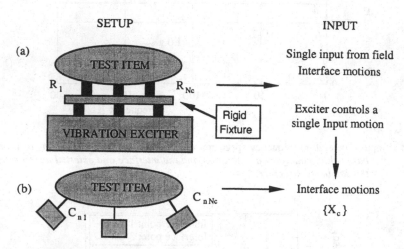

SETUP   INPUT

Figure 7   Laboratory environment calculatioin procedure: (a) Test item attached to a *rigid test fixture* and driven by a *single vibration exciter*; (b) Test item connected to multiple exciters, *each with a different input to match the field environment.*

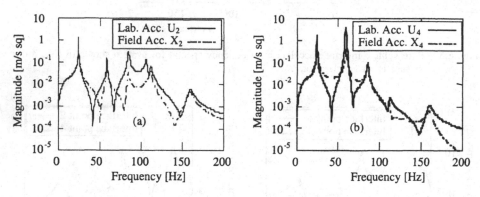

Figure 8   Magnitudes of test item external acceleration frequency spectra for laboratory simulation employing *a single exciter with a rigid test fixture and input acceleration matching field motion* X_1: (a) External acceleration U_2 and X_2; (b) External acceleration U_4 and X_4.

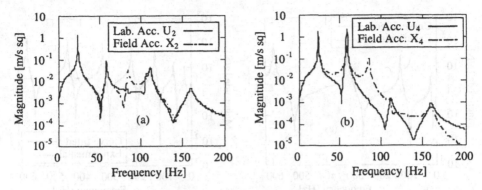

Figure 9   Magnitudes of test item external acceleration frequency spectra for laboratory simulation employing *a single exciter with a rigid test fixture and input acceleration matching the average of field motions X_1 and X_3*: (a) External acceleration U_2 and X_2; (b) External acceleration U_4 and X_4.

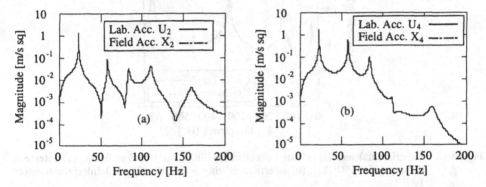

Figure 10   Magnitudes of test item external acceleration frequency spectra for laboratory simulation employing N_c *exciters attached to the test item interface points. Each Vibration exciter is controlled to match the corresponding field interface accelerations*: (a) External acceleration U_2 and X_2; (b) External acceleration U_4 and X_4.

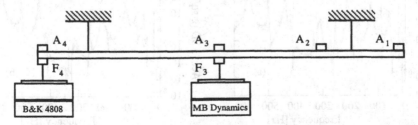

Figure 11   Setup used in experimental study. Steel beam suspended by flexible cords and driven by two exciters using random excitation. Output motion ASDs and CSDs are measured at four locations.

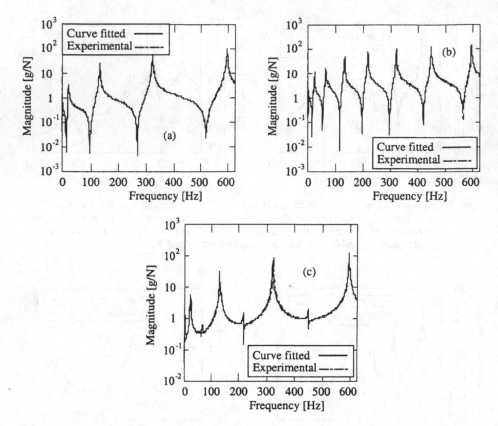

Figure 12   Experimental and curve fitted accelerance FRFs for free-free beam: (a) Interface driving point FRF T_{33}; (b) Interface driving point FRF T_{44}; (c) Interface transfer FRF T_{34}.

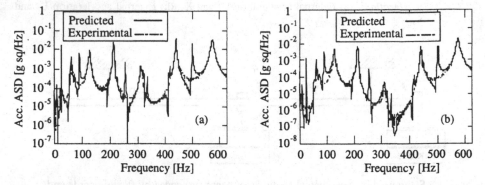

Figure 13   Experimental and predicted results for external acceleration ASDs Gu_1 and Gu_2. *The curved fitted accelerance FRFs are used in this case*: (a) External acceleration ASD Gu_1; (b) External acceleration ASD Gu_2.

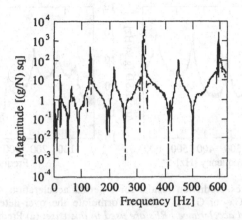

Figure 14   Determinant of the free-free beam interface acceleration FRF matrix as a function of frequency. *Solid line: curve fitted FRF; Dash-dot line: experimental.*

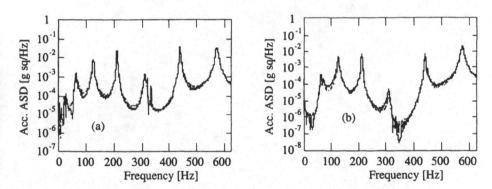

Figure 15   Least squares prediction results for external acceleration ASDs when either the measured Gx_2 or Gx_1 are used to formulate the over-determined problem. *The curve fitted acceleration FRFs are used in this case:* (a) Predicted acceleration ASD Gu_1 when Gx_2 is used as input data in the least squares problem; (b) Predicted acceleration ASD Gu_2 when Gx_1 is used as input data in the least squares problem. *Solid line: Predicted ASD , Dot-dashed line: Measured ASD.*

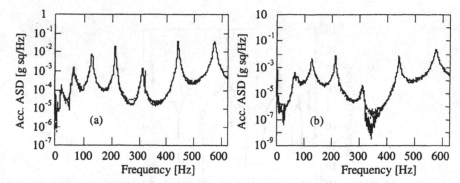

Figure 16   Least squares prediction results for external acceleration ASDs when either the measured Gx_2 or Gx_1 is used to formulate the over-determined problem. *The experimental accelerance FRFs are used in this case*: (a) Predicted acceleration ASD Gu_1 and Gx_2 is used as input data in the least squares problem; (b) Predicted acceleration ASD Gu_2 when Gx_1 is used as input data in the least squares problem; *Solid line*: *Predicted ASD , Dot-dashed line*: *Measured ASD.*

RULES FOR THE EXCHANGE AND ANALYSIS OF DYNAMIC INFORMATION

Part VI: Current Practice and Standards

P. S. VAROTO
Dept de Engenharia Mecanica
Escola De Engenharia De São Carlos,
USP
São Carlos – SP – 13560-250, Brasil

K. G. McCONNELL
Dept. Aerospace Engineering and
Engineering Mechanics
Iowa State University
Ames, IA 50011 USA

Abstract

While in the field environment, the test item is connected to the vehicle at several interface points. The combined structure is then subjected to a variety of field loads that produce motions at the test item interface and external (non-interface) points. A commonly employed laboratory simulation procedure consists in attaching the test item to a single vibration exciter through a test fixture. Choice of appropriate test fixtures and definition of suitable laboratory test item inputs is of major importance in obtaining realistic simulation of a field environment. MIL – STD810D recommends that: (i) The test item be attached to the exciter through a *rigid* test fixture; (ii) When available, *Field data* should be used to define the test item inputs in the laboratory environment. In this case, enveloping techniques are frequently used with the field data in order to define the input that will be applied in the laboratory. The objectives of this paper is to show the consequences of using these current laboratory simulation procedures on the test item dynamic response. The test item, the vehicle, and the test fixture are modeled by a multi degree of freedom (MDOF) lumped systems. Random excitation and response signals are used in the simulation process. The test item laboratory inputs are defined in terms of an acceleration auto spectral density (ASD) that is obtained by enveloping field data as recommended in MIL – STD810D. The test item accelerations obtained in the laboratory when the test item is attached to either a rigid or flexible test fixture are compared with the actual motions that occur when the test item is attached to the vehicle in the field. It is shown that both the rigid and the flexible test fixtures seriously affect the resulting test item accelerations and that the current enveloping techniques expose the test item to excessively high dynamic strain levels.

Nomenclature

A_{pq}	Accelerance FRF	C_q	q^{th} test item and vehicle
Cn_1, Cn_3	Connection points between test item and vehicle		damping coefficients

J.M.M. Silva and N.M.M. Maia (eds.), Modal Analysis and Testing, 209–226.
© *1999 Kluwer Academic Publishers. Printed in the Netherlands.*

K_q	qth test item and vehicle spring constants	$[T_{ce}]$	Test item connector-external accelerance FRF matrix
$j = \sqrt{-1}$	Imaginary number	$[TV]$	Test item-vehicle combined
$[Gx_{cc}]$	Connector acceleration ASD		matrix
$[Gx_{ee}]$	External acceleration ASD	$[M]$	Mass matrix
$[Gz_{cc}]_e$	Bare text fixture acceleration ASD	X_{pq}	Relative displacement ASD between p and q points
M_q	qth test item and vehicle mass coefficients	**Greek**	
N_t	Degrees of freedom for test item	Δf	Frequency resolution in Hz
N_v	Degrees of freedom for vehicle	$\{\phi\}_r$	r^{th} mode shape
N_c	Number of connectors	$\Gamma(\omega)$	Acceleration transmissibility
T_{pq}	Test item accelerance FRF	FRF	
$[T]$	Test item accelerance FRF matrix	ω	Frequency in rad/s
		ω_r	r^{th} natural frequency in rad/s
$[T_{cc}]$	Test item connector accelerance FRF matrix	ζ_r	r^{th} modal damping ratio

1. Introduction

The process of simulating field vibration conditions of a test item attached to a vehicle as shown in Fig. 1a requires appropriate test conditions in the laboratory if the test item true dynamic behavior is to occur. In previous papers [1,2,3,4], we have examined the requirements to achieve reasonable simulations of field conditions in the laboratory for deterministic and random type of loadings. These requirements showed that we need to (*i*) be aware of the effects of connector and external forces, (*ii*) use multiple exciters that are controlled to match either the field interface and external forces or the field interface motions and external forces, and (*iii*) these exciters must satisfy appropriate auto spectral density (ASDs) and cross spectral densities (CSDs) relationships. In this paper, we shall contrast this complexity with current practice and standards.

A common practice [5,6,7] is to use a single vibration exciter where the test item is attached to the exciter's table through a test fixture as shown in Fig. 1b. We shall assume for the purposes of this paper that there are several attachments between the test fixture and the test item. In addition, test standards, such as MIL-STD810D [5], usually require that the *test fixture* be *rigid* in the sense that it has no natural frequencies in the frequency range of the test. It is apparent that such a test fixture should significantly affect the dynamic responses of the test item in terms of natural frequencies and mode shapes. Finally, we see that we are replacing several interface motions by a common input motion. So, naturally, we need to decide what that motion should look like.

MIL-STD810D [5] recommends that we use actual field data when it is available. There are several field data variables that can be used. First, there are the interface accelerations and forces. Second, there are the bare vehicle interface accelerations. Third, there are the test item external accelerations. Among these choices, the bare vehicle interface motions are selected most often for, as Smallwood [6] points out, they are larger than those that occur when the test item is attached to the vehicle. When a single vibration exciter is employed, the vibration standards usually recommend that the interface acceleration ASD be enveloped by a flat ASD that is slightly higher than the highest resonant peak.

It is our goal in this paper to examine the consequences of following these standard procedures. The test item, vehicle, and test fixture are modeled using two lumped four DOF systems that are connected at points 1 and 3 as shown in Fig. 2. Random excitation is applied at the base of the vehicle or test fixture and is used in all simulated cases. Two *field environments* are simulated that consist of a bare vehicle without the test item and a combined test item-vehicle system. Two *laboratory environments* are simulated that consist of the test item mounted on a *rigid* test fixture and a *flexible* test fixture. We have shown in previous papers [3,4] that external forces can have significant effects on the test item's dynamic behavior. In order to keep this analysis as simple as possible, *external forces are assumed to be negligible*. Now, we look at how we can simulate this common testing situation and evaluate the amount of over-testing that may occur.

2. Response models of test structures

Figure 2 displays the two four DOF models that are used in these simulation studies to represent the test item and vehicle. The physical parameters used for test item and vehicle in the field simulations are shown in Table 1. These parameters were chosen such that the resulting natural frequencies for both the test item and vehicle fall with the 0-200 Hz frequency range.

Table 1: Physical parameters of test structures

Mass	Test Item (Kg)	Vehicle (Kg)
M_1	0.20	0.50
M_2	0.20	0.50
M_3	0.25	0.30
M_4	0.15	0.50
Stiffness	Test Item N/m x 10^4	Vehicle N/m x 10^4
K_1	5.00	6.00
K_2	4.00	14.00
K_3	2.00	10.00

The test item and vehicle modal properties (natural frequencies, mode shapes, modal masses, and modal damping ratios) are calculated by employing standard modal analysis procedures [8]. The vehicle structure is grounded by mass M_2, as shown in Fig. 2 while the test item is modeled as a free structure in space. Table 2 shows the resulting modal parameters obtained for the test item, the vehicle, and the combined structure.

Table 2: Modal properties for test structures

Natural Frequency (Hz)	Test Item	Vehicle	Combined
f_1	0	30.00	23.86
f_2	52.60	86.17	58.94
f_3	85.70	164.72	85.05
f_4	131.00	–	112.00
f_5	–	–	160.21
Damping Ratio (%)	Test Item	Vehicle	Combined
ζ_1	0	0.28	0.24
ζ_2	0.50	0.81	0.56
ζ_3	0.81	1.55	0.91
ζ_4	1.23	–	1.06
ζ_5	–	–	1.54
Modal Masses	Test Item	Vehicle	Combined
m_1	0.20	0.43	0.38
m_2	0.17	0.48	0.17
m_3	0.21	0.37	0.55
m_4	0.20	–	0.21
m_5	–	–	0.34

Accelerance FRFs are calculated for both structures according to [8]

$$A_{pq}(\omega) = -\omega^2 \sum_{r=1}^{N} \frac{\phi_{pr}\,\phi_{qr}}{m_r\left(\omega_r^2 - \omega^2 + j\,2\zeta_r\,\omega_r\,\omega\right)} \qquad (1)$$

where ϕ_{pr} and ϕ_{qr} are the pth and the qth elements of the rth mode shape, ω_r is the rth natural frequency (rad/s), ζ_r is the rth modal damping, m_r is the rth modal mass and ω (rad/s) is the excitation frequency.

Similarly, for input motion at a single coordinate, the transmissibility FRF is obtained by [2].

$$\{\Gamma(\omega)\} = \omega^2 \sum_{r=1}^{N} \frac{\{\phi\}_r\,\{\phi\}_r^T\,[M]\,\{1\}}{m_r\left(\omega_r^2 - \omega^2 + j\,2\zeta_r\,\omega_r\,\omega\right)} + \{1\} \qquad (2)$$

where $[M]$ is the structure's mass matrix. Equations 1 and 2 define response models for viscously damped structures.

3. Field environment

Two field environments are analyzed in order to generate the data that is used in the laboratory simulations. The *first field environment corresponds to the bare vehicle simulation* where the test item is not attached to the vehicle. The field data obtained in this case consists of the vehicle interface accelerations caused by external forces acting on the vehicle only; and hence, they are called the *bare* vehicle interface accelerations.

In the *second field environment, the test item is attached to the vehicle at several locations*, forming the combined structure. This simulation is performed in order to obtain the actual test item interface and external accelerations as well as relative motions (strains) between adjacent points. This field data will be used as a reference for comparison to the laboratory results.

A random acceleration ASD Gy_{22} is used as the vehicle input source in both field simulations. This acceleration ASD is applied to the vehicle base mass M_2 and is shown in Fig. 3. This spectrum increases with a slope of 40 dB/decade from 0 to 10 Hz and becomes constant at 0.00517 g^2/Hz up to 200 Hz. This amplitude level is chosen such that an overall 1.0 g_{RMS} input vibration level is obtained. *This base motion is the only input used in the field environment. No external inputs are applied to the test item in this example.*

3.1. FIELD SIMULATION 1: BARE VEHICLE INTERFACE MOTIONS

The test item is not attached to the vehicle in this case. Thus, the interface motions are given in terms of the bare vehicle interface acceleration spectral density matrix $[Gy_{cc}]_e$ that is obtained by Eq. 14 of Ref. [4] as

$$[Gy_{cc}]_e = \left[: \Gamma_{pq} :\right]^* [Gy_{qq}] \left[: \Gamma_{pq} :\right]^T \tag{3}$$

where index $p = 1, ..., N_c$ covers all interface points while $q = 2$ is a fixed coordinate where the input motion Gy_{22} is applied. The qth column of the $N_c \times N_c$ matrix $[: \Gamma_{pq} :]$ contains the acceleration transmissibility FRFs as defined by Eq. 2, and all remaining entries are zero. The $N_c \times N_c$ diagonal matrix $[Gy_{qq}]$ contains zeros at all diagonal entries except at (2,2) where the input base motion Gy_{22} is applied. The *diagonal elements* of the resulting spectral density matrix $[Gy_{cc}]_e$ *correspond to the bare vehicle interface acceleration* ASDs while the *off-diagonal entries* are the *cross spectral densities* (CSDs) among the vehicle interface points.

3.2. FIELD SIMULATION 2: COMBINED STRUCTURE

This field simulation corresponds to the situation where the test item and vehicle are connected in the field. Since no external forces are applied to the test item, the resulting test item interface and external accelerations depend only on the bare vehicle interface spectral density matrix $[Gy_{cc}]_e$. These motions are obtained from Eqs. 9 and 10 of Ref. [4] by setting $[Gx_{cc}]_e = 0$, $[Gx_{ee}]_e = 0$, and $[Gx_{ce}]_e = 0$, since these terms correspond to the motions caused by external forces applied to the test item. The resulting expressions for the test item $N_c \times N_c$ interface spectral density matrix $[Gx_{cc}]$ and the $N_e \times N_e$ external spectral density matrix $[Gx_{ee}]$ are respectively

$$[Gx_{cc}] = [T_{cc}]^* [TV]^* \left[[Gy_{cc}]_e\right] [TV] [T_{cc}] \tag{4}$$

$$[Gx_{ee}] = [T_{ec}]^* [TV]^* \left[[Gy_{cc}]_e\right] [TV] [T_{ec}] \tag{5}$$

where the diagonal entries in $[Gx_{cc}]$ and in $[Gx_{ee}]$ represent the interface and external accelerations ASDs and the off-diagonal entries are the interface and external CSDs, respectively. The $N_c \times N_c$ matrix $[TV] = [[T_{cc}] + [V_{cc}]]^{-1}$, is the test item-vehicle combined matrix and is given by the inverse of the matrix formed by the sum of the test item and vehicle connector accelerance FRF matrices, $[T_{cc}]$ and $[V_{cc}]$, respectively as outlined in [1].

Interface and external acceleration ASDs are calculated from Eqs. 4 and 5 and they are shown in Fig. 4. Figures 4a and c show the interface ASDs Gx_{11} and Gx_{33} for points 1 and 3 along with the bare vehicle interface acceleration ASDs Gy_{11} and Gy_{33} which are plotted for comparison purposes. Some interesting observations can be made about these motions. *First*, the bare vehicle (dashed-dotted line) and the combined structure (solid line) do not peak at the same frequencies. This is expected since the bare vehicle interface acceleration ASDs involve the dynamics of the vehicle only, while the combined structure involve the dynamics of the vehicle and the test item. *Second*, the combined structure interface accelerations ASDs levels are lower than the corresponding levels obtained from the bare vehicle interface acceleration ASDs. This can be explained by the fact that when the test item is attached to the vehicle, mass is being added at the interface DOFs where coupling of test item and vehicle occurs. The attachment of the test item increases the modal mass so that the resulting interface acceleration levels decrease relative to the corresponding bare vehicle interface acceleration levels [6].

4. Considerations for laboratory simulation

Often, standard testing practice is to use a single vibration exciter and test fixture arranged as shown in Fig. 1b where the test item is attached to the test fixture at N_c points. We need to examine how this arrangement affects the test outcome. Previously , we have shown [3,4] that certain requirements must be satisfied in order to have a successful simulation when there are multiple interface connections. *First*, multiple exciters must be used, one for each connection point. *Second*, each exciter must be controlled such that either the corresponding interface motion or force occurs at each interface point that has the correct phasing relative to the other inputs. *Third*, the external loads must be properly applied to the test item. Now the question is, "can we specify the test fixture and exciter input so that the desired results are obtained?" To begin to answer this question, we return to Eqs. 4 and 5.

4.1. TEST FIXTURE

Equations 4 and 5 require that the test fixture have the same driving and transfer point accelerance as the vehicle so that the [TV] matrices are the same in the field and laboratory environments. This matching of driving and transfer point accelerances is a very difficult requirement to achieve in the first place and extremely expensive to implement if such test conditions can be achieved. Thus, there is a little chance of meeting this requirement in general.

Current practice requires that the test fixture be *rigid* over the range of frequencies tested. In this study, the test fixture is simulated as a four DOF system as shown in Fig.

2. Two test fixture designs are used where *one is rigid* and *the other has a single resonance in the middle of the test frequency range*. These two different test fixture conditions are achieved by changing the spring constants in Table 1. The actual spring constant values are given in Table 3 along with the corresponding test fixture natural frequencies.

Table 3: Parameters for Test Fixtures

Stiffness (N/m)	Rigid x10^8	Flexible x10^8
K_1	0.09	0.005
K_2	1.40	14.00
K_3	1.00	10.00
Natural Freq. (Hz)	Rigid x10^2	Flexible x10^2
f_1	4.10	1.00
f_2	24.64	76.57
f_3	51.61	162.98

4.2. EXCITER INPUT

Equations 4 and 5 require that there be multiple inputs, but the test situation in Fig. 1b shows only a single input and in Fig. 2, we have only one input, Gy_{22}. Hence, we have an automatic question of what to do when using a single exciter with a test fixture and multiple connector or interface points. So, how is this question answered in test standards?

MIL-STD810D recommends that field data is used if it is available. It also recommends that the highest interface motion peaks be used in drawing a flat envelope (similar to the one shown in Fig. 3) over all bare vehicle peaks. For this simulation, the bare vehicle interface acceleration ASDs from Figs. 4a and 4c were enveloped with the same shape as that shown in Fig. 3. This means that we multiplied the ASD in Fig. 3 by a scale factor of $140^2 = 1.96 \times 10^4$.

It is immediately clear from this discussion and Fig. 1b that the requirements for an adequate simulation are not being met by using a single test fixture and input motion. Now we shall look at the consequences of using these test standard recommendations for this test setup.

5. Acceleration ASD simulation results

First, the bare test fixture acceleration ASDs must be calculated from Eq. 3 for the laboratory environment when we use the test fixture characteristics given in Table 3 and the desired input acceleration ASD that envelopes the bare vehicle interface ASD curves. Then Eq. 3 becomes

$$\left[Gz_{cc}\right]_e = \left[:\Gamma_{pq}:\right]^* \left[Gz_{qq}\right]\left[:\Gamma_{pq}:\right]^T \tag{6}$$

where $[Gz_{cc}]e$ represents the bare fixture interface acceleration spectral density matrix and pq represents the test fixture acceleration transmissiblity FRFs. The diagonal Nc x Nc matrix $[Gz_{pq}]$ has zeros at all entries except at q = 2 where the test fixture input acceleration ASD that is obtained by enveloping the bare vehicle interface ASD curves is used. The test item acceleration spectral density matrices are obtained by rewriting Eqs. 4 and 5 as

$$[Gu_{cc}] = [T_{cc}]^{*} [TF]^{*} [[Gz_{cc}]_{e}] [TF] [T_{cc}] \qquad (7)$$

$$[Gu_{ee}] = [T_{ec}]^{*} [TF]^{*} [[Gz_{ec}]_{e}] [TF] [T_{ec}] \qquad (8)$$

where $[Gu_{cc}]$ and $[Gu_{ee}]$ represent the test item connector and external acceleration spectral densities that include both ASDs on the diagonal and CSDs on the off diagonals. When we employ Eq. 6 with Eqs. 7 and 8, we can calculate the test item acceleration ASD responses for both the *rigid* and *flexible* test fixture cases that are shown in Fig. 5. Figures 5a and 5c show the interface acceleration ADSs Gu_{11} and Gu_{33}, respectively, while Figs. 5b and 5d show the external acceleration ASDs Gu_{22} and Gu_{44}, respectively for both the *rigid* and *flexible* test fixture cases. The original field ASDs are shown by the dotted lines, but these curves were multiplied by a factor of 10^{4} in order to plot in the same range as the laboratory data. This indicates that the acceleration ASDs are significantly in error in terms of levels of magnitude.

The two interface motions for the *rigid* case (Gu_{11} and Gu_{33}) are depicted by the dashed-dotted line and are essentially flat with a value near 100 as required to envelope the bare vehicle input motions shown in Fig. 4. This is the result we would expect and serves as a good indication that the inputs were correctly calculated. In the *rigid* test fixture case, each of the two test item external points show a single resonance with one at 58.0 Hz in the Gu_{44} ASD where mass M_4 resonates on spring K_4 and one at 106.2 Hz in the Gu_{22} ASD where mass M_2 resonates on springs K_2 and K_3 since interface points 1 and 3 have the same motion due to test fixture rigidity.

The *flexible* test fixture case is shown by the solid line curves in Fig. 5. In this case, three resonances are present in each ASD. These resonances occur at 56.0 Hz and 119.5 Hz and correspond to the natural frequencies of the combined test fixture and test item.

It is clearly evident in Fig. 5 that we failed to achieve a good estimate of either the test item acceleration levels or its natural frequencies. The excessive vibration levels are due to using a flat input envelope that equals or exceeds the highest bare vehicle interface acceleration ASD peaks. The resulting inaccurate natural frequencies is due to the use of a *rigid* or *nearly rigid* test fixture that does not match the vehicle driving point and transfer accelerances as required. Hence, we suspect that a significant over-test occurred in this case. Now, we need to estimate just how significant of an over-test actually occurred.

6. Exploring the amount of over-testing

A direct measure of the amount of over-test that occurred during this laboratory simulation can be determined by comparing the strain levels that occur in the test item when in the field environment to those obtained in the laboratory environment when both *rigid* and *flexible* test fixtures are employed. The basic definition of strain is the relative displacement of two adjacent points divided by the original distance between the points. Hence, we can use relative displacements between adjacent points in the test item while in the field and laboratory environments as a direct measure of test item strain.

The relative motion frequency spectra between two adjacent points p and q, X_{pq}, can be obtained from the acceleration ASDs Gx_{pp} and Gx_{qq} for each point and the real part of the acceleration CSD Gx_{pq} by the relationship

$$X_{pq} = \frac{g_0}{\omega^2} \left[\Delta f \left(Gx_{pp} + Gx_{qq} - 2 \operatorname{Re} \left(Gx_{pq} \right) \right) \right]^{1/2} \tag{9}$$

where g_0 is the acceleration of gravity in units of 9.81 m/s^2 /g, Δf is the frequency component spacing in Hz, and ω is the corresponding frequency. Equation 9 is used to calculate the relative displacements X_{12}, X_{23}, and X_{34} for the field and laboratory environments which are plotted in Fig. 6. There are three things to note about the field relative displacements. First, they must be multiplied by a factor of 100 in order to plot in the same range as the laboratory relative displacements. Second, they tend to drop off with a slope of about $1/\omega^3$ while the laboratory relative displacements have a much shallower slope. Third, there are five resonant peaks in each relative field motion while the laboratory relative motions have fewer resonant peaks. It is clear that laboratory and field strains are significantly different.

The *rigid* test fixture case shows only one resonant frequency for X_{12} and X_{23} at 106 Hz while X_{34} shows a single resonance at 58 Hz. The *flexible* test fixture case shows the same three resonant peaks in each relative displacement since the test fixture motion of point 3 allows the motion of point 4 to communicate with the motion of points 1 and 2 in this case compared to the *rigid* test fixture case. It is difficult to compare each of these relative displacements and decide the relative damage potential of each since many variables contribute to fatigue damage estimates that are dependent on the fatigue theory employed. Among these variables are the distribution of peak strains in terms of amplitude and frequency in the frequency domain as well as the probability density distribution of peaks in the time domain. For our purposes, we can make a single comparison of peak amplitudes and frequencies for each relative motion as shown in Tables 4 for X_{12}, 5 for X_{23}, and 6 for X_{34}.

In addition, we propose a simple *Figure of Merit* (FOM) for comparing the potential fatigue damage of the three cases cited by using the product of each relative displacement FRF peak amplitude in times its corresponding frequency in *Hz*. This FOM is a measure of how fast damage is accumulated at each peak. In addition, we can add up each FOM from each resonant frequency and compare this sum between test item environments.

6.1. COMPARING RELATIVE DISPLACEMENTS X_{12}

It is clear from Table 4 that the first natural frequency is the dominant strain damage element in the field environment with a relative amplitude of 302 x 10^{-6} m at 23.9 Hz while other significant damage situation occurs at the third resonance (85 Hz at 7.52 x 10^{-6} m). The other field peaks should contribute little to field fatigue in this case. For the *rigid* test fixture case, there is only one resonant component at 107 Hz with an amplitude of $7,411 \times 10^{-6}$ m. In the *flexible* test fixture case, there are three strain peaks. The 80 Hz peak of $14,020 \times 10^{-6}$ m is the largest with the other two being $2,340 \times 10^{-6}$ m and $4,450 \times 10^{-6}$ m at 56 Hz and 119.5 Hz, respectively.

The corresponding FOM results in Table 4 shows that the total FOM for the field case is 8.33×10^{-3} mHz while the *rigid* test fixture has a total FOM of 791×10^{-3} mHz and the flexible test fixture has a FOM of 1.785×10^{-3}. The ratio of the total FOM to that in the field shows that the *rigid* test fixture is about 95 times more severe and the *flexible* test fixture case is about 214 times more severe than the field environments.

Table 4: Analysis of relative displacement X_{12}

Peak No.	Freq.[Hz] x 10^0	Amp.[m] x 10^{-6}	FOM [mHz] x10^{-3}	Ratio
		FIELD		
1	23.9	302	7.22	
2	58.9	2.36	0.139	
3	85.0	7.52	0.639	
4	112.0	2.52	0.282	
5	160.2	0.28	0.045	
		TOTAL	8.330	1.0
		RIGID		
1	106.7	7,410	791	
		TOTAL	791	95.0
		FLEXIBLE		
1	56.0	2,330	131.0	
2	80.0	14,020	1,120	
3	119.5	4,450	532	
		TOTAL	1.780	214

6.2. COMPARING RELATIVE DISPLACEMENTS X_{23}

The results for the relative displacement X_{23} is shown in Table 5. In this case, the field total FOM is 7.39×10^{-3} mHz while the *flexible* test fixture case has a total FOM of 1.785×10^{-3} mHz. The ratio of the total FOM to that in the field shows that the *rigid* test fixture is about 107 times more severe and the *flexible* test fixture case is about 242 times more severe than the field environment.

6.3. COMPARING RELATIVE DISPLACEMENTS X_{34}

The results for the relative displacement X_{34} is shown in Table 6. In this case, the field total FOM is 15.88×10^{-3} mHz while the *rigid* test fixture case has a total FOM of $2,530 \times 10^{-3}$ mHz and the *flexible* test fixture case has a total FOM of $6,440 \times 10^{-3}$ mHz. The ratio of the total FOM to that in the field shows that the *rigid* test fixture is about

159 times more severe and the *flexible* test fixture case is about 405 times more sever than the field environment.

Table 5: Analysis of relative displacement X_{23}

Peak No.	Freq.[Hz] x 10^0	Amp.[m] x 10^{-6}	FOM [mHz] x10^{-3}	Ratio
FIELD				
1	23.9	214	5.11	
2	58.9	14.12	0.831	
3	85.0	13.40	1.139	
4	112.0	2.40	0.269	
5	160.2	0.23	0.037	
		TOTAL	7.390	1.0
RIGID				
1	106.7	7,400	790	
		TOTAL	790	107
FLEXIBLE				
1	56.0	2,330	130.5	
2	80.0	14,020	1,122	
3	119.5	4,450	532	
		TOTAL	1.785	242

Table 6: analysis of relative displacement X_{34}

Peak No.	Freq.[Hz] x 10^0	Amp.[m] x 10^{-6}	FOM [mHz] x10^{-3}	Ratio
FIELD				
1	23.9	336	8.03	
2	58.9	103.5	6.10	
3	85.0	14.81	1.69	
4	112.0	0.27	0.03	
5	160.2	0.13	0.02	
		TOTAL	15.88	1.0
RIGID				
1	58.0	43,100	2,500	
2	106.7	292	31.2	
		TOTAL	2,530	159
FLEXIBLE				
1	56.0	79,200	4,400	
2	80.0	23,200	1,856	
3	119.5	1,182	141	
		TOTAL	6,440	405

6.4. COMPARING THE DIFFERENT RELATIVE DISPLACEMENT FOM RESULTS

It is clear that the displacement FOM values for X_{12} and X_{23} are nearly equal (8.30×10^{-3} mHz) when in the field environment. This implies that fatigue failure should occur due to X_{34} strains. Similarly, a comparison of the FOM for the *rigid* test fixture case we see that the FOM has increased about 100 times for relative displacements X_{12} and X_{23} and about 160 times for relative displacement X_{34}. Now, a comparison of the FOM for the *flexible* test fixture case we see that the FOM has increased about 214 to 241 times for relative displacements X_{12} and X_{23} and about 405 times for X_{34}. These results imply that failure should occur in the correct location in the test item structure but at a much faster rate than would occur in the field.

7. Summary and Conclusion

This paper discusses the implications of using a single vibration exciter, a single rigid or flexible test fixture, and an enveloping of bare vehicle interface acceleration ASD field data for determining a single exciter input on laboratory test results. A simple four mass MDOF model is used to model the vehicle, the test item, and test fixtures respectively where two interface connections are used. The results from a numerical simulation of the various system arrangements shows

- that the test fixture dramatically alters the dynamic response characteristics of the test item when compared to its field characteristics when attached to the vehicle,
- that enveloping of the bare vehicle interface acceleration ASD crates excessive input levels that are certainly conservative,
- that fatigue is accumulated at significantly faster levels when using rigid test fixtures and enveloped inputs,
- that flexible test fixtures can produce excessively high fatigue levels when compared with rigid test fixtures, and
- that a detailed study of fatigue accumulation for various test fixture arrangements should be made

In addition, it may be more reasonable to use variable flat levels on the input ASD where the area under a given segment is the same for the field and the laboratory, or at least something proportional to the input.

Acknowledgements

The authors would like to thank the Iowa State University Engineering College and the Aerospace Engineering and Engineering Mechanics Department for supporting this research. Mr. Paulo S. Varoto, from Universidade de São Paulo – São Carlos – Brasil was financially sponsored by CNPq – Brazil during his PhD program.

8. References

1. Varoto, P.S., and McConnell, K.G. (1999), Rules for the Exchange and Analysis of Dynamic Information: Part I – Basic definitions and test scenarios, in J.M.M. Silva and N.M.M. Maia (eds.), *Modal Analysis & Testing, Kluwer Academic Publishers, NATO Series*, Dordrecht, 65-81.
2. Varoto, P.S. and McConnell, K.G. (1999), Rules for the Exchange and Analysis of Dynamic Information, Part II: Numerically simulated results for a deterministic excitation with no external forces, in J.M.M. Silva and N.M.M. Maia (eds.), *Modal Analysis & Testing, Kluwer Academic Publishers, NATO Series*, Dordrecht, 83-100.
3. Varoto, P.S. and McConnell, K.G. (1999), Rules for the Exchange and Analysis of Dynamic Information, Part III: Numerically simulated and experimental results for a deterministic excitation with external loads, in J.M.M. Silva and N.M.M. Maia (eds.), *Modal Analysis & Testing, Kluwer Academic Publishers, NATO Series*, Dordrecht, 101-135.
4. Varoto, P.S. and McConnell, K.G. (1999), Rules for the Exchange and Analysis of Dynamic Information, Part IV – Numerically simulated and experimental results for a random excitation, in J.M.M. Silva and N.M.M. Maia (eds.), *Modal Analysis & Testing, Kluwer Academic Publishers, NATO Series*, Dordrecht, 137-177.
5. "Environmental test methods and engineering guidelines" (1983), *MIL-STD-810D*.
6. Smallwood, D. (1989), An analytical study of a vibration test method using extremal control of acceleration and force, *Proceedings of the IES*, 263-271.
7. Scharton, T. (1990), Motion and force controlled vibration testing, *Proceedings of the IES*, 77-85.
8. Ewins, D.J. (1994), Modal Testing: Theory and Practice, *Research Studies Press*.

222

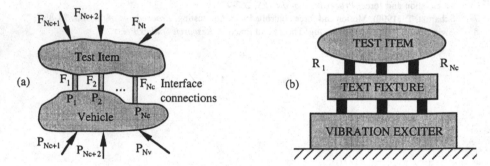

Figure 1   Test item in the filed and laboratory environments: (a) Attached to the vehicle in the filed; (b) Attached to the test fixture and vibration exciter in the laboratory.

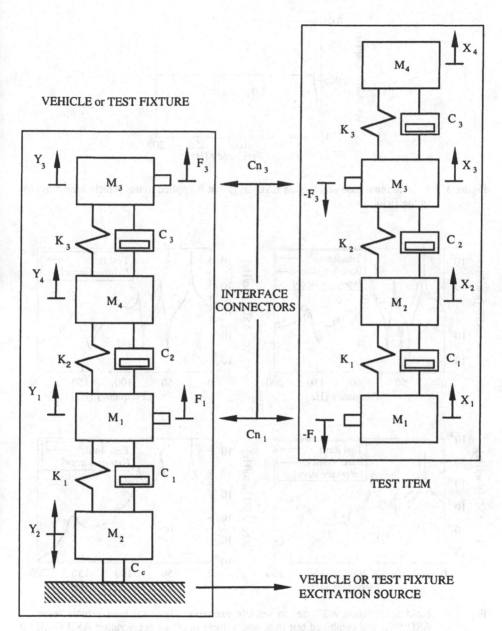

Figure 2   Simulation models: (a) Vehicle for field and test fixture for laboratory simulations;
(b) Test item.

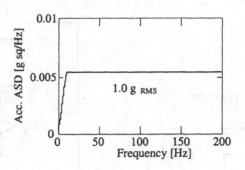

Figure 3   Vehicle input base acceleraion ASD Gy_{22} that is applied to the vehicle base mass M_2 in the field.

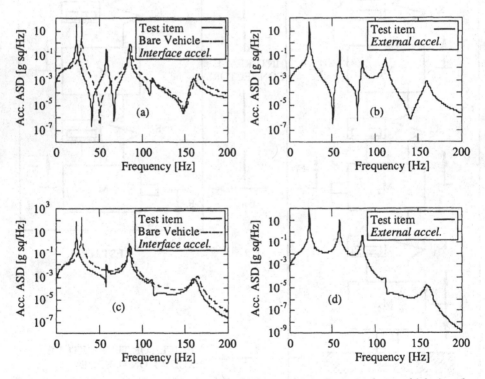

Figure 4   Field acceleration ASD due to vehicle excitation Gy_{22}: (a) bare vehicle interface ASD Gy_{11} and combined test item and vehicle interface acceleration ASD Gx_{11}; (b) test item external acceleration ASD Gx_{22} for combined system; (c) bare vehicle interface acceleration ASD Gy_{33} and combined test item and vehicle interface acceleration ASD Gx_{33}; (d) test item external acceleration ASD Gx_{44} for combined system.

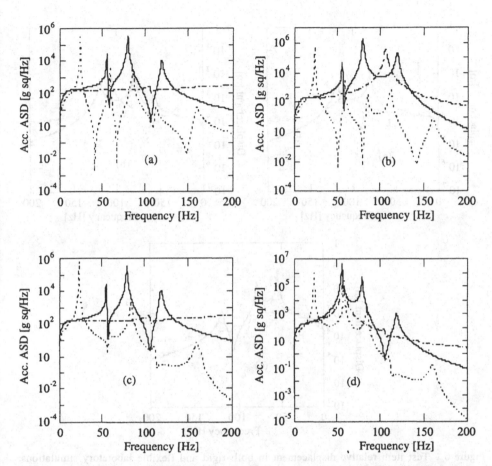

Figure 5   Laboratory acceleration ASD due to envelope excitation ASD Gz_{22} that is applied to rigid and flexible test fixtures: (a) interface acceleration ASD Gu_{11}; (b) external acceleration ASD Gu_{22}; (c) interface acceleration ASD Gu_{33}; (d) external acceleration ASD Gu_{44}; **Dotted**: Field x 10^4, **Dashed-dot**: Rigid fixture, **Solid**: Flexible test fixture.

226

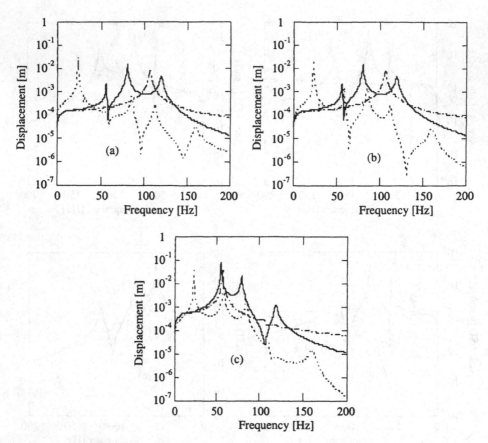

Figure 6   Test item relative displacement in both rigid and flexible laboratory simulations: (a) X_{12}; (b) X_{23}; (c) X_{34}; *Dotted*: Field x 10^2 , *Dashed-dot*: Rigid fixture, *Solid*: Flexible test fixture.

THEORETICAL MODELS FOR MODAL ANALYSIS

D. J. INMAN
Center for Intelligent Material Systems and Structures
Department of Mechanical Engineering, Virginia Tech
Blacksburg VA 24061, USA

1. Introduction

This chapter presents a summary of the analytical tools required to understand the theoretical aspects of modal analysis. From the mathematical point of view, modal analysis results because of the ability to decouple coupled sets of ordinary differential equations. This ability is normally set in the context of linear algebra and the theory of matrices. Hence, much of what follows is based upon manipulations of vectors and matrices. The orientation of this chapter is to provide the elementary background required for the following chapters.

2. Basic models

The basic assumptions in modal analysis are that the structure or machine of interest, can be adequately modeled by linear elements and a lumped mass domain. The equations of motion for lumped parameter systems with n degrees-of-freedom can be put in the following form

$$A_1 \ddot{q} + A_2 \dot{q} + A_3 q = f(t) \tag{1}$$

which is a vector differential equation. Here, $q = q(t)$ is an n vector of time-varying elements representing the displacements of the masses in the lumped mass model. The vectors $\ddot{q}$ and $\dot{q}$ represent the acceleration and velocities, respectively. The overdot means that each element of q is differentiated with respect to time. The vector q could also represent a generalized coordinate that may not be an actual physical coordinate or position but is related, usually in a simple manner, to the physical displacement. The coefficients A_1, A_2, and A_3 are n square matrices of constant real elements representing the various physical parameters of the system. The n vector $f = f(t)$ represents applied external forces and is also time-varying.

The matrices A_i will have different properties depending on the physics of the problem under consideration. The mathematical properties of these matrices will determine the physical nature of the solution $q(t)$, just as the properties of the scalars m, c, and k determined the nature of the solution to a single degree of freedom system.

227

J.M.M. Silva and N.M.M. Maia (eds.), Modal Analysis and Testing, 227–240.
© 1999 *Kluwer Academic Publishers. Printed in the Netherlands.*

With these definitions in mind, equation (1) can be rewritten as

$$M\ddot{q} + (D+G)\dot{q} + (K+H)q = f \tag{2}$$

where q and f are as before but

$\begin{aligned} M &= M^T & \textit{mass, or inertia matrix} \\ D &= D^T & \textit{viscous damping matrix} \text{ (sometimes denoted by C)} \\ G &= G^T & \textit{gyroscopic matrix} \\ K &= K^T & \textit{stiffness matrix} \\ H &= -H^T & \textit{circulatory matrix} \text{ (constraint damping)} \end{aligned}$

Some physical systems may also have asymmetric mass matrices. The mass matrix arises from the intertial forces in the system, the damping matrix arises from the intertial forces in the system, the damping matrix arises from dissipative forces proportional to velocity, and the stiffness matrix arises from elastic forces proportional to displacement. The symmetric matrices and the physical systems described by (2) can be further characterized by defining the concept of the definiteness of a matrix. In many cases, M, D, and K will be positive definite, a condition that ensures stability and follows from physical considerations, as it does in the single degree of freedom case.

3. Classifications of systems

This section lists the various classifications of systems modeled by equation (2) commonly found in the literature. These classifications are useful in verbal communication of the assumptions made when discussing a vibration problem. In the following, each word in italics is defined to imply the assumptions made in modeling the system under consideration.

The phrase *conservative system* usually refers to systems of the form

$$M\ddot{q} + Kq = f \tag{3}$$

where M and K are both symmetric and positive definite (with semi-definite implying a rigid body mode). However, the system

$$M\ddot{q} + G\dot{q} + Kq = f \tag{4}$$

where G is skew-symmetric is also conservative, in the sense of conserving energy, but is referred to as a *conservative gyroscopic system*, or an *undamped gyroscopic system*. Such systems arise naturally when spinning motions are present, such as in a gyroscope or spinning satellite.

Systems of the form

$$M\ddot{q} + D\dot{q} + Kq = f \tag{5}$$

where M, D, and K are all positive definite, are referred to as *damped nongyroscopic systems* and are also considered to be damped conservative systems in some instances, although they certainly do not conserve energy. Systems with symmetric and positive definite coefficient matrices are sometimes referred to as *passive* systems.

Classification of systems with asymmetric coefficients is not as straightforward as the classification depends on more detailed matrix theory. However, systems of the form

$$M\ddot{q} + (K + H)q = f \tag{6}$$

are referred to as *circulatory systems*. In addition, systems of the more general form

$$M\ddot{q} + D\dot{q} + (K + H)q = f \tag{7}$$

result from dissipation inferred to as *constraint damping* as well as external damping in rotating shafts. Combining all of these effects provides motivation to study the most general system of the form of equation (4), i.e.,

$$M\ddot{q} + (D + G)\dot{q} + (K + H)q = f \tag{8}$$

This expression is the most difficult analyze. To be complete, however, it is appropriate to mention that this model of a structure does not account for time-varying coefficients or nonlinearities, which are often present. Physically the expression represents the most general forces considered in the majority of modal analysis applications.

4. Modal analysis of undamped systems

The solution of equation (3) for $f = 0$ proceeds by assuming a harmonic solution of the form $x = e^{j\omega t}u$, where u is a nonzero vector of constants (and ω will become the natural frequency). Substituting this into equation (3) yields,

$$-\omega^2 M u + K u = 0 \tag{9}$$

or

$$K u = \lambda M u \tag{10}$$

where $\lambda = \omega^2$. Equation (10) is a statement of the *generalized eigenvalue problem*. The important feature of this formulation is that M and K are generally sparse, banded matrices, that are symmetric and positive definite. These features are very useful in numerically solving for the n eigenvalues λ_i and for the n eigenvectors u_i. The eigenvalues λ_i are just the squares of the natural frequencies ω_i^2. The solution of equation (3) is of the form

$$x(t) = \sum_{i=1}^{n} c_i \sin(\omega_i t + \phi_i) u_i \tag{11}$$

so that the eigenvectors u_i are also the vibration mode shapes. Here c_i and ϕ_i are constants to be determined by initial conditions. Because M and K are symmetric, it is known from matrix theory that λ_i and u_i are real valued and that u_i form a complete set which can be made orthonormal with respect to the matrix M.

Next consider multiplying equation (3) by the matrix M^{-1}. Again assuming a solution of the form $x = e^{j\omega t}u$ yields

$$-\omega^2 u + M^{-1} K u = 0 \tag{12}$$

for the homogeneous case so that

$$\left(M^{-1}K\right)u = \lambda u \tag{13}$$

This is the standard *algebraic eigenvalue problem*. The matrix $M^{-1}K$ is neither symmetric nor banded. Again there are n eigenvalues λ_i, which are the squares of the natural frequencies ω_i^2, and n eigenvectors u_i. The solution of equation (9), $x(t)$, is again of the form

$$x(t) = \sum_{i=1}^{n} c_i \sin(\omega_i t + \phi_i) u_i \tag{14}$$

where c_i and ϕ_i are constants to be determined by the initial conditions. Hence the eigenvectors u_i are also the mode shapes. Since $M^{-1}K$ is not symmetric, the solution of the algebraic problem could yield complex values for the eigenvalues and eigenvectors. However, they are known to be real valued because of the generalized eigenvalue problem formulation, equation (10), which has the same eigenvalues and eigenvectors.

Next consider the vibration problem obtained by substitution of the coordinate transformation $x = M^{-1/2} q$ into equation (3) and multiplying by $M^{-1/2}$. This yields the form

$$\ddot{q}(t) + M^{-1/2} K M^{-1/2} q(t) = 0 \tag{15}$$

where the matrix $M^{-1/2} K M^{-1/2}$ is symmetric but not necessarily sparse or banded unless M is diagonal. The solution of equation (15) is also obtained by assuming a solution of the form $q = e^{j\omega t}v$ where v is a nonzero vector of constants. Substituting this form into equation (15) yields

$$-\omega^2 v + M^{-1/2} K M^{-1/2} v = 0 \tag{16}$$

or

$$\left(M^{-1/2} K M^{-1/2}\right) v = \lambda v \tag{17}$$

where again $\lambda = \omega^2$. This is the *symmetric eigenvalue problem* and again results in n eigenvalues λ_i, which are the squares of the natural frequencies ω_i^2 and n eigenvectors v_i forming a complete set. The solution of equation (15) becomes

$$q(t) = \sum_{i=1}^{n} c_i \sin(\omega_i t_i + \phi_i) v_i \tag{18}$$

where c_i and ϕ_i are again constants of integration. The solution in the original coordinate system x is obtained from this last expression by multiplying by the matrix $M^{-1/2}$.

$$x = M^{-1/2} q = \sum_{i=1}^{n} c_i \sin(\omega_i t + \phi_i) M^{-1/2} v_i \qquad (19)$$

Hence, in this case, the mode shapes are the vectors $M^{-1/2} v_i$, where v_i are the eigenvectors of the symmetric matrix $M^{-1/2} K M^{-1/2}$. Since the eigenvalue problem here is symmetric, it is known that the eigenvalues and eigenvectors are real valued, as are the mode shapes, and that both form a complete set allowing the expansion in (19). The numerical advantage here is that the eigenvalue problem is symmetric, so more efficient numerical algorithms can be used. The numerical disadvantage is in calculating the matrix $M^{-1/2} K M^{-1/2}$ which requires computing the square root of a matrix. The alternative is to compute the eigenvectors of K and then normalize them with respect to the mass matrix M, which is simpler unless damping is present.

5. State space modal analysis

Again consider the vibration problem defined in equation (3). This equation can be transformed to a first-order vector differential equation by defining two new $n \times 1$ vectors, y_1 and y_2, by $y_1 = x$ and $y_2 = \dot{x}$. Note that y_1 is the vector of displacements and y_2 is the vector of velocities. Differentiating these two vectors yields

$$\dot{y}_1 = \dot{x} = y_2$$
$$\dot{y}_2 = \ddot{x} = -M^{-1} K \dot{y}_1 + M^{-1} f \qquad (20)$$

where the equation for $\dot{y}_2$ has been expanded by solving equation (23) for $\ddot{x}$. Equations (20) can be recognized as the first-order vector differential equation

$$\dot{y} = A y + b \qquad (21)$$

Here

$$A = \begin{bmatrix} 0 & I \\ -M^{-1} & 0 \end{bmatrix}, \quad y = \begin{bmatrix} y_1 \\ y_2 \end{bmatrix} = \begin{bmatrix} x \\ \dot{x} \end{bmatrix}, \quad b = \begin{bmatrix} 0 \\ M^{-1} f \end{bmatrix} \qquad (22)$$

where A is called the $2n \times 2n$ state-matrix, the 0 denotes an $n \times n$ matrix of zeros, I denotes the $n \times n$ identity matrix, and the state vector y is defined to be the $2n \times 1$ vector of displacements and velocities. The solution of equation (21) proceeds by assuming the exponential form $y = z e^{\lambda t}$, where z is a nonzero vector of constants and λ is a scalar. Upon substitution into equation (21) yields $\lambda z = A z$ or

$$A z = \lambda z \quad z \neq 0 \qquad (23)$$

This is again the standard algebraic eigenvalue problem. While the matrix A has many zero elements, it is now a $2n \times 2n$ eigenvalue problem. It can be shown that the 2n eigenvalues λ_i again corresponds to the n natural frequencies ω_i by the relation $\lambda_i = \omega_i j$, where $j = \sqrt{-1}$. The extra n eigenvalues are $\lambda_i = -\omega_i j$, so that there are still only n natural frequencies, ω_i. The 2n eigenvectors, z of the matrix A, however, are of the form

$$z_i = \begin{bmatrix} u_i \\ \lambda_i u_i \end{bmatrix}$$
(24)

where u_i are the mode shapes of the corresponding vibration problem. The matrix A is not symmetric and the eigenvalues λ_i and eigenvectors z_i would therefore be complex. The eigenvalues in this case are imaginary numbers of the form $\lambda_i = \omega_i j$. State space methods offer the advantage because numerical analysts and mathematicians have devoted decades to the study of the eigensolution of the single matrix A.

6. Damped systems

For large-order systems, computing the eigenvalues using equation (23) becomes numerically more difficult because it is of order 2n rather than n. The main advantage of the state-space form is in numerical simulations and in solving the damped multiple-degree-of-freedom vibration problem, which is discussed next. Now consider the damped vibration problem given in equation (5) where D represents the viscous damping in the system and is assumed to be symmetric and positive semidefinite. The state-matrix approach and related standard eigenvalue problem of equation (23) can also be used to describe the nonconservative vibration problem of equation (5). Multiplying (5) by the matrix M^{-1} yields

$$\ddot{x} + M^{-1} D \dot{x} + M^{-1} K x = 0$$
(25)

Again it is useful to rewrite this expression in a first-order or state-space form by repeating the procedure of the previous section. In this damped case, the state matrix A becomes

$$A = \begin{bmatrix} 0 & I \\ -M^{-1}K & -M^{-1}D \end{bmatrix}$$
(26)

The eigenvalue analysis for the damped system proceeds directly as for the undamped state-matrix system of equation (21).

A solution of equation (21) is again assumed of the form $y = z\,e^{\lambda t}$ and substituted into equation (21) to yield the eigenvalue problem $A z = \lambda z$. This again defines the standard eigenvalue problem of dimension $2n \times 2n$. In this case, the solution again yields 2n values λ_i, which may be complex valued. The 2n eigenvectors z_i described in equation (29) may also be complex valued (if the corresponding λ_i is complex). This, in turn, causes the physical mode shape u_i to be complex valued as well as the free-response vector $v(t)$.

The physical time response $x(t)$ is simply taken to be the real part of the first n coordinates of the vector $y(t)$ computed from

$$\mathbf{x}(t) = \sum_{i-1}^{2n} c_i \, \mathbf{y}_i \, e^{\lambda_i t} \tag{27}$$

The physical interpretation of the complex eigenvalues λ_i is taken directly from the complex numbers from the solution of underdamped single-degree-of-freedom system. In particular, the complex eigenvalues λ_i will appear in complex conjugate pairs in the form

$$\lambda_i = -\zeta_i \omega_i - \omega_i \sqrt{1 - \zeta_i^2} \, j$$

$$\lambda_{i+1} = -\zeta_i \omega_i + \omega_i \sqrt{1 - \zeta_i^2} \, j \tag{28}$$

where $j = \sqrt{-1}$, ω_i is the undamped natural frequency of the ith mode and ζ_i is the *modal damping ratio* associated with the ith mode. For the underdamped system (see Inman, 1989, for conditions), the numerical solution of the eigenvalue problem for the state matrix A of equation (26) produces a set of complex numbers of the form $\lambda_i = \alpha_i + \beta_i j$, where $\mathrm{Re}(\lambda_i) = \alpha_i$ with $\mathrm{Im}(\lambda_i) = \beta_i$. Comparing these expressions with equations (28) reveals the following formulas for calculating natural frequencies and damping ratios:

$$\omega_i = \sqrt{\alpha_i^2 + \beta_i^2} = \sqrt{\mathrm{Re}(\lambda_i)^2 + \mathrm{Im}(\lambda_i)^2} \tag{29}$$

$$\zeta_i = \frac{-\alpha_i}{\sqrt{\alpha_i^2 + \beta_i^2}} = \frac{-\mathrm{Re}(\lambda_i)}{\sqrt{\mathrm{Re}(\lambda_i)^2 + \mathrm{Im}(\lambda_i)^2}} \tag{30}$$

Equations (29) and (30) also provide a connection to the physical notions of natural frequency and damping ratios for the underdamped case. However it should be noted that the natural frequencies computed by equation (29) are not in general the same as the natural frequencies calculated by equation (13), unless the damping matrix is zero or proportional.

The complex-valued mode shape vectors $\mathbf{u}_i$ also appears in complex conjugate pairs and are referred to as *complex modes*. The physical interpretation of a complex mode is that each element describes the relative magnitude and phase of the motion of the degree of freedom associated with that element when the system is excited at that mode only. In the undamped real mode case, the mode shape vector is real and indicates the relative positions of each mass at any given instant of time at a single frequency. The difference between the real mode case and the complex mode case is that if the mode is complex, the relative position of each mass can also be out of phase by the amount indicated by the complex part of the mode shape's entry. The state-space formulation of the eigenvalue problem for the matrix A given by equation (26) is related to the most general linear vibration problem. It also forms the most difficult computational eigenvalue problem of those discussed above.

A symmetric generalized eigenvalue problem can also be formulated in state space for systems with damping in such a way as to avoid calculating the inverse of the matrix M. Note that if the mass matrix M has a very large element and a very small element, it could be computationally difficult to calculate M^{-1}. Consider substituting a new set of

234

variables $y_1 = x$ and $y_2 = \dot{x}$ into the damped system of equation (5). This yields the expression

$$M\dot{y}_2 = -D y_2 = K y_1 \tag{31}$$

Next consider the identity (recall that $\dot{y}_1 = y_2$)

$$-K\dot{y}_1 = K y_2 \tag{32}$$

Combining equations (31) and (32) into a single equation in the vector $y^T = \begin{bmatrix} y_1^T & y_2^T \end{bmatrix}$ yields

$$\begin{bmatrix} -K & 0 \\ 0 & M \end{bmatrix}\begin{bmatrix} \dot{y}_1 \\ \dot{y}_2 \end{bmatrix} = \begin{bmatrix} 0 & -K \\ -K & -D \end{bmatrix}\begin{bmatrix} y_1 \\ y_2 \end{bmatrix} \tag{33}$$

This expression can be written as

$$A\dot{y} = B y \tag{34}$$

where

$$A = \begin{bmatrix} -K & 0 \\ 0 & M \end{bmatrix} \quad B = \begin{bmatrix} 0 & -K \\ -K & -D \end{bmatrix} \quad y = \begin{bmatrix} x \\ \dot{x} \end{bmatrix} \tag{35}$$

In this form, both matrices A and B are symmetric. Substitution of $y = z e^{\lambda t}$ into equation (34) yields the symmetric generalized eigenvalue problem

$$\lambda A z = B z \tag{36}$$

where $\lambda_i = \zeta_i \omega_i \pm \omega_i \sqrt{1 - \zeta_i^2} j$ relates the solution of the symmetric generalized eigenvalue problem to the calculation of natural frequencies and damping ratios.

7. Lambda matrices

The most natural treatment of the general system of equation (1), is through the notion of a Lambda matrix as introduced to vibration analysis by Lancaster (1966). One alternative is to place equation (1) into the state space form and analyze it in the context of matrix pencils of the standard eigenvalue problem. In fact, many numerical algorithms do exactly that. However, the second-order form does retain more of the physical identity of the problem and hence is worth developing.

Again, assume solutions of (1) of the form $q(t) = u e^{\lambda t}$, where u is a vector of constants. Then equation (1) becomes

$$\left(\lambda^2 A_1 + \lambda A_2 + A_3\right)u e^{\lambda t} = 0$$

or, since $e^{\lambda t}$ is never zero

$$\left(\lambda^2 A_1 + \lambda A_2 + A_3\right)u = 0 \quad \text{or} \quad D_2(\lambda)u = 0 \tag{37}$$

where $D_2(\lambda)$ is referred to as a *lambda matrix* and u is referred to as a *latent vector*. In fact, in this case, u is called the right latent vector.

For the existence of nonzero solutions of (37), the matrix $D_2(\lambda)$ must be singular, so that

$$\det\left(D_2(\lambda)\right) = 0 \tag{38}$$

The solutions to this 2n-degree polynomial in λ are called *latent roots*, eigenvalues, or characteristic values and contain information about the system's natural frequencies and damping ratios given by equation (28). Note that the solution of (38) and the solution of $\det (A - \lambda I) = 0$, where A is the state matrix given by

$$A = \begin{bmatrix} 0 & I \\ -A_1^{-1}A_3 & -A_1^{-1}A_2 \end{bmatrix} \tag{39}$$

are the same. Also, the eigenvectors of A are just $\left[u_i^T \ \lambda_i u_i^T\right]^T$, where u_i are the latent vectors of (37) and λ_i are the solutions of (38).

An $n \times n$ lambda matrix, $D_2(\lambda)$, is said to be *simple* if A_1^{-1} exists and if for each eigenvalue (latent root) λ_i satisfying (37), the rank of $D_2(i)$ is $n - \alpha_i$; where α_i is the multiplicity of the eigenvalue λ_i. If this is not true, then $D_2(\lambda)$ is said to be *degenerate*. If each of the coefficient matrices are real and symmetric and if $D_2(\lambda)$ is simple, the solution of equation (1) is given by

$$q(t) = \sum_{i=1}^{2n} c_i \, u_i \, e^{\lambda_i t} \tag{40}$$

Here the c_i are 2n constants to be determined from the initial conditions, q is $n \times 1$, and the u_i are the right eigenvectors (latent vectors) of $D_2(\lambda)$. Note that in the undamped case ($A_2 = 0$), this equation collapses to the eigenvalue problem of a matrix.

Since, in general, u_i and λ_i are complex, the solution $q(t)$ will be complex. The physical interpretation is as follows: The displacement is the real part of $q(t)$, and the velocity is the real part of $\dot{q}(t)$. The terms *modes* and *natural frequencies* can be used if care is taken to interpret their meaning properly. The damped natural frequencies of the system are again related to the λ_i in the sense that if the initial conditions $q(0)$ and $\dot{q}(0)$ are chosen such that $c_i = 0$ for all values of i except $i = 1$, each coordinate $q_i(t)$ will oscillate (if underdamped) at a frequency determined by λ_i. The elements of u_i indicate the relative displacement and phase of each mass when the system vibrates at that frequency.

In many situations, the coefficient matrices are symmetric and the damping matrix D is chosen to be a form that allows the solution (1) to be expressed as a linear combination of the normal modes, or eigenvectors of the matrix $M^{-1/2} K M^{-1/2}$, which, of course, are real. In this case, the matrix of eigenvectors decouples the equations of motion. Consider the symmetric damped system of equation (5) and note the following:

1. If $D = \alpha M + \beta K$, where α and β are any real scalars, then the eigenvectors of (5) are the same as the eigenvectors of the undamped case.

2. If $D = \sum_{i=1}^{n} \beta_{i-1} K^{i-1}$, where β_i are real scalars, then the eigenvectors of (5) are the same as the eigenvectors of the undamped system.

3. The eigenvectors of (5) are the same as those of the undamped system if and only if $D M^{-1} K = K M^{-1} D$.

Systems satisfying any of the above rules are said to be *proportionally damped*, to have *Rayleigh damping*, or to be *normal mode systems*. Such systems can be decoupled by the modal matrix associated with the matrices M and K. Of the cases just mentioned, 3 is the most general and includes the other two as special cases.

As a generic illustration of a normal mode system, let S_m be the matrix of eigenvectors of K normalized with respect to the mass matrix M (i.e., $S_m = M^{-1/2} P$) so that

$$S_m^T M S_m = I$$
$$S_m^T K S_m = \Lambda_K = \text{diag}\left[\omega_i^2\right] \tag{41}$$

where ω_i^2 are the eigenvalues of the matrix $M^{-1} K$. If condition 3 holds, then the damping matrix is also diagonalized by the transformation S_m, so that

$$S_m^T D S_m = \text{diag}\left[2\zeta_i \omega_i\right] \tag{42}$$

where ζ_i are called the modal damping ratios. Then equation (5) can be transformed into a diagonal system via the following: Let $q(t) = S_m y(t)$ in (5) and premultiply by S_m^T to get

$$\ddot{y}_i(t) + 2\zeta_i \omega_i \dot{y}_i(t) + \omega_i^2 y_i(t) = f_i, \quad i = 1, 2, ..., n \tag{43}$$

where $y_i(t)$ denotes the ith component of the vector $y(t)$. Each of the n equations (43) is a scalar, which can be analyzed by single degree of freedom methods and f_i is the ith modal force.

8. Decoupling conditions and modal analysis

An alternative approach to solving for the response of a system by transform or matrix exponent methods is to use the eigenvalue and eigenvector information from the free response as a tool for solving the forced response. Approaches based on the eigenvectors of the system are referred to as *modal analysis* and also form the basis for understanding modal test methods.

If the state matrix A has a diagonal Jordan form (which happens when it has distinct eigenvalues), for example, then the matrix A can be diagonalized by its modal matrix. In this circumstance equation (21) can be reduced to 2n independent first-order equations. To see this, let u_i be the eigenvectors of the state matrix A with eigenvalues λ_i. Let $P = [u_1$

$u_2...u_{2n}]$ be the modal matrix of the matrix of A. Then substitute $x = P z$ into equation (21) to obtain

$$P\dot{z} = A P z + b \tag{44}$$

Premultiplying by P^{-1} yields the decoupled vector equation:

$$\dot{z} = P^{-1} A P z + P^{-1} b \tag{45}$$

each element of which is of the form

$$\dot{z}_i = \lambda_i z_i + F_i \tag{46}$$

where z_i is the ith element of the vector z and F_i is the ith element of the vector $F = P^{-1} b$. Equation (46) can now be solved using scalar integration of each of the first order equations subject to the initial condition $z_i(0) = [P^{-1} x(0)]_i$. In this way the vector $z(t)$ can be calculated, and the solution $x(t)$ in the original coordinates becomes simply $x(t) = P z(t)$. The amount of effort required to calculate the solution via this method is comparable to that required to calculate the solution via (40).

Alternately, if $D = 0$ and if K is assumed to be positive definite, each ω_i^2 is a positive number. Denoting the initial conditions by $y_i(0)$ and $\dot{y}_i(0)$, the solutions of (43) are calculated by the method of variation of parameters to be

$$y_i(t) = \frac{1}{\omega_i} \int_0^t f_i(t-\tau) \sin(\omega_i\tau) d\tau + y_i(0) \cos(\omega_i t) + \frac{\dot{y}_i(0)}{\omega_i} \sin(\omega_i t) \tag{47}$$

for $i = 1, 2, 3..., n$, (Boyce and DePrima, 1986).

If K is semidefinite, one or more values of ω_i^2 might be zero. Then equation (43) would become

$$\ddot{y}_i(t) = f_i(t) \tag{48}$$

Integrating (48) then yields

$$y_i(t) = \int_0^t \left[\int_0^t f_i(s) ds \right] dt + y_i(0) + \dot{y}_i(0)t \tag{49}$$

which represents a rigid body motion.

This method is referred to as *modal analysis* and differs from the state space modal approach in that the computations involve matrices and vectors of size n rather than 2n. The result is a solution for the position vector rather than the 2n dimensional state vector. The coordinates defined by the vector y are called *modal coordinates, normal coordinates, decoupled coordinates*, and (sometimes) *natural coordinates*.

The key to using modal analysis to solve for the forced response of systems with velocity-dependent terms is whether or not the system can be decoupled. As in the case of the free response, this will happen for symmetric systems if and only if $K M^{-1} D = D M^{-1} K$.

If it is assumed at $4\tilde{K} - \tilde{D}^2$ is positive definite (Inman, 1989), then $0 < i < 1$, and the solution for the damped case of (70) is (assuming all initial conditions are zero)

$$y_i(t) = \frac{1}{\omega_{di}} \int_0^t e^{-\zeta\omega_i\tau} f_i(t-\tau)\sin(\omega_{di}t)d\tau, \quad i = 1,2,3,...,n \tag{50}$$

where $\omega_{di} = \omega_i\sqrt{1-\zeta_i^2}$ and $\zeta_i = \lambda_i(D)/2\omega_i$, $\lambda_i(D)$ denoting the eigenvalues of the matrix D. This is the underdamped case.

In addition to the forced response given by equation (50), there will be a transient response, or homogeneous solution, due to any nonzero initial conditions. If this response is denoted by y_i^H, then the total response of the system in the decoupled coordinated system is the sum $y_i(t) + y_i^H(t)$.

For systems in which the coefficient matrices do not commute, i.e., for which $DM^{-1}K \neq KM^{-1}D$ in equation (5), modal analysis of a sort is still possible without resorting to state space. Let u_i be the eigenvectors of the lambda matrix $D_2(\lambda)$ defined by equation (37) with associated eigenvalue λ_i. Let n be the number of degrees of freedom (so there are 2n eigenvalues), let 2s be the number of real eigenvalues, and let $2(n-s)$ be the number of complex eigenvalues. Assuming that $D_2(\lambda)$ is simple and that the u_i are normalized so that

$$u_i^T(2M\lambda_i + D)u_i = 1 \tag{51}$$

Then a particular solution of (5) is given by Lancaster (1966) in terms of the generalized modes u_i to be

$$q(t) = \sum_{k=1}^{2s} u_k u_k^T \int_0^t e^{-\lambda_k(t+\tau)} f(\tau)d\tau + \sum_{k=2s+1}^{2n} \int_0^t Re\left\{e^{\lambda_k(t-\tau)}u_k u_k^T\right\}f(\tau)d\tau \tag{52}$$

This expression is more difficult to compute but does offer some insight into the form of the solution that is useful in modal testing. Also note that the normalization of the modes as given by equation (51) is very different from that of the undamped case.

9. Frequency response methods

This section extends the concept of frequency response for single degree of freedom systems to multiple degree of freedom systems. In so doing, the material in this section makes the connection between analytical modal analysis and experimental modal analysis. The development starts by considering the response of a structure due to a harmonic or sinusoidal input, denoted by $f(t) = f_0 e^{j\omega t}$. In the undamped case, an oscillatory solution of (3) of the form given by $q(t) = ue^{j\omega t}$ is assumed. This is equivalent to the frequency response theorem: if the system is harmonically excited, the response will consist of a steady-state term that oscillates at the driving frequency with different amplitude and phase.

Substitution of the assumed oscillatory solution into equation (3) yields

$$\left(K - \omega^2 M\right)u\, e^{j\omega t} = f_0\, e^{j\omega t} \tag{53}$$

Dividing through by the nonzero scalar $e^{j\omega t}$ and solving for $\mathbf{u}$ yields

$$\mathbf{u} = \left(K - \omega^2 M\right)^{-1} \mathbf{f}_0 \tag{54}$$

Note that the matrix inverse of $(K - \omega^2 M)$ exists as long as ω is not one of the natural frequencies of the structure. The matrix coefficient of (54) is defined to be the *receptance matrix*, denote by $\alpha(\omega)$, i.e.,

$$\alpha(\omega) = \left(K - \omega^2 M\right)^{-1} \tag{55}$$

Equation (54) can be thought of as the *response model* of the structure because it yields the vector $\mathbf{u}$, which, in turn yields the steady-state repose of the system to the input force $\mathbf{f}(t)$. Each element of the response matrix can be related to a single-frequency response function by examining the definition of matrix multiplication. In particular if all the elements of the vector $\mathbf{f}_0$, denoted by f_i, except the kth element are set equal to zero, then the ik^{th} element of $\alpha(\omega)$ is just the receptance transfer function between u_i, the ith element of the response vector $\mathbf{u}$, and f_k. That is

$$\alpha_{ik}(\omega) = \frac{u_i}{f_k}, \quad f_i = 0, \quad i = 0, ..., n, \quad i \neq k \tag{56}$$

An alternative to computing the inverse of the matrix $(K - \omega^2 M)$ is to use the modal decomposition of $\alpha(\omega)$. Recalling equations (41) the matrices M and K can be rewritten as $M = S_m^T S_m^{-1}$ and $K = S_m^{-T} \text{diag}\,(\omega_r^2) S_m^{-1}$ where ω_i are the natural frequencies of the system and S_m is the matrix of modal vectors normalized with respect to the mass matrix. Substitution of these "modal" expressions into (55) yields

$$\alpha(\omega) = S_m \left[\text{diag}\left(\omega_r^2 - \omega^2\right)\right]^{-1} S_m^T \tag{57}$$

Equation (57) can also be written in summation notation by considering the ik^{th} element of $\alpha(\omega)$ and partitioning the matrix S_m into columns, denoted by s_r. This yields

$$\alpha(\omega) = \sum_{r=1}^{n} \left(\omega_r^2 - \omega^2\right)^{-1} s_r s_r^T \tag{58}$$

The ik^{th} element of the receptance matrix becomes

$$\alpha_{ik}(\omega) = \sum_{r=1}^{n} \left(\omega_r^2 - \omega^2\right)^{-1} \left[s_r s_r^T\right]_{ik} \tag{59}$$

where the matrix element $[s_r s_r^T]_{ik}$ is identified as the *modal constant* or *residue* for the rth mode and the matrix $s_r s_r^T$ is called the *residue matrix*. Note that the right-hand side of (59) can also be rationalized to form a single fraction consisting of the ratio of two polynomials in ω^2. Hence $[s_r s_r^T]_{ik}$ can also be viewed as the matrix of constants in the partial fraction expansion of (56).

240

Next, consider the same procedure applied to the damped case. As always, consideration of damped systems results in two cases: those systems that decouple and those that do not.

First consider the damped case such that $DM^{-1}K = KM^{-1}D$, so that the system decouples and the system eigenvectors are real. In this case the eigenvectors of the undamped system are also eigenvectors for the damped system. The definition of the receptance matrix takes on a slightly different form to reflect the damping in the system. Under the additional assumption that the system is underdamped, (54) becomes

$$\mathbf{u} = \left(K + j\omega D - \omega^2 M \right)^{-1} \mathbf{f}_0 \tag{60}$$

Because the system decouples, the matrix D can be written as $D = S_m^{-T} \operatorname{diag}(2\zeta_r \omega_r) S_m^{-1}$ and (57) becomes

$$\mathbf{u} = S_m \left[\operatorname{diag}\left(\omega_r^2 + 2k\zeta_r \omega_r \omega - \omega^2 \right)^{-1} \right] S_m^T \mathbf{f}_0 \tag{61}$$

This expression defines the complex receptance matrix given by

$$\alpha(\omega) = \sum_{r=1}^{n} \left(\omega_r^2 + 2j\zeta_r \omega_r \omega - \omega^2 \right)^{-1} \mathbf{s}_r \, \mathbf{s}_r^T \tag{62}$$

Next consider the nonproportionally damped case. In this case the eigenvectors $\mathbf{s}_r$ are complex and the receptance matrix is given (see Lancaster 1966) as

$$\alpha(\omega) = \sum_{r=1}^{n} \left\{ \frac{\mathbf{s}_r \mathbf{s}_r^T}{j\omega - \lambda_r} + \frac{\mathbf{s}_r^* \mathbf{s}_r^{*T}}{j\omega - \lambda_r^*} \right\} \tag{63}$$

Here the asterisk denotes the conjugate; the λ_r are the complex system eigenvalues and the $\mathbf{s}_r$ are the system eigenvectors. The expressions for the receptance matrix and the interpretation of an element of the receptance matrix given by equation (56) form the background for modal testing.

10. Summary

The methods of theoretical modal analysis have been reviewed for linear lumped mass systems. Various classifications of linear systems have been discussed and the issue of complex versus real modes has been summarized.

11. References

1. Boyce, E. D. and DePrima, R. C. (1986), Elementary Differential Equation and Boundary Value Problem, 4th ed., New York: John Wiley.
2. Inman, D. J. (1989), Vibration with Control, Measurement and Stability, *Prentice Hall*, Englewood Cliffs, New Jersey.
3. Lancaster, P. (1968), Lambda Matrices and Vibrating Systems.

FUNDAMENTALS OF TIME DOMAIN MODAL IDENTIFICATION

S. R. IBRAHIM
Professor of Mechanical Engineering
Old Dominion University
Norfolk, VA, U.S.A.

1. Introduction

Identification of modal parameters in the time domain offers an attractive alternative to the more classical approaches in the frequency domain. The elimination of the need to perform frequency transformation on the inputs and responses and the associated errors of leakage, truncation, frequency resolution and lengthy time records, among other errors, enhances the accuracy and practicality of the time domain approaches.

Two classes of time domain identification exist. The first class are those techniques that requires measuring or identifying inputs. These have proven more complex and their computational requirements are prohibitive. A good reason in support of the second class of time domain identification techniques which are based on using the systems free response thus eliminates the complexities of dealing with measuring or identifying inputs; a powerful advantage that balances the inability of these techniques to compute the modal participation factors needed in computing modal residuals and modal masses. These quantities are only available if the free responses are in fact the unit impulse responses.

Free decays time domain identification techniques coupled with random decrement transformation offer a valuable tool for identifying the modal parameters of structures from operational or ambient responses. When structures are subjected to some unmeasurable or unknown random inputs, the random decrement transformation can reduce these responses to equivalent free decay responses or correlation functions. These random decrement functions can be used directly to identify structures' modal parameters.

2. Free-decay and unit impulse responses

Here the structure under consideration is assumed to be linear with general viscous damping. If [m], [c] and [k] are the system's N x N mass, damping and stiffness matrices, then the governing equations of motion are:

$$[m]\{\ddot{y}\} + [c]\{\dot{y}\} + [k]\{y\} = \{0\} \tag{1}$$

The solution to such a system is known to be of the form:

J.M.M. Silva and N.M.M. Maia (eds.), Modal Analysis and Testing, 241–250.
© 1999 *Kluwer Academic Publishers. Printed in the Netherlands.*

$$\{y\} = \sum_{i=1}^{2N} A_i \{\psi\}_i \, e^{\lambda_i t} \qquad (2)$$

where $\{\psi\}_i$ is the system's i^{th} complex mode shape and λ_i is the system's i^{th} characteristic root which contains information on natural frequencies and damping factors. The A_i is the initial condition constant.

If in some cases the structures' unit impulse responses $h_{pq}(t)$, which is the inverse Fourier Transform of the systems transfer function $H_{pq}(\omega)$ at location p due to input at location q, is available, then it can be used as free decay response. The equation for the unit impulse response is:

$$h_{pq}(t) = \sum_{i=1}^{2N} A_{pqi} \, e^{\lambda_i t} \qquad (3)$$

where A_{pqi} are the residuals.

3. Idealized theory of time domain identification

In this section, to simplify the derivation, it is assumed that the structure's response is measured at N locations, contains N modes and is free of noise. Then in the following section the issues pertaining to incomplete and noisy measurements are systematically addressed.

Thus from equation (2) and incorporating the constants A_i with mode shape vector $\{\psi\}_i$, the response vector at time t_k can be written as

$$\{y(t_k)\} = \{y_k\} = \begin{bmatrix} \psi_1 & \psi_2 & \cdots & \psi_{2N} \end{bmatrix} \begin{Bmatrix} e^{\lambda_1 t_k} \\ e^{\lambda_2 t_k} \\ \vdots \\ e^{\lambda_{2N} t_k} \end{Bmatrix} \qquad (4)$$

Measuring the response vector $\{y\}$ at 2N time instants and form the N x 2N response matrix [y], the following equation results:

$$[y] = \begin{bmatrix} y_1 & y_2 & \cdots & y_{2N} \end{bmatrix} = [\psi][\Lambda] \qquad (5)$$

where the [ψ] is the N x 2N modal matrix and [Λ] is 2N x 2N matrix of the form:

$$[\Lambda] = \begin{bmatrix} e^{\lambda_1 t_1} & \cdots & e^{\lambda_1 t_{2N}} \\ \vdots & & \vdots \\ e^{\lambda_{2N} t_1} & \cdots & e^{\lambda_{2N} t_{2N}} \end{bmatrix} \qquad (6)$$

It is important to note here that there were no constraints or conditions on the selection of times $t_1, t_2 \dots t_{2N}$.

Equation (5) forms the basis of the identification. However, efforts to deal directly with the equation in its form proved futile due to the nonlinearities of the $[\Lambda]$ matrix. To eliminate the nonlinearities of equations, equation (5) is repeated for responses delayed some time shift from those responses in equation (5).

To establish some relation between a response vector and the delayed response vector, consider a response vector at time $t_k + \Delta t_3$ (other time shifts Δt_1 and Δt_2 are to be addressed later):

$$
\begin{aligned}
\left\{ y(t_k + \Delta t_3) \right\} &= \sum_{i=1}^{2N} \{\psi\}_i \, e^{\lambda_i (t_k + \Delta t_3)} \\
&= \sum_{i=1}^{2N} \{\psi\}_i \, e^{\lambda_i \Delta t_3} e^{\lambda_i t_k} \\
&= \sum_{i=1}^{2N} \{\overline{\psi}\}_i \, e^{\lambda_i t_k}
\end{aligned}
\tag{7}
$$

where

$$
\{\overline{\psi}\}_i = \{\psi\}_i \, e^{\lambda_i \Delta t_3}
\tag{8}
$$

Thus time shift of the response vector simply multiplies the mode shape $\{\psi\}_i$ by a scalar $e^{\lambda_i \Delta t_k}$.

Now for an N x 2N response matrix delayed Δt_3 from those responses of equation (5):

$$
[\overline{y}] = [\overline{\psi}][\Lambda]
\tag{9}
$$

Combining equations (5) and (9) gives:

$$
\begin{bmatrix} [y] \\ [\overline{y}] \end{bmatrix} = \begin{bmatrix} [\psi] \\ [\overline{\psi}] \end{bmatrix} [\Lambda]
\tag{10}
$$

or

$$
[\phi] = [\Psi][\Lambda]
\tag{11}
$$

Applying a time shift Δt_1 to equation (11) results in a new 2N x 2N response matrix $[\hat{\phi}]$ of the form:

$$
[\hat{\phi}] = \begin{bmatrix} \hat{y} \\ \hat{\overline{y}} \end{bmatrix} = \begin{bmatrix} [y(t_k + \Delta t_1)] \\ [y(t_k + \Delta t_3 + \Delta t_1)] \end{bmatrix}
\tag{12}
$$

and

$$
[\hat{\phi}] = [\Psi] \, \text{diag}[\alpha][\Lambda]
\tag{13}
$$

where

$$
\alpha_i = e^{\lambda_i \Delta t_1}
\tag{14}
$$

Eliminating $[\Lambda]$ from equations (11) and (13) gives

$$[\hat{\phi}][\phi]^{-1}[\Psi] = [\Psi]\,\text{diag}[\alpha]$$

If

$$[A] = [\hat{\phi}][\phi]^{-1} \tag{15}$$

then,

$$[A]\{\Psi\} = \alpha\,\{\Psi\} \tag{16}$$

which is an eigenvalue problem.

3.1.   CALCULATION OF CHARACTERISTIC ROOTS

For an eigenvalue α_i of equation (16), which is usually complex, let:

$$\alpha_i = \beta_i + j\gamma_i \tag{17}$$

If α_i corresponds to the characteristic root λ_i, where

$$\lambda_i = a_i + jb_i \tag{18}$$

and since according to equation (14)

$$\alpha_i = e^{\lambda_i \Delta t_1}$$

then,

$$a_i = \frac{1}{2\Delta t_1}\ln\!\left(\beta_i^2 + \gamma_i^2\right) \tag{19}$$

and

$$b_i = \frac{1}{\Delta t_1}\tan^{-1}\frac{\gamma_i}{\beta_i} \tag{20}$$

In this case, for the linear structure, b_i is the damped natural frequency $(\omega_d)_i$:

$$b_i = (\omega_d)_i = (\omega_n)_i\,\sqrt{1 - \zeta_i^2} \tag{21}$$

and

$$a_i = -(\omega_n)_i\,\zeta_i \tag{22}$$

where ζ_i is the ith damping factor. Then:

$$(\omega_n)_i = \sqrt{a_i^2 + b_i^2} \tag{23}$$

and

$$\zeta_i = \frac{a_i}{\sqrt{a_i^2 + b_i^2}} \tag{24}$$

3.1.1. *Condition on Δt_1*

To ensure a unique value for b_i from equation (20), the following condition is imposed:

$$b_{max} \cdot \Delta t_1 < \pi \qquad (25)$$

but

$$f_{max} = \frac{b_{max}}{2\pi}$$

then

$$(1/\Delta t_1) < 2 f_{max} \qquad (26)$$

Usually f_{max} is known from the setting of the law pass filters. It is advisable to sample at a higher rate than $2f_{max}$ in order to obtain sufficient response records for identification since we are dealing with decaying response.

3.1.2. *Zooming or On-Purpose Aliasing*

For structures being tested in a higher frequency range f_{min} to f_{max}, sampling at a rate higher than $2f_{max}$ may prove impractical. In this case zooming or on-purpose aliasing can be achieved by limiting the angles of equation (20) to within $2\pi k$ and $2\pi k + \pi$ thus:

$$b_{min} \, \Delta t_1 > 2\pi k \qquad (27\text{-}a)$$

$$b_{max} \, \Delta t_1 < 2\pi k + \pi \qquad (27\text{-}b)$$

Thus, the condition on Δt_1 becomes

$$\frac{2 f_{max}}{2k+1} < \frac{1}{\Delta t_1} < \frac{f_{min}}{k} \qquad (28)$$

From equation (28) k can be chosen and the maximum possible k is:

$$k_{max} = integer\left(\frac{f_{min}}{2(f_{max} - f_{min})}\right) \qquad (29)$$

To illustrate the above, let a structure be tested in the 900-1000 Hz range. Thus

$$k_{max} = integer\left(\frac{900}{200}\right) = 4$$

Then, applying equation (28) for different choice of k give:

$$
\begin{array}{llll}
\text{for} & k = 4 & 222.2 < & f_s < 225 \\
& = 3 & 286 < & f_s < 300 \\
& = 2 & 400 < & f_s < 450 \\
& = 1 & 667 < & f_s < 900 \\
& = 0 & 2000 < & f_s < \infty
\end{array}
$$

After k is chosen, then equation 20 takes the form,

$$b_i = \frac{1}{\Delta t_1}\left(\tan^{-1}\frac{\gamma_i}{\beta_i} + 2k\pi\right) \tag{30}$$

4. Addressing practical considerations

In the previous section, the basic theory of time domain modal identification was derived under ideal or theoretical conditions. As any other inverse problem, the complexities of modal identification arise from:

1. The number of modes contributing to the responses is unknown.
2. The number of measurements is unequal to the "unknown" number of modes.
3. The measurements are corrupted with noise.

These problems are addressed in this section.

4.1. PSEUDO MEASUREMENTS

In most applications, the number of available measurements, p, is smaller than the number of modes. The measurements vector $\{y\}$ of length p is of the form,

$$\{y(t)\} = \sum_{i=1}^{2N} \{\psi\}_i e^{\lambda_i t} \quad , N > p \tag{31}$$

The concept of time shift is here utilized to create a new "pseudo" set of measurements. For a time shift Δt_2,

$$\{y(t+\Delta t_2)\} = \sum_{i=1}^{2N} \{\psi\}_i e^{\lambda_i(t+\Delta t_2)}$$

$$= \sum_{i=1}^{2N} e^{\lambda_i \Delta t_2}\{\psi\}_i e^{\lambda_i t}$$

or

$$\{\bar{y}(t)\} = \sum_{i=1}^{2N} \{\bar{\psi}\}_i e^{\lambda_i t} \tag{32}$$

This additional set of measurements, combined with the original set of measurements, give a set measurements containing 2p measurements. This process can be repeated until the response vector is of length N.

4.1.1. *Choice of Δt_2*
Considering the fact that the measured response is a free decay response, large time shifts will utilize higher noise to signal responses. On the other hand, too small Δt_2 will make the rows of the response matrices $[\phi]$ and $[\hat{\phi}]$ closely dependent thus adversely affect the accuracy of matrix inversion. One period of the center frequency is suggested as a guideline.

4.2. LEAST SQUARES SOLUTION FOR NOISE REDUCTION

In order to reduce the effects of noise on the matrix [A], the response matrices [φ] and [φ̂] are constructed to be 2N x 2r, r > N. This produces an overdetermined system of equations from which [A] can be computed as:

$$[A] = [\hat{\phi} \, \phi^T][\phi \, \phi^T]^{-1} \tag{33}$$

It is recommended that r be larger than 2N.

Least squares solution for [A] produce biased estimates for damping. Singular Value Decomposition and QR algorithms can as well be used to solve for the [A] matrix.

4.3. OVERSIZED IDENTIFICATION MODELS

This concept has proven to be the primary reason for the recent developments in modal identification in general. The noisy responses of structures are of the form:

$$\{y(t)\} = \sum_{i=1}^{2N} \{\psi\}_i e^{\lambda_i t} + \{n(t)\} \tag{34}$$

Here the noise {n(t)} is modeled in the same form as the structural response, e.g.

$$\{y(t)\} = \sum_{i=1}^{2N} \{\psi\}_i e^{\lambda_i t} + \sum_{i=2N+1}^{2M} \{R\}_i e^{\eta_i t}$$

$$= \sum_{i=1}^{2M} \{\psi\}_i e^{\lambda_i t} \tag{35}$$

Thus the noise is modeled as "computational" or "noise" modes to be identified together with the structural modes.

This does not only improve identification accuracy but as well solves the need to accurately determine N, the number of modes in the responses. Only a rough estimate for N is necessary. Then an identification model with M modes, M >> N, is used.

4.4. MODAL CONFIDENCE FACTOR (MCF)

An oversized identification model identifies M modes. Out of these M modes, N are structural modes and (M-N) are computational mode. The MCF has proven to be quite effective in differentiating between the two types.

If Q_{ij} is the identified j^{th} mode displacement at measurement i and $\overline{Q}_{ij}$ is that for same measurement delayed Δt_3, then according to the linear theory of vibration the expected value for $\overline{Q}_{ij}$ should be:

$$\left(\overline{Q}_{ij}\right)_{expected} = Q_{ij} \, e^{\lambda_j \Delta t_3} \tag{36}$$

Then by defining the (MCF)$_{ij}$ as:

$$(MCF)_{ij} = \overline{Q}_{ij} \Big/ \Big(Q_{ij} \, e^{\lambda_j \Delta t_3} \Big) \tag{37}$$

If this particular mode j was a structural mode, then $(MCF)_{ij} = 1.0 + j0.0$, otherwise it is a complex quantity. A user set criterion can be used to distinguish structural modes from computational ones. A magnitude of 0.9 and a phase of $\pm$ 10.0° is an example of such a criterion.

5. Excitation methods

The required free-decay responses can be generated in several ways. Among them are:

1. Impact hammer or general impact. Here the input need not to be measured. Response is collected right after the impact diminishes.

2. Quick release.

3. Free decay following a stopped wide-band excitation.

4. Inverse FFT of transfer functions.

5. Random Decrement Transformation.

5.1. RANDOM DECREMENT TRANSFORMATION

Since its introduction by H. A. Cole in 1968 the RD technique has been extensively investigated and used as a powerful tool in modal identification. The technique is basically a time domain ensemble, or record, averaging process which converts random responses — due to unknown or unmeasurable stationary random inputs — to free decay responses.

When introduced by Cole, the RD technique was intended to extract the "signature" of one particular vibration mode for a single narrow-band filtered random response measurement. The time records to be averaged are started when the response reaches a pre-specified level. Such a condition is referred to as a triggering condition and in this case a constant level is used for triggering. The resulting free decay in this case a resembles a step response or auto correlation function. Other triggering conditions are possible as shown in equations (40)-(43).

For use in modal identification, the RD technique was extended by Ibrahim in 1977 to cover multi-mode multi-measurement random response. In order to maintain phase relations between measurements, triggering is applied to only one of the measurements and the entire response vector is averaged accordingly.

The mathematical basis of the RD technique can be explained by writing the equations of the response of a linear structure as

$$x(t) = e^{A(t-t_0)} x(t_0) + \int_{t_0}^{t} h(t - \tau) \, f(\tau) \, d\tau \tag{38}$$

The first part of equation (38) is the homogeneous solution of the governing differential equations of motion of the system, which are dependent on system dynamic characteristics and not on the input. The second part is the convolution integral dependent on system transfer function and input. If the inputs to the system are stationary random, then averaging segments of the responses will cause the random part of the response (second part of equation (38)) to average out. To ensure that the homogeneous part of the response (first part of equation (38)) does not also average out, the triggering condition is applied to one of the measurements. Thus the RD functions can be computed from equation (38) as

$$D_{xx_j}(\tau) = \frac{1}{N} \sum_{i=1}^{N} x(t_i + \tau) \Big| T_{x_j(t_j)}$$ (39)

where D_{xx} contains the RD functions of measurement vector x with measurement x_j as the triggering measurement, N is the number of averages and $T_{x_j}(t_i)$ is the chosen triggering condition. D_{xx} is the estimate of the mean value of a stochastic vector process on the condition T_{x_j}, provided that the measurement x are realizations of an ergodic stochastic vector process. Well-known triggering conditions are:

$$T_{x_j}(t) = \left\{ x_j(t) = a \right\}$$ (40)

$$T_{x_j}(t) = \left\{ a \leq x_j(t) < b \right\}$$ (41)

$$T_{x_j}(t) = \left\{ x_j(t) = 0, \dot{x}_j(t) > 0 \right\}$$ (42)

$$T_{x_j}(t) = \left\{ a \leq x_j(t) < b, \dot{x}_j(t) = 0 \right\}$$ (43)

The triggering conditions in equations (40)-(43) are referred to as level crossing, positive point, zero crossing with positive slope and local extremum, respectively. The reference measurement, x_j, can be any arbitrary measurement or the RD computation can be repeated as many times as the number of measurements, changing the reference measurement every time. The RD function obtained from the measurement x_j is referred to as an auto RD function, whereas other RD functions are referred to as cross RD functions. This is analogous to auto correlation and cross correlation functions. The auto RD functions are in general known to have less noise than the cross RD functions. One of the reasons is that the initial conditions of the auto RD functions are strictly given by the triggering condition, whereas the initial condition of the cross RD functions is the result of an averaging process. Another explanation is that the rms of cross RD functions is usually lower than the rms of auto RD functions. This is due to a possible low correlation between the responses at two different locations of a structure. It would be preferable to base an identification on auto RD functions solely, but this causes the loss of phase information and thus the possibility to identify mode shapes. If n measurements are available n different sets of RD functions, which all contain 1 auto and n - 1 cross RD functions could be estimated. Only a single one of these sets of RD functions is necessary in order to estimate modal parameters. The problem is to select a proper measurement for triggering beforehand. On the other hand all sets of RD functions could be used in the modal parameter extraction procedure. For several measurements, say 8 to 16, recorded simultaneously, the estimation time for all sets of RD functions will be high, but the accuracy of the modal parameters will also be high, since all possible RD functions are used. The above considerations illustrate that the RD technique can be an accurate technique with relatively high estimation time or an extremely fast technique with corresponding less accuracy.

6. Concluding remarks

Modal parameter identification in the time domain has been summarized in a generic manner. The presentation is limited to those techniques which use free decay response. Several other time domain techniques has emerged over the past twenty years. They all share the same concepts especially that of using oversized identification models to model the measurements noise as computational modes. The combination of the free decay time domain identification and the Random Decrement Transformation offer a powerful tool in identifying structures from their operational or ambient responses.

7. Bibliography

1. Ibrahim, S. R. and Mikulcik, E. C. (1977), A method for the direct identification of vibration parameters from the free response, *The Shock and Vibration Bulletin* 47, 183-196.
2. Ibrahim, S. R. (1977), Random decrement technique for modal identification of structures, *Journal of Spacecraft and Rockets* 14, 696-700.
3. Ibrahim, S. R. (1978), Modal confidence factor in vibration testing, *Journal of Spacecraft and Rockets* 15, 313-316.
4. Ibrahim, S.R. and Pappa, R. S. (1982), Large modal survey testing using the Ibrahim time domain identification technique, *Journal of Spacecraft and Rockets* 19, 459-465.
5. Ibrahim, S. R. (1984), A modal identification algorithm for higher accuracy requirements, *Proceedings of the 25th Structures, Structural Dynamics and Materials Conference*, Paper No. 84-0928, 117-122.
6. Hanks, B. R., Miserentino, R., Ibrahim, S. R., Lee, S H., and Wada, B. K. (1978), Comparison of modal test methods on the voyager payload, *Transactions of the American Society of Mechanical Engineers* 87.
7. Andrew, L. V. (1981), An automated application of Ibrahim time domain method to responses of the space shuttle, *AIAA Proceedings of the 22nd Structures, Structural Dynamics and Materials Conference*, Paper No. 81-0526, 155-165.
8. Carney, T. G., Martinez, D. R., and Ibrahim, S. R. (1983), Modal identification of a rotating blade system, *Proceedings of the 24th Structures, Structural Dynamics and Materials Conference*, AIAA Paper No. 83-0815.
9. Kauffman, R. R. (1984), Application of the Ibrahim time domain algorithm to spacecraft transient responses, *Proceedings of the 25th Structures, Structural Dynamics and Materials Conference*, AIAA Paper 84-0946.
10. Vold, H., Kundrat, J., Rocklin, G. T., and Russel, R. (1982), A multi-input modal estimation algorithm for minicomputers, SAE Paper No. 820194.
11. Vold, H. and Crowley, Jr. (1983), A modal confidence factor for the polyreference method, *Proceedings of the 4rd International Modal Analysis Conference*, Sponsored by Union College, Schenectady, N.Y., 305-310.
12. Smith, W.R. (1981), Least squares time-domain method for simultaneous identification of vibration parameters from multiple free-response records, *Proceedings of 22nd AIAA Structures, Structural Dynamics, and Materials Conference*, Atlanta, Georgia, 194-201.
13. Smith, K. E. (1984), An evaluation of a least squares time domain parameter identification method for free response measurement, *Proceedings of the 2nd International Modal Analysis Conference* (Union College).
14. Ibrahim, S. R. (1984), Time domain quasi-linear identification of nonlinear systems, *American Institute of Aeronautics and Astronautics Journal* 22, 817-823.
15. Ibrahim, S. R. (1978), An upper Hessenberg sparse matrix algorithm for modal identification on minicomputers, *Journal of Sound and Vibration* 113 (1), 46-57.
16. Ibrahim, S. R., Asmussen, J. S., and Brincker, R. (1998), Vector triggering random decrement technique for higher identification accuracy, *ASME Journal of Vibration and Acoustics*, to appear.

MODAL IDENTIFICATION METHODS IN THE FREQUENCY DOMAIN

N. M. M. MAIA
Instituto Superior Técnico, Dep. Engª. Mec.,
Av. Rovisco Pais, 1049-001 Lisboa, Portugal

1. Introduction

As for time domain methods, there are two main categories of identification methods in the frequency domain: indirect methods and direct methods. The former are based on the modal model, i.e, on the modal parameters (natural frequency, damping ratio and modal constants), while the latter are based on the spatial model, i.e, they allow for the evaluation of the mass, stiffness and damping matrix coefficients.
There is still a third category, the tuned-sinusoidal methods. This is a very particular case, that will be briefly mentioned in section 4.

Indirect methods can be based either on a single-degree-of-freedom (SDOF) or multiple-degree-of-freedom (MDOF) approach. Historically, indirect methods using an SDOF approach were the first to appear, mainly due to the big limitations in computer capacity (or in their absolute absence!). They are much richer in phisical content and form the basis for the more complicated ones, mostly not much more than just numeric and algebraic extensions.

There is still another classification concerning the number of inputs and outputs. Methods that analyse one FRF at a time are called SISO (single-input-single-output). When they process several FRFs simultaneously taken with the excitation always at the same point, they are called SIMO (single-input-multiple-output). Finally, there are MIMO methods (multiple-input-multiple-output), i.e., they have into account FRFs taken with the exciter in different locations and measurements also in different co-ordinates. Here, we shall give a brief overview of some identification methods. Details can be found in [1].

2. Indirect methods

2.1. SDOF

The Peak Amplitude method is the simplest of all, where the natural frequencies are taken from the observation of the peaks of the amplitude of the response. The damping ratios are evaluated from the sharpness of those peaks and the mode shapes from the ratios of the peak amplitudes at various points along the structure. As the natural frequencies and

J.M.M. Silva and N.M.M. Maia (eds.), Modal Analysis and Testing, 251–264.

damping ratios are global properties of the structure, they must be the same for each mode, so averaged values from the various FRFs can be made.

For lightly damped structures with well-separated modes, this method may provide quite reasonable results. Figure 1 illustrates this method.

The use of this method may constitute a good exercise for someone giving the first steps into modal identification. This is also true for most of the early methods, as they are very much based on the physical interpretation of the dynamic behaviour of structures and in the way the graphical representation of the different quantities are presented. For example, the display of the real part of the receptance frequency response function (FRF) exhibits zero crossings along the frequency axis. It is not difficult to realise that these crossings correspond to the undamped natural frequencies of the structure. At those same frequencies, the imaginary part of the response exhibits minima (see figure 2).

These characteristics of the response, allowing for the identification of the natural frequencies, are the basis of the Quadrature Response and Maximum Quadrature Component methods. These methods in themselves may not be very accurate in most situations, but their strong physical liaison help to consolidade the understanding of the link between the identification procedure and the physical phenomena behind it.

More than just mathematical techniques, most of the SDOF identification methods are based on the physical interpretation of the dynamic responses of structures. One of the most relevant and acknowledged contributions were those of Kennedy and Pancu [2], about fifty years ago. They realised that the display of the real part of the receptance versus its imaginary part (Nyquist plot) being a circle around each natural frequency could be used as a useful tool to identify the modal parameters associated to each resonance of a structure. The particular circular shape is not the real key issue here. The important fact is that the response around the natural frequency is amplified and even if the accuracy of the measurements is not very good, we know that the data should theoretically lay along a particular curve, a circle in that case (see figure 3).

Therefore, it is not very important to have a very accurate response exactly at the resonance. Moreover, it was found that the natural frequency was located at the point where the rate of change of arc length with frequency attained a maximum. The half-power-points were used to evaluate the damping ratio, and the mode shapes were estimated from the ratios of the diameters of the circles, fitted around each natural frequency for various output responses.

The Kennedy-Pancu method suffered various refinements along the years, until it has been consolidated and settled as one of the most popular modal analysis methods: the Circle-Fitting method. In contrast with the more primitive methods mentioned before, which relied heavily on the simple observation of the responses, the Circle-Fitting is, in fact, the first identification method, comprising a series of numerical techniques that permit - in an interactive way - the correct establishment of a mathematical model to support the physical evidence.

The receptance of an N d.o.f system with hysteretic damping is given by:

$$\alpha_{jk}(\omega) = \sum_{r=1}^{N} \frac{{}_rC_{jk}}{\omega_r^2 - \omega^2 + i\eta_r\omega_r^2} \tag{1}$$

where η_r and $_rC_{jk}$ are the hysteretic damping ratio and the complex modal constant $_rC_{jk} = (C_r\, e^{i\phi_r})_{jk}$, respectively, associated with each mode r.

The SDOF approach implies that we study eq.(1) around each mode, as:

$$\alpha_{jk}(\omega) = \frac{_rC_{jk}}{\omega_r^2 - \omega^2 + i\eta_r\omega_r^2} + _rD_{jk} \tag{2}$$

where $_rD_{jk}$ represents the contribution of the modes besides the one under consideration, known as the "residual".

In the Circle-Fitting method the first term on the r.h.s. of (2) represents a circle in the complex plane amplified and rotated by $_rC_{jk}$. $_rD_{jk}$ is a complex constant that displaces the circle (see figure 4).

After a first identification of each of the modes individually, the analysis is repeated for each mode, this time subtracting from the original FRF data the contribution of the modes (besides the one under study) that have already been identified. Mathematically, this can be expressed as:

$$\tilde{\alpha}_r = \tilde{\alpha} - \sum_{\substack{s=1 \\ s \neq r}}^{N} \alpha_s \tag{3}$$

where $\tilde{\alpha}$ is the initially measured FRF data, $\tilde{\alpha}_r$ is the resulting FRF of the mode under consideration and α_s is the regenerated FRF contribution of each mode already analysed.

In practice, there is a limited frequency range for which experimental data are collected. After the identification of all the modes is completed, it is still necessary to account for the contribution of the terms situated outside the experimental frequency range (figure 5). This adjustment is made through a frequency dependent term, of the form:

$$D_{jk}(\omega) = \frac{R_{jk}}{\omega^2} + S_{jk} \tag{4}$$

where R_{jk} and S_{jk} are complex terms. R_{jk}/ω^2 represents a "massic" term, whereas S_{jk} represents a "stiffness" term. If $\tilde{\alpha}_{jk}$ and α_{jk} are the measured and identified FRFs, respectively,

$$\tilde{\alpha}_{jk} = \alpha_{jk} + D_{jk} \tag{5}$$

Two data points (complex) are necessary to identify the two complex unknowns, although more data points can be considered.

Another important contribution in the field of SDOF methods is the Inverse method. Basically, is the inverse of the Circle-fitting method. The identification is based on the inverse of the first term on the r.h.s. of (2). The graphical representation of the real and imaginary parts of that term versus the frequency (squared) gives straight lines (see figure 6), although they may appear distorted by neighbouring modes.

To analyse and fit straight lines is simpler than to analyse or fit circles and for systems with well separated modes the Inverse method can be more advantageous and quicker. In the case of close modes an iterative procedure based on equation (3) is necessary. In general, it may be visually easier to distinguish close modes with the Inverse method (figure 7).

In the Nyquist plot, two very close modes may look just like a single circle, while in the inverse representation one may still observe variations in the slopes of the straight lines. Moreover, when some modes are very lightly damped, a circle fit may be difficult, due to the lack of points. In such a case, a straight-line fitting is preferable and presents no problem. Figures 8 a) and b) illustrate this point.

Some SDOF methods take automatically into consideration the contribution of neighbouring modes along the identification process. These are Dobson's method [3] and the Characteristic Response Function (CRF) [4]. From equation (2), both methods consider the residual $_rD_{jk}$ as a constant and so, taking differences between the receptance at different frequencies, the residual term is eliminated. In Dobson's method, we have:

$$\Delta\alpha = \alpha(\omega) - \alpha(\Omega) =$$

$$C\left[\frac{\omega^2 - \Omega^2}{\left(\omega_r^2 - \omega^2\right)\left(\omega_r^2 - \Omega^2\right) - \eta_r^2\omega_r^4 + i\eta_r\omega_r^2\left(2\omega_r^2 - \omega^2 - \Omega^2\right)}\right] \tag{6}$$

from which a function Δ is defined as:

$$\Delta = \frac{\omega^2 - \Omega^2}{\alpha(\omega) - \alpha(\Omega)} \tag{7}$$

This procedure leads to a series of straight lines and from their slopes and intercepts it is possible to identify the modal parameters. In [5], it is shown that Dobson's method can be viewed as a generalization of the Inverse method. Figure 9 shows a typical display of this method for theoretically generated data.

The CRF method takes a third frequency point and through one more subtraction, elliminates the information about the modal constant. The result is an expression that depends only on the natural frequencies and damping ratios and therefore is the same no matter where we excite and measure along the structure, i.e., it is a characteristic of the structure:

$$\beta_j = \frac{\dfrac{\left(\alpha_j - \alpha_k\right)\left(\omega_k^2 - \omega_\ell^2\right)}{\left(\alpha_k - \alpha_\ell\right)\left(\omega_j^2 - \omega_k^2\right)} - 1}{\omega_j^2 - \omega_\ell^2} = \frac{1}{\omega_r^2 - \omega_j^2 + i\eta_r\omega_r^2} \tag{8}$$

The identification process is once again based on an inverse procedure [6].

In summary, one can say that there are essencially two basic SDOF modal analysis methods: the Circle-fitting and the Inverse method, and if used in conjunction enable us to solve a wide range of practical problems. For better accuracy, we can adopt Dobson's method instead of the Inverse method and for more confidence and interpretation of the number of existing modes we can use the CRF, superimposing several curves taken along the structure, as they should always provide the same answer. In figure 10 a practical example is shown.

2.2. MDOF

Probably the simplest MDOF method of all is the Ewins-Gleeson method, for lightly damped structures. In fact, in a first stage the damping is neglected and we have:

$$Re(\alpha(\omega)) = \sum_{r=1}^{N} \frac{C_r}{\omega_r^2 - \omega^2} \tag{9}$$

Because the damping is small, the FRF peaks are sharp and the natural frequencies are easy to identify directly (as in the Peak Amplitude method). Therefore, the only unknowns are the modal constants. So, by taking N frequency points, a linear system is built to calculate the modal constants:

$$
\begin{Bmatrix} C_1 \\ C_2 \\ \vdots \\ C_N \end{Bmatrix} =
\begin{bmatrix}
\frac{1}{\omega_1^2 - \Omega_1^2} & \frac{1}{\omega_2^2 - \Omega_1^2} & \cdots & \frac{1}{\omega_N^2 - \Omega_1^2} \\
\frac{1}{\omega_1^2 - \Omega_2^2} & \frac{1}{\omega_2^2 - \Omega_2^2} & \cdots & \frac{1}{\omega_N^2 - \Omega_2^2} \\
\vdots & \vdots & \cdots & \vdots \\
\frac{1}{\omega_1^2 - \Omega_N^2} & \frac{1}{\omega_2^2 - \Omega_N^2} & \cdots & \frac{1}{\omega_N^2 - \Omega_N^2}
\end{bmatrix}^{-1}
\begin{Bmatrix} Re(\tilde{\alpha}_1) \\ Re(\tilde{\alpha}_2) \\ \vdots \\ Re(\tilde{\alpha}_N) \end{Bmatrix} \tag{10}
$$

where Ω are the chosen frequencies and $\tilde{\alpha}$ the corresponding experimental results. Later, the damping factors are evaluated at each peak and the residuals are also estimated, as explained before.

One of the methods worth to mention is th Global method [7, 8], which is a single-input-multi-output (SIMO) method. It has the same philosophy as Dobson's in the sense that it takes differences of the receptances ($\Delta\alpha$) and mobilities ($\Delta\dot{\alpha}$) to elliminate the residual term, taken as a constant. The development of this method leads to the following eigenvalue problem:

$$\left[[\Delta\dot{\alpha}]' - s_r [\Delta\alpha]' \right]\{z_r\} = \{0\} \tag{11}$$

where $[\Delta\dot{\alpha}]'$ and $[\Delta\alpha]'$ are obtained from the initial ones $[\Delta\dot{\alpha}]$ and $[\Delta\alpha]$ after a refinement process to obtain the number of effective existing modes, through an SVD analysis.

Another very popular method is the Rational Fraction Polynomial method (RFP) [9]. Considering the viscously damped model, it starts from

$$\alpha(\omega) = \sum_{r=1}^{N} \frac{A_r + i\omega B_r}{\omega_r^2 - \omega^2 + i2\xi_r\omega_r\omega} \tag{12}$$

which is written in rational fraction form:

$$\alpha(\omega) = \frac{\displaystyle\sum_{k=0}^{2N-1} a_k(i\omega)^k}{\displaystyle\sum_{k=0}^{2N} b_k(i\omega)^k} \tag{13}$$

Establishing an error functional between $\alpha(\omega)$ and the measured values $\tilde{\alpha}(\omega)$, and minimizing it, leads to a linear system to be solved for a_k and b_k:

$$\begin{bmatrix} [Y] & [X] \\ [X]^T & [Z] \end{bmatrix} \begin{Bmatrix} \{a\} \\ \{b\} \end{Bmatrix} = \begin{Bmatrix} \{G\} \\ \{F\} \end{Bmatrix} \tag{14}$$

As the matrices involved have elements with increasing powers of frequency, they tend easily to be ill-conditioned. The problem is circumvented by reformulating it in terms of orthogonal polynomials. The coefficients of those polynomials are the new unknowns, and at the end a_k and b_k are recovered and the modal parameters are evaluated.

An extension of this method to analyse a global set of FRFs was later developed (the SIMO version), which is the Global Rational Fraction Polynomial method (GRFP) [10, 11].

The Complex Exponential Frequency Domain [12] is a single-input-single-output method, corresponding to the Ibrahim Time Domain method [13, 14]. From the Impulse-Response-Fuction (IRF),

$$h(t) = \sum_{r=1}^{2N} A_r \, e^{s_r t} \tag{15}$$

and so,

$$\alpha(\omega) = \sum_{r=1}^{2N} \frac{A_r}{i\omega - s_r} \tag{16}$$

For N shifted responses, $h_k(t) = h(t + kT)$, and so, $\alpha_k(\omega) = \sum_{r=1}^{2N} \frac{A_r \, e^{s_r kT}}{i\omega - s_r}$.

This means that this method is based on the analysis of series of FRFs obtained from a series of shifted values of the same IRF. At the end of the process, we have an eigenvalue problem to solve, formally similar to the one found in Ibrahim's method, from which the modal parameters are estimated.

3. Direct methods

In this category, various methods can be mentioned, such as the Spectral [15], the SFD [16,17], the ISSPA [18] and the Multi-Matrix [19-21]. The objective is to identify directly from the measurements the matrices of the system. In very general terms (see [19] for more details), they take the MDOF equilibrium equation (in terms of amplitudes):

$$\left[-\omega^2[M] + i\omega[C] + [K] \right]\{\overline{Y}\} = \{F\} \tag{17}$$

Taking as many frequencies, responses and forces as wished, it is possible to derive expressions of the type:

$$\left[-\omega_1^2[M] + i\omega_1[C] + [K]\right]\left\{\overline{Y}_1\right\} = \{F\}$$
$$\left[-\omega_2^2[M] + i\omega_2[C] + [K]\right]\left\{\overline{Y}_2\right\} = \{F\}$$
$$\vdots \qquad \vdots \qquad (18)$$
$$\left[-\omega_L^2[M] + i\omega_L[C] + [K]\right]\left\{\overline{Y}_L\right\} = \{F\}$$

From these, it is possible to obtain estimates for [K], [M] and [C] matrices.

4. Tuned-sinusoidal methods

This is a very special category of identification methods, based on the tuning of real modes, by exciting a structure at each natural frequency by a set of exciters appropriately distributed in space and time (and phase). They require an a priori aproximate location of the natural frequencies. The philosophy is to apply various forces so that they balance the dissipative ones. This happens when the measured displacements are in quadrature with the applied forces. When such a tuning is achieved, the undamped modes and natural frequencies are known.

5. Final remarks

There are many available modal identification methods in the frequency domain. The simplest ones usually imply a loger and more interactive analysis, but they give more insight into the physics of the problem under study. The more sophisticated ones are faster but often are not very much flexible.

Nowadays, the existing commercial software for industrial applications tends to provide very automatic and quick methods of analysis rather than slow interactive techniques, more suited for the development and research environments. However, those packages usually offer means for controlling and assessing the quality of the analyses, using indicators, charts, various graphical possibilities associated to the re-synthesised curves, etc. One of the most popular means to help the user in taking decisions when doing an analysis is the so-called stabilisation chart. This is often active during global curve fitting of various FRFs. The analysis is repeated several times, taking different frequency points or time intervals, and the quality of the analysis is assessed from the percentage variation of the results obtained. For instance, the user may impose that the estimates on the natural frequencies and damping ratios do not vary more than 1% and 3%, respectively, and then observes the repeatability or stabilisation of those properties in some modes, while a scatter is seen in others (figure 11 illustrates an example). In this way, the user can obtain a picture of which are the reliable answers and which ones are just the result of errors due to the mathematical process used in the analysis or due to some other origin, like noisy data or nonlinear behaviour.

A final conclusion is that it is not possible to say a priori that "this" or "that" method is better for a particular situation. The choice of the method to use depends on the available resources, on the available time, on the objectives of the study and on the personal experience of the user.

258

6. References

1. Maia, N. M. M., Silva, J. M. M. *et al.* (1997), Theoretical and Experimental Modal Analysis, *Ed.: Maia & Silva, Publ. Research Studies Press, Distrib. John Wiley & Sons.*

2. Kennedy, C. C., Pancu, C. D. P. (1947), Use of vectors in vibration measurement and analysis, *Journal of the Aeronautical Sciences*, Vol. 14, No. 11, 603-625.

3. Dobson, B. J. (1987), A straight-line technique for extracting modal properties from frequency response data, *Mechanical Systems and Signal Processing*, Vol. 1, No. 1, 29-40.

4. Maia, N. M. M., Ribeiro, A. M. R., Silva, J. M. M. (1994), A new concept in modal analysis: the Characteristic Response Function (CRF), *International Journal of Analytical and Experimental Modal Analysis*, Vol. 9, No. 3, 191-202.

5. Maia, N. M. M. (1991), Reflections on some sdof modal analysis methods, *International Journal of Analytical and Experimental Modal Analysis*, Vol. 6, No. 2, 69-80.

6. Maia, N. M. M., Silva, J. M. M., Ribeiro, A. M. R. (1995), Identification of structural dynamic properties with modal constant consistency, *Proceedings of Vibration and Noise*, Venice, Italy, 366-374.

7. Fillod, R., Lallement, G., Piranda, J., Raynaud, J. L. (1984), Méthode globale d'identification modale, *Mécanique, Matériaux, Électricité.*

8. Fillod, R., Lallement, G., Piranda, J., Raynaud, J. L. (1985), Global method of modal identification, *Proceedings of the 3rd International Modal Analysis Conference (IMAC III)*, Vol. II, Orlando, Florida, U. S. A., 1145-1151.

9. Richardson, M. H., Formenti, D. L. (1982), Parameter estimation from frequency response measurements using rational fraction polynomials, *Proceedings of the 1st International Modal Analysis Conference (IMAC I)*, Orlando, Florida, U. S. A., 167-181.

10. Richardson, M. H., Formenti, D. L. (1985), Global curve-fitting of frequency response measurements using the rational fraction polynomial method, *Proceedings of the 3rd International Modal Analysis Conference (IMAC III)*, Orlando, Florida, U. S. A., 390-397.

11. Richardson, M. H. (1986), Global frequency and damping estimates from frequency response measurements, *Proceedings of the 4th International Modal Analysis Conference (IMAC IV)*, Los Angeles, California, U. S. A., 465-470.

12. Schmerr, L. W. (1982), A new complex exponential frequency domain technique for analysing dynamic response data, *Proceedings of the 1st International Modal Analysis Conference (IMAC I)*, Orlando, Florida, U. S. A., 183-186.

13. Ibrahim, S. R., Mikulcik, E. C. (1973), A time domain modal vibration test technique, *The Shock and Vibration Bulletin*, Vol. 43, No. 4, 21-37.

14. Ibrahim, S. R., Mikulcik, E. C. (1976), The experimental determination of vibration parameters from time responses, *The Shock and Vibration Bulletin*, Vol. 46, No. 5, 187-196.

15. Klosterman, A. (1971), On the experimental determination and use of modal representation of dynamic characteristics, *Ph.D. Thesis*, University of Cincinnati.

16. Coppolino, R. N. (1981), A simultaneous frequency domain technique for estimation of modal parameters from measured data, *SAE Technical Paper Series*, No. 811046.

17. Coppolino, R. N., Stroud, R. C. (1986), A global technique for estimation of modal parameters from measured data, *Proceedings of the 4th International Modal Analysis Conference (IMAC IV)*, Los Angeles, California, U. S. A., 674-681.

18. Link, M., Vollan, A. (1978), Identification of structural system parameters from dynamic response data, *Z. Flugwiss. Weltraumforsch*, Vol. 2, No. 3, 165-174.

19. Leuridan, J. (1984), Some direct parameter model identification methods applicable for multiple modal analysis, *Ph.D. Dissertation*, Department of Mechanical and Industrial Engineering, University of Cincinnati.

20. Vold, H., Leuridan, J. (1982), A generalized frequency domain matrix estimation method for structural parameter identification, *7th International Seminar on Modal Analysis*, Katholieke Universiteit Leuven, Belgium.

21. Leuridan, J. M., Kundrat, J. A. (1982), Advanced matrix methods for experimental modal analysis - a multi-matrix method for direct parameter excitation, *Proceedings of the 1st International Modal Analysis Conference (IMAC I)*, Orlando, Florida, U. S. A., 192-200.

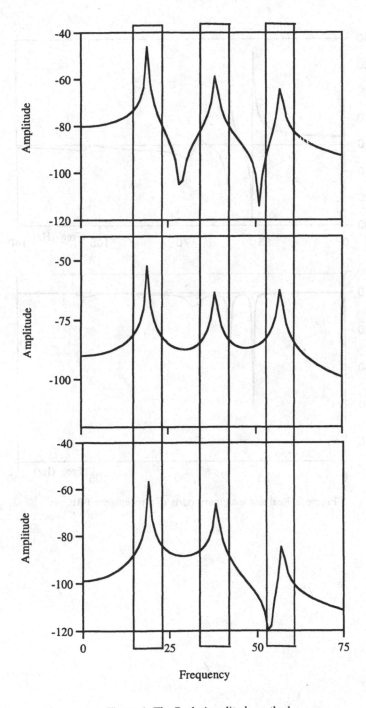

Figure 1 The Peak Amplitude method.

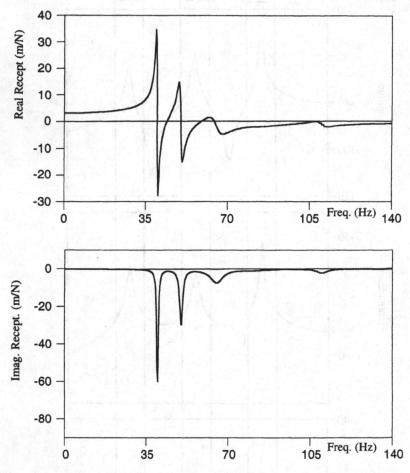

Figure 2 Real and imaginary parts of a receptance FRF.

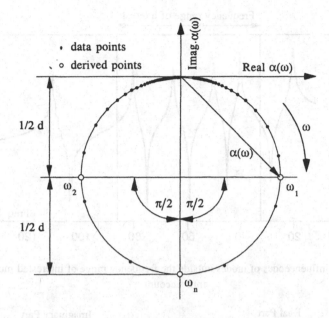

Figure 3 Nyquist plot of an SDOF receptance FRF.

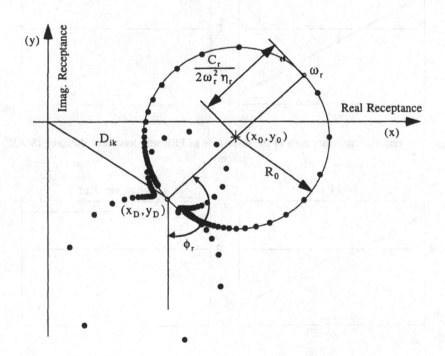

Figure 4 Nyquist plot of the receptance, showing the SDOF circle-fit approach.

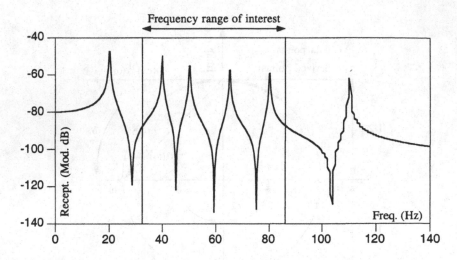

Figure 5 The influenceodes of modes outside the frequency range of interested must be taken into account.

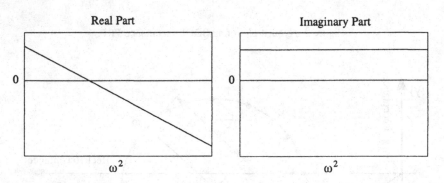

Figure 6 Real and imaginary parts of the inverse of an FRF with hysteretic damping (SDOF system).

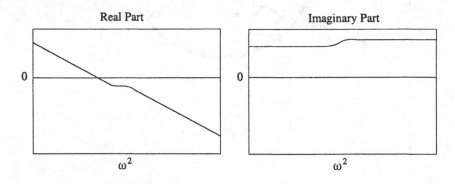

Figure 7 Real and imaginary parts of the inverse of an FRF showing two close modes.

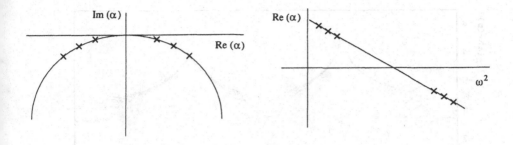

Figure 8 a)   Representation of the
               receptance on the Nyquist plot.

Figure 8 b)   Representation of the real part
               of the inverse of receptance.

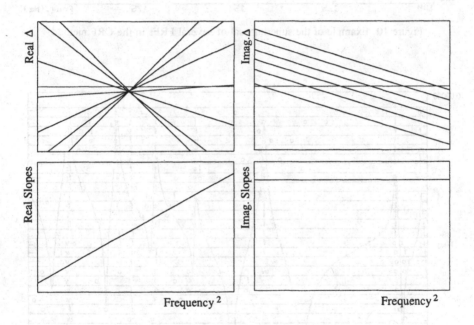

Figure 9 Example of analysis using Dobson's method (theoretical data).

264

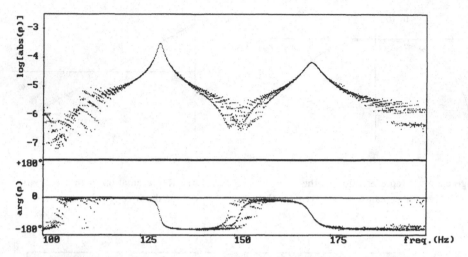

Figure 10 Example of the superimposal of several FRFs in the CRF method.

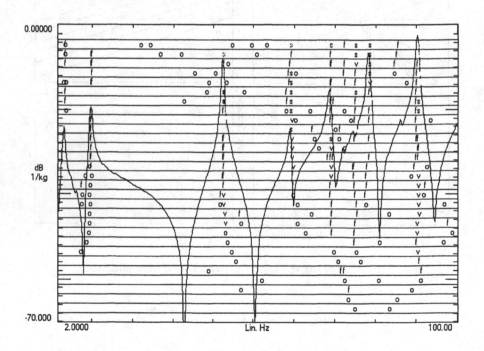

Figure 11 Example of an analysis using a stabilisation chart.

PARAMETRIC IDENTIFICATION BASED ON PSEUDO-TESTS

G. LALLEMENT, S. COGAN
Applied Mechanics Laboratory, UMR 6604 CNRS-UFC
24 rue de l'Epitaphe, 25030 Besançon Cedex, France

Abstract

A strategy is developed to reduce the poor conditioning of estimation equations used in the parametric identification of finite element models for linear elastodynamic structures. It is based on the simultaneous exploitation of synthesized responses resulting from different boundary condition configurations obtained by introducing fictive constraints. This approach is then applied to an updating procedure employing partial output residuals constructed from the eigensolutions of the model and structure. Two numerical examples illustrate the potential of this method.

1. Introduction

The recent developments in the field of parametric identification of mechanical structures are motivated by the necessity of improving the quality of knowledge models such as those provided by finite element methods. These models are used for classical dynamic analyses, such as predicting of the effects of structural modifications, as well as for parametric and shape optimization. The primary difficulties encountered in the parametric updating of complex structures are the result of the ill-posedness of this type of inverse problem and include:

- incoherencies between the structure and the corresponding model form;

- the high dimensionality in the space of uncertain parameters;

- the limited knowledge space, that is to say, structure is observed with a relatively small number of sensors (narrow spatial window) and limited frequency band (narrow frequency window).

A methodology is proposed which limits the effect of the last two sources of ill-posedness, thus allowing the poor conditioning of the estimation equations to be reduced a priori. A first solution consists in exploiting additional information resulting from complementary structural tests, namely:

- static and dynamic tests under diverse configurations of linearly independent excitations;

J.M.M. Silva and N.M.M. Maia (eds.), Modal Analysis and Testing, 265–280.
© 1999 *Kluwer Academic Publishers. Printed in the Netherlands.*

- tests on decoupled structures;

- tests on structures after introduction of known modifications such as parameters [1]; [2], subdomains, connectivities and boundary conditions [3] - [5].

These *real* complementary tests have the disadvantage of being excessively costly and are thus generally unacceptable in an industrial environment. Moreover, even known modifications are frequently accompagnied by additional unknowns, for example, the stiffness of the connections between the structure and added inertias or the residual effects of a non-ideally grounded degree of freedom.

A second solution exploits the additional data obtained from *fictive* complementary tests. These tests are the same as above but the structural behavior is predicted based on a more limited experimental data base. By hypothesis, the principle of superposition is valid for the structure under study. They allow the construction of complex loading conditions which cannot be realized in practice. More generally, they open the door to both input optimization and selective sensitivity approaches.

In order to illustrate the last points, three examples can be cited where the concept of fictive tests can be applied:

- Construction of fictive inputs corresponding to outputs having imposed properties;

- Determination of the spatial distribution of inputs, based on a set of observed frequency response functions, which lead to a response function which is colinear to an eigenvector of the associated conservative structure to the dissipative system [6];

- Construction of outputs having special properties resulting from fictive inputs;

- Transformation of a parametric identification problem having a large number of unknowns to a series of decoupled estimation problems having a reduced number of unknowns (selective sensitivity [7]);

- A priori regularisation;

- Regularization of the system of estimation equations by exploitation of outputs resulting from fictive grounding of degrees of freedom (abr. dof).

Only the principles concerning this last aspect will be developed in the following sections.

2. Enlargement of the knowledge space

It is assumed in what follows that the parameter estimation problem leads to a procedure-specific linear (or linearized) system of estimation equations based on the initial boundary condition configuration of the structure, denoted by the superscript $^{(0)}$, and having the form:

$$A^{(0)} x = b^{(0)} \tag{1}$$

where:

$A^{(0)}$ known observation matrix of the initial configuration;

x   unknown vector of correction coefficients;

$b^{(0)}$ known vector characterizing the distance between the model and the structure in their initial configurations.

We can associate to (1) a series of q systems of estimation equations corresponding to modified structures, denoted by (u) , u=1 to q, defined by different fictive groundings of one or several dof. Each of these systems has the form:

$$A^{(u)} x = b^{(u)} \quad , u = 1 \text{ to } q \qquad (2)$$

where:

$A^{(u)}$ known observation matrix of the uth boundary configuration

$b^{(u)}$ known vector characterizing the *distance* between the model and the structure in their u^{th} configuration

Regrouping (1) with the q systems (2) leads to the global system:

$$A x = b \qquad (3)$$

where:

$$A^T = \left[A^{(0)\ T}; ...; A^{(u)\ T}; ...; A^{(q)\ T} \right]$$

$$b^T = \left[b^{(0)\ T}; ...; b^{(u)\ T}; ...; b^{(q)\ T} \right]$$

The global system (3), general largely overdetermined, can be condensed before proceeding to its solution. This condensation consists in selecting *a posteriori* the subset of identifiable parameters with the best precision as well as the best corresponding subset of estimation equations. In the following paragraph, a method for performing such a selection will be presented. A procedure for constructing the system (2), in the particular case of linearized output residuals based on eigensolutions, will be presented further on.

3. Selection of parameters and equations

The procedure comprises two successive steps based on the properties of the QR factorization.

3.1. SELECTION OF THE BEST IDENTIFIABLE PARAMETERS

The global system (3) is assumed to be overdetermed: $A \in \mathbb{R}^{M,N}$, M > N. Applying the QR factorization with pivoting to the matrix A, its columns are ordered by the orthogonal permutation matrix $\Pi \in \mathbb{R}^{N,N}$ according to their best linear independence:

$$A \Pi = Q R \qquad (4)$$

where: $\Pi^T \Pi = \Pi \Pi^T = I_N$; $Q \in \mathbb{R}^{M,M}$, $Q^T Q = Q Q^T = I_N$. The matrix $R \in \mathbb{R}^{M,N}$ is an upper triangular matrix in which:

$$|R_{11}| \geq |R_{22}| \geq ... |R_{nn}| \geq |R_{NN}|$$

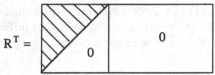

Equation (4) becomes: $R y = z$, where:

$$y \overset{\Delta}{=} \Pi^T x; \quad z \overset{\Delta}{=} Q^T b \tag{5}$$

Expression (5) can be partitioned into submatrices and leads to a selection of the n, $1 \leq n \leq N$, most identifiable parameters based on the observation matrix A.

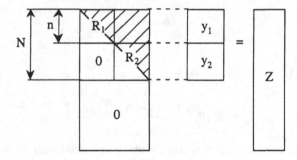

These parameters are defined by the subvector $y_1 \in \mathbb{R}^n$ corresponding to the triangular submatrix R_1. Their position in the vector x is then obtained from the relation $x = \Pi y$ which selects the n parameters retained and regrouped in $x_1 \in \mathbb{R}^n$ and the corresponding n columns of A regrouped in $A_1 \in \mathbb{R}^{M,n}$.

At the end of this first step, the system $A x = b$ has been replaced by the reduced system:

$$A_1 x_1 = b \tag{6}$$

Differents criteria allow the value of n to be determined. For example, two classical criteria include:

1. $\dfrac{R_{nn}}{R_{11}} \geq \alpha$, where α is a positive scalar;

2. $\dfrac{\sigma_1}{\sigma_n} \geq \beta$, where β is a positive scalar and σ_i, i = 1 to N are the singular values of A.

3.2. SELECTION OF THE BEST ESTIMATION EQUATIONS

In analogy with the optimal selection of parameters, the rows of the matrix A_1 are ordered according to their best linear independence by the permutation matrix $\hat{\Pi} \in \mathbb{R}^{M,M}$:

$$A_1^T \hat{\Pi} = \hat{Q}\,\hat{R}$$

where :

$$\hat{Q} \in \mathbb{R}^{n,n}; \ \hat{R} \in \mathbb{R}^{n,M}$$

$$\hat{\Pi}^T \hat{\Pi} = \hat{\Pi}\,\hat{\Pi}^T = I_M; \ \hat{Q}^T \hat{Q} = \hat{Q}\,\hat{Q}^T = I_n$$

The n first columns of $A_1^T\,\hat{\Pi}$ contain the n most linearly independent equations among the N rows of the matrix A1. In order to preserve the advantages of an overdetermined system, the selection of equations is extended to the m (m = 2 to 3n) first columns of the matrix $A_1^T\,\hat{\Pi}$. This selection leads to the following reduced system of estimation equations:

$$\hat{A}\,x_1 = b_1 \quad , \ \hat{A} \in \mathbb{R}^{m,n} \tag{7}$$

The construction of the global system (3) requires a knowledge of the dynamic characteristics of the modified structures defined by a fictive grounding of one or more dof. The properties of the modified behaviors are described in the following two paragraphs. Paragraph 4 establishes the relationship between the modified behaviors and the notion of a classical or generalized antiresonance while paragraph 5 develops the methods for evaluating the modified eigensolutions of the model and structure.

4. Zeros of a Frequency Response Function

In order to simply the formulation, the following developments only concern non dissipative mechanical structures. The extension to weakly dissipative structures is given in [8]. It is assumed that the eigensolutions λ_v ; y_v of the initial structure satisfy the dynamic equiilibrium equation:

$$\left[K - \lambda_v M \right] y_v = 0 \quad , v = 1 \text{ to } N$$

where:

$K = K^T \in \mathbb{R}^{N,N}$, semi-positive definite;

$M = M^T \in \mathbb{R}^{N,N}$, positive definite.

The simultaneous diagonalization of the matrices M; K is expressed by the relations:

$$Y^T M Y = I_N; \ Y^T K Y = \Lambda$$

$$Y = [\ \dots\ ;\ y_v\ ;\ \dots\ ];\ \Lambda = \text{Diag}\{\lambda_v\},\ Y;\ \Lambda \in \mathbb{R}^{N,N}$$

Let f_j be a control force applied to the jth dof of the initial structure and whose amplitude is defined to be a linear function of the displacement on the ith dof:

$$f_j(t) = -k_{ij}\, y_i(t) = -k_{ij}\, e_i^T\, y(t)$$

where $e_i \in \mathbb{R}^N$ is the ith column of the unit matrix I_N.

The equilibrium equation of the modified structure is expressed by:

$$M\, \ddot{y}(t) + K\, y(t) = -k_{ij}\, e_j\, e_i^T\, y(t)$$

This equation leads to a homogeneous generalized eigenvalue problem:

$$\left[K + \Delta K - {}^{ij}\lambda_v\, M \right]\, {}^{ij}y_v = 0 \quad ,\ v = 1 \text{ to } N \tag{8}$$

which is Hermitian for $i = j$ and non-Hermitian ($\Delta K^T \neq \Delta K = k_{ij}\, e_j\, e_i^T$) otherwise.

In (8), ΔK is a rank one singular matrix. This property allows the characteristic equation of the homogeneous problem (8) to be expressed as a function of the element $\Gamma_{ij}(\lambda)$ of the dynamic flexibility matrix $\Gamma(\lambda) \in \mathbb{R}^{N,N}$ for the initial system.

Indeed, for $\lambda \neq \lambda_v$, $v = 1$ to N:

$$\Gamma(\lambda) = [K - \lambda M]^{-1} = Y [\Lambda - \lambda I_N]^{-1} Y^T\ , \tag{9}$$

and if $\Gamma\left({}^{ij}\lambda_v\right)$ is regular, (8) can be rewritten:

$$\left[I_N + k_{ij}\, \Gamma\left({}^{ij}\lambda_v\right) e_j\, e_i^T \right]\, {}^{ij}y_v = 0 \tag{10}$$

Premultiplying (10) by e_i^T yields the characteristic equation:

$$\left[1 + k_{ij}\, e_i^T\, \Gamma\left({}^{ij}\lambda_v\right) e_j \right] e_i^T\, {}^{ij}y_v = 0$$

If the eigenvectors ${}^{ij}y_v$ of the constrained problem (8) satisfy the conditions:

$$e_i^T\, {}^{ij}y_v \neq 0 \quad ,\ v = 1 \text{ to } N \tag{11}$$

then the new eigenvalues ${}^{ij}\lambda_v$, solutions to the preceding characteristic equation, are given by:

$$\left(k_{ij}\right)^{-1} = -\Gamma_{ij}\left(^{ij}\lambda_{\nu}\right) \tag{12}$$

where $\Gamma_{ij}(\lambda) = e_i^T \Gamma(\lambda) e_j$. If it assumed that the row vector $e_i^T Y$ does not contain any zero elements, then it is evident that the condition (11) has a simple physical interpretation. For $k_{ij} \to \infty$, the eigenvalues $^{ij}\lambda_{\nu}$ of the modified structure, noted in what follows as structure "i; j", are the zeros of $\Gamma_{ij}(\lambda)$:

$$\lim_{k_{ij} \to \infty} \Gamma_{ij}(\lambda) = 0 \tag{13}$$

This result is the generalization of the case $i = j$ corresponding to a classical antiresonance on the dof i. To summarize, the following three cases can be distinguished.

4.1. CASE 1: CLASSICAL ANTIRESONANCE

Classical antiresonance denoted "i; i":

- $i = j$; $k_{ii} \to \infty$

- $^{ii}\lambda_{\nu}$: squares of the real angular frequencies at which a finite applied force fi applied to the ith dof provokes a zero displacement on the ith dof;

- $^{ii}y_{\nu}$: deformation vector for $\lambda = {^{ii}\lambda_{\nu}}$, $\nu = 1$ to $N - 1$
  (10) leads to the classical interlacing condition [10] for the eigenvalues:

$$\lambda_{\nu} \leq {^{ii}\lambda_{\nu}} \leq \lambda_{\nu+1} , \nu = 1 \text{ to } N - 1$$

where the equality corresponds to the case where the condition (11) is not satisfied.

4.2. CASE 2: GENERALIZED ANTIRESONANCE

Generalized antiresonance denoted "i; j":

- $i \neq j$; $k_{ij} \to \infty$

- $^{ij}\lambda_{\nu}$: squares of the angular frequencies (real or complex) at which a finite force f_j applied to the jth dof provokes a zero displacement on the ith dof;

- $^{ij}y_{\nu}$: deformation vector (real or complex) at $\lambda = {^{ij}\lambda_{\nu}}$, $\nu = 1$ to $N - 1$

- In general, it is possible that 0 to 2 real eigenvalues $^{ij}\lambda_{\nu}$ lie on the interval defined by two successive eigenvalues.

4.3. CASE 3: GENERALIZED MULTIPLE ANTIRESONANCE

The notion of generalized multiple antiresonances are introduced below and are denoted by "i, k, ...; j, ℓ, ...". They correspond to the simultaneous zeroing of the displacements on the a dof i, k,... due to a control forces applied to the dof j, ℓ,...

5. Eigensolutions of the constrained system

5.1. EIGENSOLUTIONS OF THE MODIFIED AUTONOMOUS MODEL

Given a modified model of type "i; j", we are interested in calculating the eigenvalues (real or complex) $\lambda = \omega^2$ for which the displacements of the dof i are zeroed due to the application of a control force f_j at the dof j.

Partitioning the equilibrium equations according to these two types of dof yields:

$$
\begin{array}{c}
\quad\quad i \quad\quad\quad\quad\quad j \\
\vdots \quad\quad\quad\quad\quad \vdots
\end{array}
$$

$$
i \cdots
\begin{bmatrix}
Z_{1i} & Z_{11} \overset{\Delta}{=} {}^{ij}Z \\
Z_{ji} & Z_{j1}
\end{bmatrix}
\begin{bmatrix}
y_i \\
y_1
\end{bmatrix}
=
\begin{bmatrix}
0 \\
f_j
\end{bmatrix}
\tag{14}
$$

where: $Z = K - \lambda\,M$, K and M being respectively the stiffness and mass matrices of of the initial model under its natural boundary conditions.

Seeking f_j and λ such that $y_i = 0$ yields:

$$
{}^{ij}Z(\lambda) \cdot y_1 = 0 \tag{15}
$$

$$
Z_{j1}(\lambda)\, y_1 = f_j \tag{16}
$$

where ${}^{ij}Z \in \mathbb{R}(\mathbb{C})^{N-1,N,1}$ is symmetric only if $i = j$. In (15), the solutions ${}^{ij}\lambda_v$; ${}^{ij}y_{1v}$, $v = 1$ to $N-1$, real or complex, are respectively the eigenvalues and right eigenvectors of the modified model.

Since the matrices K; M of the model are known, the sought eigensolutions can be obtained by evaluating the solutions of (15).

<u>Remarks</u>:

- Properties of the eigenvector ${}^{ij}y_{1v} \in \mathbb{R}(\mathbb{C})^{N-1}$
  By definition, the initial stiffness and flexibility matrices, $Z(\lambda)$ and $\Gamma(\lambda)$ respectively, satisfy the condition:

$$
Z\Gamma = I_N
$$

or again, in terms of the partitioned dof:

$$
\begin{bmatrix}
Z_{1i} & {}^{ij}Z \\
Z_{ji} & Z_{j1}
\end{bmatrix}
\begin{bmatrix}
\Gamma_{i1} & \Gamma_{ij} \\
\Gamma_{11} & \Gamma_{1j}
\end{bmatrix}
=
\begin{bmatrix}
I_{N-1} & \\
& 1
\end{bmatrix}
$$

Among the four matrix relations which result from this partitioning, the following equation is particularly useful:

$$Z_{1i}\,\Gamma_{ij} + {}^{ij}Z\Gamma_{1j} = 0$$

The antiresonance condition $\Gamma_{ij}\left({}^{ij}\lambda_v\right) = 0$ yields:

$$^{ij}Z\left({}^{ij}\lambda_v\right)\Gamma_{1j}\left({}^{ij}\lambda_v\right) = 0 \, , \, v = 1 \text{ to } N-1 \qquad (17)$$

The right eigenvector ${}^{ij}y_{1v}$ is thus colinear to the column corresponding to the dof j of the dynamic flexibility matrix Γ evaluated for $\lambda = {}^{ij}\lambda_v$;

- Use of the orthonormality relations for the modified model imply an evaluation of the eigenvectors of the adjoint system.
  The right hand eigenvectors of the modified model satisfy the relation:

$$\left[\,{}^{ij}K^T - {}^{ij}\lambda_v\,{}^{ij}M\right]\,{}^{ij}y_{1v} = 0 \quad , \, v = 1 \text{ to } N-1 \qquad (18)$$

while the eigensolutions ${}^{ij}\bar{\lambda}_v$; ${}^{ij}y_{1v}^*$ satisfy the equilibrium equations of the adjoint system given by:

$$\left[\,{}^{ij}K^T - {}^{ij}\bar{\lambda}_v\,{}^{ij}M^T\right]\,{}^{ij}y_{1v}^* = 0 \quad , \, v = 1 \text{ to } N-1 \qquad (19)$$

where $\bar{\lambda}$ denotes the complex conjugate of λ.
Moreover, the two sets of eigenvectors satisfy the orthonormality relations:

$$^{ij}y_{1v}^{T*}\left[\,{}^{ij}M\right]\,{}^{ij}y_{1v} = \delta_{\sigma v}$$

$$^{ij}y_{1v}^{T*}\left[\,{}^{ij}K\right]\,{}^{ij}y_{1v} = {}^{ij}\lambda_v\delta_{\sigma v}$$

with $\sigma, v = 1, 2, \ldots$ and where it is easily demonstrated that: ${}^{ij}\bar{y}_{1v}^* = {}^{ji}y_{1v}$;

- In order to simplify the procedure, only the real eigensolutions of the modified model which are contained in the observed frequency band of the structure are exploited. Under the condition that i = j, the state matrices of the modified model are symmetric, hence:

$$^{ij}\lambda_v \in \mathbb{R}\,, \, {}^{ii}y_{1v} = {}^{ii}y_{1v}^* \in \mathbb{R}^{N-1}\,, \, v = 1 \text{ to } N-1$$

- The generalization of these developments to the case of generalized multiple antiresonances is immediate.

5.2. FREQUENCY RESPONSES AND EIGENSOLUTIONS OF THE MODIFIED STRUCTURE

The following developments are presented for the case of a generalized antiresonance "i; j", denoting the zeroing of the displacement on dof i due to a control force applied to dof

j. It is assumed that the dynamic flexibility matrix $\Gamma(\omega) \in \mathbb{C}^{c,c}$ of the initial structure is available on the c observed dof in the frequency band of interest. Let f_2 be a exterior force vector applied to the dof 2 (i and j $\neq$ 2) The frequency response functions for the dof i, j and 2, as well as the (c - 3) remaining dof regrouped in the vector y_1, can be written:

$$y_i(\omega) = \Gamma_{ij}(\omega)f_j + \Gamma_{i2}(\omega)f_2$$
$$y_j(\omega) = \Gamma_{jj}(\omega)f_j + \Gamma_{j2}(\omega)f_2$$
$$y_2(\omega) = \Gamma_{2j}(\omega)f_j + \Gamma_{22}(\omega)f_2$$
$$y_1(\omega) = \Gamma_{1j}(\omega)f_j + \Gamma_{12}(\omega)f_2$$

The condition $y_i(\omega) = 0$ allows f_j to be expressed as a function of f_2 and leads to the FRF of the modified structure:

$$\begin{Bmatrix} y_j \\ y_2 \\ y_1 \end{Bmatrix} = \left(-\Gamma_{ij}^{-1}\Gamma_{i2} \begin{bmatrix} \Gamma_{jj} \\ \Gamma_{2j} \\ \Gamma_{1j} \end{bmatrix} + \begin{bmatrix} \Gamma_{j2} \\ \Gamma_{22} \\ \Gamma_{12} \end{bmatrix} \right) f_2 \tag{20}$$

$$\overset{\Delta}{=} \hat{\Gamma}(\omega) f_2$$

Remarks:

- The eigensolutions (eigenvalues and right hand eigenvectors) of the associated conservative structure are obtained by applying a modal identification procedure to the series of frequency response vectors obtained from (20);

- The left hand eigenvectors are obtained by permuting the indices i and j in (20);

- The generalization of these principles to the simultaneous grounding of a dof using a control forces is immediate. It should be noted that a minimum (a + 1) columns of the dynamic flexibility matrix of the initial structure must be known in order to ground the displacements of a dof. However, in practice the rank deficiency of the matrix Γ_{ij} often precludes a physically meaningful inversion for a > 2;

- The above relations have been extended to weakly dissipative structures in [8]; [9].

6. Application to linearized eigensolution residuals

This type of residual is motivated by the fact that it uses data which are very commonly available. Moreover, the corresponding observation matrix A is strictly evaluated from model-based data. It is thus independent of the diverse uncertainties which affect the observed data.

6.1. ESTIMATION EQUATIONS FOR THE INITIAL MODEL

In the formulation based on the linearized output residuals formed from the eigensolutions, the system (1): $A^{(0)} x = b^{(0)}$ results in the minimization of an objective function $J_{(a)}^{(0)}$ introducing a measure of the distance between the calculated and identified eigensolutions.

The following notation will be employed:

e vector of initial parameter estimations;

a = e + x vector of corrected parameter values;

x correction vector;

$\lambda_v^{(s)}$: v^{th} eigenvalue of the initial structure;

$H y_v^{(s)} \in \mathbb{R}^c$ restriction of the vth eigenvector of the initial structure to the c observed dof, $v = 1$ to m;

$\lambda_v(e)$, $H y_v(e)$, $v = 1$ to m, eigensolutions of the initial model matched to the identified eigensolutions of the initial structure;

W_i diagonal weighting matrix, positive definite, i = 1 to 3;

Following a linearization with respect to the vector x, the objective function $J^{(0)}(a)$ is given by:

$$J^{(0)}(x) = r_1(x)^T W_1 \, r_1(x) + r_2(x)^T W_2 \, r_2(x) + x^T W_3 \, x \qquad (21)$$

The residuals r_1 and r_2 are defined by:

$$r_1(x) \stackrel{\Delta}{=} \Delta\lambda - S_\lambda \, x \quad \in \mathbb{R}^m$$

$$r_2(x) \stackrel{\Delta}{=} \Delta y - S_y \, x \quad \in \mathbb{R}^{cm}$$

where:

$$\Delta\lambda = \left\{ ...; \Delta\lambda = \lambda_v^{(s)} - \lambda_v^{(e)}; ... \right\}^T$$

$$\Delta y = \left\{ ...; \Delta y_v^T; ... \right\}^T$$

$$\Delta y_v = H\left(y_v^s - y_v(e) \right)$$

S_λ ; S_y are the Jacobian matrices of the eigenvalues and eigenvectors, respectively.

The necessary condition for the minimization of $J^{(0)}(x)$ leads to the non homogeneous system (1) denoted earlier by: $A^{(0)} x = b^{(0)}$ and whose detailed expression is given by:

$$S_\lambda^T W_1 \left[\Delta\lambda - S_\lambda \, x \right] + S_y^T W_2 \left[\Delta y - S_y \, x \right] + W_3 x = 0 \qquad (22)$$

6.2.   ESTIMATION EQUATIONS FOR THE MODIFIED MODEL

The objective function $J^{(u)}(a)$ for the uth modified model has a form analogous to (21) and thus also leads to a linearized system resembling (22). The developments in the paragraph 5 specify the necessary data for the model and structure modified by the introduction of grounded dof, in particular:

- The eigensolutions of the modified structure are evaluated from the dynamic flexibility relations of the initial structure;

- The eigensolutions of the modified model are determined numerically from the dynamic stiffness relations of the initial model.

7. Numerical examples

The two academic examples presented below show that a significant enrichment in the knowledge space can be obtained by implementing the proposed methodology, namely:

- reduction of the condition number of the observation matrix A;

- increase in the modal sensitivity of a parameter to be identified.

These examples are based on an application of the proposed methodology to estimation equations formed from the linearized eigensolution residuals described in the preceeding section.

7.1. IMPROVED PARAMETER SEPARABILITY

The truss structure in Figure 1 is composed of 27 members and each member is modelled by two parallel Euler-Bernouilli beam elements, each comprising 2 nodes and 3 dof by node. The total number of dof is 42. Each member is characterized by two physical parameters, namely the cross-sectional area S and the quadratic moment of inertia I, yielding a total of 54 parameters.

The initial structure is simulated from the model by modifying the cross-sections and inertia of the 6 members indicated in Figure 1. A total of 8 dof have been observed in the frequency band of [0 ; 30] Hz.

The values of the modifications are reported in Table 1. Table 2 specifies the distances between the 5 first eigensolutions of the initial model and structure.

The evolution of the condition numbers has been examined for two systems of estimation equations, constructed from partial modal residuals (see paragraph 6), during the iterations of a parametric updating:

1. condition numbers of the matrix $A^{(0)} \in \mathbb{R}^{36,54}$ evaluated by exploiting the first 4 matched eigensolutions of the model and structure under their natural boundary conditions; (36 = 4 eigenvalues + (4 x 8) = 32 eigenvectorscomponents)

2. condition numbers of the matrix $A \in \mathbb{R}^{163,54}$ formed from the matrix $A^{(0)}$ as well as four matrices $A^{(u)}$. The latter are constructed from the eigensolutions of the model and structure (in the frequency band [0 ; 30] Hz) obtained from the successive fictive grounding the following four dof: 2x; 2y; 6x; 6y (grounding of the "i; i" type).

Figure 2 shows the evolution of Cond ($A^{(0)}$) and Cond (A) during the first 17 iterations of a parametric updating and demonstrates that a significant improvement in the numerical conditioning of the observation matrix can be obtained by the simultaneous exploitation of the different boundary condition configurations.

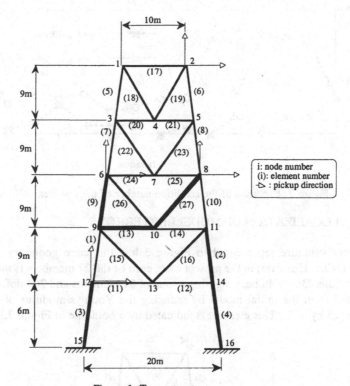

Figure 1 Truss structure

Table 1 Values of the parametric modifications

Element	$S = \alpha \, S^0$	$I = \beta \, I^0$
(2)	0.64	0.41
(3)	2.25	5.06
(4)	3.24	10.50
(9)	1.44	2.07
(13)	0.49	0.24
(27)	0.25	0.06

Table 2 Differences between the eigensolutions of the modeland the structure

	Model	Structure		
Mode	Freq. (Hz)	Freq. (Hz)	$\Delta f_v/f_v$ (%)	MAC_v (%)
1	5.85	7.05	17.0	93
2	11.77	16.56	28.9	84
3	20.42	21.96	7.0	97
4	28.51	24.86	14.7	95
5	36.81	35.07	5.0	44

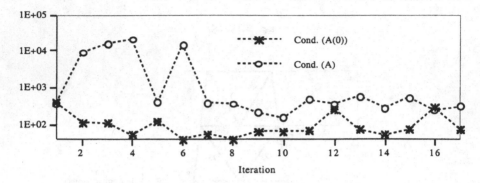

Figure 2 Evolution of the condition number of the matrices $A^{(0)}$ and A.

7.2. LOCALIZATION OF MODELLING ERRORS

The truss structure represented in Figure 3 has the same geometry as that used in paragraph 7.1. However, in the present case, each of the 27 members is modelled by three parallel Euler-Bernoulli beams giving a total of 81 elements and 204 dof. The structure is simulated from the initial model by reducing the Young's modulus of one element of member 25 by 90%. This element is indicated by a bold line in Figure 3.

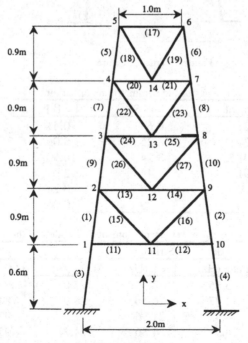

Figure 3 Truss structure

The eigenvectors of the structure are observed at 14 nodes along the x and y directions giving a total of 28 pickups. The eigensolutions exploited in the estimation equations are those contained in the frequency interval [0;310] Hz. This band contains the first 5 eigenfrequencies of the initial model and structure. Table 3 specifies the following distances between the model and the structure for the mode v, $v = 1$ to 5 based on the following comparison criteria.

Eigenfrequency error: $\Delta f_v / f_v$

Eigenvector MAC: $1 - MAC_v$

Eigenvector error: $\| \Delta y_v \| / \| y_v \|$

Table 3 Distances between the model and the structure

Mode	Model (Hz)	Structure (Hz)	$\Delta f_v / f_v$ %	$1 - MAC_v$ %	$\| \Delta y_v \| / \| y_v \|$ %
1	47.2	47.1	0.4	10^{-3}	1
2	102.9	102.3	0.6	6×10^{-2}	2.5
3	155.8	155.4	0.3	8×10^{-2}	2.8
4	275.7	274.5	0.5	1.5	12.4
5	308.8	306.9	0.6	28	58

It can be noted that these distances are very small and do not allow the damaged member to be localized based on the matrix $A^{(0)}$. In order to sensitize the effects of the introduced damage of unknown position, a series of 14 matrices $A^{(u)}$, $u = 1$ to 14, are constructed based on the differences between the eigensolutions of the model and structure following the simultaneous fictive grounding of the x and y dof of each of the nodes 1 to 14 successively. The average model-structure frequency differences are then evaluated for each of these 14 boundary condition configurations in order to select the most sensitizing ones. Figure 4 reports these average values as a function of the grounded node. The grounding of nodes 3, 8, 12 and 13 all have an significant effect and it may be noted that they are all in the neighborhood of the effectively damaged member. An observation matrix formed from the five matrices $A^{(0)}$; $A^{(3)}$; $A^{(8)}$; $A^{(12)}$; $A^{(13)}$ allows the structural damage to be both localized and quantified.

8. Conclusion

A strategy for the enrichment of the observation matrix used in inverse estimation procedures have been proposed which is based on the responses of the model and structure under fictive multiple boundary configurations. This enrichment is due to the introduction of information resulting from kinetic and potential energy distributions which are linearly independent of those available based on data obtained from only the initial boundary condition configuration and is the result of the mode contributions outside of the analysed frequency band. Although the information contained in the configurations "i ; i" and "i ; j" are present in the initial frequency response functions, a deliberate procedure is required to amplify and filter this data. The efficiency of the proposed methodology depends

280

strongly on the uncertainties affecting the observed frequency response functions which will limit the maximum number of dof that can be simultaneously grounded.

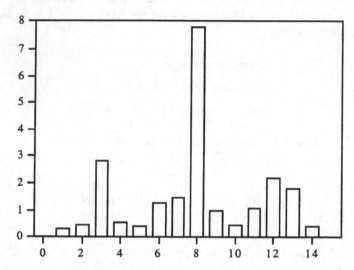

Figure 4   Average model-structure frequency differences following the simultaneous fictive grounding of the x and y dof of each of the nodes 1 to 14 successively

9 . References

1.   Nalitolela, N.G., Penny, J.E. and Friswell, M.I. (1990), Updating structural parameters of a finite element model by adding mass or stiffness to the system, *Proceedings of the 8th International Modal Analysis Conference*, 836-842.
2.   Salvini, P. and Sestieri, A. (1992), Predicting the new frequency response functions of a structure when adding constraints, *Proceedings of the 10th International Modal Analysis Conference*, 284-290.
3.   Wada, B.K., Kuo, C.P. and Glaser R.J. (1986), Multiple boundary condition tests for verification of large space structure, *AIAA Paper 86-0905*.
4.   Berman, A. and Fuh, J. (1990), Structural System identification using multiple tests, *AIAA Paper 90-1200*.
5.   Shepard, G.D. and Milani, J. (1990), Frequency based localisation of structural discrepancies, *Mechanical Systems and Signal Processing, Vol. 4, No. 2*, 173-184.
6.   Piranda, J. and Ratsifandrihana, L. (1996), Modal identification of non-linear structures by coupling appropriation and curve fitting, *Euromech - 2nd European Nonlinear Oscillation Confrence*, Prague, 353-356.
7.   Cogan, S., Lallement, G., Ayer, F. and Ben-Haim, Y. (1995), Updating linear elastic models with modal selective sensitivity, *Inverse Problems in Engineering, Vol. 2*, 29-47.
8.   Rade, D.A. (1994), Parametric correction of finite element models: enlargment of the knowledge space (in French), *Ph.D., No. 403*, Université de Franche-Comté, Besançon, France.
9.   Rade, D.A. and Lallement, G. (1996), Vibration analysis of structures subjected to boundary condition modifications using experimental data, *Journal of the Brazilian Society of Mechanical Sciences, Vol XVIII, No. 4, 374-382*.
10.   Lancaster, P. and Tismenetsky, M. (1985), The Theory of Matrices, *2nd Edition, Academic Press Limited*, London.

UPDATING OF ANALYTICAL MODELS – BASIC PROCEDURES AND EXTENSIONS

M. LINK
Light Weight Structures and Structural Mechanics Laboratory
University of Kassel, Germany

Abstract

In the paper basic procedures for computational updating of analytical model parameters are presented. The procedures have been investigated thoroughly in recent years with respect to

* the numerical estimation techniques for solving the updating equations;

* the influence of different model parametrisations defining the type and the location of the erroneous parameters;

* the type of the residuals formed by the test/analysis differences to be minimised;

* the requirements to be posed on the initial analysis model.

The residuals presented are formed by force and response equation errors, by eigenfrequency and mode shape errors and by frequency response errors. The procedures have been derived to handle incomplete test vectors, where the number of measured degrees of freedom (DOF) is much less than the DOF no. of the computational model. Finally an example of updating a laboratory test structure is reported including some recommendations and experiences.

1. Introduction

The validation of analytical models in practise is mainly based on comparing experimental modal analysis results with the analytical predictions. Despite the high sophistication of analytical (Finite Element) modelling practical applications often reveal considerable discrepancies between analytical and test results. In recent times some effort has therefore been spent in the development of mathematical procedures for updating analytical mass and stiffness matrices using dynamic test data. The books of Natke [1] and Friswell / Mottershead [2] represent comprehensive monographs containing the most relevant techniques.

The requirement for updating design parameters selected by the analyst like local mass and stiffness parameters results in iterative methods using non-linear optimisation in conjunction with least square procedures. The success of these methods is governed not

J.M.M. Silva and N.M.M. Maia (eds.), Modal Analysis and Testing, 281–304.

only by the skill of the analyst to assume an appropriate initial analysis model but also the source and the location of the erroneous parameters to be corrected. In practical applications the source and location of the errors can be manifold resulting in non-unique updated matrices all of them fulfilling the mathematical minimisation criteria. For example, using a physical design parameter like a bending stiffness to update a discretisation error caused by a coarse finite element mesh would not be consistent with the real error source and would therefore destroy the physical significance of the design parameters. The updated model plays the role of a substitute model which at least has to fulfil the requirement of reproducing the test data used for updating. Generally we call all updating approaches where the real error source and location is not consistent with the assumed error source and location an <u>inconsistent</u> updating approach. The practical applicability of any localisation and updating procedure requires its ability to handle

(1) incomplete test data where the no. of measurement DOF's is less than the no. analytical DOF,

(2) local and global physical modelling errors related to parameters like the stiffness or the mass of a single or a group of finite elements,

(3) inconsistent error assumptions,

(4) measured data polluted with random noise and unavoidable (small) systematic errors.

2. Mathematical background of model parameter estimation

Parameter estimation techniques aim at fitting the parameters of a given initial analytical model in such a way that the model behaviour corresponds as close as possible to the measured behaviour. The resulting parameters represent estimated values rather than true values since the test data are unavoidably polluted by unknown random and systematic errors. Also the mathematical structure of the initial analysis is not unique depending on the idealisations made by the analyst for the real structure. The method of extended weighted least squares is summarised in the following since it represents the most important estimation technique.

The first step in parameter estimation is the definition of a residual containing the difference between analytical and measured structural behaviour, for example the difference between analytical and measured eigenfrequencies. The weighted least squares technique requires to define a weighting matrix $\mathbf{W}_v$ accounting for the importance of each individual term in the residual vector ε:

$$\varepsilon_w = \mathbf{W}_v \varepsilon = \mathbf{W}_v (\mathbf{v}_M - \mathbf{v}(p_i)) \tag{1}$$

$\mathbf{v}_M$ represents the measured and $\mathbf{v}(p)$ the corresponding analytical vector which is a function of the parameters p_i ($i = 1,...np$ = number of correction parameters). The weighted squared sum of the residual vector yields the objective function

$$J = \varepsilon_w^T \varepsilon_w = \varepsilon^T \mathbf{W} \varepsilon \quad \rightarrow \quad \min , \quad \mathbf{W} = \mathbf{W}_v^T \mathbf{W}_v \tag{2}$$

whose minimisation yields the unknown parameters. In general the model vector $\mathbf{v}$ represents a non- linear function of the parameters resulting in a non- linear minimisation problem. One of the techniques to solve this non- linear optimisation problem is to expand the model vector into a Taylor series truncated after the linear term according

$$\mathbf{v}(p) = \mathbf{v}_a + \mathbf{G}\,\Delta\mathbf{p} \qquad (3)$$

where $\mathbf{v}_a = \mathbf{v}\big|_{p=p_a}$ represents the model vector at the linearization point $\mathbf{p} = \mathbf{p}_a$.

$\mathbf{G} = \dfrac{\partial \mathbf{v}}{\partial \mathbf{p}}\bigg|_{p=p_a}$ represents the sensitivity matrix (order m, np with m = no. of measurements

and np = no. of parameters) and $\Delta\mathbf{p} = \mathbf{p} - \mathbf{p}_a$ represents the vector of the parameter changes. Eq.(3) introduced into eq.(2) yields the linear residual

$$\varepsilon_w = \mathbf{W}_v \varepsilon = \mathbf{W}_v(\mathbf{v}_M - \mathbf{v}_a - \mathbf{G}\,\Delta\mathbf{p}) = \mathbf{W}_v(\mathbf{r}_a - \mathbf{G}\,\Delta\mathbf{p}) \qquad (4)$$

where $\mathbf{r}_a = \mathbf{v}_M - \mathbf{v}_a$ contains the residual at the linearization point. Of course, this formulation includes the special case when the model vector is a linear function of the parameters resulting in a constant sensitivity matrix. The stepwise calculated minimum of the objective function with respect to the parameter changes is obtained from the derivative of the objection function $\partial J / \partial \Delta\mathbf{p} = 0$ yielding the linear system of equations

$$\mathbf{W}_v\,\mathbf{G}\,\Delta\mathbf{p} = \mathbf{W}_v\,\mathbf{r}_a \qquad (5)$$

with the solution

$$\Delta\mathbf{p} = \underbrace{(\mathbf{G}^T\mathbf{W}\mathbf{G})^{-1}\mathbf{G}^T\mathbf{W}}_{\mathbf{Z}^T}\mathbf{r}_a = \mathbf{Z}^T\mathbf{r}_a \qquad (6)$$

The condition of the sensitivity matrix $\mathbf{G}$ plays an important role for the accuracy and the uniqueness of the solution. It is clear that in the case when less measurements than parameters $(m < n_p)$ are available eq.(4) leads to an underdetermined system whose solution is not unique. Even if a minimum norm or a minimum parameter change solution is selected the resulting parameters will in general not retain their physical meaning. In parameter updating the number of measurements should always be made larger than the number of parameters $(m > n_p)$ which yields overdetermined equation systems. If in practical applications it is not possible to increase the number of measurements it is recommended to reduce the number of parameters by applying parameter localisation techniques like those described in [3]- [5] in order to retain only the most erroneous and sensitive parameters.

Instead of decomposing $(\mathbf{G}^T\mathbf{W}\mathbf{G})$ in eq.(6) the numerical solution of the overdetermined system is preferably done via QR or singular value decomposition in order to check the condition of $\mathbf{G}$. The singular value decomposition (SVD) of the matrix $\mathbf{G}$ is defined by (see e.g.[6])

$$\mathbf{G}_{(m,n_p)} = \mathbf{U}_{(m,m)}\,\Sigma_{(m,n_p)}\,\mathbf{V}^T_{(n_p,n_p)} \qquad (7)$$

U and V represent orthogonal matrices with the properties $U^T U = V^T V = V V^T = I_{n_p}$ and $\Sigma = \text{diag}(s_1 \ldots s_r \ldots s_{n_p})$. The singular values s_r are the roots of the eigenvalues of $G^T G$. If

$$\text{rank}(G) = R: \quad \Sigma = \text{diag}(s_1 \ldots s_r \ldots s_R, s_{R+1} = 0 \ldots s_{n_p} = 0) \tag{8}$$

the pseudo inverse is calculated from:

$$G^+ = V \Sigma^+ U^T \tag{9}$$

with

$$\Sigma^+ = \text{diag}(1/s_1 \ldots 1/s_R, 1/s_{R+1} = 0 \ldots 1/s_{n_p} = 0)$$

In practical cases a clear rank defect represented by zero singular values and caused by linear dependent parameter sensitivities can often not be detected, for example due to the influence of measurement noise. In this case the singular values can be truncated below a certain threshold of the ration s_R / s_1. In any case it is recommended to consider a priori only parameters exhibiting linear independent sensitivities.

With these definitions the solution to Eq.(4) can be expressed by

$$\Delta p = G_v^+ r_{av} \tag{10}$$

with

$$G_v = W_v G \quad \text{and} \quad r_{av} = W_v r_a.$$

The statistical properties of the solution are calculated from the mean values and the covariance matrix of the estimate. After substituting the unknown true vector Δp^0 into the eq.(4) we obtain

$$\varepsilon^0 = v_M - v_a - G \Delta p^0 = r_a - G \Delta p^0$$

which is a measure for the for the random measurement error. With $r_a = G \Delta p^0 + \varepsilon^0$ an estimate vector is calculated from

$$\Delta \hat{p} = Z^T r_a = Z^T (G \Delta p^0 + \varepsilon^0) = \Delta p^0 + Z^T \varepsilon^0 \tag{11}$$

The mean values of this estimate is calculated by the expectation operation

$$E(\Delta \hat{p}) = \Delta p^0 + E(Z^T \varepsilon^0) \tag{12}$$

If the matrix Z defined in eq.(6) is statistically independent of ε^0 and if it is assumed that the mean of the measurements error vector equals zero, $E(\varepsilon^0) = 0$, then $E(Z^T \varepsilon^0)$ also equals zero and

$$E(\Delta \hat{p}) = \Delta p^0 \tag{13}$$

The mean values of the estimate in this case are equal to the true values. Such an estimate is called an unbiased estimate. Often the above assumptions are not valid in particular when the sensitivity matrix is corrupted by measurement errors or when the model vector used to calculate $\mathbf{Z}$ is a non- linear function of the parameters, i.e. the estimate is biased. Procedures like the instrumental variable technique [7] directed to reduce the bias are useful in model updating, however, updating methods should be preferred where the sensitivity matrix is not directly corrupted by measurement errors.

The covariance matrix $\text{cov}(\Delta\hat{\mathbf{p}})$ represents a measure for the deviation of the estimate depending on the covariance matrix of the measurement vector $\text{cov}(\varepsilon^0) = E(\varepsilon^0\varepsilon^{0T}\varepsilon)$. With the assumption of an unbiased estimate

$$\text{cov}(\Delta\hat{\mathbf{p}}) = E[(\Delta\hat{\mathbf{p}} - \Delta\mathbf{p}^0)(\Delta\hat{\mathbf{p}} - \Delta\mathbf{p}^0)^T] = \mathbf{Z}^T E(\varepsilon^0\varepsilon^{0T})\mathbf{Z} \qquad (14)$$

Substituting the inverse of the measurement covariance matrix into eq. (14) as a weighting matrix, i.e. $\mathbf{W} = [E(\varepsilon^0\varepsilon^{0T})]^{-1}$, yields the covariance matrix of the estimate in the form

$$\text{cov}(\Delta\hat{\mathbf{p}}) = (\mathbf{G}^T\mathbf{W}\mathbf{G})^{-1} \qquad (15)$$

This result shows the importance of the numerical condition of the sensitivity matrix $\mathbf{G}$ as well as the influence of the measurement errors on the error of the parameter estimate. In the special case $E(\varepsilon^0\varepsilon^{0T}) = \sigma_v^2\mathbf{I}$, i.e. when it is assumed that all measured quantities have the same standard deviation σ_v the covariance matrix of the parameter estimate is

$$\text{cov}(\Delta\hat{\mathbf{p}}) = \sigma_v^2(\mathbf{G}^T\mathbf{G})^{-1} \qquad (16)$$

In [7] it is shown that the covariance matrix of the estimate is always a minimum when the above assumption for the weighting matrix is used, i.e. the covariance matrix is always smaller than the covariance matrix resulting from the standard unweighted LS with $\mathbf{W} = \mathbf{I}$. Such estimation procedures are called Markov estimates in estimation theory, they yield the best linear unbiased estimate.

The classical weighted LS method described above can be extended in cases where it difficult to obtain a convergent solution because of an ill- conditioned sensitivity matrix. The objective function (2) is extended by the requirement that the parameter changes $\Delta\mathbf{p}$ shall be kept minimal

$$J(\mathbf{p}) = \varepsilon^T \mathbf{W} \varepsilon + \Delta\mathbf{p}^T \mathbf{W}_p \Delta\mathbf{p} \rightarrow \min \qquad (17)$$

When the parameters are unbounded the minimisation (17), now with respect to the parameter changes $\Delta\mathbf{p}$, yields the following linear problem to be solved within each iteration step which represents the linearization point a:

$$(\mathbf{G}^T\mathbf{W}\mathbf{G} + \mathbf{W}_p) \Delta\mathbf{p} = \mathbf{G}^T\mathbf{W}\mathbf{r}_a \qquad (18)$$

Of course, any other mathematical minimisation technique could also be applied, in particular when the parameters shall be constrained by upper and lower bounds. In case of $\mathbf{W}_p = \mathbf{0}$ the solution of eq. (17) represents the standard weighted least squares solution, otherwise the solution is affected by the choice of the weighting matrix $\mathbf{W}_p$. In [8] this matrix was related to the inverse of the squared sensitivity matrix according to

$$W_p = w_p \, B \quad \text{where} \quad B = \frac{\text{mean}(g)}{\text{mean}(g^{-1})} g^{-1} \quad \text{and} \quad g = \text{diag}(G^T W G) \qquad (19)$$

This definition allows to constrain Δp according to the sensitivity of the parameters. In consequence the parameters p_k remain unchanged if their sensitivity approaches zero ($W_{pk} \to 0$). The weighting factor w_p allows to scale W_p with respect to B. $B = I$ represents the classical Tikonov regularisation [9] used to solve ill conditioned systems of equations. The question remains who to choose the regularisation parameter w_p. Several proposals for the selection of W_p have been proposed, e.g. in refs.[10]-[12]. Hansen [13] proposed to balance the norms $n_\varepsilon = \varepsilon^T W \varepsilon$ and $n_p = w_p \, \Delta p^T B \, \Delta p$ of the two terms in the extended objective function (17) with respect to minimising $n_\varepsilon + n_p$ as a function of w_p. This minimum is localised as the corner of the so-called L-curve obtained from plotting n_ε versus n_p. In [14] model updating applications of this and other regularisation methods were investigated. MATLAB algorithms for the different procedures have been developed and described by Hansen [15].

3. Definition of updating parameters

Starting point for updating are the assumptions on those model parameters defining the type and the location of the erroneous parameters to be updated in the equation motion of the finite element elastodynamic model :

$$(-\omega^2 M + K + j\omega \, D)y = \hat{K} \, y = f \qquad (20a)$$

where the system matrices M, K and D represent the mass, the stiffness and the damping matrix, ω the excitation frequency, f the excitation force vector and $y(j\omega)$ the complex frequency response vector ($j = \sqrt{-1}$).

The special case of the undamped eigenequation is also considered in this paper:

$$(-\omega_0^2 M + K)y_0 = 0 \qquad (20b)$$

with ω_0 and y_0 denoting the undamped eigenfrequency and eigenvector, respectively.

In the most popular approach first introduced by Natke [16] the system matrices are updated by substructure matrices according to

$$K = K_A + \sum \alpha_i K_i \qquad (21a)$$

$$M = M_A + \sum \beta_j M_j \qquad (21b)$$

$$D = D_A + \sum \gamma_k D_k \qquad (21c)$$

where

$[\alpha_i \; \beta_j \; \gamma_k] = [p_s] =$ unknown correction (design) parameters,

$(i = 1, 2, \dots, I; \; j = 1, 2, \dots, J; \; k = 1, 2, \dots, K; \; s = 1, 2, \dots, S)$

with

$S = I + J + K$ = no. of correction parameters

K_A, M_A, D_A = analytical (initial) stiffness, mass and damping matrix

$\mathbf{K}_i$, $\mathbf{M}_j$, $\mathbf{D}_k$ = assumed correction substructure matrices (elements or element groups) defining source and location of modelling error.

The correction sub-matrices defined above can be considered as the first derivative of the updated matrices with respect to a physical or geometrical model parameter :

$$\mathbf{K}_i = \alpha_A \, \partial\mathbf{K}/\partial\alpha_i \, , \quad \mathbf{M}_j = \beta_A \, \partial\mathbf{M}/\partial\beta_j \quad \text{and} \quad \mathbf{D}_k = \gamma_A \, \partial\mathbf{D}/\partial\gamma_k \tag{22}$$

These derivatives are constant like in the case of a beam element with α representing Young's modulus where the stiffness matrix is a linear function of the modulus. The derivatives must not be constant like in the case when α represents the shear modulus of a Timoshenko beam where the stiffness matrix terms are non-linear functions of the shear modulus. α_A , β_A and γ_A denote the initial parameters used to make the parameter changes dimensionless. Other parametrisations related to generalised elements or substructures have been proposed in [23] and [37].

The success of parameter updating is governed not only by the skill of the analyst to assume an appropriate initial analysis model but also to assume the right source and location of the erroneous parameters to be corrected. In practical applications the source and location of the errors can be manifold resulting in non-unique updated matrices all of them fulfilling the mathematical minimisation criteria. For example, using a physical design parameter like a bending stiffness to update a discretisation error caused by a coarse finite element mesh would not be consistent with the real error source and would therefore destroy the physical significance of the design parameters. The updated model plays the role of a substitute model which at least has to fulfil the requirement of reproducing the test data used for updating. Generally we call all updating approaches where the real error source and location is not consistent with the assumed error source and location an inconsistent updating approach.

4. Definition of test/analysis residuals

Another important assumption the analyst has to make is the choice of the residuals formed by the differences of the predicted analytical and the measured behaviour. In the present investigation we have considered the following residuals which have most often been applied in the past:

* eigenvalues
* mode shapes
* (weighted) input forces and
* frequency response functions (FRF).

In [17] the authors present a comprehensive selection of these and other residuals with special consideration of statistically based weighting and the statistical properties of the parameter estimates.

The linearized undamped **eigenvalue residuals** are defined by the differences between measured (index M) and analytical undamped eigenvalues at the linearization point a:

$$\varepsilon_L = \lambda_M - \lambda = r_{\lambda a} - G_\lambda \Delta p \qquad (\lambda = \omega_0^2) \qquad (23a)$$

where

$$r_\lambda = \lambda_M - \lambda_a \qquad (23b)$$

$\quad$ = residual vector containing test/analysis differences of eigenvalues and

$$G_\lambda = [...\partial\lambda / \partial p_s ...] = [...\partial\lambda/\partial\alpha_i ...\partial\lambda/\partial\beta_j ...]\Big|_{\alpha=\alpha_a, \beta=\beta_a} \qquad (23c)$$

$\quad$ = sensitivity matrix at point a.

If the undamped problem is considered G_λ can be calculated by differentiation of the undamped eigenvalue equation and by substituting the parametrisation of eqs. (21a-b) which results in

$$\partial\lambda/\partial\alpha_i = y_0^T K_i y_0 \quad \text{and} \quad \partial\lambda/\partial\beta_j = -\lambda\, y_0^T M_j y_0 \qquad (23d\text{-}e)$$

for the sensitivities with respect to the i-th stiffness and the j-th mass parameter (y_0 = real mode shape normalised to unit modal mass).

The linearized real **mode shape residuals** are obtained from the differences of the measured modes at the reduced set of $n_M < n$ measured DOF's denoted by the index c:

$$\varepsilon_y = y_{0M} - y_{0c} = r_{ya} - G_y \Delta p \qquad (24a)$$

where

$$r_{ya} = y_{0M} - y_{0ca} \qquad (24b)$$

$\quad$ = residual vector with test/ analysis differences of eigenvectors at point a and

$y_{0ca} := (y_{0c1} \cdots y_{0cr} \cdots y_{0cR})_a$ = real model modes

$\quad$ (r = 1, 2 ... R = no. of measured modes)

y_{0M} = corresponding measured modes

$$G_y = [...\partial y_{0c} / \partial p_s ...] = [...\partial y_{0c}/\partial\alpha_i ...\partial y_{0c}/\partial\beta_j ...]\Big|_{\alpha=\alpha_a, \beta=\beta_a} \qquad (24c)$$

$\quad$ = sensitivity matrix at point a.

The calculation of the mode shape sensitivity matrix involves a major numerical effort. The modal method of Fox and Kapoor [18] is widely used due to its simplicity of implementation. It is based on expanding the gradients by a weighted sum of the eigenvectors

$$\partial y/\partial p_s = \sum_r y_{0r} c_r \quad (r = 1,...R \leq n = \text{model order}) \qquad (25a)$$

which yields after substitution into the derivative of the eigenequation the gradients with respect to the stiffness parameters:

$$\frac{\partial y_o}{\partial \alpha_i} = -\sum_{s=1}^{R} y_{os} \, y_{os}^T \, K_i \, y_o \, / (\lambda_s - \lambda) \qquad \text{for } \lambda_s \neq \lambda, \; y_{os} \neq y_o$$

$$\frac{\partial y_o}{\partial \alpha_i} = 0 \qquad \text{for } \lambda_{os} = \lambda, \; y_{os} = y_o$$

(25b)

The gradients with respect to the mass parameters follow from:

$$\frac{\partial y_o}{\partial \beta_j} = -\sum_{s=1}^{R} y_{os} \, y_{os}^T \, M_j \, y_o \, / (\lambda_s - \lambda) \qquad \text{for } \lambda_s \neq \lambda, \; \lambda_s \neq \lambda$$

$$\frac{\partial y_o}{\partial \beta_j} = -0.5 \, y_{os} \, y_{os}^T \, M_j \, y_{os} \qquad \text{for } \lambda_s = \lambda, \; \lambda_s = \lambda$$

(25c)

This expansion is exact if R = n modes are used. For R < n the expansion represents an approximation depending on the number of modal terms. Corrections to this approach have been investigated by several authors [19]. Eqs. (25 b, c) also show that the convergence of the expansion will decrease for neighboured eigenvalues $\lambda_s \approx \lambda$. Lallement [20] proposed a procedure to overcome this difficulty. A recent investigation of several procedures developed in the past was given by Balmes [21] with respect to using other reduced projection bases than the modal basis in eq. (25a).

Since the sensitivity matrices are derived from the (updated) analytical model they do not contain measurement errors which is an essential prerequisite for an unbiased estimate. However it should be kept in mind that due to the iterative process the sensitivity matrices depend on the parameters from the previous iteration step calculated from the noise polluted residuals in eqs.(23b) and (24b). Another advantage stems from the fact that the mode shape residuals and the sensitivities need only to be calculated for the measured DOF's, i.e. the analytical model must neither be condensed nor must the measured mode shapes be expanded to the unmeasured DOF's. It must be noted that the modal residual have to be formed between paired mode shapes. Most often the correct mode shape correlation is checked using the modal assurance criterion MAC = $(y_M^T \, y_A)^2 /$ $(y_M^T \, y_M \, y_A^T \, y_A)$ which approaches one if the measured mode y_M and the analytical mode y_A are fully correlated. Mode pairs with MAC values smaller than a certain threshold (for example MAC < 0.7) should not be included in the residuals.

Using the deviations between the analytical and experimental mode shapes in the residual vector suffers from the disadvantage that the updated model is fitted to the test data in a least square's sense as close as possible although the test data are uncertain. In [22] and [23] we presented a technique of relaxing this requirement by applying a model based smoothing procedure to the experimental modes. This derivation is repeated here for completeness. The result is that the modal data of the updated model are not fitted as exactly as possible to the original experimental data but more to the smoothed data which allows to cancel not only random but also systematic measurement errors. The idea behind this approach is to bring simultaneously together both sides, the test data and the analytical parameters. Of course, both types of modifications must be kept within realistic bounds.

The smoothed modal matrix $\Phi = [\phi_1 \ldots \phi_R]$ is expressed by a linear combination of the originally measured mode shapes $Y_M = [y_{o1} \ldots y_{oR}]_M$ (R number of measured modes) via an unknown transformation matrix Q :

$$\Phi = Y_M\, Q \tag{26}$$

We now require that the smoothed shapes shall satisfy the orthonormality condition with respect to the updated analytical mass matrix in the form :

$$\Phi^T M\, \Phi = I \tag{27}$$

In order to carry out the product in equation (27) the measured modes must be expanded with respect to the unmeasured DOF's. This is done by expressing the expanded measured modal matrix Y_M by the analytical modal matrix $Y_A = [y_{o1} \ldots y_{oR}]_A$ according to

$$Y_M = Y_A C \tag{28}$$

After partitioning the modal matrices with respect to the measured DOF's (index c) and the unmeasured DOF's (index u) equation (28) yields

$$\begin{bmatrix} Y_{Mc} \\ Y_{Mu} \end{bmatrix} = \begin{bmatrix} Y_{Ac} \\ Y_{Au} \end{bmatrix} C \tag{29}$$

The first row of equation (29) allows to solve for the unknown transformation matrix by minimising the norm $\| Y_{Mc} - Y_{Ac}\, C \|$ which results in

$$C = Y_{Ac}^+ Y_{Mc} \tag{30}$$

("+" indicates the pseudoinverse of the analytical (c, r_A) modal sub-matrix Y_{Ac} , c = no. of measured DOF's, r_A = no. of modes). Using equation (30) the full measured modal matrix is expressed by equation (28). After introducing (26) and (28) and using the orthonormality condition, $Y_A^T M Y_A = I$, eq. (27) yields:

$$Q^T C^T C\, Q = I \tag{31}$$

Provided that the transformation matrix is symmetric $Q^T = Q$ then:

$$Q = (C^T C)^{-1/2} \tag{32}$$

Equation (32) is identical with the optimal orthogonalisation introduced by Baruch [25] which also includes the method of Targoff [24]. The smoothed shapes are finally obtained from eqs.(26) and (32) by

$$\Phi = Y_A\, C(C^T C)^{-1/2} \tag{33}$$

This method was used in [25] and [26] to orthogonalise the mode shapes, and to update all terms of the stiffness matrix, in a global manner. We use this technique here after a given number of iteration steps, which allows to reduce the influence of the measurement errors and to build a test data set that is more consistent with the analytical model. It should be noted that the presented technique combines both, the mode shape smoothing and the expansion on the basis of the orthonormality condition. Numerically it does not require the use of the full mass matrix. The original measured mode shapes in the residual in eq.(24b) can now be replaced by the smoothes shapes taken at the measured DOF's: $y_{oM} => \phi_c$.

The **input error** is given by substituting the measured frequency response into the equation of motion. Since the number of measured DOF's is generally much smaller than the number of analytical DOF's it is necessary to expand the measured vector to full model size or to condense the model order down to the number of measured DOF's.

$$\varepsilon_F = f_M - \hat{K}_c(j\omega_M, p)y_M \tag{34a}$$

where

$$\hat{K}_c(j\omega_M) = \hat{K}_{ca}(j\omega_M) + \sum_s p_s S_s \tag{34b}$$

$$= \text{ updated dynamic stiffness matrix dynamically condensed}$$
$$\text{to } N_M < N \text{ measured DOF's}$$

$$\hat{K}_{ca}(j\omega_M) = -\omega_M^2 M_{ca} + K_{ca} + j\omega_M D_{ca} \tag{34c}$$
$$= \text{condensed dynamic stiffness matrix at point a.}$$

$p_s = \alpha_i, \beta_j$ or γ_k correction parameters

$S_s = K_i, M_j$ or D_k correction submatrices

$y_M(j\omega_M) = $ complex frequency response vector measured
       at c measurement DOF's

f_M = measured harmonic exciter forces, if $f_{Mi} = 1$ at exciter DOF i , y_M represents the
     i-th column of the frequency response matrix

With eqs. (34 b, c) introduced the force residual at the measured DOF's can be expressed by

$$\varepsilon_F = r_F - G_F \Delta p \tag{35a}$$

where

$$r_F = f_M - \hat{K}_{ca}(j\omega_M, p_a)y_M \quad = \text{residual vector at point a} \tag{35b}$$

$$G_F = [\ ... -\omega_M^2 M_{cj} y_M .... K_{ci} y_M ... j\omega_M D_{ck} y_M ...]_a \quad = \text{gradient matrix} \tag{35c}$$

Two condensation methods have most extensively been investigated in the past. One is the dynamic condensation which includes the static condensation (also called the Guyan reduction) as a special case. The dynamically condensed matrices (index c) in the above equations are obtained by first introducing the parametrisations of eqs. (21) into the equation of motion and by partitioning the matrices with respect to the measured (index M) and unmeasured DOF's (index U):

$$\left(\begin{bmatrix}\hat{K}_{MM} & \hat{K}_{MU} \\ \hat{K}_{UM} & \hat{K}_{UU}\end{bmatrix}_A + \sum_s p_s \begin{bmatrix}\hat{K}_{MM} & \hat{K}_{MU} \\ \hat{K}_{UM} & \hat{K}_{UU}\end{bmatrix}_s\right) \cdot \begin{bmatrix}y_M \\ y_U\end{bmatrix} = \begin{bmatrix}f_M \\ f_U = 0\end{bmatrix} \tag{36}$$

where the vector y_u contains the unknown response vector components at the unmeasured DOF's. The second row of this equation can be used to express the unmeasured vector y_u as a function of the correction parameters and the measured vector y_M:

$$y_U = T(j\omega_M, p)y_M = -\hat{K}_{UU}^{-1}\hat{K}_{UM}y_M \tag{37}$$

The transformation matrix T is a function of the excitation frequency ω_M. For $\omega_M = 0$ this matrix represents the static condensation matrix (also called Guyan condensation).

The expanded full vector is given by

$$y = \begin{bmatrix}y_M \\ y_U\end{bmatrix} = \begin{bmatrix}I \\ T(j\omega_M, p)\end{bmatrix}y_M = \overline{T}(j\omega_M, p)y_M \tag{38}$$

where the transformation matrix $\overline{T}$ can be used to condense the system and the correction matrices according to

$$S_c = \overline{T}^T(S_A + \sum_s p_s S_s)\overline{T} = S_{Ac} + \sum_s p_s S_{cs} \tag{39}$$

$S_c = K_c, M_c$ or D_c = condensed system matrices.

Note: in the general damped case $\overline{T}$ is complex and $\overline{T}^T$ denotes complex - conjugate transpose.

This type of model reduction has several drawbacks. Looking at large order model typical for industrial applications the number of measured DOF's is much smaller than the model order: $n_M \ll n$. Since the transformation matrix depends on the current erroneous parameter estimate it becomes apparent that the sensitivity matrix in eq. (35c) will be polluted by a systematic error which can not be compensated by the LS method. In addition the numerical effort to decompose the $\hat{K}_{UU}$ matrix in eq. (31) makes its application prohibitive for updating of large order models.

The second technique of order reduction also called the modal reduction technique has already been presented in eqs.(29) and (30). When applied to the frequency response vectors the transformation between the response at the unmeasured and the measured DOF's is given by :

$$y_U = T y_M = Y_{Au}Y_{Ac}^+ y_M \tag{40}$$

It should be noted that using this transformation to condense the system matrices according to eq. (39) the eigendata of the original and the condensed system are the same.

Another disadvantage results from the fact that the gradient matrix is polluted not only by the systematic condensation errors but also by measurement errors which both result in biased parameter estimates.

The **pseudo response** error residual is obtained from transforming the input force error to an output error by multiplying the force residual of eq. (34a) with the FRF matrix of initial model (index A). This is equivalent with exciting the initial model (assumed to exhibit modal damping for simplicity) with the arbitrary frequency ω_F:

$$\varepsilon_{PR} = \hat{K}_{Ac}^{-1}(\omega_F)\,\varepsilon_F \tag{41}$$

where

$$\hat{K}_{Ac}^{-1} = (-\omega_F^2\,M_{Ac} + j\,\omega_F\,D_{Ac} + K_{Ac})^{-1} = \sum_i y_{oAi}\,y_{oAi}^T\,/(\omega_{Ai}^2 - \omega_F^2 + j\,2\,\omega_{Ai}\,\omega_F\,\xi_F) \tag{42}$$

= condensed FRF matrix expressed by modal data of initial model (index A)
ω_F = filter frequency

$\hat{K}_{Ac}^{-1}$ can also be interpreted as a dynamic filter. Applying the force residual vector on the initial system vibrating with the arbitrary filter frequency ω_F controls the magnification of the pseudo response error ε_{PR}. The idea behind this filtering is to reduce the bias of the estimate.

A **special case** is obtained by applying the residual force vector on the analytical system linearized at point 'a' using the measured excitation frequencies :

With $\hat{K}_{Ac}^{-1} = \hat{K}_{ca}^{-1}$ and $\omega_F = \omega_M$ introduced into eqs.(41) and (35a-c) the pseudo response residuals are calculated by

$$\varepsilon_{PR} = \hat{K}_{ca}^{-1}(\omega_F)\,\varepsilon_F = r_{PR} - G_{PR}\,\Delta p \tag{43a}$$

where

$$r_{PR} = y_{ca}(j\omega_M) - y_M(j\omega_M) \tag{43b}$$

= residual response at point 'a'

$$G_{PR} = \hat{K}_{ca}^{-1}\big[\ \ldots - \omega_M^2\,M_{cj}\,y_M \ldots.\ K_{ci}\,y_M \ldots j\omega_M\,D_{ck}\,y_M\ \ldots\big]_a \tag{43c}$$

= gradient matrix at measured DOF's.

The force residual and the pseudo response technique have been investigated by several authors with respect to the bias and ill-conditioning problems mentioned above [27]- [32] and with respect to the optimal choice of the excitation frequencies.

The linearized **frequency response (FR) residuals** are obtained from the differences of the measured and the analytical FR at the reduced set of $n_M < n$ measured DOF's denoted by the index c. The sensitivity matrix of the **FR** residual is derived from differentiating the equation of motion (20a) with respect to the parameters.

$$\varepsilon_R = y_M - y_c = r_{Ra} - G_R\,\Delta p \tag{44a}$$

where

$$r_R = y_M(j\omega_M) - y_{ca}(j\omega_A) \tag{44b}$$

= residual vector with test/ analysis differences of frequency response at linearization point a and at excitation frequency $\omega_A = \omega_M$.

$y_{ca}(j\omega_A) :=$ analytical complex frequency response vector

at excitation frequency $\omega_A = \omega_M$

$y_M(j\omega_M) =$ frequency response vector measured at the same DOF's

at excitation frequency ω_M

$$G_R = \left[...\partial y_c / \partial p_s...\right] = -\hat{K}^{-1} \left[...\frac{\partial \hat{K}}{\partial p_s} y ...\right]\big|_{p=p_a} \tag{44c}$$

$$= -\hat{K}^{-1} \left[... -\omega_M^2 M_j y ....K_i y ...j\omega_M D_k y ...\right]\big|_{p=p_a}$$

= sensitivity matrix at measured DOF's at linearization point a.

$$\hat{K}^{-1} = (-\omega_M^2 M + j\omega_M D + K)^{-1}\big|_{p=p_a} = \sum_i^{R<=N} y_{oci}\, y_{oci}^T /(\omega_{oi}^2 - \omega_M^2 + j2\omega_{oi}\,\omega_M\,\xi_i)\big|_{p=p_a} \tag{44d}$$

= FRF matrix of analytical model expressed by R <= N analytical modal quantities at linearization point a (proportional modal damping values ξ_i assumed for simplicity) and calculated at the measured DOF's.

Like the sensitivity matrix for the modal sensitivities the FR sensitivity matrix is not directly corrupted by measurement errors allowing (approximately) unbiased parameter estimates in contrast to the force and the pseudo response residuals in eqs.(35c) and (43c). In addition there is no need to expand the test vectors to the unmeasured DOF's. However, there is a crucial drawback in this formulation which has prevented its application to other than academic cases. Comparing measured and analytical FR functions like those of Figure 3 it may be noticed that due to the shifts of the resonance peaks caused by the mismatch of the eigenfrequencies the test/analysis differences at a given excitation frequency become extremely large in particular in such cases where the ordering of the test and analysis eigenfrequencies is not the same (also called mode crossing). The consequence is that the first order theory either fails to predict the updated model response or the rate of convergence is very low unless the test/analysis deviations are very small. Attempts have been made by several authors to overcome the problem, e.g. by using second order derivatives , artificial damping or eliminating or reducing the weight [32] of the data in the range of the resonances. Other approaches deliberately make use of the shift of the resonance peaks [33], [34].

It is advocated here to restrict updating to the resonance peaks taken at the analytical and experimental eigenfrequencies which are not identical due to mass and stiffness errors. The residuals therefore are the same as in eq. (44) except that the analytical excitation frequencies are replaced by the analytical eigenfrequencies and the experimental excitation frequencies by the experimental eigenfrequencies :

$$\omega_A \Rightarrow \omega_o \quad \text{and} \quad \omega_M \Rightarrow \omega_{Mo} \tag{45}$$

Of course, the latter assumption means to extract the experimental eigenfrequencies prior to FRF updating which is in conflict with the motivation behind using FRF's for model updating instead of modal data aimed to avoid experimental modal analysis errors. However, it is believed that in practise when FRF measurements have been taken an experimental modal analysis will be performed anyway. Of course, the experimental modal data would also be useful to validate the FRF updating results. Also is must be expected that in cases when the accuracy of the test data and/or the elastodynamic mode assumptions will not allow a successful modal extraction then FRF updating will even be less successful due to the fact that the physical model includes more error sources than the curve fitting approach used in experimental modal analysis.

One advantage of FRF updating remains: the possibility to identify damping parameters. If it is assumed that the stiffness and mass parameters have been updated before by using the undamped modal residuals of eqs. (23) and (24) (Note: undamped experimental modal parameters can either be measured directly by phase resonance testing or by phase separation techniques like ISSPA [35] or by techniques transforming complex modes to real modes [36]) then the FRF residuals of eqs.(44) under consideration of eq. (45) could be applied directly with the damping parameters $\Delta p := \Delta\gamma$ being the only unknowns in this case. However, this possibility is very restricted because the local physical damping parametrisation of eq. (21c) is not available for most structural applications. This is the reason why the global modal damping approach is widely used in practise and why it should be retained in FRF based parameter updating.

The approach described in the following is directed to update initial estimates of modal damping parameters. Using the mass/stiffness updated models to calculate the frequency response one would of course use at first the damping values resulting from experimental modal analysis. If the fit of the eigenfrequencies and modes of the updated model to their experimental counterparts was perfect then the FRF's calculated with the updated model would be the same as the experimental FRF's used to extract the modal data, i.e. any updating of the modal damping parameters would not be necessary.

Since in practical applications such a perfect fit cannot be obtained it seems desirable to have a procedure allowing to fit an initial modal damping estimate to the experimental FRF's.

Such an approach is described in the following. The residual used is the same as in eqs. (43) and (44), i.e. the excitation frequencies are taken at the eigenfrequencies differing between test and analysis :

$$\varepsilon_\xi = y_M(\omega_{Mo}) - y_c(\omega_o) = r_{\xi a} - G_\xi \Delta\xi \tag{46}$$

where

$$r_{\xi a} = y_M(j\omega_{Mo}) - y_{ca}(j\omega_o)$$

   = residual vector with test/ analysis differences of frequency response at linearization point a

$y_{ca}(j\omega_o) :=$ complex frequency response vectors at excitation frequency = analytical eigenfrequency ω_o

$y_M(j\omega_{Mo}) = $ frequency response vectors measured at the same DOF' s
at experimental excitation frequency =
experimental eigenfrequency ω_{Mo}

To calculate the sensitivity matrix of the modal damping the equation of motion (20a) is first transformed to modal co-ordinates by the superposition of undamped modes

$$y_c = \sum_r^{R \leq N} y_{ocr}\, q_r \tag{47a}$$

resulting in $r = 1, \dots R$ decoupled equations if proportional damping is assumed otherwise the modal equations are coupled by the non- diagonal modal damping matrix $D_g^T = y_o^T D\, y_o$ (the proportional damping assumption is not necessary and was made to simplify the derivation):

$$(\omega_{or}^2 - \omega_{os}^2 + j\, 2\,\omega_{os}\,\omega_{or}\,\xi_r)\, q_r = y_o^T f \tag{47b}$$

Eq.(46b,) derived with respect to the viscous modal damping parameter $\xi_r = D_{gr}/(2\omega_{or})$ yields

$$G_\xi = \left[\dots \partial y_c(\omega_{os})/\partial \xi_r = y_{ocr}\frac{\partial q_r(\omega_{os})}{\partial \xi_r} \dots \right]_{\xi=\xi_a} \tag{48a}$$

where

$$\frac{\partial q_r(\omega_{os})}{\partial \xi_r} = \frac{-j\, 2\,\omega_{os}\omega_{or}q_r(\omega_{os})}{\hat{K}_{gr}}. \tag{48b}$$

$(r, s = 1, \dots R$ number of active modes $<= N)$

5. Applications

In the following the different steps of updating a small 5 DOF laboratory test structure using measured modal residuals are presented. The stiffness parameters are updated using measured eigenfrequencies and mode shapes whereas the damping parameters are updated using measured FRF's.

The structure is shown in Figure 1. It consists of 5 cantilever steel beams carrying a tip mass and coupled by arch springs. Due to the concentrated tip masses being larger than the structural masses the beams and the arches were statically condensed to the connection DOF's. By this procedure the beams and the arches were reduced to one and 2 DOF's, respectively. The condensed elements were subsequently used to assemble the 5 DOF spring mass system shown in Figure 2. The following numerical data for the condensed element matrices were used for the initial model:

Stiffness parameters N/m :

$K_1 = K_2 = K_4 = K_5 = 27\ 216$; $K_3 = 39\ 191$; $K_6 = K_7 = K_8 = K_9 = 17\ 901$; $K_{10} = K_{11} = 15\ 312$

Tip masses kg : $M_{1-5} = [0.429\ 0.574\ 0.565\ 0.574\ 0.729\ ]$ kg

M_5 includes the mass of the force transducer used to measure the excitation force.

The condensed mass matrices of the arches are given by $M_i = m_c \begin{bmatrix} 1 & 0.5 \\ 0.5 & 1 \end{bmatrix}$ with

$m_c = 0.0083$ kg (for i = 6-9); $m_c = 0.105$ kg (for i = 10, 11)

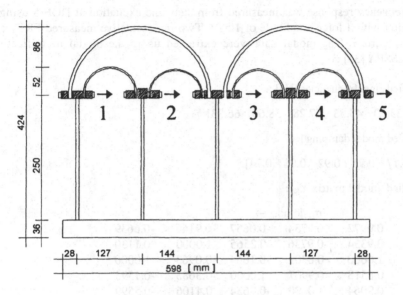

Figure 1 Test structure

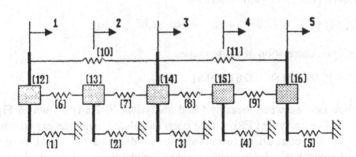

Figure 2 Spring / mass model of test structure

The results for the undamped model are:

Eigenfrequencies Hz :

$f_A = [31.82\ 41.49\ 54.89\ 59.71\ 66.08]$

Modal matrix Y_A :

DOF	mode no. $\rightarrow$				
1	0.6064	0.9879	-0.0904	-0.9087	-0.6396
2	0.6077	1.0000	-0.1593	1.0000	-0.1729
3	0.7191	0.2707	0.1530	-0.1785	1.0000
4	0.7881	-0.4219	1.0000	0.1414	-0.3822
5	1.0000	-0.8017	-0.5458	0.0247	-0.2034

The frequency response was measured from step sine excitation at DOF 5 using a modal exciter with a force amplitude of 1.5 N. Two representative measured FRF's are shown in Figure 3. The modal data were extracted using the modal identification procedure ISSPA [35] :

Identified eigenfrequencies [Hz] :

$f_M = [33.50 \quad 43.53 \quad 57.28 \quad 58.60 \quad 66.33]$

Identified modal damping [%]

$\xi = [1.17 \quad 0.94 \quad 0.92 \quad 0.91 \quad 0.74]$

Identified modal matrix Y_M :

DOF	mode no. $\rightarrow$				
1	0.9972	-0.9734	-0.0657	-0.9165	-0.6639
2	0.9334	-0.9736	0.2565	1.0000	-0.1130
3	1.0000	0.0309	-0.1664	-0.0963	1.0000
4	0.8513	0.9907	1.0000	-0.3683	-0.1593
5	0.9064	1.0000	-0.7634	0.4106	-0.5599

Test / analysis error of eigenfrequencies :

$(f_M - f_A)/f_M \ \% \ = \ [5.27 \quad 4.92 \quad 4.36 \quad -1.85 \quad 0.36]$

Test/ Analysis correlation of mode shapes :

$MAC \ \% = [0.95 \quad 0.89 \quad 0.81 \quad 0.81 \quad 0.90]$

The comparison between measured and analytical FRF's is shown in Figure 3 for DOF's 2 and 3. The analytical FRF's have been calculated using the experimental modal damping values. The comparison reveals significant discrepancies in the range of the 3rd and 4th eigenfrequency which are much closer in the test than in the analysis model.

The first step starting model updating is the definition of the updating parameters. In order to avoid the selection of parameters with small sensitivity the parameter sensitivity expressed by the column norms of the sensitivity matrix G_λ and G_y is plotted in Figure 4(a) for the 11 stiffness parameters. All 5 eigenfrequencies and modes were used to calculate the sensitivity matrix. The plot shows that in the present case the sensitivity of all the parameters is high enough if a value beyond 10 % resulting from experience is assumed to be the minimum requirement. The plots in Figures 4 (b) and (c) show the sensitivity of the parameters with respect to the individual modes. Such plots are useful in practise to decide how many modes should be measured in order to observe as many parameters as possible. For example, if only the first 3 modes were taken parameters 1, 2 and 5 - 9 would be too insensitive to be used for updating.

Updating of 11 stiffness parameters using 5 measured modes at all 5 DOF's yields the results shown in Figures 5(a)-(d). The mass parameters were assumed to be correct. The regularisation parameter was set to $w_p = 0.2$. The test/analysis deviations are shown in Figures 5 (a) and (c) for the eigenfrequencies and the MAC values. In Figure 5(d) the average of all frequency deviations over all frequencies and of the (100-MAC) - values shows an excellent convergence behaviour. The following correction parameters calculated after the 6th iteration step were used to update the model stiffness matrix :

Number:	1	2	3	4	5	6	7	8	9	10	11
α [%] =	[3.1	3.4	-11.7	1.8	52.4	-12.6	-1.0	-30.0	16.0	11.3	32.1]

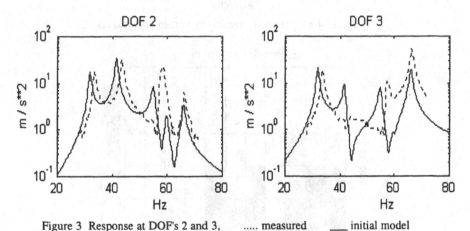

Figure 3 Response at DOF's 2 and 3, ... measured   ___ initial model

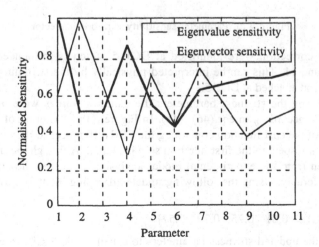

Figure 4 a) Normalised parameter sensitivity

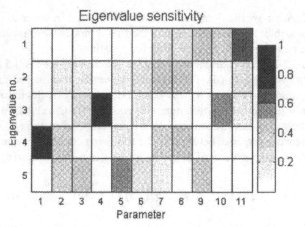

Figure 4 b) Parameter sensitivity w.r. to eigenvalues

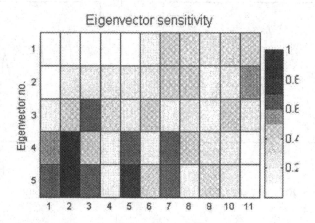

Figure 4 c) Parameter sensitivity w.r. to eigenvector

The main correction were identified at DOF 5 which is influenced by stiffness parameters 5 and 11. This can be interpreted physically by the stiffening effect of the exciter stinger rod attached at DOF 5.

After updating the stiffness parameters the damping values were updating using measured FRF's according to eqs.(46)- (48). In Figure 6 the evolution of the parameter estimates and of the cost function containing the FRF differences at the resonance peaks normalised with respect to the first iteration step is plotted. Although the initial damping estimates taken from the experimental modal analysis were taken the cost function is reduced considerably. Using the following updated damping identified after 5 iteration steps

$$\xi \, [\%] = [1.11 \quad 0.81 \quad 0.75 \quad 0.79 \quad 0.57]$$

together with the updated stiffness parameters to calculate the FRF's an excellent fit to the experimental data is obtained as can be seen in Figure 7.

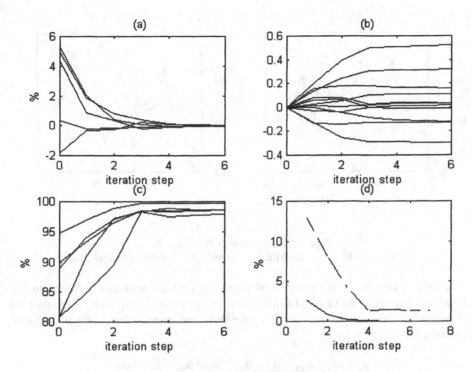

Figure 5 Results of stiffness updating
   (a)   eigenfrequency error
   (b)   correction parameters
   (c)   MAC values
   (d)   mean deviations ___frequency - - - 100-MAC
            regularisation parameter $w_p = 0.2$

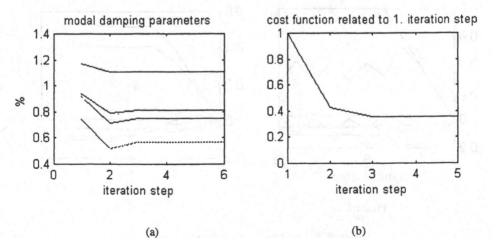

(a)   (b)
Figure 6 Results of damping parameter updating :
(a) viscous damping parameters   (b) cost function

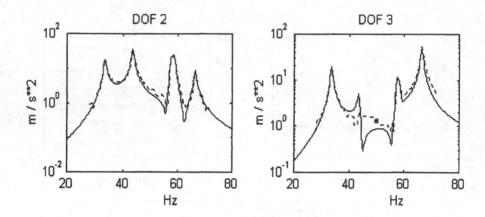

Figure 7 Response at DOF's 2 and 3
..... measured ___ analytical after stiffness and damping updating

Another example of the benefits of using regularised solutions is given in the following. Looking at the physical similarity of the structural components a reduced set of 7 parameters may also be considered a meaningful parametrisation. Using the reduced set related to the full set by

$$\bar{\alpha}_1 = \alpha_6 = \alpha_7 = \alpha_8; \quad \bar{\alpha}_2 = \alpha_9; \quad \bar{\alpha}_3 = \alpha_{10}; \quad \bar{\alpha}_4 = \alpha_{11};$$
$$\bar{\alpha}_5 = \alpha_1 = \alpha_2 = \alpha_4; \quad \bar{\alpha}_6 = \alpha_3; \quad \bar{\alpha}_7 = \alpha_5$$

the oscillating convergence behaviour of Figure 8 can be observed. The regularisation with $w_p = 0.2$ results in the smoothed convergence of Figure 9.

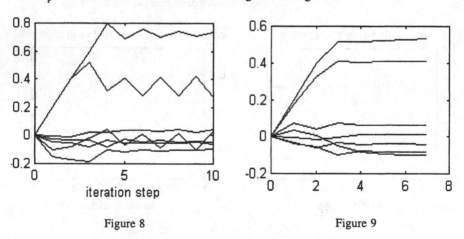

Figure 8   Figure 9

Figures 8, 9 Updating with a reduced parameter set,
(8) no regularisation
(9) with regularisation

6. Summary

In the paper a summary of current procedures for model updating are presented. At first the mathematical background of the parameter estimation is given with respect to the numerical estimation techniques for solving the updating equations based on the iterative extended least squares technique. The model parametrisations defining the type and the location of the erroneous parameters are then used to construct the residuals formed by the test/analysis differences to be minimised. The residuals presented are formed by force and pseudo response equation errors, by eigenfrequency and mode shape errors and by frequency response errors. The procedures have been derived to handle incomplete test vectors, where the number of measured degrees of freedom (DOF) is much less than the DOF no. of the computational model. Some recent extensions of the estimation techniques with respect to handle systematic mode shape errors and to updating of modal damping values have also been included. Finally an application of updating the model of a laboratory test structure was used to demonstrate the most important updating steps.

7. References

1. Natke (1992), Einführung in theorie und praxis der zeitreihen und modalanalyse, *Vieweg Verlag*, Braunschweig/Wiesbaden.
2. Friswell and Mottershead, J.E. (1995), Finite Element Model Updating in Structural Dynamics, *Kluwer Academic Publishers*, Dordrecht.
3. Lallement, G. (1988), Localisation techniques, *Proc. of Workshop "Structural Safety Evaluation Based on System Identification Approaches"*, Vieweg Verlag, Braunschweig/Wiesbaden.
4. Flores, O., Link, M. (1993), Localization techniques for parametric updating of dynamic mathematical models, *Proc. of Int. Forum On Aeroelasticity and Structural Dynamics*, Strassbourg, France, *Association Aeronautic et Astronautic de France (Hrsg)*, 75782 Paris.
5. Ahmadian, G.M.L., Gladwell and Ismail, F., Parameter selection strategies in finite element model updating, *ASME, J. of Vibration and Acoustics 119*, 37-45.
6. Demmel (1997), Applied numerical linear algebra, *SIAM*, Philadelphia, PA, U.S.A.
7. Soederstroem and Stoica, P. (1989), System Identification, *Prentice Hall Int.*, UK.
8. Link, M. (1993), Updating of analytical models- procedures and experience, *Proc. of Conf. on Modern Practice in Stress and Vibration Analysis*, J.L. Wearing (ed.), *Sheffield Academic Press*, 35-52.
9. Tikhonov and Arsenin, V.Y. (1977), Solutions of Ill-posed Problems, *J. Wiley*, New York.
10. Natke (1991), On regularization methods applied to the error localization of mathematical models, *Proc. Int. Modal Analysis Conf. IMAC IX*, Florence, Union College, Schenectady.
11. Mottershead and Foster, C.D. (1991), On the treatment of ill-conditioning in spatial parameter estimation from measured vibration data, *Mechanical Systems and Signal Processing, Vol. 5, No. 2*, 139-154.
12. Prells (1995), Eine regularisierungsmethode für die lineare fehlerlokalisierung von modellen elastomechanischer systeme, *Dissertation*, Univ. Hannover.
13. Hansen, C. (1992), Analysis of discrete ill-posed problems by means of the L - curve, *Siam Review, Vol. 34, No. 4*, 561- 580.
14. Ahmadian, Mottershead, J.E. and Friswell, M.I. (1998), Regularisation methods for finite element model updating, *Mechanical Systems and Signal Processing, Vol. 12, No. 1*.
15. Hansen, C., Regularisation Tools: a MATLAB package for analysis and solution of discrete ill-posed problems, *Technical University of Denmark*, Lyngby, Denmark
16. Natke, Collmann, D. and Zimmermann, H. (1974), Beitrag zur korrektur des rechenmodells eines elastomechanischen systems anhand von versuchsergebnissen, *VDI- Berichte 221*, 23-32.
17. Natke, Lallement, G. and Cottin, N. (1995), Properties of various residuals within updating of mathematical models, *Inverse Problems in Engineering., Vol. 1*, 329- 348.

18. Fox and Kapoor, M. (1968), Rate of change of eigenvalues and eigenvectors, *AIAA Journal, Vol. 6*, 2426- 2429.
19. Sutter, T. R., Camarda,Ch. J., Walsh, J.L. and Adelman, H. M. (1988), Comparison of several methods for calculating vibration mode shape derivatives, *AIAA J., Vol. 26, No. 12*.
20. Lallement, G. and Zhang, Q. (1989), Selective structural modifications, applications to the problems of eigensolution sensitivity and model adjustment, *Mechanical Systems and Signal Processing, Vol. 3, No. 1*.
21. Balmes, E. (1998), Efficient sensitivity analysis based on finite element model reduction, *Proc. of 16th Int. Modal Analysis Conf., IMAC XVI*, Santa Barbara, USA.
22. Link, M. and Mardorf, J. (1996), The role of finite element idealisation and test data errors in model updating, *Proc. 2nd Int. Conf. Structural Dynamics Modelling, NAFEMS*, Glasgow, 493- 504.
23. Link, M. (1998), Updating analytical models by using local and global parameters and relaxed optimisation requirements, *Mechanical Systems and Signal Processing, Vol. 12, No. 1*.
24. Targoff (1976), Orthogonality check and correction of measured modes, *AIAA Journal, Vol. 14, No. 2* ,164 - 167.
25. Baruch and Bar-Itzhack, I.Y. (1978), Optimal weighted orthogonalisation of measured modes, *AIAA Journal, Vol. 16, No. 4*, 346- 351.
26. Baruch (1982), Optimal correction of mass and stiffness matrices using measured modes, *AIAA Journal Vol. 20, No. 11*, 1623- 1626.
27. Link, M. (1991), Localisation of errors in computational models using dynamic test data, Proc. of the European Conf. on Struct. Dynamics, *EURODYN '90, Structural Dynamics*, W. B. Kraetzig et al (eds.), A. A. Balkema.
28. Link, M. (1992), Experiences with different procedures for updating structural parameters of analytical models using test data, *Proc. Int. Modal Analysis Conf., IMAC X*, San Diego, Union College, Schenectady, NY, USA.
29. Fritzen, P. and Kiefer, T. (1992), Localisation and correction of errors in finite element models based on experimental data, *Proc. Int. Modal Analysis Conf., IMAC X*, San Diego, Union College, Schenectady, NY, USA.
30. Larsson and Sas, P. (1992), Model updating based on forced vibration testing using numerically stable formulation, *Proc. Int. Modal Analysis Conf., IMAC X*, San Diego, Union College, Schenectady, NY, USA.
31. D'Ambrogio and Fregolent, A. (1998), On the use of consistent and significant information to reduce ill-conditioning in dynamic model updating, *Mechanical Systems and Signal Processing , Vol. 12, No. 1*.
32. Ibrahim, Teichert, W. and Brunner, O. (1998), Frequency response function FE model updating using multi perturbed analytical models and information density matrix, *Proc. Int. Modal Analysis Conf., IMAC XVI*, Santa Barbara, USA.
33. Pascual, Golinval, J.C. and Razeto, M. (1997), A frequency domain correlation technique for model correlation and updating, *Proc. Int. Modal Analysis Conf., IMAC 15*, Orlando, USA.
34. Cogan, S., Lenoir, D. and Lallement, G. (1996), An improved frequency response residual for model correction, *Proc. Int. Modal Analysis Conf., IMAC 14*, Dearborn, USA.
35. Link, M., Weiland, M. and Moreno-Barragan, J. (1987), Direct physical matrix identification as compared to phase resonance testing, *Proc. Int. Modal Analysis Conf., IMAC 5*, London, UK.
36. Niedbal (1984), Analytical determination of real normal modes from measured complex modes, *AIAA Paper 84-0995*.
37. Gladwell and Ahmadian, H. (1995), Generic element matrices suitable for finite element model updating, *Mechanical Systems and Signal Processing, Vol. 9*.

MODEL QUALITY ASSESSMENT AND MODEL UPDATING

M. LINK, G. HANKE
Light Weight Structures and Structural Mechanics Laboratory
University of Kassel, Germany

1. Introduction

Before using an initial finite element model for subsequent model updating the user must be aware of the modelling uncertainties arising from three main sources :

(1) Idealisation errors resulting from the assumptions made to characterise mechanical behaviour of the physical structure

(2) Discretisation errors introduced by numerical methods like those inherent in the finite element method and

(3) Erroneous assumptions for model parameters.

The parameters can affect one element or a group of elements with the same properties, i.e. a single parameter may be assigned to a substructure. Numerical procedures for model updating are restricted to correct model parameters only. When the model includes idealisation and discretisation errors it may only be updated in the sense that the deviations between test and analysis are minimised. The same happens when the selected correction parameters are not consistent with the real source and the location of the error. The parameters in such cases may lose their physical meaning after updating. A typical result of updating such *inconsistent* models will be that they may be capable of reproducing the test data but may not be useful to predict the system behaviour beyond the frequency range used in the updating, or to predict the effects of structural modifications or to serve as a substructure model to be assembled with an overall structural model.

The aim of all structural analyses to predict the structural response can only be achieved if all three kinds of modelling errors are minimised with respect to the given purpose of the structural analysis. Models which fulfil these requirements shall be called validated models. Model quality must therefore be assessed in three steps.

Step1: Assessment of idealisation and numerical method errors prior to parameter updating

Step2: Correlation of analytical model predictions and test results and selection of correction parameters

Step3: Assessment of model quality after parameter updating

J.M.M. Silva and N.M.M. Maia (eds.), Modal Analysis and Testing, 305–324.
© 1999 *Kluwer Academic Publishers. Printed in the Netherlands.*

In this paper we report about the influence of model idealisation and discretisation errors on the results of numerical parameter updating. Steps and experience are described with generating validated models for industrial type structures.

2. Consistent *versus* inconsistent models

Models which include idealisation and discretisation errors resulting from the assumptions made to characterise the mechanical behaviour of the structure and from discretisation errors introduced by numerical methods are *inconsistent* with the real physical behaviour. Such models may only be updated in the sense that the deviations between test and analysis are minimised but the parameters may lose their physical meaning after updating.

Without idealisation and discretisation errors the model uncertainties can be assigned to the parameters of the model only.

In the most popular approach the parameters of the system matrices are updated by substructure matrices according to

$$K = K_A + \sum \alpha_i K_i \tag{1a}$$

$$M = M_A + \sum \beta_j M_j \tag{1b}$$

$$D = D_A + \sum \gamma_k D_k \tag{1c}$$

where

$[\alpha_i \ \beta_j \ \gamma_k]$ = unknown correction (design) parameters,

$(i = 1, 2, \dots, I; \quad j = 1, 2, \dots, J; \quad k = 1, 2, \dots, K)$

$I + J + K$ = no. of correction parameters

K_A, M_A, D_A = analytical (initial) stiffness, mass and damping matrix

K_i, M_j, D_k = assumed correction substructure matrices (elements or element

groups) defining source and location of modelling error.

The correction sub-matrices defined above can be considered as the first derivative of the updated matrices with respect to a physical or geometrical model parameter :

$$K_i = \alpha_A \, \partial K/\partial \alpha_i \,, \quad M_j = \beta_A \, \partial M/\partial \beta_j \quad \text{and} \quad D_k = \gamma_A \, \partial D/\partial \gamma_k \tag{2}$$

These derivatives are constant like in the case of a beam element with α representing Young's modulus where the stiffness matrix is a linear function of the modulus. The derivatives must not be constant like in the case when α represents the shear modulus of a Timoshenko beam where the stiffness matrix terms are non-linear functions of the shear modulus. α_A, β_A and γ_A denote the initial parameters used to make the parameter changes dimensionless. The correction parameters α_i, β_j and γ_A are calculated by the updating algorithm whereas the sub-matrices $\mathbf{K}_i$, $\mathbf{M}_j$ and $\mathbf{D}_k$ containing the assumptions for the type and the location of the modelling error have to be defined by the user. These assumptions are important for the success of the updating.

If these assumptions are *not consistent* with the real error source and location the parameters may again minimise the difference between analytical and experimental data, however, the parameters may lose their physical significance and represent pure mathematical optimisation variables like in the case of idealisation and discretisation errors.

A typical result of updating such inconsistent models will be that they may be capable of reproducing the test data but may not be useful to predict the system behaviour beyond the frequency range used in the updating, or to predict the effects of structural modifications or to serve as a substructure model to be assembled with an overall structural model.

It is of crucial importance to avoid inconsistent models especially idealisation and numerical approximation errors in order to avoid the physical parameters to lose their physical significance. However, situations may arise where this requirement may be relaxed, for example, when the validity of the updated model shall intentionally be restricted to a specified frequency range ($\rightarrow$ equivalent model). During the preparation of the finite element model the analyst must therefore be aware of the modelling uncertainties mentioned above in order to interpret correctly the results of the parameter updating.

3. Requirements for validated models

The requirements posed on validated model depend on the purpose of its final utilisation in terms of its ability to

a) reproduce the test data used in the updating,

b) predict the system behaviour to other load cases than those used in the test,

c) predict the system behaviour beyond the frequency range and /or at other degrees of freedom than those used for model updating,

d) predict the effects of structural modifications,

e) to be utilised as a substructure model assembled within an overall structural model.

The above requirements are ranked w.r. to their increasing demand on the number and the type of eigenmodes. From the arguments above the following requirements for the initial analysis model (*pre-update* model) can be derived :

The initial analysis model as well as the assumptions defining the source and the location of the assumed modelling errors must be consistent in order that the model parameters retain their physical significance after updating. Only in the case when the quantity of experimental modal data is sufficient for the purposes (a)-(e) inconsistent initial analysis and error assumptions may be tolerated. In particular this holds when so-called *equivalent models* (EM) shall be generated.

A model is called an equivalent model when the mesh and the properties have intentionally been simplified to obtain a model of smaller order. Its properties may either be generated directly by the user or may be derived from a high order finite element model by using the modal data of the fine model as target data to be approximated in the updating procedure. After updating the equivalent models should yield a minimum deviation w.r. to the experimental modal data. As mentioned above EM's generally are inconsistent models where the source and the location of erroneous parameters and the structure of the mathematical model is not consistent with the true model. As always in the case of inconsistent models the parameters may lose their physical meaning after updating. A typical result of updating such inconsistent models will be that they may be capable of reproducing the test data but may not be useful to predict the system behaviour beyond the frequency range used in the updating, or to predict the effects of structural modifications or to serve as a substructure model to be assembled with an overall structural model *unless a sufficient number of modes* has been used for updating. Consequently the quantity of experimental data to be used for updating EM's must generally be larger for inconsistent models than for consistent models. To preserve the physical meaning of the parameters should generally be considered the primary goal in model updating.

4. A priori assessment of model quality

4.1. IDEALISATION ERRORS

The first step in model quality assessment must consider idealisation errors resulting from the assumptions made to characterise mechanical behaviour of the physical structure. Typical such errors are resulting from

a) erroneous simplification of the structure, for example, when a plate is treated like a beam, when a thick volumetric structure exhibiting a 3D non uniform stress state is simplified by a 2D model with a uniform stress distribution in one direction, i.e. lumping of stiffness properties is too coarse,

b) inaccurate assignment of mass properties, for example when distributed masses are modelled with too few lumped masses, when an existing eccentricity of a lumped mass is disregarded,

c) when the used finite element formulation neglects particular properties, for example, when the influence of transverse shear deformation or warping due to torsion in beam elements is neglected,

d) errors in the connectivity of the mesh i.e. some elements are not connected or connected to a wrong node,

e) erroneous modelling of boundary conditions, for example when an elastic foundation is assumed to be rigid,

f) erroneous modelling of joints, for example when an elastic connection is assumed to be rigid (clamped) or when an eccentricity of a beam or a plate connection possibly existing in the real structure is disregarded,

g) erroneous assumptions for the external loads,

h) erroneous geometrical shape assumptions,

i) when a non-linear structure is assumed to behave linearly.

4.2. DISCRETISATION ERRORS

The second group of model uncertainties is related to discretisation errors introduced by numerical methods like those inherent in the finite element method, for example

a) Discretisation errors when the finite element mesh is too coarse (not fully converged modal data in the frequency range of interest).

b) Truncation errors in order reduction methods like static condensation.

c) Incomplete integration order used to calculate element matrices.

d) Finite elements exhibiting spurious modes.

e) Poor convergence and apparent stiffness increase due to the element's shape sensitivity.

All the approximation errors mentioned above must be seen in the light of the final utilisation of the model. Obviously the requirements for a dynamic model aimed at predicting a few fundamental natural frequencies are different to those for a static model aimed at predicting some local stress concentrations. Concentrating on dynamic models the first important decision to be made is the selection of the active frequency range containing the active modes necessary to fulfil the requirements mentioned under 3.a)-e) above. The active modes should correspond to experimental modes within certain limits, for example, eigenfrequency deviation not more than ±20 % and corresponding MAC values not less than 70 %. In many practical cases of complicated industrial structures the initial analysis model, in particular when equivalent models are used, will not fulfil this requirement. Very often such models are not yet suited for computational updating and will be called pre-updating models in the following.

In such cases steps have to be undertaken to improve the initial analysis model in order to make it suitable for subsequent computational updating. This improved initial analysis model will be called the **pre-update** model in what follows. In order to avoid unjustified expectations it is important to understand that computational updating offers some kind of fine-tuning of the pre-update model which is not too far away from the true model in the sense of the requirements stated above. Some of the most frequent idealisation and discretisation errors of the list in Sections 4.1 and 4.2 above are discussed in the following. Some of these errors may be tolerated when equivalent models shall be used.

4.3. ILLUSTRATIVE EXAMPLES

In what follows some examples are reported to illustrate the effects on the results of parameter updating in the presence of idealisation and discretisation errors.

Starting with a classical example of an erroneous simplification of the structure three possible idealisations of the cantilever beam shown in Figure 1 are considered when calculating the first 10 bending vibration modes. The first model represents the classical Euler-Bernoulli beam model where shear deformation effects are neglected whereas in the second model shear was considered by using elements including shear deformation according to Timoshenko beam theory. The results were obtained from a 24 element division. The third model is represented by a 12 x 24 mesh of 4 - noded isoparametric membrane elements. The idealisation errors are shown for two height to length ratios h/ℓ = 1/7 (short beam) and h/ℓ = 1/14 (long beam). Figure 2 shows the deviation of the beam solution without shear versus the solution when the shear deformation is considered (Timoshenko beam model). The deviations increase with the mode order and with the height to length ratio.

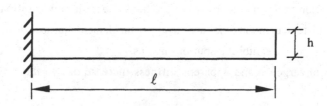

Figure 1   Cantilever beam

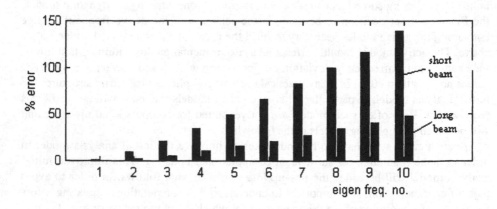

Figure 2   Error of beam solution without shear a vs. Timoshenko beam solution with shear

Of course, the Timoshenko solution is also an approximation only. Figure 3 shows the errors of the Timoshenko beam solution with respect to the results of the third model represented by the 12 x 24 mesh of membrane elements.

Shear deformation generally affects the eigenfrequencies with increasing mode orders and should not be neglected since it is often necessary to use higher modes in model updating. Any attempt to update physical parameters for example the stiffness parameters of the Timoshenko beam will always lead to non-physical results in particular when the higher order modes are used.

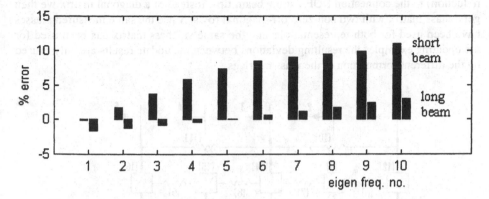

Figure 3 Error of Timoshenko beam solution with respect to 2D finite element solution

In the following example the influence of using distributed or lumped mass representations on the results of computational parameter updating was investigated. Real test data of the laboratory test structure shown in Figure 4 were used. The structure consists of 5 cantilever steel beams carrying a tip mass and coupled by 6 arch springs. With this structure the effect of two popular types of mass representation was investigated, the lumped mass and the consistent mass representation. Due to the concentrated tip masses the structure could be modeled as a discrete 5-DOF system shown in Figure 5, where the masses are lumped at the 5 DOF's resulting in a diagonal mass matrix.

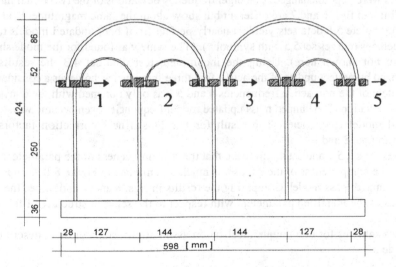

Figure 4 Laboratory test structure

The alternative approach was to use a consistent mass matrix for each arch and each beam obtained from reducing a multi-element mass matrix by static condensation (Guyan

312

reduction) to the connection DOF's at the beam tips. Instead of a diagonal matrix we then get a mass matrix with two non-zero off-diagonal rows. Since the same measured masses have been used for both representations and the same stiffness matrix has been used for the updating examples the resulting deviations between the update results are only caused by the different formulation of the mass matrices.

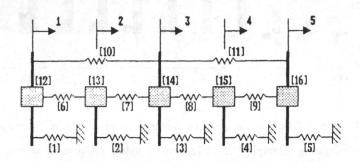

Figure 5 Discrete 5 -DOF model of laboratory test structure

The updating was performed with the inverse sensitivity approach (e.g. [1]) using two test data sets with different numbers of measured DOF's and modes: set 5,5: DOF's 1-5 and modes 1-5 (complete set) and set 4,3: DOF's 1-4 and modes 1-3 (incomplete set).

Using the incomplete set with only 3 modes for updating allows to assess the accuracy of the non-active modes 4 and 5 of the updated models. All 11 spring stiffness parameters of the discrete model in Figure 5 were used for updating. Mass matrix parameters were kept unchanged. The eigenfrequency deviations of the two initial models (curves "ini") in figs. 6 and 7 are different but show about the same magnitude. Updating with the complete 5,5 data sets yields a nearly perfect fit of both updated models to the test frequencies (curves set 5,5 with symbol *). (The same was found for the mode shapes which are not shown here). Using the incomplete test data set 4,3 the results are characterized by perfect update of the active first three frequencies, however a considerable improvement of the non-active frequencies 4 and 5 was only obtained with the consistent mass representation. The lumped mass updated model frequencies even became worse than the initial model frequencies. The result for the 11 stiffness correction factors are presented in figs. 8 and 9.

Curves for set 5,5 and set 4,3 indicate that the non-uniqueness of the parameters with respect to the completeness of the data set is much less marked in Figure 8 than in Figure 9 for the lumped mass model. Comparing the results in figs. 8 and 9 underlines the non-uniqueness of the identified parameters with respect to the selected structure of the mass model.

Other examples for non-unique idealisation assumptions, some of them described in [8], include :

-   Uncertain boundary conditions which should not be assumed to be rigid unless they have been proved not to affect the effective modes. Instead elastic springs should be introduced with stiffness parameters to be updated later.

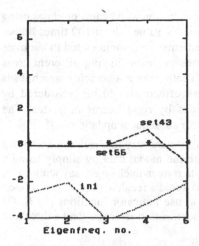

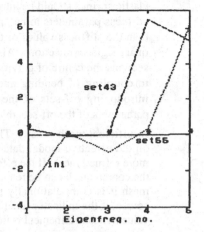

Figure 6   Frequency errors before
            and after updating,
            consistent masses

Figure 7   Frequency errors before
            and after updating,
            lumped masses

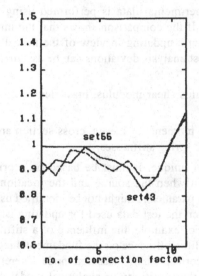

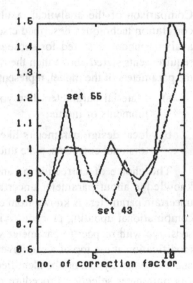

Figure 8   Correlation factors after update
            with consistent mass

Figure 9   Correlation factors after
            update with lumped mass

- Erroneous modelling of joints, for example, when a connection between beams or plates is assumed to be rigid or when an eccentricity of a beam or a plate connection possibly existing in the real structure is disregarded. Unless stiffened by ribs even welded connections of thin walled beams do not behave rigidly due to local deformations of the cross sections. Connections should not be assumed to be rigid unless they have been proved not to affect the effective modes. Instead elastic springs should be introduced to connect beams with beams or plates using stiffness parameters to be updated later. Initially values about 100 times higher than the stiffness values of the adjacent elements are recommended to simulate quasi rigid connections. Also in connections of beams having different cross sections the centre of gravity and the shear centre may not coincide which leads to coupling of bending and torsion. These effects should be considered by introducing off-sets of the connected nodes by rigid beams or plates. The parameters of the off-sets may also be open for subsequent updating.

- Insufficient mesh density. The most straightforward way to check this influence on the effective modal data is to recalculate the modal data by simply using a more refined mesh. If the effective modal data remain unchanged, say within 1%, the coarse mesh can be retained . In order to avoid a recalculation with a refined mesh it is computationally more efficient to use indicator functions [2, 3]. Of course, if the indicator functions indicate discretisation errors beyond a threshold value a mesh refinement is unavoidable.

5. Correlation of analytical model predictions and test results and selection of correction parameters

Comparison of the analytical results with experimental data is performed using the correlation techniques described elsewhere [4]. If the comparison shows that the initial analysis model is suited for subsequent automatic updating in view of the modelling requirements stated above then the remaining test/analysis deviations can be assigned to the parameters of the model, represented by

a) Material parameters like young's modulus shear modulus, mass density, mass moments of inertia,

b) local design parameters like beam area moments of inertia, cross section areas torsional constants, plate thicknesses and spring stiffnesses.

The choice of correction parameters is not unique. It must be based on a priori knowledge about parameter uncertainties. Even when the source and the location of uncertain parameters is known from evidence the parameters might not be identified using computational updating procedures in cases when the test data used for updating is not sensitive with respect to parameter variations. For example, the influence of a stiffness modification at the top of a cantilever beam has little influence on the fundamental mode and consequently can not be identified from the corresponding measured mode. Therefore any parameter selection procedure must check the sensitivity of analytical modal data (eigenfrequencies and mode shapes) with respect to the updating parameters. A simple means to assess the variation of an eigenvalue λ with respect to the (small) parameter changes $\Delta\alpha_i$ and $\Delta\beta_j$ of the stiffness and mass matrix is given by the relations :

$$\Delta\lambda = \sum_i \partial\lambda / \partial\alpha_i \, \Delta\alpha_i + \sum_j \partial\lambda / \partial\beta_j \, \Delta\beta_j \qquad (3a)$$

where the sensitivities

$$\partial\lambda / \partial\alpha_i = \mathbf{X}^T \mathbf{K}_i \mathbf{X} \quad \text{and} \quad \partial\lambda / \partial\beta_j = -\lambda \, \mathbf{X}^T \mathbf{M}_i \mathbf{X} \qquad (3b)$$

represent the modal energies of the ith and j^{th} stiffness and mass substructure to be updated (i = 1,2, ..., no. of stiffness, j = 1,2, ..., no. of mass correction parameters, **X** = eigenvector). These relations state that correction substructures located in areas where the mode shape components are small have little influence on the corresponding eigenvalue. Existing software contains graphical tools to visualise these sensitivities spatially on the structure. Of course the most straightforward but tedious way to check the sensitivity of the modal data is to recalculate them using the pre-update model for a given parameter change of say 20 %.

Frequently springs are used to model elastic boundary conditions and joints. In order to make the eigenvalues and mode shapes sensitive with respect to variations of these spring stiffnesses their initial stiffness value should be related to the stiffness of the adjacent elastic elements.

The information of the sensitivity study can be used to tune the initial analysis model into the vicinity of the experimental data until it fulfils the minimum requirements for initial model quality.

Computational procedures to identify the most effective correction parameters have been described in the literature [5, 6].

6. A posteriori assessment of model quality

Considering the non- unique correction parameter sets caused by the unavoidable idealisation and discretisation errors (not to forget the influence of random and systematic measurement errors which are discussed elsewhere, e.g. [1]) the problem of finding criteria to select the best parameter estimate must be solved . From the viewpoint of the subsequent utilisation of the models an assessment of the model quality must be based on checks whether the requirements listed in Section 3 are met.

In the following this procedure is illustrated by an industrial application example. The aim was to validate the models of two aero- engine components (HPT, high pressure turbine casing and RBSS, rear bearing support structure, highlighted in Figure 10) as parts of the carcass of an aero-engine. Due to the complexity of such structures non-consistent model structures and parametrisations as well as experimental errors and incomplete test data can generally not be avoided.

Main goal of this industrial project was to improve the confidence in the predictions for an extremely complex model of the whole engine. The validation concept is based on updating the FE- models of engine components using experimental modal data of the components which allows to restrict the number of uncertain updating parameters of the whole engine model.

The aim of model improvement for each component as stated in Section 3, point e) is to verify that the updated single models can improve the eigendynamics predictions of the assembly.

6.1. UPDATE OF HPT COMPONENT MODEL

The HPT cross section design as shown in Figure 11 was modelled according to Figure 12. Twenty mode shapes could be correlated with their experimental counterparts with an average MAC - value of 91.1 %. The average eigenfrequency error was − 4 %, the

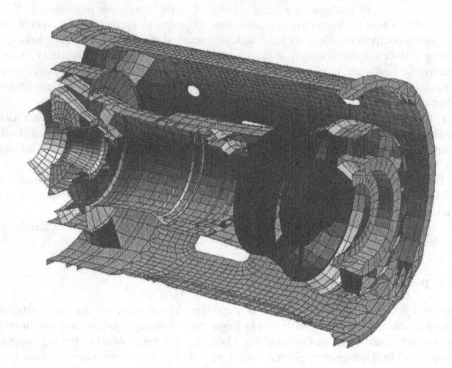

Figure 10 Aero- engine carcass model with high pressure turbine casing (HPT) and rear bearing support structure (RBSS)

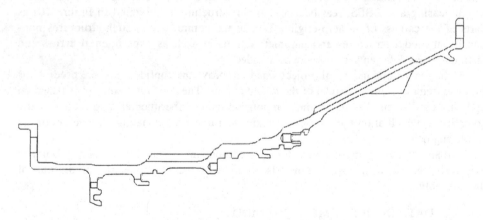

Figure 11 HPT cross section as designed

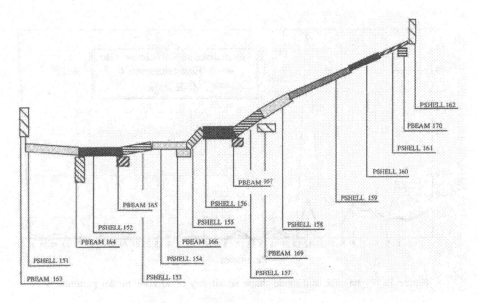

Figure 12 FE model (NASTRAN) of cross section

maximum error -6.6 % and the error of the fundamental frequency was -2.5 %. Usually this degree of accuracy can be considered to be sufficient for practical applications. Since due to the variety of parameters further improvement of the results by manual adjustment is no more feasible the numerical updating software UPDATE_N developed at the university of Kassel for updating NASTRAN FE models was applied.

In the first step the optimisation parameters have to be selected. This selection is based on the sensitivity of the mode shapes and the eigenfrequencies with respect to small parameter variations.

All parameters which are suspect of being uncertain should be considered in the sensitivity analysis.

The related sensitivity matrix is calculated within the NASTRAN FE code and is imported into UPDATE_N. Figure 13 shows the sensitivity of 37 parameters assumed to be uncertain. The first 12 parameters represent the shell thicknesses, the following 24 parameters the beam area moments of inertia and the 37th parameter represents the modulus of elasticity. One curve shows the column norm of that partition of the sensitivity matrix in eq. 3 related to eigenfrequency variations, the other curve the norm of the sensitivity matrix with respect to the mode shape variations.

From this sensitivity analysis it was found that the beam parameters were quite insensitive compared with the shell thicknesses and Young's modulus. The latter were therefore finally retained for the automatic updating process.

The results of this last step are shown in Figures. 14-15. Now 21 analytical mode shapes are correlated with the experimental shapes with an average MAC value of 92% and an average frequency error of -0.2%. The convergence of the frequency deviations, the

318

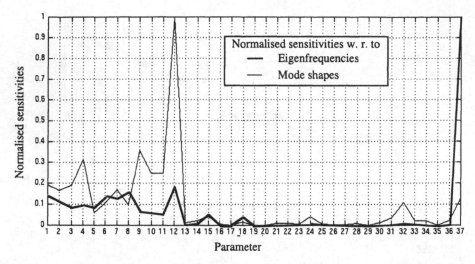

Figure 13 Eigenvalue and mode shape sensitivity of 37 HPT model parameters

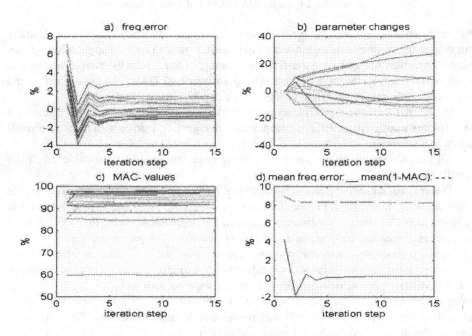

Figure 14 Evolution of eigenfrequency errors, parameter changes, MAC - values and objective
function during iteration of HPT update

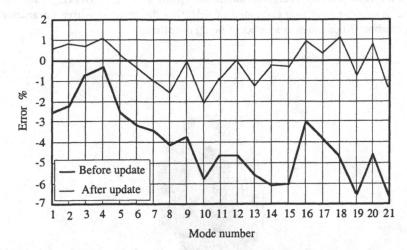

Figure 15 Eigenfrequency errors before and after numerical updating of HPT

parameter changes, the MAC values and the objective function expressed by the mean frequency error and the mean MAC error during the iterations is presented in Figure 14 a-d. It should be noted that in contrast to the objective function some parameters in Figure 14b have not yet converged. This means that the diverging parameters may be changed without changing the modal behaviour in the frequency range since a convergence of the objective function can be stated from Figure 14d.

The physical significance of such type of parameters must be questioned. In the present case these parameters represent shell thicknesses. Changing a thickness means to increase the membrane stiffness and the mass distribution simultaneously which tends to cancel the effect of the thickness change on the eigenfrequencies. It should also be noted that even when the parameters have not converged each parameter set at a given iteration step represents an input data set which yields a valid FE solution. Of course, the solution is not unique in such cases. Other criteria can help to find the best solution. In the present case that solution was selected where the overall model mass and the measured mass was nearly equal.

Figure 15 shows the significant reduction of the 21 eigenfrequency errors before and after updating. The quality of the final correlation results should be considered as above what can be expected from large scale industrial applications. In the present case this quality can be attributed to the quality of the initial modelling and also to the accuracy of the experimental modal data obtained under scientific laboratory conditions [7].

6.2. UPDATE OF RBSS COMPONENT MODEL

The initial FE model of the second component , the rear bearing support structure (RBSS), is shown in Figure 16. 14 analytical and experimental modes could be correlated with an average of 81.8% for the MAC and -2.8% for the eigenfrequencies. However, the frequency error range between -20% for the fundamental frequency and a maximum 6.3%

was considered too large. Therefore, automatic updating using UPDATE_N software was applied. Again the first step consisted of selecting the most sensitive parameters as a subset of about 70 parameters suspect of being uncertain.

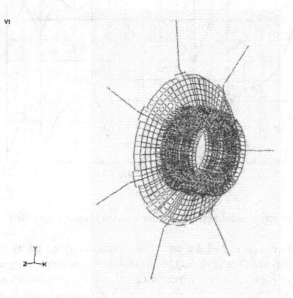

Figure 16 Initial pre- update model of rear bearing support structure (RBSS)

Finally 9 parameters were defined by the area and torsional moments of inertia of the spokes and of the outer ring and of the stiffness of the transition area between the circular plate and the cylinder represented by an equivalent Young's modulus of the shell elements in that area.

The results of automatic updating are presented in Figure 17. The correlation of 14 experimental modes could be increased to an average of 88.9% for the MAC and of 3% for the eigenfrequencies. The frequency error range was reduced to between -5.9% to 4.9%. The convergence behaviour of the updating process was even more satisfactory than that for the HPT component in view of Figure 17.

6.3. ASSEMBLY OF HPT & RBSS COMPONENTS

The main goal of the present study was to improve the prediction capability of the assembly model by using updated component models. The correlation of the test and analysis results of the assembly model where the initial not updated component models were used showed poor correlation.

As mentioned before the next step of assembling the *updated component* models represents the most important goal of the present study. With the assumption that the updating parameters were consistent with the source and the location of the real modelling uncertainties and with the assumption that the mathematical model structure was appropriate one can expect that the assembly of the two updated components would yield an improved model and should improve its prediction capability without further updating of the component parameters except for the uncertain parameters introduced at the bolted connection areas.

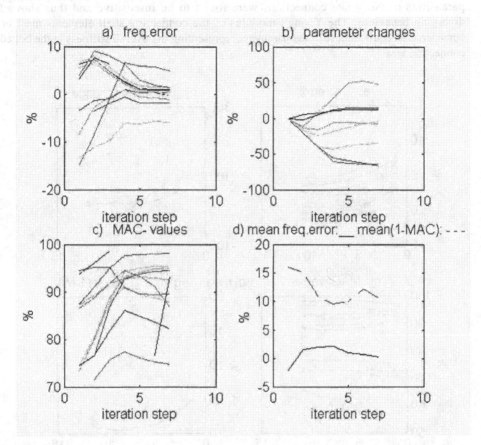

Figure 17 Evolution of frequency errors (a), parameters (b), MAC values (c) and objective function (d) for computational updating of RBSS component

The results of using the *updated component* models confirmed that the prediction accuracy has been improved significantly compared with that of using the initial component models. 21 modes were correlated with an average of 85.8% for the MAC and 5.5% for the eigenfrequencies. However, the frequency error range between 0.44% for the fundamental frequency and a maximum 10.15% showed a positive bias which could further reduced by computational updating of the whole assembly. The correction parameters in this case could be restricted to 7 parameters located in the bolted joints areas and represented by translational and rotational spring stiffnesses and the Young modulus of elasticity for the shell elements in the HPT/ spoke connection areas.

The results of updating the assembly model are shown in Figure 18. 27 modes , that means 6 more modes compared to the previous case were correlated with an MAC average

322

of 82.9% and an average frequency error of 0.9%. The average MAC was a little bit lower than in the previous case which however can be tolerated in view of the fact that 6 more modes could be correlated. The xx and zz rotational and the x translational spring stiffness parameters of the spoke connections were found to be insensitive and thus showed diverging behaviour. The Young's modulus of the connecting shell elements must be considered an equivalent stiffness parameter representing the overall stiffness in the bolted connection area.

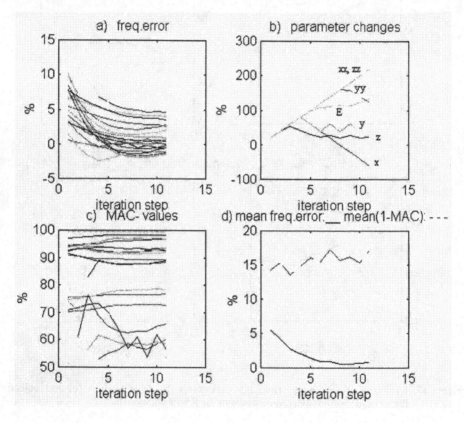

Figure 18   Evolution of frequency errors, parameters , MAC values and objective function for computational updating of HPT & RBSS assembly

7. Conclusions

Updating of physical design parameters like bending stiffnesses for minimizing test/analysis discrepancies leads to non-unique solutions which depend on the assumptions contained in the initial structure of the analysis model, on the amount and the accuracy of the test data, on the type of residuals in the objective function to be minimised in computational parameter updating as well as on the error assumptions in terms of type, numbers and location of correction parameters. However, the aim of retaining the physical meaning of the estimated parameters requires that the parameters are invariant with respect to these influences. Each solution may nevertheless produce an adequate fit of the experimental data used for parameter updating. It is therefore necessary to *reduce* the degree of non-uniqueness although unique solutions cannot be expected because in practical applications the above errors are unavoidable. In the ideal case different idealisations and parametrisations should be investigated. The selection of the best suited initial model is mainly governed by the specified use of the final model according to the requirements listed in Section 3.

For all these requirements it is important to determine the no. of active modes, i.e. the minimum number and the type of modes necessary for accurate predictions. For substructure coupled analysis presented in Section 6.3 it is additionally important to improve not only the active modes but also the interface flexibility which is influenced by the higher modes. These considerations lead to the obvious conclusion, that it makes no sense to use an updated model predicting only n_{act} modes accurately when $n > n_{act}$ modes are necessary to fulfil the requirements.

As shown in Figure 7 for the inconsistent mass model it may happen that the non-active mode predictions are worse than for the initial model. On the other hand when the number of active modes is high enough for the individual application the corresponding updated model may be interpreted as a special purpose model with limited validity. In the ideal case the updated model will also improve the predictions in the frequency range *beyond* the active range, which seems to be a good and reliable criterion to check the suitability of the initial model structure. This behaviour was demonstrated in Figure 10, where the non-active frequencies (set 4,3) of the updated model were improved for the consistent mass only but not for the lumped mass model and also in Section 6.3 where the prediction of the assembly dynamics was improved.

Acknowledgement

Part of this work was supported by BMW Rolls-Royce, Dahlewitz, Germany.

8 . References

1. Link, M. (1999), Updating of analytical models - basic procedures and extensions, in J.M.M. Silva and N. M. M. Maia (eds.), *Modal Analysis & Testing, Kluwer Academic Publishers, NATO Series*, Dordrecht, 281-304,
2. Zienkiewicz, O.C., Zhu, J.Z. (1987), A simple error estimator and adaptive procedure for practical engineering analysis, *International Journal for Numerical Methods in Engineering, Vol. 24*, 337-357,
3. Link, M. and Mardorf, J. (1996), The role of finite element idealization and test data errors in model updating, *Proc. of Int. Conf. on Structural Dynamics Modelling- Test, Analysis, Correlation and Updating*, Lake Windermere, UK.
4. Ewins, D.J. (1998), Correlation methods for experimental and analytical models, *Proc. of NATO Advanced Study Institute, Modal Analysis & Testing*, Sesimbra, Portugal, 271-288
5. Lallement, G. (1988), Localisation techniques, *Proc. of Workshop Structural Safety Evaluation Based on System Identification Approaches, Vieweg Verlag*, Braunschweig/Wiesbaden.
6. Flores, O., Link, M. (1993), Localisation techniques for parametric updating of dynamic mathematical models, *Proc. of Int. Forum On Aeroelasticity and Structural Dynamics*, Strassbourg, France *Association Aeronautic et Astronautic de France (Hrsg.)*, 75782 Paris.

324

7. Schedlinski, C. (1997), Modal testing and experimental modal analysis of bmw rolls-royce aeroengine components, *FG Leichtbau Bericht, internal report, 4.6.*
8. Link, M. (1992), Requirements for the structure of analytical models used for parameter identification, *IUTAM Symposium on Inverse Problems in Engineering Mechanics*, Springer-Verlag, Heidelberg, Tokyo.

DAMAGE DETECTION AND EVALUATION I

J. HE
Victoria University of Technology
Melbourne, Australia

1. Introduction

Structural damage detection using non-destructive vibration test data has received considerable attention in recent years. The analysis is usually based on the assumption that damage will change the structural (mass, stiffness or damping) properties which further lead to changes in the dynamic characteristics such as the natural frequencies, damping loss factors and mode shapes. Since the changes on the dynamic characteristics can be measured and studied, it is possible to trace what structural changes have caused the dynamic characteristics to change, thus identifying the damage.

Early research relating to damage detection was focused on using the information of natural frequency changes [Cawley, Adam 1979]. This focus was extended with the help of fracture mechanics to define the changes in modal frequencies as functions of damage characteristics [Ju, Minovich 1986]. Natural frequency changes usually require significant extent of damage. This brings certain degrees of difficulty in damage detection. Recent discovery, however, shows that natural frequency changes are less susceptible to random error sources than other modal parameters [Farrar, *et al* 1997][Doebling *et al* 1997].

The strong research interests in modal analysis community for analytical model updating has lent damage detection with many useful tools and methods. A significant part of research in structural damage detection was very similar to model updating in approach. This is because damage detection and model updating are intrinsically linked. Both aim to determine the difference between two mathematical models using some incomplete data. For model updating, the difference is the modelling errors, while for damage detection, it is structural damage. The model updating methods which are relevant to damage detection can be categorised as 'optimal matrix', 'eigenstructure assignment', 'sensitivity analysis' and 'application of FRF data'. The goal of the optimal matrix approach is to find an updated matrix that is closest to the original matrix and that produces the measured natural frequencies and mode shapes. One of the earliest works for 'optimal matrix' approach was that of Rodden [1967] who used vibration test data to determine the structural influence coefficients of a structure. Brock [1968] addressed the question of finding a matrix that satisfies a set of measurements as well as symmetry and positive definiteness. Berman and Flannely [1971] discussed the question of deriving system matrices with incomplete vibration data. Later, a formulation based on a minimised Frobenius norm was proposed by Baruch and Bar Itzhack [1978]. The result of

J.M.M. Silva and N.M.M. Maia (eds.), Modal Analysis and Testing, 325–344.
© *1999 Kluwer Academic Publishers. Printed in the Netherlands.*

this formulation is a stiffness matrix constructed only from experimental modal data that is mathematically 'closest' to the measured data. The drawback is that such a matrix is not able to preserve physical connectivity of a structure for which the matrix is derived. The same idea was later pursued by Kabe [1985] who derived the system stiffness matrix from experimental modal data that preserves correct physical connectivity of the system. Kabe's method was later reformulated and improved by Kammer [1988].

When used for structural damage detection, methods in the category of 'optimal matrix' obtain varying degrees of success. Baruch's method may be able to identify elements on the stiffness matrix which are most affected by damage, but, because of the absence of correct physical connectivity, will be inadequate in pinpointing damage location from a structure. Kabe's method utilises a percentage change in the stiffness value cost function and preserves the connectivity of the structure. It is therefore possible to use it to study the extent of damage [Smith and Hendricks 1987]. Zimmerman and Kaouk, in a series of publications [Zimmerman and Kaouk 1994][Kaouk and Zimmerman 1994a, 1994b][Zimmerman etc 1995], promoted an optimal matrix approach that minimises the rank (rather than the norm) of the stiffness damage matrix as perturbation to the original stiffness matrix. The same approach is also pursued by Doebling [1996].

Eigenstructure assignment is another approach for damage detection. This technique was used in control engineering in order to dictate the forced response of a structure [Andry etc 1983]. For model updating and damage detection, it aims to determine a pseudo-control that would be required to produce the measured modal properties with the initial Finite Element model. The pseudo-control is then translated into matrix adjustments applied to the initial FEM. Earlier work using this approach for model updating and damage detection includes that by Inman and Minas [1990]. Zimmerman and Widengren [1990] used symmetry preserving eigenstructure assignment theorem where the information regarding eigenparameters of the damaged structure was incorporated in the spatial model of the undamaged structure. This algorithm used the solution of a generalised algebraic Riccati equation whose dimension was defined solely by the number of measured modes. Lim and Kashangaki [1994] further developed the idea and proposed a method by which measured vibration modes can be used to determine the location and magnitude of damage in a space truss structure. The damage is located by computing the Euclidean distances between the measured mode shapes and the best achievable eigenvectors which are the projection of the measured mode shapes onto the subspace defined by the analytical model of the structure.

Another approach of damage detection relies on sensitivity analysis of modal parameters with respect to physical variables. The mathematical base of sensitivity analysis is derivatives. This means that this approach is unable to deal with 'large' physical variable changes. Nevertheless, the approach has been used for damage detection [Hajela and Soeiro 1990][Soeiro 1990][Jung and Ewins 1992][Adelman and Haftka 1986][Wolff and Richardson 1989]. Matrix perturbation methods are similar to the sensitivity approach. They usually involve an investigation of the effect of change in the system matrices on the eigenvalues [Law and Li 1993] [Dong etc 1994].

In addition to the aforementioned methods, Liu and Yao [15] considered the development of the probabilistic methodology for the prediction of multiple crack distribution in a structure of beam elements. The probabilistic measure of crack distribution could then be used for probabilistic diagnosis of crack location and extent. Several other authors' work have also shown that crack locations could be identified by using the concept of fracture mechanics along with information about change in natural frequencies due to damage. Penny, Wilson, and Friswell [1993] used a statistical method of identification based on generalised least square theory to detect structural damage.

Model updating has also been attempted in two steps, firstly by locating elements from system matrices that need to be updated, and secondly by quantifying these elements [He and Ewins 1986][Ibraham 1987]. This approach is also used for damage detection. Chen and Garba [1988] developed a three step damage location procedure that initially used residual force vectors to locate potential damage areas; then a least squares approach was adopted to determine scalars for the appropriate element stiffness matrices and finally, damage was located in structural members where the calculated element scalars were less than unity. Ricles and Kosmatka [1992] used measured modal data along with an analytical model to locate damaged regions using residual force vectors and to conduct a weighted sensitivity analysis to assess the extent of variations, where damage was characterised by stiffness reduction.

Most of the damage detection methods mentioned are based on the assumption that a structure's dynamic properties are described by system matrices. As a result, damage detection will be shown in terms of global coordinates and matrix elements. A different approach is also pursued that makes use of the concept of fracture mechanics. In this approach, a damage characteristic is quantified by certain parameters such as a crack depth. The changes in modal data (mostly natural frequencies) are linked to the parameters for damage [Chondros and Dimarogonas 1980][Ju and Minovich 1986].

Although many damage detection algorithms are focused merely on detecting the existence and location of damage, there are other methods that can be used to determine the extent of damage. For example, a 'submatrix' approach [Lim 1990] for model updating can be used to determine the extent of damage once it is located. However, the success of damage evaluation will be greatly enhanced if the information of structural connectivity in system matrices and that of elemental composition in an FE model can be used. Attempts to treat a stiffness matrix of a structural system as simply another mathematical matrix neglects vital information about the connectivity and composition of the system which is essential in determining the extent of damage.

In recent years, a different approach was proposed to use measured frequency response function (FRF) data directly for model updating. [Lin and Ewins, 1990]. This approach is a departure from the traditional approach of using modal analysis data and is quickly adopted for damage detection [Choudhury and He 1993] [Zimmerman 1995][Choudhury and He 1996] [Zimmerman and Smith 1992]. Measured FRF data are obtained directly from the structure. Thus, they do not carry numerical errors modal data may have from curve fitting analysis. Considerable numerical efforts to process measured data in order to derive modal data has been saved. However, the most significant difference of using measured FRF data over derived modal data lies in the fact that FRF data provide abundant information on the dynamic behaviour of a structure, while modal data lose much of the information FRF data have, due to the necessary numerical process to extract them.

2. Sensitivity of damage on modal and FRF data

Structural damage causes changes on the mass, stiffness and damping properties. These physical property changes will alter the modal properties, namely natural frequencies, mode shapes, and damping loss factors, of the structure. It is customary to assume that damage is more likely to result in mainly stiffness changes. Hence, it is useful for damage detection to study the link between stiffness changes of a structure and the changes of its modal and FRF properties.

The sensitivity of a natural frequency to a change in a physical parameter 'p_i' such as the stiffness of an elastic element can be described by the following derivative:

$$\frac{\partial \omega_r^2}{\partial p_i} = \{\phi\}_r^T \frac{\partial [K]}{\partial p_i} \{\phi\}_r \tag{2.1}$$

For an individual element in stiffness matrix, this sensitivity can be written as:

$$\frac{\partial \omega_r^2}{\partial k_{ij}} = \begin{cases} 2\phi_{ir}\phi_{jr} & i \neq j \\ \phi_{ir}^2 & i = j \end{cases} \tag{2.2}$$

Likewise, the sensitivity of a mode shape to a change in a physical parameter 'p_i' such as the stiffness of an elastic element and to an individual element in stiffness matrix can be described respectively by the following derivatives:

$$\frac{\partial \{\phi\}_r}{\partial p_i} = \sum_{j=1; j \neq r}^{N} \frac{\{\phi\}_r \{\phi\}_r^T}{\omega_s^2 - \omega_r^2} \frac{\partial [K]}{\partial p_i} \{\phi\}_r \tag{2.3}$$

$$\frac{\partial \{\phi\}_r}{\partial k_{ij}} = \sum_{s=1}^{n} \gamma_s \{\phi\}_s \quad \text{where} \quad \gamma_s = \begin{cases} \dfrac{\phi_{is}\phi_{ir}}{\omega_s^2 - \omega_r^2} & \text{for } r \neq s \quad i = j \\ 0 & \text{for } r = s \quad i = j \\ \dfrac{\phi_{js}\phi_{ir} + \phi_{is}\phi_{jr}}{\omega_s^2 - \omega_r^2} & \text{for } r \neq s \quad i \neq j \end{cases} \tag{2.4}$$

The sensitivity of the FRF matrix to structural damage caused by a system parameter 'p_i' that affect stiffness properties can only be derived similarly to that of modal data as:

$$\frac{\partial [\alpha(\omega)]}{\partial p_r} = -[\alpha(\omega)] \frac{\partial [K]}{\partial p_r} [\alpha(\omega)] \tag{2.5}$$

Although this sensitivity matrix quantifies the direction of changes of each FRF at a given frequency with respect to the change of system parameter 'p', its application in damage detection is limited. This is because such sensitivity becomes incorrect near resonances due to the discontinuity of an FRF at resonances.

Direct comparison of natural frequencies, mode shapes or FRF data before and after damage occurs is a primitive way of identifying damage. However, some methods have evolved from sensitivity analysis and direct comparison. They are usually aimed at locating damage rather than estimating its extent.

Another simple approach of using sensitivity analysis in damage detection is to calculate a hybrid type of sensitivity using undamaged system matrices and measured mode shapes:

$$\frac{\partial e_{ij}}{\partial p} = \{\phi_D\}_i^T \frac{\partial [K]_U}{\partial p} \{\phi_D\}_j \qquad (2.6)$$

This sensitivity corresponds to the system parameter 'x'. It shows the changes of the energy if the undamaged structure is forced to vibrate in the measured mode shape $\{\phi_D\}_j$. The significance of the sensitivity would indicate possible system parameters responsible for damage.

Usually, mode shapes of a damaged structure are derived from its experimental vibration data. Methods relying on mode shapes therefore have to deal with numerical errors introduced during the identification of these modes.

3. Damage detection using experimental data only

Damage detection without a readily available spatial model of the undamaged structure is an approach departed completely from model updating path. Usually, the data available are the experimental data before and after damage occurred. As a result, we are dealing with two sets of modal (or FRF) data. The comparison of these two sets should yield the information about the existence and location of damage. The main question here is how to relate the differences between modal and FRF data before and after damage to the spatial stiffness changes that resulted in the differences.

3.1. DERIVATION OF A SPATIAL MODEL FROM EXPERIMENTAL DATA

One method that has been explored is to derive the spatial model of a structure before and after damage using modal or FRF data and subsequently to determine the physical changes occurred (if any) from the differences between the two derived spatial models. The idea of this method is simple and numerical execution is straightforward but it requires very accurate experimental data. A spatial model can be derived from either modal or FRF data. The former uses a pseudo inverse technique to derive the following [Mannan and Richardson 1990]:

$$[K]_U = [\phi]_U^{+T} \left[`k_\cdot \right]_U [\phi]_U^+ \qquad (3.1)$$

$$[M]_U = [\phi]_U^{+T} \left[`m_\cdot \right]_U [\phi]_U^+ \qquad (3.2)$$

$$[C]_U = [\phi]_U^{+T} \left[`c_\cdot \right]_U [\phi]_U^+ \qquad (3.3)$$

$$[K]_D = [\phi]_D^{+T} \left[`k_\cdot \right]_D [\phi]_D^+ \qquad (3.4)$$

$$[M]_D = [\phi]_D^{+T} \left[\, \dot{} \, m_{\cdot} \right]_D [\phi]_D^{+} \tag{3.5}$$

$$[C]_D = [\phi]_D^{+T} \left[\, \dot{} \, c_{\cdot} \right]_D [\phi]_D^{+} \tag{3.6}$$

The damage can then be determined from:

$$[\Delta K] = [K]_U - [K]_D \tag{3.7}$$

Similar to using modal data, the spatial model of a structure can also be derived from its measured FRF data [Chen etc 1996].

$$[[A] \quad [B]] \begin{bmatrix} [M] \\ [K] \end{bmatrix} = [E] \tag{3.8}$$

$$[A] = - \left[\Omega_1^2 [\alpha(\Omega_1)] \quad \Omega_2^2 [\alpha(\Omega_2)] \quad \dots \quad \Omega_m^2 [\alpha(\Omega_m)] \right]^T \tag{3.9}$$

$$[B] = \left[[\alpha(\Omega_1)] \quad [\alpha(\Omega_2)] \quad \dots \quad [\alpha(\Omega_m)] \right]^T \tag{3.10}$$

$$[E] = \left[[I] \quad [I] \quad \dots \quad [I] \right]^T \tag{3.11}$$

Thus, two sets of spatial models can be derived, one for undamaged and the other for damaged. The differences between these models should reveal structural damage.

The approach of deriving a spatial model from experimental FRF or modal data can be useful for damage detection provided a few issues are clearly thought over and dealt with. Due to the nature of matrix inverse or similar mathematical operation, the accuracy required from experimental data is very high. When modal data are used, it is usually unrealistic to expect many vibration modes to be available from the experiment. This may result in significantly inaccurate stiffness matrices, as higher frequency modes, which are more important for constructing a stiffness matrix, are usually not as available as lower modes. In addition, a stiffness matrix constructed from incomplete modal data will not exhibit correct structural connectivity. This may hinder proper identification of damage from the difference between two stiffness matrices.

As for using FRF data, issues related to modal data approach largely remain. There is, however, one notable advantage, i.e. the abundant information on structural spatial properties inherent in FRF data may be made full use in order to derive a stiffness matrix with correct structural connectivity and possibly greater accuracy.

3.2. DAMAGE DETECTION USING EXPERIMENTAL MODAL DATA

An obvious application of damage detection using modal data is to compare intelligently mode shapes (as well as natural frequencies) before and after damage has occurred. Modal Assurance Criterion (MAC) [Alllemang and Brown 1982] can serve for this purpose. Originally designed for matching modes from experimental identification and computer simulation, MAC can be used to determine if the changes on experimental mode shapes have occurred with two sets of data. Presumably the changes would be due to structural damage. However, it is not always possible to detect the location of damage from mode shape changes. This is because significant mode shape differences do not spatially correspond to significant physical changes. For example, the stiffness changes at the fixed end of a cantilever would be reflected mostly at the free end of the beam.

For damage detection, MAC can be written as:

$$\left[\text{MAC}(\{\phi_U\}_p, \{\phi_D\}_q)\right] = \left[\frac{\left|\{\phi_U\}_p^T \{\phi_D\}_q\right|^2}{\{\phi_U\}_p^T \{\phi_U\}_p \{\phi_D\}_q^T \{\phi_D\}_q}\right]_{m \times m} \tag{3.12}$$

where 'm' is the number of vibration modes used. Although MAC value may show disparity between mode shapes before and after damage, it will not explicitly convey the location of damage. An alternative is to use Co-ordinate Modal Assurance Criterion (COMAC) [Lieven and Ewins 1988]:

$$\{\text{COMAC}(j)\} = \left[\frac{\sum_L \left|(\{\phi_U\}_L)_j (\{\phi_D\}_L)_j\right|^2}{\sum_L (\{\phi_U\}_L)_j^2 \sum_L (\{\phi_D\}_L)_j^2}\right] \tag{3.13}$$

where 'L' is the number of vibration mode pairs. The COMAC values reflect the discrepancy between mode shapes of original and damaged structure in terms of co-ordinates. Therefore, it may be indicative of damage location.

By further processing mode shape data and nodal line data, it is possible to determine not only the existence of damage but also its location. West [1984] used MAC to locate structural damage of the Space Shuttle Orbiter body. Wolff and Richardson proposed a method [Wolff and Richardson 1989] to rank mode shape differences and nodal point differences. This method can be used on mode shape data from a set of measurements of a structure over time in order to capture changes on physical properties from mode shape changes.

An alternative of using MAC and COMAC to detect damage is to use the mode shape curvature. This curvature can be derived either from numerical derivative of a vibration mode shape or from the measured bending strain of a structure. The former requires stringent accuracy from measured mode shapes which real modal testing may have difficulty to satisfy. Strain mode shapes can be measured directly using special strain transducers with suitable sensitivity and dynamic response characteristics.

3.3. DAMAGE DETECTION USING THE FRF CURVATURE

A different approach of using FRF data directly for damage detection does not rely on deriving a spatial model. Instead, it looks into a specially defined curvature of a structure from its FRF data and aims to localise damage [Silva *et al* 1998]. The curvature is defined as a finite difference estimated from three adjacent FRFs as:

$$\alpha''_{ij}(\omega) = \frac{\alpha_{(i+1)j}(\omega) - 2\alpha_{ij}(\omega) + \alpha_{(i-1)j}(\omega)}{h^2} \tag{3.14}$$

where 'h' is the distance between consecutive coordinates. For a given frequency value, this curvature can be defined for FRFs of all measurement coordinates (with excitation applied at co-ordinate 'j').

$$\left\{\alpha''(\omega)\right\}_j = \left\{\alpha''_{1j}(\omega) \quad \alpha''_{2j}(\omega) \quad \alpha''_{Nj}(\omega)\right\}^T \tag{3.15}$$

The curve defined by the estimated curvature values from all coordinates should be a structural property sensitive to damage. Thus, without system matrices, the difference between the curvature vector defined in Equation (3.14) reveals the location of damage:

$$\left\{\Delta\alpha''(\omega)\right\}_j = \left\{\alpha''_D(\omega)\right\}_j - \left\{\alpha''_U(\omega)\right\}_j \tag{3.16}$$

4. Damage detection using modal data and analytical model

Damage detection using modal data is an approach that was largely adopted from model updating. Its algorithm aims to determine damage by using the modal data from a damage structure and an analytical model for its undamaged counterpart. Although many model updating methods utilising modal data can be adopted directly for damage detection, one particular method based on modal force vector is most convenient and efficient. Model updating using modal force vector firstly appeared for adjusting a stiffness matrix [He and Ewins, 1986]. The idea was explored and it developed into an effective method for model updating [Ojalvo and Pilon 1988]. In order for the method to work, the measured vibration mode shapes need to be expanded to the set of DoFs compatible to the analytical model for undamaged structure. There are a number of analytical techniques designed for the purpose of expanding measured vibration mode shapes. Gysin (1990) provided a useful review on these techniques. The modal force vector method was subsequently developed to become an effective algorithm for structural damage detection [Zimmerman and Kaouk 1994]. It commences with the eigenvalue problem of the undamaged and damaged structures:

$$\left([K]_U - \omega_k^2 [M]_U\right)\{\phi_U\}_k = \{0\} \tag{4.1}$$

$$\left([K]_D - \omega_k^2 [M]_D\right)\{\phi_D\}_k = \{0\} \tag{4.2}$$

Considering the damage caused by stiffness changes, then;

$$\left([K]_U - [\Delta K] - \omega_k^2 [M]_U\right)\{\phi_D\}_k = \{0\} \tag{4.3}$$

Equation (4.3) reveals:

$$\{d\}_k = [\Delta K]\{\phi_D\}_k \qquad (4.4)$$

or

$$\{d\}_k = \left([K]_U - \omega_k^2 [M]_U\right)\{\phi_D\}_k \qquad (4.5)$$

Damage detection vector $\{d\}$ is indicative of location of damage. If damage exists at co-ordinate 'r', then the rth element in vector $\{d\}$ will be non-zero. Otherwise, theoretically the elements will be zero.

This idealised pattern of zeros and non-zeros from damage detection vector $\{d\}$ never occurs for a real problem. There are two main reasons for this. The first is that the modal data used are not error-free and the second that experimental mode shape $\{\phi_D\}_k$ usually needs to be expanded in order to match the co-ordinates used in system matrices before calculation can proceed. This expansion generates numerical errors as well as proliferates measurement ones [Lieven and Ewins 1990]. The errors in experimental mode shape $\{\phi_D\}_k$ may be due to both measurement noise and numerical errors while deriving it from measured vibration data. The mode shape errors, when weighted by large elements in the mass and stiffness matrices ($[M_U],[K_U]$), will jeopardise the ability of vector $\{d\}$ to indicate damage reflected in matrix $[\Delta K]$. To overcome this, Zimmerman and Kaouk (1994) suggested an alternative of using the scalar values in vector $\{d\}$. Instead, they defined an angle from vector $\{d\}$ such that its deviation from 90 degrees will be a better indicator of damage location.

An alternative method using modal data for damage detection was proposed by Chen and Garba [1988]. It assumes that damage of a structural system is represented by a stiffness perturbation matrix $[\Delta K]$ so that:

$$[K]_D = [K]_U + [\Delta K] \qquad (4.6)$$

According to eigenvalue properties of the system,

$$[\Delta K]\{\phi_D\}_k = \left([K]_U - \omega_k^2 [M]\right)\{\phi_D\}_k \qquad (4.7)$$

If the connectivity within stiffness matrix is known and is not altered by stiffness perturbation, the perturbation matrix can be expressed in terms of unknown elements $[\Delta K_{ij}]$ so that:

$$[\Delta K]\{\phi_D\}_k = [C]_k\{\Delta K_{ij}\} \qquad (4.8)$$

$$[C]_k\{\Delta K_{ij}\} = \left([K]_U - \omega_k^2 [M]\right)\{\phi_D\}_k = \{A\}_k \qquad (4.9)$$

Here, vector $\{\Delta K_{ij}\}$ contains all unknowns in stiffness matrix $[\Delta K]$ and matrix $[C]_k$ is a connectivity matrix derived from the kth measured mode shape $\{\phi_D\}_k$. The dimension of $\{\Delta K_{ij}\}$ is m x 1 where 'm' is the number of unknowns and the dimension of $[C]_k$ is n x m where 'n' is the number of DoFs. Usually, $n < m$. In order to solve for unknowns in vector $\{\Delta K_{ij}\}$, more than one mode is needed such that:

$$
\begin{bmatrix} [C]_1 \\ [C]_2 \\ .... \\ [C]_h \end{bmatrix} \{\Delta K_{ij}\} = \begin{Bmatrix} \{A\}_1 \\ \{A\}_2 \\ .... \\ \{A\}_h \end{Bmatrix} \qquad (4.10)
$$

where 'h' is the number of modes used. This equation represents a set of nh linear simultaneous equations with 'm' unknowns. The optimal solution can be sought if nh ≠ m. The success of this method relies on the accuracy of mode shape data and selection of effective modes that are more affected by the damage.

5. Damage detection using FRF data

Using measured FRF data for damage detection has many advantages over the traditional methods using modal analysis data. First, any numerical errors inherent in modal analysis results caused by inaccurate curve fitting and unavailable residual terms are avoided. Second, no more effort is needed to process measured FRF data in order to derive modal data. Finally, the most significant advantage of using measured FRF data over derived modal analysis data lies in the fact that FRF data provide abundant information on the dynamic behaviour of a structure. In comparison, modal analysis data, due to the necessary numerical process to extract them, lose much of the information that FRF data have to offer.

The general concept of structural damage detection using FRF data can be explained using a simple beam example as illustrated below.

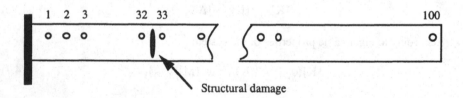

Structural damage

Assume that the beam can be discretised using 100 nodes and its dynamic characteristics are described analytically at the transverse co-ordinates of these nodes. Suppose structural damage exists between nodes 32 and 33. If FRF data are measured from nodes 1, 10, 20, ..., 100 at a space of 10 nodal positions only, then these data can be numerically expanded to all 100 nodes and the novel approach developed by the Chief Investigators can be used to detect and locate the damage at a position between nodes 30 and 40 - the two measured nodes closest to the damage to the best available spatial resolution. Further, if more FRF data are acquired say from node 35, then it is possible to further determine the location of damage between nodes 30 and 35. The technique uses the FRF data directly, thus circumventing numerical or computational errors other methods have to endure. With FRF data are available from more co-ordinates, the detection of structural damage can succeed with greater spatial accuracy.

Damage detection using measured FRF data and an FE model is based on a simple mathematical fact that the dynamic stiffness matrix $[Z(\omega)]$ and the receptance FRF matrix $[\alpha(\omega)]$ of a multi-degree-of-freedom system are orthogonal to each other:

$$[Z(\omega)][\alpha(\omega)] = ([K] - \omega^2[M])[\alpha(\omega)] = [I] \qquad (5.1)$$

Consider the fact that usually only one column of FRF data are available from a modal testing measurement, we may have equations:

$$[Z(\omega)]_U \{\alpha_U(\omega)\}_r = \{\delta\}_r \qquad (5.2)$$

$$[Z(\omega)]_D \{\alpha_D(\omega)\}_r = \{\delta\}_r \qquad (5.3)$$

Therefore,

$$[Z_U(\omega)] (\{\alpha_D(\omega)\} - \{\alpha_U(\omega)\}) = [\Delta Z(\omega)]\{\alpha_D(\omega)\} = \{d(\omega)\} \qquad (5.4)$$

or

$$[Z_U(\omega)] \{\Delta\alpha(\omega)\} = [\Delta Z(\omega)] \{\alpha_D(\omega)\} = \{d(\omega)\} \qquad (5.5)$$

Vector $\{\alpha_U(\omega)\}$ can be derived from the FE model and has a full set of co-ordinates. Since measurement usually adopts a subset of those co-ordinates, only a subset of vector $\{\alpha_D(\omega)\}$ (denoted as $\{\alpha_D(\omega)\}^R$) will be made available. As a result, the complete vector $\{\alpha_D(\omega)\}$ has to be derived if damage detection using FRF data can proceed. This can be done by expanding measured FRF vector to the full set of coordinates, as described later.

Vector $\{d(\omega)\}$, which is the product of $[\Delta Z(\omega)]\{\alpha_D(\omega)\}$, should indicate any stiffness changes since any nonzero dynamic stiffness elements would be reflected in vector $[\Delta Z(\omega)]\{\alpha_D(\omega)\}$ as nonzero elements. Although the product $[\Delta Z(\omega)]\{\alpha_D(\omega)\}$ cannot be estimated due to unknown matrix $[\Delta Z(\omega)]$, the product $[Z_U(\omega)](\{\alpha_D(\omega)\} - \{\alpha_U(\omega)\})$ can be readily computed once $\{\alpha_D(\omega)\}$ becomes complete after FRF data expansion.

An alternative to using equations (5.4) and (5.5) is to derive stiffness matrix $[K]_D$ using the undamaged stiffness matrix $[K]_U$ and measured FRF data [Choudhury 1996]. It is customary to assume in damage detection that structural damage does not change physical connectivity exhibited in stiffness matrix. Thus, using measured FRF data at certain frequency points, matrix $[K]_D$ can be derived (see Appendix). The comparison of matrix $[K]_D$ and $[K]_U$ reveals structural damage.

6. Coordinate incompatibility and expansion of FRF data

Damage detection using measured FRF data and an FE model is based on equation (5.5). However, since measured FRF data only adopts a subset of those co-ordinates the FE model uses, only a subset of vector $\{\alpha_D(\omega)\}$ (denoted as $\{\alpha_D(\omega)\}^R$) will be available from measurement. As a result, the complete vector $\{\alpha_D(\omega)\}$ has to be derived if equation (5.5) can be used for damage detection. A feasible solution is to expand measured FRF vector $\{\alpha_D(\omega)\}^R$ into its full set $\{\alpha_D(\omega)\}$ using the existing FE model.

Partition the FRF vector $\{\alpha_D(\omega)\}$ into a subset related to measured co-ordinates and another subset related to unmeasured co-ordinates. This partition involves re-ordering of elements in the FRF vector to locate all measured co-ordinates on the upper half of the vector and unmeasured the lower half.

$$\{\alpha_D(\omega)\} = \begin{Bmatrix} \alpha_D^{mc}(\omega) \\ \alpha_D^{uc}(\omega) \end{Bmatrix} \tag{6.1}$$

Here, the unmeasured FRF data in vector $\{\alpha_D(\omega)\}$ is denoted as $\{\alpha_D^{uc}(\omega)\}$ (uc stands for unmeasured co-ordinates), and the measured subset is denoted as $\{\alpha_D^{mc}(\omega)\}$ (mc stands for measured co-ordinates). The objective is to determine a linear transformation matrix $[T]$ that relates vectors $\{\alpha_D^{mc}(\omega)\}$ with $\{\alpha_D^{uc}(\omega)\}$. This is a reasonable proposition at a low frequency range.

$$\{\alpha_D^{uc}(\omega)\} = [T]\{\alpha_D^{mc}(\omega)\} \tag{6.2}$$

Like FRF data in vector $\{\alpha_D(\omega)\}$, the dynamic stiffness matrix in equation (5.5) can be re-arranged and partitioned in terms of measured and unmeasured co-ordinates. It can be recast as:

$$\begin{bmatrix} [Z_{11}(\Omega)] & [Z_{12}(\Omega)] \\ [Z_{21}(\Omega)] & [Z_{22}(\Omega)] \end{bmatrix}_U \begin{Bmatrix} \alpha_U^{mc}(\Omega) - \alpha_D^{mc}(\Omega) \\ \alpha_U^{uc}(\Omega) - \alpha_D^{uc}(\Omega) \end{Bmatrix} = [\Delta Z(\Omega)] \begin{Bmatrix} \alpha_D^{mc}(\Omega) \\ \alpha_D^{uc}(\Omega) \end{Bmatrix} \tag{6.3}$$

Without knowing matrix $[\Delta Z(\Omega)]$, this equation offers little help for expanding measured FRF data to unmeasured co-ordinates. However, it is analytically possible to assume first that matrix $[\Delta Z(\Omega)]$ is a null matrix. This will enable the derivation of the sub-vector $\{\alpha_D^{uc}(\omega)\}$ from $\{\alpha_D^{mc}(\omega)\}$ using equation (6.2). The consequence of this derivation is that the FRF on the expanded co-ordinates will not be able to indicate stiffness changes in matrix $[\Delta Z(\Omega)]$ since they are derived by assuming matrix $[\Delta Z(\Omega)]$ is a null one. Nevertheless, an expanded vector $\{\alpha_D(\omega)\}$ now becomes available for damage detection. Only those measured co-ordinates will be able to locate damage in matrix $[\Delta Z(\Omega)]$. Equation (6.3) can be written as two separate matrix equations:

$$[Z_{11}(\Omega)]_U\{\Delta\alpha^{mc}(\Omega)\} + [Z_{12}(\Omega)]_U\{\Delta\alpha^{uc}(\Omega)\} = \{0\} \tag{6.4}$$

$$[Z_{21}(\Omega)]_U\{\Delta\alpha^{mc}(\Omega)\} + [Z_{22}(\Omega)]_U\{\Delta\alpha^{uc}(\Omega)\} = \{0\} \tag{6.5}$$

These two equations enable the solution of the subset vector $\{\Delta\alpha^{uc}(\Omega)\}$ to be found from $\{\Delta\alpha^{mc}(\Omega)\}$. However, different algorithms are possible to implement that solution and they will arrive at different answers, although all will be of the form given in equation (6.2). Table 1 summarises these possible solutions.

Table 1 Dynamic expansion methods for expanding measured FRF data

Method	Analysis	Remarks
Expansion A	$[T] = -\left[[K_{22}]_u - \Omega^2[M_{22}]_u\right]^{-1}\left[[K_{21}]_u - \Omega^2[M_{21}]_u\right]$	Matrix inversion
Expansion B	$[T] = -\left[[K_{12}]_u - \Omega^2[M_{22}]_u\right]^{+}\left[[K_{11}]_u - \Omega^2[M_{11}]_u\right]$	Generalised inverse
Expansion C	$[A_1] := \begin{bmatrix} [K_{11}]_u - \Omega^2[M_{11}]_u \\ [K_{21}]_u - \Omega^2[M_{21}]_u \end{bmatrix}$ $[A_2] := \begin{bmatrix} [K_{12}]_u - \Omega^2[M_{12}]_u \\ [K_{22}]_u - \Omega^2[M_{22}]_u \end{bmatrix}$ $[T] = [A_2]^{+}[A_1]$	The whole dynamic stiffness matrix is used
Hybrid FRF vector	$\{\alpha_d(\omega)\} = \begin{Bmatrix} \{\alpha_d^{mc}(\Omega)\} \\ \{\alpha_u^{uc}(\Omega)\} \end{Bmatrix}$	Mix measured and unmeasured FRF data together

In addition to the dynamic expansion methods, a method called System equivalent reduction expansion process (SEREP) [O'Callahan, J etc 1989] can be used for expanding measured FRF data. SEREP was originally designed for the expansion of measured vibration mode shapes. Initial study suggested that FRF data expanded using dynamic expansion are more suitable for damage detection [Choudhury and He 1996].

7. Concluding remarks

As an application of modal analysis, structural damage detection using measured vibration data has been a subject of much practical interests and research efforts. Structural damage usually results in identifiable stiffness changes. Damage detection and analytical model updating share similar objectives, both aiming to determine the difference between two mathematical models using incomplete modal or FRF data. Methods have been available to develop spatial models of a structure before and after damage occurs using its measured modal or FRF data. Damage location can then be accomplished by correlation of two derived spatial models. This approach requires strict accuracy from measured data. The stability and repeatability of the derived spatial model needs to be ascertained. Methods adopted from model updating using modal data often determines the damaged stiffness matrix that is closest to the original stiffness matrix and that also produces the measured natural frequencies and mode shapes. Direct application of these methods locates and quantifies damage simultaneously but may result in a stiffness matrix that loses physical connectivity of a structure. Some refined methods have overcome this problem. Other methods aim to locate structural damage using measured modal data first and then use well designed mathematical techniques such as submatrix approach to quantify damage.

Sensitivity analysis is based on the derivatives of the modal properties of a structure to changes in material or physical parameters. The sensitivity can be used to determine the parameter changes needed in order for the modal data to match the measured ones. This approach can be used together with other methods to determine the location of damage and then to estimate it. Perturbation methods are similar to the sensitivity approach.

Damage detection using measured FRF data has shown many advantages over the methods using modal analysis data. FRF data possess abundant information about the modal and spatial properties of the measured structure. To maximise the use of the abundant information in measured FRF data lend significant versatility to damage detection. A number of methods have been developed to use measured FRF data to locate damage and estimate damage. One significant issue to be addressed in using FRF data is the uniqueness of damage detection. This is not exclusive in this case but is more obvious than the case of using modal data because of the abundance of FRF data points compared with the number of modes. Damage detection using FRF data may lead to different damage locations with different FRF data points used. However, the abundance of FRF data points also provides an opportunity to judiciously use data points to conduct accurate damage detection.

Co-ordinate incompatibility between the analytical model and measured modal or FRF data needs to be addressed before damage detection can proceed. It is now evident that Guyan Reduction or similar reduction methods tend to create a reduced spatial model with no meaningful physical connectivity. The loss of the physical connectivity information will impair damage detection attempt. Expansion of measured mode shapes has been studied with useful results. The expanded mode shapes can provide accurate damage detection results and the spatial accuracy depends on the measured co-ordinates rather than on the expanded set of co-ordinates. Methods have also been studied for the expansion of measured FRF data. The expanded data facilitate damage detection that takes full advantage of the abundant modal and spatial information in measured FRF data.

Structural damage detection using an FE model as the baseline assumes that any found difference between the experimental data and the model is due to damage. In applications, this assumption may become a significant undertaking, as an accurate FE model itself needs immense efforts to establish before any damage detection occurs.

The minimum structural damage identifiable using various damage detection methods is a question usually not addressed by those methods themselves. It is relatively easy to demonstrate in numerical analysis the workability of new methods with sizeable simulated damage or with small damage in the absence of measurement or numerical errors. However, real damage detection has to be able to distinguish between the damage and results caused by measurement errors. Some statistics based efforts have been reported in literature recently. Significant progress has to be made before damage detection can become a robust and effective tool for real structures.

8. References

1.   Adelman, H M and Haftka, R T (1986), Sensitivity analysis of discrete structural system, *AIAA Journal*, Vol.24/5, 823-832.
2.   Allemang, R J and Brown, D L (1982), A correlation coefficient for modal vector analysis, *Proceedings of 1ˢᵗ IMAC*, 110-5.

3. Andry, A N, Shapiro, E Y and Chung, J C (1983), Eigenstructure assignment for linear systems, *IEEE Transactions of Aerospace and Electric Systems, Vol AES-19, No 5,* 711-29.
4. Baruch,M, and Bar Itzhack,I Y (1978), Optimum weighted orthogonalization of measured modes, *AIAA Journal, Vol.16, No.4,* 346-351.
5. Berman,A, and Flanelly,W G (1971), Theory of incomplete models of dynamic structures, *AIAA Journal, Vol.9, No.8,* 1481-1487.
6. Berman,A, and Nagy,E J (1983), Improvement of a large analytical model using test data, *AIAA Journal, Vol.21, No.8,* 1168-1173.
7. Brock,J E (1968), Optimal matrices describing linear systems, *AIAA Journal, Vol.6, No.7,* 1292-1296.
8. Cawley,P, and Adams, R D (1979), The location of defects in structures from measurements of natural frequencies, *Journal of Strain Analysis, Vol. 14(2),* 49-57.
9. Chen.J.C., and Garba.J.A (1988), On-orbit damage assessment for large space structures, *AIAA Journal, Vol.26, No.9,* 1119-1126.
10. Chen, S Y, Ju, M S and Tsuei, Y G (1996), Estimation of mass, stiffness and damping matrices from frequency response functions, *Transactions of ASME, Vol 118,* 78-82.
11. Chondros,T.G., and Dimarogonas,A.D.(1980), Identification of cracks in welded joints of complex structures, *Journal of Sound & Vibration, Vol. 69, No. 4,* 531-538.
12. Choudhury, A (1996), Damage detection in structures using measured frequency response function data, *Ph.D thesis,* Department of Mechanical Engineering, Victoria University of Technology, Melbourne, Australia.
13. Choudhury, R. and He, J. (1993), Expansion of measured frf data for model updating, *Proceedings of the Asia-Pacific Vibration Conference,* Japan, 677-682.
14. Choudhury, R and He, J (1996), Structural damage detection using expanded measured frequency response function data, *Proceedings of the 14th International Modal Analysis Conference,* Dearborn, USA, 934-942.
15. Doebling, S W (1996), Minimum-rank optimal update of elemental stiffness parameters for structural damage identification, *AIAA Journal, Vol. 34, No. 12,* 2615-21.
16. Doebling, S. W., Farrar, C. R., and Goodman, R. S., (1997), Effects of measurement statistics on the detection of damage in the Alamosa Canyon bridge, *Proceedings 15th International Modal Analysis Conference,* Orlando, FL, 919-29.
17. Dong.C., Zhang.P., Feng.W., and Huang.T.(1994), Sensitivity of the modal parameters of a cracked beam, *Proceedings of 12th Int. Modal Analysis Conference,* Hawaii, 98 -104.
18. Farrar, C. R., Doebling, S. W., Cornwell, P. J., and Straser, E. G., (1997), Variability of modal parameters measured on the Alamosa Canyon bridge, *Proceedings 15th International Modal Analysis Conference,* Orlando, FL, 257-63.
19. Gysin, H (1989), Comparison of expansion methods for FE modelling error localisation, *Proceedings of the 8th International Modal Analysis Conference,* Kissimmee, FL, 195-204.
20. Hajela,P., and Soeiro,F.(1990), Structural damage detection based on static and modal analysis, *AIAA Journal, Vol.28, No.6,* 1110-5.
21. He.J., and Ewins.D.J.(1986), Analytical stiffness matrix correction using measured vibration modes, *International Journal of Analytical and Experimental Modal Analysis, Vol.1, No.3,* 9-14.
22. Inman,D.J., and Minas,C.(1986), Matching analytical models with experimental modal data in mechanical systems, *Control and Dynamics Systems, Vol.37,* 327-63.
23. Ju, F.D., and Minovich, M. (1986), Modal frequency method in diagnosis of fracture damage in structures, *Proceedings of the 4th International Modal Analysis Conference,* LA, 1168-74.
24. Jung.H., and Ewins.D.J. (1992), Error sensitivity of the inverse eigensensitivity method for model updating, *Proceedings of 10th International Modal Analysis Conference,* SanDiego, California, 992- 8.
25. Kabe,A.M. (1985), Stiffness Matrix Adjustment Using Mode Data, *AIAA Journal, Vol.23, No.9,* 1431-36.
26. Kammer, D C (1988), Optimum approximation for residual stiffness in linear system identification, *AIAA Journal, Vol 26, No 1,* 104-12.
27. Kaouk, M and Zimmerman, D C, (1994a), Structural damage assessment using a generalized minimum rank perturbation theory, *AIAA Journal, Vol. 32, No. 4,* 836–842.
28. Kaouk, M. and Zimmerman, D C, (1994b), Structural damage detection using measured modal data and no original analytical model, *Proceedings of the 12th International Modal Analysis Conference,* 731-7.

340

29. Law.S., and Li.X. (1993), Structural damage detection based on higher order analysis, *Proceedings of Asia Pacific Vibration Conference*, Kitakyushu, Japan, 640-3.
30. Lim,T.W. (1990), Submatrix approach to stiffness matrix correction using modal test data, *AIAA Journal, Vol.28, No.6*, 1123-30.
31. Lim, T W and Kashangaki, T A L (1994), Structural damage detection of space truss structures using best achievable eigenvectors, *AIAA Journal, Vol 32, No 5*, 1049-57.
32. Lieven, N A J and Ewins, D J (1988), Spatial correlation of mode shapes, the co-ordinate modal assurence criterion, *Proceedings of 6th IMAC*, Kissimmee, FL, 690-5.
33. Lieven, N A J and Ewins, D J (1990), Expansion of modal data for correlation, *Proceedings of the 8th IMAC*, Orlando, FL, 605-9.
34. Lin.R.M., and Ewins.D.J. (1990), Model updating using FRF data, *Proceedings of 15th International Seminar on Modal Analysis*, Leuven.
35. Liu,S.C., and Yao,J.T.P. (1978), Structural identification concept, *ASCE Journal of the Structural Division, Vol.104, No.ST12*, 1845-58.
36. Mannan, M A and Richardson M H (1989), Detection and location of structural cracks using FRF measurements, *Proceedings of 7th International Modal Analysis Conference*, Las Vagas, Nevada, 652-7.
37. O'Callahan, J, Avitable, P and Riemer, R (1989), System Equivalent Reduction Expansion Process (SEREP), *Proceedings of the 7th International Modal Analysis Conference*, Las Vagas, Nevada, 29-37.
38. Penny.J., Wilson.D., and Friswell.M.I. (1993), Damage location in structures using vibration data, *Proceedings of 11th International Modal Analysis Conference*, Kissimmee, Florida, 861-7.
39. Ricles.J.M., and Kosmatka,J.B. (1992), Damage detection in elastic structures using vibratory residual forces and weighted sensitivity, *AIAA Journal, Vol.30, No. 9*, 2310-6.
40. Rodden,W.P.(1967), A method for deriving structural influence coefficients from ground vibration tests, *AIAA Journal, Vol.5, No.5*, 991-1000.
41. Salawu.O.S., and Williams.C. (1993), Structural damage detection using experimental modal analysis- A comparison of some methods, *Proceedings of 11th International Modal Analysis Conference*, Kissimme, Florida, .254-260.
42. Silva, J M M, Maia, N M M and Sampaio, R P C (1998), Localisation of damage using FRF's curvatures: assessment and discussion in experimental cases, *Proceedings of 16th International Modal Analysis Conference*, Santa Barbara, CA, 1587-90.
43. Smith,S.W., and Hendricks,S.L. (1987), Evaluation of two identification methods for damage detection in large space trusses, *Proceedings of the 6th VPI&SU/AIAA Symposium on Dynamics and Controls of Large Space Structure*, Virginia Polytechnic Inst. and State Univ., Blacksburg, VA, 127-42.
44. Soeiro,F. (1990), Structural damage assessment using identification techniques, *Ph.D Dissertation*, AeMES Department., University of Florida.
45. Wolff.T., and Richardson.M (1989), Fault detection in structures from changes in their modal parameters, *Proceedings of 7th International Modal Analysis Conference*, SanDiego, California.
46. Zimmerman, D C and Kaouk, M (1994), Structural damage detection using a minimum rank update theory, *Journal of Vibration and Acoustics, Vol. 116*, 222-31.
47. Zimmerman.D., Simmermacher.T., and Kaouk.M. (1995), Structural Damage Detection Using Frequency Response Functions, *Proceedings of 13th International Modal Analysis Conference*, Nashville, USA, 179-84.
48. Zimmerman, D C and Smith, SW (1992), Model refinement and damage location for intelligent structures, *Intelligent Structural Systems, Kluwer Academic Publishers*, 403-452.
49. Zimmerman,D.C.,and Widengren,W. (1990), Model correction using a symmetric eigenstructure assignment technique, *AIAA Journal, Vol.28, No.9*, 1670-6.

Appendix

Assume in damage detection that structural damage does not change physical connectivity. Thus:

$$[K]_D = [K]_U \otimes [Q_K] \qquad (A1)$$

Here the operator $\otimes$ defines element by element multiplication between two matrices. The matrix $[Q_K]$ is denoted as damage quantification coefficient matrix due to changes in stiffness. The use of the operator $\otimes$ in equation (A1) ensures that all elements of dynamic stiffness matrix with values of zero prior to damage occurrence remain zero afterwards. Equation (A1) can be written as:

$$[\Delta K] = [K]_U \otimes ([U] - [U] \otimes [Q_K]) \quad \text{where} \quad U_{ij} = \begin{cases} 1 & \text{if} & K_{U,ij} \neq 0 \\ 0 & \text{if} & K_{U,ij} = 0 \end{cases} \quad (A2)$$

Define an error function as the norm of matrix $([U] - [U] \otimes [Q_K])$ that can be denoted as:

$$e = \| [U] - [U] \otimes [Q_K] \| \qquad (A3)$$

or

$$e = \sum_{i=1}^{n} \sum_{j=1}^{n} (U_{ij} - U_{ij} Q_{K,ij})^2 \qquad (A4)$$

Assume that the kth column of receptance FRF data $\{\alpha_D(\Omega)\}_k$ is available from measurement, it is possible to find a matrix $[Q_K]$ which minimises the norm defined in equation (A3) and also satisfies the constraints given by equations (A5) and (A6) below:

$$[Q_K] - [Q_K]^T = [0] \qquad (A5)$$

$$([K]_U \otimes [Q_K] - \Omega^2 [M]) \{\alpha_D(\Omega)\}_k - \{\delta\}_k = \{0\} \qquad (A6)$$

Physically, constraint equation (A5) guarantees that the derived dynamic stiffness matrix of the damaged structure is symmetrical while constraint equation (A6) ensures that the dynamic stiffness derived is orthogonal to any column of the receptance matrix of the damaged structure.

The constraint presented in equation (A6) can be written for 'n' frequency points from g_1 to g_2 as:

$$([K]_U \otimes [Q_K]) [\{\alpha_D(\Omega_1)\}_k \quad \cdots \quad \{\alpha_D(\Omega_n)\}_k] - $$
$$[\Omega_1^2 [M] \{\alpha_D(\Omega_1)\}_k \quad \cdots \quad \Omega_n^2 [M] \{\alpha_D(\Omega_n)\}_k] - [\{\delta\}_k \quad \cdots \quad \{\delta\}_k] = \{0\} \qquad (A7)$$

or,

$$([K]_U \otimes [Q_K]) [A_D(\Omega)] - [M][A_D(\Omega)] [\cdot \Omega_r^2 \cdot] - [\delta_k] = [0] \qquad (A8)$$

Frequency Ω_i (i = 1, 2, ..., n) is measured frequency point for receptance FRF data and

$$[A_D(\Omega)] = \left[\{\alpha_D(\Omega_1)\}_k \quad \cdots \cdots \quad \{\alpha_D(\Omega_n)\}_k\right] \tag{A9}$$

Using the method of Lagrange multipliers to incorporate the two constraints, a Lagrange function can be defined as:

$$L = e + \lambda g_1 + \mu g_2 \tag{A10}$$

where λ and μ are the Lagrange multipliers, g_1 and g_2 are the given constraints in equations (A5) and (A6) and L is the Lagrange function. It can be written as

$$L = e + \sum_{i=1}^{n}\sum_{j=1}^{s}\lambda_{ij}\left(\sum_{l=1}^{n}K_{U,il}Q_{K,il}A_{D,lj}\right) - Y + \sum_{i=1}^{n}\sum_{j=1}^{n}\mu_{ij}(Q_{K,ij} - Q_{K,ji}) \tag{A11}$$

The term Y represents the second and third terms at the left hand side of equation (A8) which are not a function of $Q_{K,ij}$ and need not be defined. They do not contribute to the derivative of L with respect to $Q_{K,ij}$. Taking the partial derivative of L with respect to $Q_{K,ij}$ and setting them equal to zero yields equations that $Q_{K,ij}$ have to satisfy in order for L to be minimal:

$$2\left([U] - [Q_K]\right) + [K]_U \otimes \left([\lambda][A_D]^T\right) + [\mu] - [\mu]^T = [0] \tag{A12}$$

Adding the transpose of equation (A12) to itself yields and multiplying the resultant equation by 1/4 ($[K]_U$) will lead to:

$$[K]_U \otimes [Q_K] = [K]_U - \frac{1}{4}[K]_U^{(2)} \otimes \left([\lambda][A_D]^T + [A_D][\lambda]^T\right) \tag{A13}$$

where $[K]_U^{(2)} = [K]_U \otimes [K]_U$.

Substituting equations (A13) into (A8) leads to:

$$[D] + \left([K]_U^{(2)} \otimes [\lambda][A_D]^T\right)[A_D] + \left([K]_U^{(2)} \otimes [A_D][\lambda]^T\right)[A_D] = [0] \tag{A14}$$

where

$$[D] = 4\left([K]_U[A_D] - [M]_U[A_D]\left[\cdot\Omega_r^2\cdot\right]\right) - \left[\{\delta\}_k \quad \cdots \quad \{\delta\}_k\right] \tag{A15}$$

The damaged stiffness matrix $[K]_D$ can be found from equation (A13) once matrix $[\lambda]$ is determined from equation (A14).

Nomenclature

$[\ ]_D$ matrix for the damaged structure

$[\ ]_U$	matrix for the undamaged structure
$[\ ]^+$	pseudo inverse
$[\ \diagdown\ ]$	a diagonal matrix
x'	the first derivative of x
$\|x\|$	Euclidean norm of x
$\otimes$	element-wise matrix multiplication
$[C]_k$	connectivity matrix of $\{\phi_D\}_k$
$[\ \diagdown c \diagdown\ ]$	modal damping matrix
$\{d\}$	damage detection vector
$\{d(\omega)\}$	damage detection vector
e	Euclidean norm
$[I]$	unity matrix
$[K]$	stiffness matrix
$[\Delta K]$	stiffness difference matrix
$[\ \diagdown k \diagdown\ ]$	modal Stiffness matrix
L	Lagrange function
$[M]$	mass matrix
$[\ \diagdown m \diagdown\ ]$	modal mass matrix
$[Q_K]$	damage quantification coefficient matrix
$Q_{K,ij}$	the ij^{th} element of $[Q_K]$
$[Z(\omega)]$	dynamic stiffness matrix
$[\Delta Z(\omega)]$	dynamic stiffness difference matrix
$[Z_{11}(\omega)]$	partitioned dynamic stiffness matrix
$[\alpha(\omega)]$	receptance FRF matrix at frequency ω
$\alpha_{ij}(\omega)$	the ij^{th} element in matrix $[\alpha(\omega)]$
$\{\alpha_D^{uc}(\omega)\}$	unmeasured FRF data in $\{\alpha_D(\omega)\}$
$\{\alpha_D^{mc}(\omega)\}$	measured FRF data in $\{\alpha_D(\omega)\}$
$\{\delta\}_r$	the rth Kronecker delta vector: $\{\delta\}_r$
$(\phi)_j$	the jth element of mode shape $\{\phi\}$
$\{\phi_D\}_j$	the jth of mode shape of the damaged structure

344

$\{\phi_U\}_j$ the jth of mode shape of the undamaged structure

ϕ_{jr} the jth element of mode shape $\{\phi\}_r$

ω_r the rth natural frequency

DAMAGE DETECTION AND EVALUATION II

Field Applications to Large Structures

C. R. FARRAR, S. W. DOEBLING
Los Alamos National Laboratory
MS P946
Los Alamos, New Mexico, USA, 87545

1. Introduction

In the most general terms damage can be defined as changes introduced into a system that adversely effect the current or future performance of that system. Implicit in this definition is the concept that damage is not meaningful without a comparison between two different states of the system, one of which is assumed to represent the initial, and often undamaged, state. This discussion is focused on the study of damage identification in structural and mechanical systems. Therefore, the definition of damage will be limited to changes to the material and/or geometric properties of these systems, including changes to the boundary conditions and system connectivity, which adversely effect the current or future performance of that system.

The interest in the ability to monitor a structure and detect damage at the earliest possible stage is pervasive throughout the civil, mechanical and aerospace engineering communities. Current damage-detection methods are either visual or localized experimental methods such as acoustic or ultrasonic methods, magnetic field methods, radiograph, eddy-current methods and thermal field methods (Doherty, 1987). All of these experimental techniques require that the vicinity of the damage is known *a priori* and that the portion of the structure being inspected is readily accessible. Subjected to these limitations, these experimental methods can detect damage on or near the surface of the structure. The need for quantitative global damage detection methods that can be applied to complex structures has led to the development and continued research of methods that examine changes in the vibration characteristics of the structure.

The basic premise of vibration-based damage detection is that the damage will significantly alter the stiffness, mass or energy dissipation properties of a system, which, in turn, will alter the measured dynamic response of that system. Although the basis for vibration-based damage detection appears intuitive, its actual application poses many significant technical challenges. The most fundamental challenge is the fact that damage is typically a local phenomenon and may not significantly influence the lower-frequency global response of structures that is typically measured during vibration tests. This challenge is supplemented by many practical issues associated with making accurate and repeatable vibration measurements at a limited number of locations on structures often operating in adverse environments.

In an effort to emphasize the extent of the research efforts in vibration-based damage detection a brief summary of applications that have driven developments in this field over

345

J.M.M. Silva and N.M.M. Maia (eds.), Modal Analysis and Testing, 345–378.
© 1999 *Kluwer Academic Publishers. Printed in the Netherlands.*

the last thirty years is first presented. Recent research has begun to recognize that the vibration-based damage detection problem is fundamentally one of statistical pattern recognition and this paradigm is described in detail. Current damage detection methods are then summarized in the context of this paradigm.

Finally, this paper will discuss the application of this technology to large civil engineering structures. The destructive tests performed on the I-40 Bridge over the Rio Grande in Albuquerque, NM are used as the example. This structure was chosen because numerous investigators have independently applied a variety of damage detection methods to the data obtained from these tests. To the authors' knowledge, this represents one of the most studied data sets from a test specifically aimed at detecting damage in an *in situ* structure from changes in its vibration characteristics. A study of the various damage detection methods that have been applied to this structure provides a better understanding of the methods themselves and many issues associated with the actual implementation of this technology. The paper concludes by describing a damage detection investigation of concrete columns in terms of the statistical pattern recognition paradigm.

2. Historical developments

It is the authors' speculation that damage or fault detection, as determined by changes in the dynamic properties or response of systems, has been practiced in a qualitative manner, using acoustic techniques, since modern man has used tools. More recently, this subject has received considerable attention in the technical literature and a brief summary of the developments in this technology over the last thirty years is presented below. Specific references are not cited; instead the reader is referred to (Doebling, et al. 1996) for a review of literature on this subject.

The development of vibration-based damage detection technology has been closely coupled with the evolution, miniaturization and cost reductions of Fast Fourier Transform (FFT) analyzer hardware and computing hardware. To date, the most successful application of vibration-based damage detection technology has been for monitoring rotating machinery. The rotating machinery application has taken an almost exclusive non-model based approach to damage detection. The detection process is based on pattern recognition applied to time histories or spectra generally measured on the housing of the machinery during normal operating conditions. Databases have been developed that allow specific types of damage to be identified from particular features of the vibration signature. For these systems the approximate location of the damage is generally known making a single channel FFT analyzer sufficient for most periodic monitoring activities. Today, commercial software integrated with measurement hardware is marketed to help the user systematically apply this technology to operating equipment.

During the 1970s and 1980s the oil industry made considerable efforts to develop vibration-based damage detection methods for offshore platforms. This damage detection problem is fundamentally different from that of rotating machinery because the damage location is unknown and because the majority of the structure is not readily accessible for measurement. To circumvent these difficulties, a common methodology adopted by this industry was to simulate candidate damage scenarios with numerical models, examine the changes in resonant frequencies that were produced by these simulated changes, and correlate these changes with those measured on a platform. A number of very practical problems were encountered including measurement difficulties caused by platform machine noise, instrumentation difficulties in hostile environments, changing mass caused by marine growth and varying fluid storage levels, temporal variability of foundation conditions, and the inability of wave motion to excite higher modes. These issues prevented adaptation of this technology and efforts at further developing this technology for offshore platforms were largely abandoned in the early 1980s.

The aerospace community began to study the use of vibration-based damage detection during the late 1970's and early 1980's in conjunction with the development of the space shuttle. This work has continued with current applications being investigated for the National Aeronautics and Space Administration's space station and reusable launch vehicle. The Shuttle Modal Inspection System (SMIS) was developed to identify fatigue damage in components such as control surfaces, fuselage panels and lifting surfaces. These areas were covered with a thermal protection system making these portions of the shuttle inaccessible and, hence impractical for conventional local non-destructive examination methods. This system has been successful in locating damaged components that are covered by the thermal protection system. All orbiter vehicles have been periodically subjected to SMIS testing since 1987. Space station applications have primarily driven the development of experimental/analytical damage detection methods. These approaches are based on correlating analytical models of the undamaged structure with measured modal properties from both the undamaged and damaged structure. Changes in stiffness indices as assessed from the two model updates are used to locate and quantify the damage. Since the mid 1990's, studies of damage detection for composite materials have been motivated by the development composite fuel tank for a reusable launch vehicle.

The civil engineering community has studied vibration based damage assessment of bridge structures since the early 1980's. Modal properties and quantities derived from these properties such as mode-shape curvature and dynamic flexibility matrix indices have been the primary features used to identify damage in bridge structures. Environmental and operating condition variability present significant challenges to the bridge monitoring application. Regulatory requirements in eastern Asian countries, which mandate the companies that construct the bridges to periodically certify their structural health, are driving current research and development of vibration-based bridge monitoring systems.

In summary, the review of the technical literature presented by (Doebling et al. 1996) shows an increasing number of research studies related to vibration-based damage detection. These studies identify many technical challenges to the adaptation of vibration-based damage detection that are common to all applications of this technology. These challenges include better utilizing the nonlinear response characteristics of the damaged system, development of methods to optimally define the number and location of the sensors, identifying the features sensitive to small damage levels, the ability to discriminate changes in features cause by damage from those caused by changing environmental and/or test conditions, the development of statistical methods to discriminate features from undamaged and damaged structures, and performing comparative studies of different damage detection methods applied to common data sets. These topics are currently the focus of various research efforts by many industries including defense, automotive, and semiconductor manufacturing where multi-disciplinary approaches are being used to advance the current capabilities of vibration-based damage detection.

3. Vibration-based damage detection and structural health monitoring

The process of implementing a damage detection strategy is referred to as *structural health monitoring*. This process involves the observation of a structure over a period of time using periodically spaced measurements, the extraction of features from these measurements, and the analysis of these features to determine the current state of health of the system. The output of this process is periodically updated information regarding the ability of the structure to continue to perform its desired function in light of the inevitable aging and degradation resulting from the operational environments. Figure 1 shows a chart summarizing the structural health-monitoring process. The topics summarized in this figure are discussed below.

3.1. OPERATIONAL EVALUATION

Operational evaluation answers two questions in the implementation of a structural health monitoring system:

1.   What are the conditions, both operational and environmental, under which the system to be monitored functions?

2.   What are the limitations on acquiring data in the operational environment?

Operational evaluation begins to set the limitations on what will be monitored and how the monitoring will be accomplished. This evaluation starts to tailor the damage detection process to features that are unique to the system being monitored and tries to take advantage of unique features of the postulated damage that is to be detected.

3. 2. DATA ACQUISITION AND CLEANSING

The data acquisition portion of the structural health monitoring process involves selecting the types of sensors to be used, the location where the sensors should be placed, the number of sensors to be used, and the data acquisition/storage/transmittal hardware. This process will be application specific. Economic considerations will play a major role in making these decisions. Another consideration is how often the data should be collected. In some cases it may be adequate to collect data immediately before and at periodic intervals after a severe event. However, if fatigue crack growth is the failure mode of concern, it may be necessary to collect data almost continuously at relatively short time intervals.

Because data can be measured under varying conditions, the ability to normalize the data becomes very important to the damage detection process. One of the most common procedures is to normalize the measured responses by the measured inputs. When environmental or operating condition variability is an issue, the need can arise to normalize the data in some temporal fashion to facilitate the comparison of data measured at similar times of an environmental or operational cycle. Sources of variability in the data acquisition process and with the system being monitored need to be identified and minimized to the extent possible. In general, all sources of variability can not be eliminated. Therefore, it is necessary to make the appropriate measurements such that these sources can be statistically quantified.

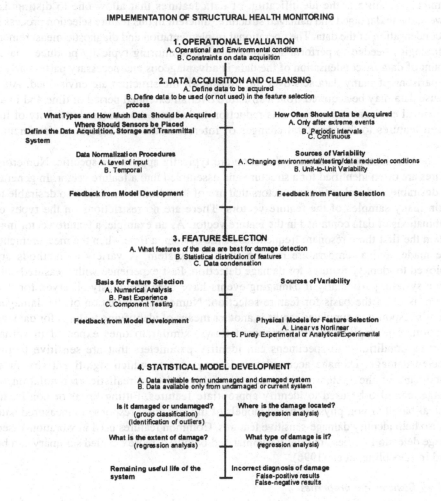

IMPLEMENTATION OF STRUCTURAL HEALTH MONITORING

1. OPERATIONAL EVALUATION
A. Operational and Enviromental conditions
B. Constraints on data acquisition

2. DATA ACQUISITION AND CLEANSING
A. Define data to be acquired
B. Define data to be used (or not used) in the feature selection process

What Types and How Much Data Should be Acquired
Where Should Sensors be Placed
Define the Data Acquisition, Storage and Transmittal System

How Often Should Data be Acquired
A. Only after extreme events
B. Periodic intervals
C. Continuous

Data Normalization Procedures
A. Level of input
B. Temporal

Sources of Variability
A. Changing environmental/testing/data reduction condtions
B. Unit-to-Unit Variability

Feedback from Model Devdopment

Feedback from Feature Selection

3. FEATURE SELECTION
A. What features of the data are best for damage detection
B. Statistical distribution of features
C. Data condensation

Basis for Feature Selection
A. Numerical Analysis
B. Past Experience
C. Component Testing

Sources of Variability

Feedback from Model Development

Physical Models for Feature Selection
A. Linear vs Nonlinear
B. Purely Experimental or Analytical/Experimental

4. STATISTICAL MODEL DEVELOPMENT
A. Data available from undamaged and damaged system
B. Data available only from undamaged or current system

Is it damaged or undamaged?
(group classification)
(Identification of outliers)

Where is the damage located?
(regression analysis)

What is the extent of damage?
(regression analysis)

What type of damage is it?
(regression analysis)

Remaining useful life of the system

Incorrect diagnosis of damage
False-positive results
False-negative results

Figure 1 Flow chart for implementing a structural health monitoring program.

Data cleansing is the process of selectively choosing data to accept for, or reject from, the feature selection process. The data cleansing process is usually based on knowledge gained by individuals directly involved with the data acquisition. Finally, is should be noted that the data acquisition and cleansing portion of a structural health-monitoring process should not be static. Insight gained from the feature selection process and the statistical model development process will provide information regarding changes that can improve the data acquisition process.

3. 3.   FEATURE SELECTION

The area of the structural damage detection process that receives the most attention in the technical literature is the identification of data features that allow one to distinguish between the undamaged and damaged structure. Inherent in this feature selection process is the condensation of the data. The operational implementation and diagnostic measurement technologies needed to perform structural health monitoring typically produce a large amount of data. A condensation of the data is advantageous and necessary particularly if comparisons of many data sets over the lifetime of the structure are envisioned. Also, because data may be acquired from a structure over an extended period of time and in an operational environment, robust data reduction techniques must retain sensitivity of the chosen features to the structural changes of interest in the presence of environmental noise.

The best features for damage detection are typically application specific. Numerous features are often identified for a structure and assembled into a feature vector. In general, it is desirable to develop feature vectors that are of low dimension. It is also desirable to obtain many samples of the feature vectors. There are no restrictions on the types or combinations of data contained in the feature vector. As an example, a feature vector may contain the first three resonant frequencies of the system, a time when the measurements were made, and a temperature reading from the system. A variety of methods are employed to identify features for damage detection. Past experience with measured data from a system, particularly if damaging events have been previously observed for that system, is often the basis for feature selection. Numerical simulation of the damaged system's response to simulated inputs is another means of identifying features for damage detection. The application of engineered flaws, similar to ones expected in actual operating conditions, to specimens can identify parameters that are sensitive to the expected damage. Damage accumulation testing, during which significant structural components of the system under study are subjected to a realistic accumulation of damage, can also be used to identify appropriate features. Fitting linear or nonlinear, physical-based or non-physical-based models of the structural response to measured data can also help identify damage-sensitive features. Common features used in vibration-based damage detection studies are briefly summarized below. A more detailed summary can be found in (Doebling, et al., 1996).

3.3.1.   *Basic modal properties*
The most common features that are used in vibration-based damage detection, and that represent a significant amount of data condensation from the actual measured quantities, are the common modal properties of resonant frequencies and mode-shape vectors. These features are identified from measured response time-histories, most often absolute acceleration, or spectra of these time-histories. The technology required to accurately make these measurements is summarized in (McConnell, 1995). Often these spectra are normalized by spectra of the measured force input to form frequency response functions. Well-developed experimental modal analysis procedures are applied to these functions or to the measured-response spectra to estimate the system's modal properties (Ewins, 1995, and Maia and Silva, 1997).

The amount of literature that uses resonant frequency shifts as a data feature for damage detection is quite large. The observation that changes in structural properties cause changes in vibration frequencies was a primary impetus for developing vibration-based damage identification technology. In general, changes in frequencies cannot provide spatial information about structural changes. For applications to large civil engineering structures the somewhat low sensitivity of frequency shifts to damage requires either very precise measurements of frequency change or large levels of damage. An exception to this limitation occurs at higher modal frequencies, where the modes are associated with local responses. However, the practical limitations involved with the excitation and identification of the resonant frequencies associated with these local modes, caused in part by high modal density and low participation factors, can make them difficult to identify.

Damage detection methods using mode shape vectors as a feature generally analyze differences between the measured modal vectors before and after damage. Mode shape vectors are spatially distributed quantities; therefore, they provide information that can be used to locate damage. However, large number of measurement locations can be required to accurately characterize mode shape vectors and provide sufficient resolution for determining the damage location.

3.3.2. *Mode shape curvature changes*
An alternative to using mode shapes to obtain spatially distributed features sensitive to damage is to use mode shape derivatives, such as curvature. Mode shape curvature can be computed by numerically differentiating the identified mode shape vectors twice to obtain an estimate of the curvature. These methods are motivated by the fact that the second derivative of the mode shape is much more sensitive to small perturbations in the system than is the mode shape itself. Also, for beam- and plate-like structures changes in curvature can be related to changes in strain energy, which has been shown to be a sensitive indicator of damage. A comparison of the relative statistical uncertainty associated with estimates of mode shape curvature, mode shape vectors and resonant frequencies showed that the largest variability is associated with estimates of mode shape curvature followed by estimates of the mode shape vector. Resonant frequencies could be estimated with least uncertainty (Doebling, Farrar and Goodman, 1997).

3.3.3. *Dynamically measured flexibility*
Changes in the dynamically measured flexibility matrix indices have also been used as damage sensitive features. The dynamically measured flexibility matrix is estimated from the mass-normalized measured mode shapes and measured eigenvalue matrix (diagonal matrix of squared modal frequencies). The formulation of the flexibility matrix is approximate because in most cases all of the structure's modes are not measured. Typically, damage is detected using flexibility matrices by comparing the flexibility matrix indices computed using the modes of the damaged structure to the flexibility matrix indices computed using the modes of the undamaged structure. Because of the inverse relationship to the square of the modal frequencies, the measured flexibility matrix is most sensitive to changes in the lower-frequency modes of the structure.

3.3.4. *Updating structural model parameters*
Another class of damage identification methods is based on features related to changes in mass, stiffness and damping matrix indices that have been correlated such that the

numerical model predicts as closely as possible to the identified dynamic properties (resonant frequencies, modal damping and mode shape vectors) of the undamaged and damaged structures, respectively. These methods solve for the updated matrices (or perturbations to the nominal model that produce the updated matrices) by forming a constrained optimization problem based on the structural equations of motion, the nominal model, and the identified modal properties (Friswell and Mottershead, 1995). Comparisons of the matrix indices that have been correlated with modal properties identified from the damaged structure to the original correlated matrix indices provide an indication of damage that can be used to quantify the location and extent of damage. Degree of freedom mismatch between the numerical model and the measurement locations can be a severe limitation for performing the required matrix updates.

3.3.5. *Nonlinear methods*
Identification of the previously described features is based on the assumption that a linear modal can be used to represent the structural response before and after damage. However, in many cases the damage will cause the structure to exhibit nonlinear response. Therefore, the identification of features indicative of nonlinear response can be a very effective means of identifying damage in a structure that originally exhibited linear response. The specific features that indicate a system is responding in a nonlinear manner vary widely. Examples include the generation of resonant frequency harmonics in a cracked beam excited in a manner such that the crack opens and closes (Prime and Shevitz, 1996). For extreme event such as earthquake, the normalized arias intensity provides an estimate of kinetic energy of the structure and has been successfully used to identify the onset of nonlinear response buildings subject to damaging earthquake excitations (Straser, 1998). Deviations from a Gaussian probability distribution function of acceleration response amplitudes for a system subjected to a Gaussian input have been used successfully to identify that loose parts are present in a system. Temporal variation in resonant frequencies as identified with canonical variate analysis is another method to identify the onset of damage (Hunter, 1999). In general, features based on the nonlinear response of a system have only been used to identify that damage has occurred. Few methods have been described that locate the source of the nonlinearity. Because all systems will exhibit some degree of nonlinearity, a challenge becomes to establish a threshold at which changes in the nonlinear response features are indicative of damage. The statistical model building portion of the structural health monitoring process is essential for establishing such thresholds.

3.4. STATISTICAL MODEL DEVELOPMENT

The portion of the structural health monitoring process that has received the least attention in the technical literature is the development of statistical models to enhance the damage detection. Almost none of the hundreds of studies summarized in (Doebling, et al, 1996) make use of any statistical methods to assess if the changes in the selected features used to identify damaged are statistically significant. Statistical model development is concerned with the implementation of the algorithms that operate on the extracted features and unambiguously determine the damage state of the structure. The algorithms used in statistical model development usually fall into three categories and will depend on the availability of data from both an undamaged and damaged structure. The first category is group classification, that is, placement of the features into respective "undamaged" or "damaged" categories. Analysis of outliers is the second type of algorithm. When data from a damaged structure are not available for comparison, do the observed features indicate a significant change from the previously observed features that can not be explained by extrapolation of the feature distribution? The third category is regression analysis. This analysis refers to the process of correlating data features with particular types, locations or extents of damage. All three algorithm categories analyze statistical distributions of the measured or derived features to enhance the damage detection process.

The statistical models are used to answer the following questions regarding the damage state of the structure, (Rytter, 1993): 1. Is there damage in the structure (existence)?; 2. Where is the damage in the structure (location)?; and 3. How severe is the damage (extent)? Answers to these questions in the order presented represents increasing knowledge of the damage state. Experimental structural dynamics techniques can be used to address the first two questions. Analytical models are usually needed to answer the third question unless examples of data are available from the system (or a similar system) when it exhibits varying level of the damage. Statistical model development can also determine the type of damage that is present. To identify the type of damage, data from structures with the specific types of damage must be available for correlation with the measured features.

Finally, an important part of the statistical model development process is the testing of these models on actual data to establish the sensitivity of the selected features to damage and to study the possibility of false indications of damage. False indications of damage fall into two categories: 1.) False-positive damage indication (indication of damage when none is present), and 2). False-negative damage indications (no indication of damage when damage is present). Although the second category is usually very detrimental to the damage detection and can have serious life-safety implications, false-positive readings can also erode confidence in the damage detection process.

This paper will now summarize a damage detection study performed on an *in situ* bridge. This summary is followed by the application of methods from statistical pattern recognition and machine learning to the vibration-based damage detection study of concrete columns. A damage detection experiment performed on concrete bridge columns will be described in terms of the statistical-pattern-recognition damage-detection paradigm that has just been summarized.

4. I-40 Bridge damage detection study

The Interstate 40 (I-40) Bridges over the Rio Grande in Albuquerque, New Mexico were scheduled to be demolished in the summer of 1993 as part of a construction project to widen the bridges for increased traffic flow. Prior to this demolition, damage was incrementally applied to the bride in a controlled manner. Vibration tests were performed on the undamaged bridge and after each level of damage had been introduced. The primary purpose of these tests was to study the ability to identify and locate damage in a large structure based on changes in its measured dynamic response.

4.1. TEST STRUCTURE GEOMETRY

The I-40 Bridges formerly consisted of twin spans (there are separate bridges for each traffic direction) made up of a concrete deck supported by two welded-steel plate girders and three steel stringers. Loads from the stringers were transferred to the plate girders by floor beams located at 6.1-m (20-ft) intervals. Cross-bracing was provided between the floor beams. Figure 2 shows an elevation view of the portion of the bridge that was tested. The cross-section geometry of each bridge is shown in Figure 3. Each bridge was made up of three identical sections. Each section had three spans; the end spans were of equal length, approximately 39.9 m (131 ft), and the center span was approximately 49.4-m (163 ft) long. All subsequent discussions of the I-40 Bridge will refer to the bridge carrying eastbound traffic, particularly the three eastern spans, which were the only ones tested.

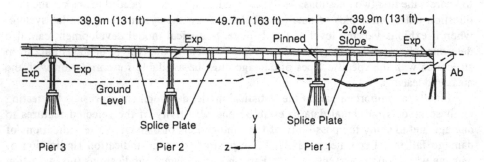

Figure 2 Elevation view of the portion of the I-40 Bridge that was tested.

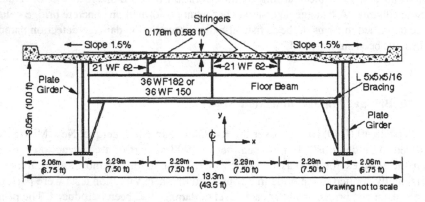

Figure 3 Cross-section geometry of the I-40 Bridge.

4.2. DAMAGE SCENARIOS

The damage that was introduced was intended to simulate fatigue cracking that has been observed in plate-girder bridges. This type of cracking results from out-of-plane bending of the plate girder web and usually begins at welded attachments to the web such as the seats supporting the floor beams. Four levels of damage were introduced to the middle span of the north plate girder close to the seat supporting the floor beam at mid-span. Damage was introduced by making various torch cuts in the web and flange of the girder. A major drawback of introducing damage in this manner is that the torch cuts produce crack much wider than an actual fatigue crack. Hence, these cracks do not open and close under the loading that was applied during this study.

The first level of damage, designated E-1, consisted of a 0.61-m-long (2-ft-long), 10-mm-wide (3/8-in-wide) cut through the web centered at mid-height of the web. Next, this cut was continued to the bottom of the web to produce a second level of damage designated E-2. For the third level of damage, E-3, the flange was then cut halfway in from either side directly below the cut in the web. Finally, the flange was cut completely through for damage case E-4 leaving the top 1.2 m (4 ft) of the web and the top flange to carry the load at this location. The various levels of damage are shown in Figure 4. Photographs of the damage can be found in (Farrar, et al., 1994).

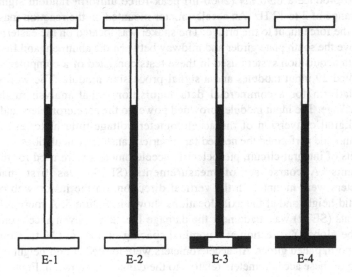

Figure 4 Damage Scenarios

4.3 OPERATIONAL EVALUATION

Safety concerns did not allow traffic on the bridge after damage had been introduced. Therefore, all testing in the damaged condition required a shaker to be used as the excitation source. Safety and time constraints dictated that the shaker could only be setup in one location and only on the span closest to the abutment. The data acquisition system used limited the number of channels that could be acquired simultaneously to 32. Access to the girders under the bridge by catwalks limited the locations that accelerometers could be placed to the two main plate girders. Considerable extraneous vibration inputs were caused by traffic on the adjacent spans. Because the contractor was anxious to proceed with the demolition they began to tear down the bridge being tested at its other end and to remove fill near the abutment that was being tested. These activities add additional sources of variability into the testing that are difficult to quantify. The testing was constrained by budget and scheduling concerns to be completed during daylight hours over a two-week time window during the hottest time of the year. Electric power was not available during these tests.

4.4. DATA ACQUISITION AND CLEANSING

The experimental procedures used to obtain vibration response data from the undamaged and damaged structure are described in this section. To obtain these data, a forced vibration test was conducted on the undamaged bridge. Forced vibration tests similar to those done on the undamaged structure were repeated after each level of damage had been introduced. A detailed summary of the experimental procedures can be found in (Farrar et al., 1994).

Engineers from Sandia National Laboratory (SNL) provided a hydraulic shaker that generated the measured force input (Mayes and Nusser, 1994). A random-signal generator was used to produce a 8.90 kN (2000-lb) peak-force uniform random signal over the frequency range of 2 to 12 Hz. An accelerometer mounted on the reaction mass was used to measure the force input to the bridge. The shaker was located on the eastern-most span directly above the south plate girder and midway between the abutment and first pier.

The data acquisition system used in these tests consisted of a computer workstation that controlled 29 input modules and a signal-processing module. The workstation was also the platform for a commercial data-acquisition/signal-analysis/modal-analysis software package. The input modules provided power to the accelerometers and performed analog-to-digital conversion of the accelerometer voltage-time histories. The signal-processing module performed the needed fast Fourier transform calculations.

Two sets of integral-circuit, piezoelectric accelerometers were used for the vibration measurements. A coarse set of measurements (SET1) was first made. These accelerometers were mounted in the vertical direction, on the inside web of the plate girder, at mid-height and at the axial locations shown in Figure 5. A more refined set of measurements (SET2) was made near the damage location. Eleven accelerometers were placed in the global Y direction at a nominal spacing of 4.88 m (16 ft) along the mid-span of the north plated girder. All accelerometers were located at mid-height of the girder. The spacing of these accelerometers relative to the damage is shown in Figure 6.

The only form of data cleansing that was performed was to use the overload reject feature in the data acquisition system. All accelerometers and their corresponding wiring were checked before and after each test to verify that they were working properly. In instances when a faulty accelerometer or wire was detected, the problem was corrected and the test was repeated.

4.5. FEATURE EXTRACTION AND DATA COMPRESSION

Basic modal parameters (resonant frequencies, mode shapes and modal damping values) were used as features as well as properties derived from these quantities. Also, stiffness indices from updated finite element models were used as damage sensitive features. The reduction of the measured accelerometer time histories to modal parameters represents a significant amount of data compression. A typical time-history for the I-40 Bridge test had 2048 data points, and if measurements are made at 26 points plus an input, there are 55,296 pieces of information regarding the current state of the structure. Through system identification procedures commonly referred to as experimental modal analysis this volume of data was reduced to six resonant frequencies, mode shapes and modal damping values. This data compression was done because the modal quantities are easier to

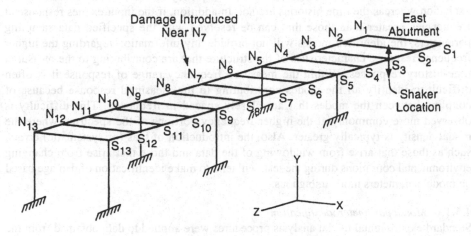

Figure 5 Set1 (coarse) accelerometer locations.

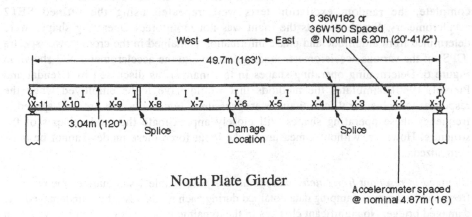

North Plate Girder

Figure 6 Set2 (refined) accelerometer locations.

visualize, physically interpret, and interpret in terms of standard mathematical modeling of vibrating systems than are the actual time-history measurements. Therefore, the 55,296 pieces of information were reduced to 324 pieces of information (6 modes made up of 26 amplitude and phase values, 6 resonant frequencies and 6 modal damping values).

Intuitively, information about the current state of the structure must be lost in this data reduction and system identification process. The loss of information occurs primarily from the fact that for a linear system the modal properties are independent of the excitation signal characteristics (amplitude and frequency content) and the location of the excitation whereas the time histories are not. In addition, if the input excites response at frequencies greater than those that can be resolved with the specified data sampling parameters, the identified modes will not provide any information regarding the higher frequency response characteristics of the structure that are contributing to the measured time-history responses. Within the measured frequency range of response it is often difficult to identify all the modes contributing to the measured response because of coupling between the modes that are closely spaced in frequency. This difficulty is observed more commonly at the higher frequency portions of the spectrum where the modal density is typically greater. Also, the introduction of bias (or systematic) errors, such as those that arise from windowing of the data and those that arise from changing environmental conditions during the test, will tend to make identification of damage based on modal parameters more ambiguous.

4.5.1. *Modal parameter identification*
Standard experimental modal analysis procedures were applied to data obtained from the SET1 accelerometers during the forced vibration tests to identify the modal parameters of the bridge in its damaged and undamaged condition. A rational-fraction polynomial, global, curve-fitting algorithm in a commercial modal analysis software package was used estimate the modal properties. Figure 7 shows the first three modes of the undamaged bridge identified from these data. By measuring the input force and the corresponding driving point acceleration, these mode shapes can be unit-mass normalized.

Immediately after the forced vibration tests with the SET1 accelerometers were complete, the random excitation tests were repeated using the refined SET2 accelerometers. For these tests the input was not monitored. Operating shapes were determined from amplitude and phase information contained in the cross-power spectra (CPS) of the various accelerometer readings relative to the accelerometer X-3 shown in Figure 6. Determining operating shapes in this manner, as discussed by (Bendat and Piersol, 1980), simulates the methods that would have to be employed when the responses to ambient traffic excitations are measured. For modes that are well-spaced in frequency these operating shapes will closely approximate the mode shapes of the structure. However, without a measure of the input force these modes cannot be mass normalized.

Changed in resonant frequencies and modal damping. Table 1 summarizes the resonant frequency and modal damping data obtained during each modal test of the undamaged and damaged bridge. No significant changes in the dynamic properties can be observed until the final level of damage was introduced. At the final level of damage the resonant frequencies for the first two modes have dropped to values 7.6 and 4.4 percent less, respectively, than those measured during the undamaged tests. For modes where the damage was introduced near a node for that mode (modes 3 and 5) no significant changes in resonant frequencies can be observed.

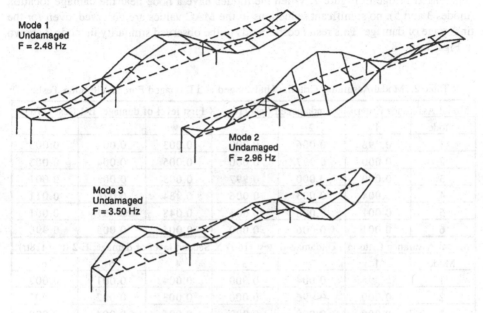

Mode 1
Undamaged
F = 2.48 Hz

Mode 2
Undamaged
F = 2.96 Hz

Mode 3
Undamaged
F = 3.50 Hz

Figure 7 First three modes measured on the undamaged structure.

Table 1   Resonant Frequencies and Modal Damping Values Identified from Undamaged and Damaged Forced Vibration Tests Using SET1 Accelerometers

Test Designation	Damage Case	Mode 1 Freq. (Hz)/ Damp. (%)	Mode 2 Freq. (Hz)/ Damp. (%)	Mode 3 Freq. (Hz)/ Damp. (%)	Mode 4 Freq. (Hz)/ Damp. (%)	Mode 5 Freq. (Hz)/ Damp. (%)	Mode 6 Freq. (Hz)/ Damp. (%)
t16tr	Undamaged	2.48/ 1.06	2.96/ 1.29	3.50/ 1.52	4.08/ 1.10	4.17/ 0.86	4.63/ 0.92
t17tr	E-1 cut at center of web	2.52/ 1.20	3.00/ 0.80	3.57/ 0.87	4.12/ 1.00	4.21/ 1.04	4.69/ 0.90
t18tr	E-2 extend cut to bottom flange	2.52/ 1.33	2.99/ 0.82	3.52/ 0.95	4.09/ 0.85	4.19/ 0.65	4.66/ 0.84
t19tr	E-3 bottom flange cut half way	2.46/ 0.82	2.95/ 0.89	3.48/ 0.92	4.04/ 0.81	4.14/ 0.62	4.58/ 1.06
t22tr	E-4 bottom flange cut completely	2.30/ 1.60	2.84/ 0.66	3.49/ 0.80	3.99/ 0.80	4.15/ 0.71	4.52/ 1.06

Changes in mode shapes. A modal assurance criterion (MAC), sometimes referred to as a modal correlation coefficient (Ewins, 1985), was calculated to quantify the correlation between mode shapes measured during different tests. Table 2 shows the MAC values that are calculated when mode shapes from tests t17tr (damage level E-1), t18tr (damage level E-2), t19tr (damage level E-3), and t22tr (damage level E-4) are compared to the modes measured on the undamaged forced vibration test, t16tr. The MAC values show no change in the mode shapes for the first three stages of damage. When the final level of damage is introduced, significant drops in the MAC values for modes 1 and 2 are noticed. These two modes are shown in Figure 8 and can be compared to similar modes identified for the undamaged bridge in Figure 7. When the modes have a node near the damage location (modes 3 and 5), no significant reductions in the MAC values are observed, even for the final stage of damage. This result corresponds to the observed similarity in mode 3 shown in Figs. 7 and 8.

Table 2. Modal Assurance Criteria: Undamaged and Damaged Forced Vibration Tests

Modal Assurance Criteria	Undamaged (test t16tr) X First level of damage, E-1 (test t17tr)					
Mode	1	2	3	4	5	6
1	0.996	0.006	0.000	0.003	0.001	0.003
2	0.000	0.997	0.000	0.005	0.004	0.003
3	0.000	0.000	0.997	0.003	0.008	0.001
4	0.004	0.003	0.006	0.984	0.026	0.011
5	0.001	0.008	0.003	0.048	0.991	0.001
6	0.001	0.006	0.000	0.005	0.005	0.996
Modal Assurance Criteria	Undamaged (test t16tr) X Second level of damage, E-2 (test t18tr)					
Mode	1	2	3	4	5	6
1	0.995	0.004	0.000	0.004	0.001	0.002
2	0.000	0.996	0.000	0.003	0.002	0.002
3	0.000	0.000	0.999	0.006	0.004	0.000
4	0.003	0.006	0.005	0.992	0.032	0.011
5	0.001	0.006	0.008	0.061	0.997	0.004
6	0.002	0.004	0.000	0.005	0.005	0.997
Modal Assurance Criteria	Undamaged (test t16tr) X Third level of damage, E-3 (test t19tr)					
Mode	1	2	3	4	5	6
1	0.997	0.002	0.000	0.005	0.001	0.001
2	0.000	0.996	0.001	0.003	0.002	0.002
3	0.000	0.000	0.999	0.006	0.006	0.000
4	0.003	0.005	0.004	0.981	0.032	0.011
5	0.001	0.006	0.004	0.064	0.995	0.003
6	0.002	0.002	0.000	0.004	0.009	0.995
Modal Assurance Criteria	Undamaged (test t16tr) X Fourth level of damage, E-4 (test t22tr)					
Mode	1	2	3	4	5	6
1	0.821	0.168	0.002	0.001	0.000	0.001
2	0.083	0.884	0.001	0.004	0.001	0.002
3	0.000	0.000	0.997	0.005	0.007	0.001
4	0.011	0.022	0.006	0.917	0.010	0.048
5	0.001	0.006	0.003	0.046	0.988	0.002
6	0.005	0.005	0.000	0.004	0.009	0.965

From the observed changes in modal parameters it is clear that damage can only be definitively identified from change in the modal properties after the final cut was made in the bridge. Prior to the final cut, one could not say that the changes observed were caused by damage or were within the repeatability of the tests. In two tests at increasing levels of damage (t17tr and t18tr) the resonant frequencies were actually found to increase slightly from that of the undamaged case. These slight increases in frequency were measured independently by other researchers studying this bridge at the same time (Farrar, et al. (1994)) and are assumed to be caused by changing test conditions. The examination of changes in the basic modal properties (resonant frequencies and mode shapes) demonstrates the need to identify data features more sensitive to damage.

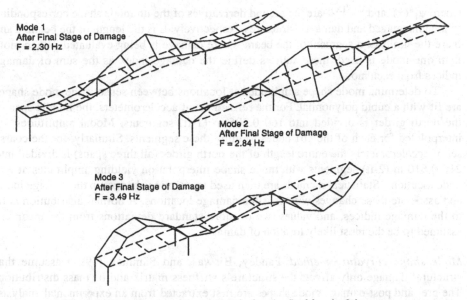

Mode 1
After Final Stage of Damage
F = 2.30 Hz

Mode 2
After Final Stage of Damage
F = 2.84 Hz

Mode 3
After Final Stage of Damage
F = 3.49 Hz

Figure 8 First three modes measured after the final level of damage.

4.5.2. *Additional features*

For comparative purposes, five linear modal-based damage identification algorithms were applied to data obtained from the I-40 Bridge in its undamaged and damaged condition. These algorithms require mode shape data (in some cases unit-mass normalized mode shape data) and resonant frequencies. A summary of these algorithms and their implementation for the study reported herein can be found in (Farrar and Jauregui, 1996). All five methods are based on the observation that in the vicinity of damage there will be a local increase in the structure's flexibility. This increase will alter the mode shapes of the structure in the damage vicinity, hence a comparison of mode shape data processed by these five different methods, before and after damage, should reveal the location of the damage.

Damage index method. The Damage Index Method was developed by Stubbs and Kim (1994) to locate damage in structures given their characteristic mode shapes before and after damage. For a structure that can be represented as a beam, a damage index, β, is developed based on the change in strain energy stored in the structure when it deforms in its particular mode shape. For location j on the beam this change in the ith mode strain energy is related to the changes in curvature of the mode at location j. The damage index for this location and this mode, β_{ij}, is defined as

$$\beta_{ij} = \frac{\left(\int_a^b [\psi_i^{*}{''}(x)]^2 dx + \int_0^L [\psi_i^{*}{''}(x)]^2 dx \right)}{\left(\int_a^b [\psi_i{''}(x)]^2 dx + \int_0^L [\psi_i{''}(x)]^2 dx \right)} \cdot \frac{\int_0^L [\psi_i{''}(x)]^2 dx}{\int_0^L [\psi_i^{*}{''}(x)]^2 dx}, \tag{1}$$

where $\psi_i{''}(x)$ and $\psi_i^{*}{''}(x)$ are the second derivatives of the ith mode shape corresponding to the undamaged and damaged structures, respectively. L is the length of the beam. a and b are the limits of a segment of the beam where damage is being evaluated. When more than one mode is used, these authors define the damage index as the sum of damage indices from each mode.

To determine mode shape amplitudes at locations between sensors, the mode shapes are fit with a cubic polynomial. For the refined set of accelerometers, the middle span of the north girder is divided into 160 0.305-m (1-ft) segments. Modal amplitudes are interpolated for each of the 161 nodes forming these segments. Similarly, for the coarse set of accelerometers the entire length of the north girder (all three spans) is divided into 210 0.610 m (2-ft) segments with mode shape interpolation yielding amplitudes at 211 node locations. Statistical methods are then used to examine changes in the damage index and associate these changes with possible damage locations. A normal distribution is fit to the damage indices, and values two or more standard deviations from the mean are assumed to be the most likely location of damage.

Mode shape curvature method. Pandey, Biswas, and Samman (1991) assume that structural damage only affects the structure's stiffness matrix and its mass distribution. The pre- and post-damage mode shapes are first extracted from an experimental analysis. Curvature of the mode shapes for the beam in its undamaged and damaged conditions can then be estimated numerically from the displacement mode shapes with a central difference approximation or other means of differentiation. Given the before- and after-damage mode shapes, the authors consider a beam cross section at location x subjected to a bending moment M(x). The curvature at location x along the length of the beam, $v''(x)$, is

$$v''(x) = M(x)/(EI), \tag{2}$$

where E = the modulus of elasticity, and I = the moment of inertia of the section.

From Eq. 2, it is evident that the curvature is inversely proportional to the flexural stiffness, EI. Thus, a reduction of stiffness associated with damage will, in turn, lead to an increase in curvature. Differences in the pre- and post-damage curvature mode shapes will, in theory, be largest in the damaged region. For multiple modes, the absolute values of change in curvature associated with each mode can be summed to yield a damage parameter for a particular location.

Change in flexibility method. Pandey and Biswas (1994) show that for the undamaged and damaged structure, the flexibility matrix, [F], can be approximated from the unit-mass-normalized modal data as follows

$$[F] \approx \sum_{i=1}^{n} \frac{1}{\omega_i^2} \{\phi_i\}\{\phi_i\}^T \text{ and} \tag{3}$$

$$[F]^* \approx \sum_{i=1}^{n} \frac{1}{\omega_i^{*2}} \{\phi_i\}^* \{\phi_i\}^{*T}, \tag{4}$$

where ω_i = the ith modal frequency, ϕ_i = the ith unit-mass-normalized mode, n = the number of measured modes, and the asterisks signify properties of the damaged structure. Equations 3 and 4 are approximations because fewer modes are typically identified than the total numbers of measurement points or degrees of freedom. From the pre- and post-damage flexibility matrices, a measure of the flexibility change caused by the damage can be obtained from the difference of the respective matrices. The column of the flexibility matrix corresponding to the largest change is indicative of the degree of freedom where the damage is located.

Change in uniform load surface curvature. The coefficients of the ith column of the flexibility matrix represent the deflected shape assumed by the structure with a unit load applied at the *i*th degree of freedom. The sum of all columns of the flexibility matrix represent the deformed shape assumed by the structure if a unit load is applied at each degree of freedom, and this shape is referred to as the uniform load surface. Zhang and Aktan (1995) state that the change in curvature of the uniform load surface can be used to determine the location of damage. In terms of the curvature of the uniform load surface, F", the curvature change at location 1 is evaluated as follows

$$\Delta F_l^{''} = \left| F_l^{*''} - F_l^{''} \right|, \tag{5}$$

where $\Delta F''$ represents the absolute curvature change. The curvature of the uniform load surface can be obtained with a central difference operator as suggested by these authors.

Change in stiffness method. Zimmerman and Kaouk (1994) have developed a damage detection method based on changes in the stiffness matrix that is derived from measured modal data. The eigenvalue problem of an undamaged, undamped structure is

$$\left(\lambda_i [M] + [K] \right) \{\psi_i\} = \{0\}, \tag{6}$$

where [M] is the system mass matrix, [K] is the system stiffness matrix, and λ_i is the squared natural frequencies corresponding to the modal vector $\{\psi_i\}$.

The eigenvalue problem of the damaged structure is formulated by first replacing the pre-damaged eigenvectors and eigenvalues with a set of post-damaged modal parameters (indicated by an asterisk) and, second, subtracting the perturbations in the mass and stiffness matrices caused by damage from the original matrices. Letting ΔM_d and ΔK_d

represent the perturbations to the original mass and stiffness matrices, two forms of a damage vector, $\{D_i\}$, for the ith mode are then obtained by separating the terms containing the original matrices from those containing the perturbation matrices. Hence,

$$\{D_i\} = \left(\lambda_i^*[M]+[K]\right)\{\psi_i\}^* = \left(\lambda_i^*[\Delta M_d]+[\Delta K_d]\right)\{\psi_i\}^*. \tag{7}$$

To simplify the investigation, damage is considered to alter only the stiffness of the structure (i.e. $[\Delta M_d] = [0]$). Therefore, the damage vector reduces to

$$\{D_i\} = [\Delta K_d]\{\psi_i\}^*. \tag{8}$$

In a similar manner as the modal-based flexibility matrices previously defined, the approximations to the stiffness matrices, before and after damage, can be used to determine $[\Delta K_d]$ as

$$[\Delta K_d] = [K]-[K]^* \approx \sum_{i=1}^{n}\omega_i^2\phi_i\phi_i^T - \sum_{i=1}^{n}\omega_i^{*2}\phi_i^*\phi_i^{*T}. \tag{9}$$

A scaling procedure discussed by these authors was used to avoid spurious readings at stiff locations near supports where the signal- to-noise ratio of the measured responses is lower.

Modifications made to the methods to facilitate direct comparison. The primary modification made to these methods was the adaptation of the cubic polynomial interpolation scheme to approximate mode shape amplitudes at locations between sensors as suggested by Stubbs and Kim (1994). This interpolation effectively introduces artificial degrees of freedom into the experimental measurements. Also, the cubic polynomial can be directly differentiated to obtain the needed mode shape curvatures thus avoiding the finite difference scheme for evaluating the mode shape derivatives suggested by some authors.

Two of the damage detection methods require only consistently normalized mode shapes, namely the Damage Index Method and the Mode Shape Curvature Method. The Change in Flexibility Method, the Change in Uniform Load Surface Curvature Method, and the Change in Stiffness Method require the resonant frequency for each mode and mass-normalized mode shape vectors. When the methods that require unit-mass normalized mode shapes were applied to the mode shape data obtained with the SET2 instruments, the mass along the length of the beam was considered constant and these mode shapes were normalized using an identity matrix. Because these damage detection methods are only concerned with changes in flexibility or stiffness matrices rather than their absolute values, it was assumed that this method of mode shape normalization would not introduce significant errors into the damage detection process.

Tables 3 and 4 summarize the results from applying the five damage detection algorithms to the experimental modal data from the SET1 and SET2 instruments, respectively. In this study, the Damage Index Method was found to have performed the best. All methods were able to definitively locate the damage for the final case, E-4. For the intermediate damage cases mixed results were obtained.

Table 3 Summary of damage detection results using experimental modal data (SET1)

Damage ID Method	Damage Cases			
	E-1	E-2	E-3	E-4
Damage Index Method	**	**	**	*
Mode Shape Curvature Method	**	*	**	*
Change in Flexibility Method	O	O	**	*
Change in Uniform Load Surface Curvature Method	O	O	O	*
Change in Stiffness Method	**	**	**	*
* Damage located, ** Damage located using only 2 modes, O Damage not located				

Table 4 Summary of damage detection results using experimental modal data (SET2)

Damage ID Method	Damage Cases			
	E-1	E-2	E-3	E-4
Damage Index Method	●	●	●	●
Mode Shape Curvature Method	● ● ●	● ●	●	●
Change in Flexibility Method	O	O	O	●
Change in Uniform Load Surface Curvature Method	O	● ● ●	●	●
Change in Stiffness Method	O	O	O	●
● Damage located, ● ● Damage narrowed down to two locations, ● ● ● Damage narrowed down to three locations, O Damage not located				

Model updating study. Simmermacher (1996) applied a finite element model updating procedure to the I-40 Bridge data. The damage detection process essentially parallels that of the change in stiffness method previously discussed. However, the stiffness matrix was obtained from a finite element model that had been correlated with the measured modal parameters. Three correlation procedures were used. The minimum rank perturbation method (MRPT) was applied to update only the stiffness matrix and then to update both the mass and stiffness matrix. Next, a method for updating the stiffness matrix proposed by Baruch and Bar Itzhack (1978) was also investigated. Two different finite element models of different complexity were studied. Both models required considerable degree of freedom (DOF) reduction so that the model DOF corresponded to the measurement DOF. Four different reduction procedures were studied and that proposed by Guyan (1965) was found to produce the best results.

For a model with 78 DOF all methods were able to locate damage for Case E-4. The Baruch method was able to locate damage for case E-3. Neither method could locate the damage for case E-2. The authors point out that for this model the Baruch method has the advantage that it produces minimal changes to the stiffness matrix such that the model agrees with the test data. The MRPT methods do not preserve the proper physical stiffness distribution in the updated model. For the model with 1422 DOF, which required 99% of the DOF to be removed in the reduction process, the MRPT method could locate the damage for Case E-4. The Baruch method did not locate any damage cases properly. In this case the original model was so poor that large changes to the stiffness matrix were required. The Baruch method did not adjust the model sufficiently to account for these necessary changes.

Finally, a study of the effects of noise on the damage detection process was undertaken by adding noise to the modal vectors. The sensitivity of the various measurement locations to damage was studied. It was found that the locations at the center of the middle span were the most sensitive to noise and one of these locations correspond to the point where damage was introduced. Subsequent applications of the MRPT method using the noisy modes revealed that the addition of noise actually enhanced the damage detection process because the damage was located at a point very susceptible to the influence of noise.

4.6. STATISTICAL MODEL BUILDING

Exclusive of the procedure used in the damage index method to identify the locations that show the largest change in the damage index statistical models were not employed in this study. The distinct need for statistical analysis of the data and quantification of the environmental effects on the measured modal properties is evident when one considers the small changes in measured dynamic response that are being examined. The topic of variability in modal properties and statistical analysis of the I-40 Bridge data are discussed in more detail in (Doebling and Farrar, 1998) and (Doebling and Farrar, 1997). Performing statistical analysis of the dynamic data is imperative if one is to establish that changes brought about by damage are greater than the test-to-test repeatability. Methods for performing such statistical analyses are summarized in (Doebling, Farrar, and Goodman, 1997) and (Farrar, Doebling and Cornwell, 1998)

Although the number of papers reporting vibration-based damage detection results from a variety of structures has greatly increased in recent years, very few of the articles examine the variability in the modal properties that can arise from changes in environmental conditions or from random and systematic errors inherent in the data acquisition/data reduction process. A thorough study of the modal parameter variability must be conducted before vibration-based damage id algorithms can be applied with confidence.

4.7. IN HIND SIGHT, THINGS THAT SHOULD HAVE BEEN DONE DIFFERENT

Based on subsequent analysis and observations related to the I-40 bridge tests (Doebling and Farrar, 1997 and 1998), subsequent tests on another bridge (Doebling Farrar and Goodman, 1997) and (Farrar Doebling and Cornwell, 1998), interactions with other researchers in the field (particularly those at the Univ. of Cincinnati and Drexel Univ.), and review of the technical literature related to bridge testing, there are several things that should have been done during these tests (and have been done on subsequent tests) to improve the confidence in the damage ID results. These improvements are listed below.

4.7.1. *Perform a more thorough pre-test visual inspection*
In addition to vibration tests visual inspection is needed to ascertain the initial condition of the structure. Particular attention should be paid to boundary conditions and changes to the neighboring vicinity of the test structure. During the tests visual inspection revealed that the east end of the top portion of the south girder were not in contact with the concrete at the top of the abutment, but very close to it. During earlier tests observations

of the end condition were not made and it was speculated that during cooler weather the girders could have actually been in contact with the abutment.

4.7.2. *Perform linearity and reciprocity checks*

The system identification portion of the experimental modal analysis procedure typically relies on the assumption that the structure is linear. Linearity can be checked, to some degree, by exciting the structure at different levels and overlaying the measured FRFs for a particular point. Ideally, with the thought of an on-line monitoring system in mind, these different excitation levels would span the range of loading observed during ambient traffic vibration measurements. Also, a change in the linearity properties can in itself be an indication of damage. In addition to the assumption of linearity, the system identification portion of the experimental modal analysis typically relies on the assumption that the structure will exhibit reciprocity. Performing a reciprocity check is much more involved when a large shaker is being used because of the setup time involved in relocating the shaker. Also, to check the reciprocity of the structure alone, one must relocate the accelerometers and cables as well as the shaker (Farrar, Doebling, Cornwell and Straser, 1997). Without moving the instrumentation, the reciprocity check will involve reciprocity of the electronics as well as that of the structure. Because, in general, the electronics will not be moved once the test has started, the latter test is more representative of the reciprocity of the system.

4.7.3. *Perform as many environmental and testing procedure sensitivity studies as possible*

Sensitivity of modal test results to environmental conditions and test procedures should be quantified to the extent possible. Subsequent tests (after the potential damage has occurred) should be performed under similar environmental conditions using similar test procedures, if possible. Figure 9 shows the change in modal frequencies measured on another bridge as a function of the temperature differential across the deck (Farrar, Doebling, Cornwell and Straser, 1997). Changes in the first mode frequency of five-percent are noted over a 24 Hr time span. These changes are comparable to the changes in resonant frequency resulting from the final level of damage in the I-40 Bridge tests. Also, a baseline noise measurement should be made for the data acquisition system.

4.7.4. *Perform false-positive studies*

As a means of quantifying the effects observed in the sensitivity studies, sets of data from the undamaged structure should be analyzed with the damage ID algorithm to demonstrate that the algorithm will not falsely predict damage when in fact none has occurred.

4.8. WHAT CAN BE DONE WITH ONLY INITIAL MEASUREMENTS AND FEM?

Once a finite element analysis has been benchmarked or correlated against the measure modal properties simulated damage scenarios can be introduced into the model and either an eigenvalue analysis can be performed or, to better simulate an actual modal test, a time-history analysis can be performed. Mode shape data can then be obtained from either type of analysis and the various damage ID methods can be applied to the observed changes in the modal properties. If a statistical analysis has been applied to the measured modal parameters of the baseline or undamaged structure, then it can be established that

368

the changes in the monitored modal properties such as mode shape curvature resulting from the simulated damage are greater than the variations that can be attributed to experimental repeatability. In addition, the statistical variations calculated for the measured modal properties on the undamaged structure can be assumed to apply to the numerical results from the damaged structure. The use of statistical variations measured on the undamaged structure and assumed for the numerical simulation of the damaged structure can then be used to establish the threshold damage level that can be reliably detected.

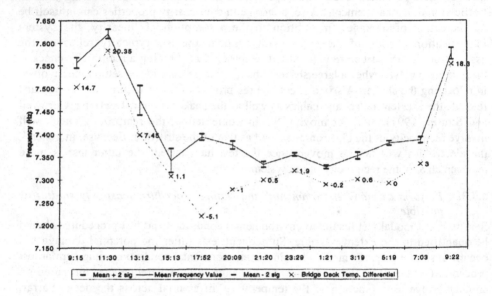

Figure 9 Change in the first mode frequency during a 24 hr time period

5. Application of vibration-based damage detection to concrete bridge columns

A damage detection study was conducted on two concrete columns that were quasi-statically loaded to failure in an incremental manner. The focus of this study was to establish a relatively simple feature vector coupled with a simple statistical model that would unambiguously identify that the columns had been damaged.

5.1. TEST STRUCTURE GEOMETRY

The test structures consisted of two 61-cm-dia (24-in-dia) concrete bridge columns that were subsequently retrofitted to 91-cm-dia (36-in-dia) columns. Figure 10 shows the test structure geometry. The first column tested, labeled Column 3, was retrofitted by placing forms around the existing column and placing additional concrete within the form. The second column, labeled Column 2, was extended to the 91-cm-diameter by spraying concrete in a process referred to as shotcreting. Column 2 was then finished with a trowel to obtain the circular cross-section.

The 91-cm-dia. portion of both columns were 345 cm (136 in) in length. The columns were cast on top of a 142-cm-sq. (56-in-sq.) concrete foundation that was 63.5-cm (25-in) high. A 61-cm-sq. concrete block that had been cast integrally with the column extends 46-cm (18-in.) above the top of the 91-cm-dia. portion of the column. This block was used to attach the hydraulic actuator to the columns for quasi-static cyclic testing and to attach the electro-magnetic shaker used for the experimental modal analyses. As is typical of actual retrofits in the field, a 3.8-cm-gap (1.5-in-gap) was left between the top of the foundation and the bottom of retrofit jacket. Therefore, the longitudinal reinforcement in the retrofitted portion of the column did not extend into the foundation. The concrete foundation was bolted to the 0.6-m-thick (2-ft-thick) testing floor in the University of California-Irvine structural-testing laboratory during both the static cyclic tests and the experimental modal analyses. The structures were not moved once testing was initiated.

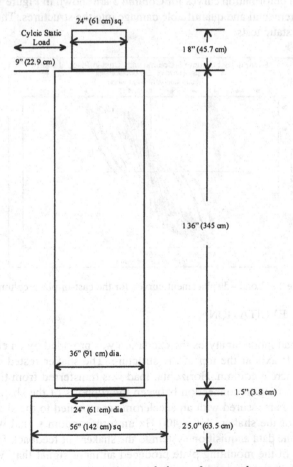

Figure 10 Column Dimensions and photo of an actual test structure.

5.2. QUASI-STATIC LOADING

Prior to applying lateral loads, an axial load of 400 kN (90 kips)was applied to simulate gravitational loads that an actual column would experience. Next, a hydraulic actuator was used to apply lateral load to the top of the column in a cyclic manner. The loads were first applied in a force-controlled manner to produce lateral deformations at the top of the column corresponding to $0.25\Delta y_T$, $0.5\Delta y_T$, $0.75\Delta y_T$ and Δy_T. Here Δy_T is the lateral deformation at the top of the column corresponding to the theoretical first yield of the longitudinal reinforcement. The structure was cycled three times at each of these load levels.

Based on the observed response, a lateral deformation corresponding to the actual first yield, Δy, was calculated and the structure was cycled three times in a displacement-controlled manner to that deformation level. Next, the loading was applied in a displacement-controlled manner, again in sets of three cycles, at displacements corresponding to $1.5\Delta y$, $2.0\Delta y$, $2.5\Delta y$, etc. until the ultimate capacity of the column was reached. Load deformation curves for Column 3 are shown in Figure 11. This manner of loading put incremental and quantifiable damage into the structures. The axial load was applied during all static tests.

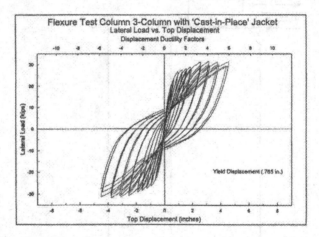

Figure 11 Load –displacement curves for the cast-in-place column.

5.3. DYNAMIC EXCITATION

For the experimental modal analyses the excitation was provided by an electro-magnetic shaker mounted off-axis at the top of the structure. The shaker rested on a steel plate attached to the concrete column. Horizontal load was transferred from the shaker to the structure through a friction connection between the supports of the shaker and the steel plate. This force was measured with an accelerometer mounted to the sliding mass (0.18 lb-s^2/in (31 Kg)) of the shaker. A 0 - 400 Hz uniform random signal was sent from a source module in the data acquisition system to the shaker but feedback from the column and the dynamics of the mounting plate produced an input signal that was not uniform over the specified frequency range.

5.4.   OPERATIONAL EVALUATION

Because the structure being tested was a laboratory specimen, operational evaluation was not conducted in a manner that would typically be applied to an *in situ* structure. However, the vibration tests were not the primary purpose of this investigation. Therefore, compromises had to be made regarding the manner in which the vibration tests were conducted. The primary compromise was associated with the mounting of the shaker. These compromises are analogous to operational constraints that may occur with *in situ* structures. Environmental variability was not considered an issue because these tests were conducted in a laboratory setting. The available measurement hardware and software placed the only constraints on the data acquisition process.

5. 5.   DATA ACQUISITION AND CLEANSING

Forty accelerometers were mounted on the structure as shown in Figure 12. These locations were selected based on the initial desire to measure the global bending, axial and torsional modes of the column. Note that the accelerometers at locations 2, 39 and 40 had a nominal sensitivity of 10 mV/g and were not sensitive enough for the measurements being made. As part of the data cleansing process, data from these channels were not used in subsequent portions of the damage detection process. Locations 33, 34, 35, 36, and 37 were accelerometers with a nominal sensitivity of 100 mV/g. All other channels had accelerometers with a nominal sensitivity of 1 V/g.

A commercial data acquisition system was used to record and digitize all accelerometer signals. Data acquisition parameters were specified to obtain 8-s-duration time-histories discretized with 8192 points. Only one average was measured. A uniform window was specified for these data, as the intent was to measure a time history only.

5. 6.   FEATURE SELECTION

Typically, systematic differences between time series from the undamaged and damaged structures are nearly impossible to detect by eye. Therefore, other features of the measured data must be examined for damage detection. Originally, damage detection features were to be based on common modal properties as have been done in many previous studies. However, the feedback from the structure and mounting system to the shaker produced an input that did not have a uniform power spectrum over the frequency range of interest as previously discussed. This input form coupled with the nonlinear response observed at higher levels of damage made it extremely difficult to track changing modal properties through the various levels of damage. Therefore, other features were sought for the damage detection process.

The alternate features were selected based on previous experience from speech pattern recognition where auto- regressive models have been used to estimate the transfer function of the human vocal track (Morgan and Scofield, 1992). The time series were modeled using a common method of auto-regressive estimation also referred to as Linear Predictive Coding (LPC) (Rabiner and Shafe, 1978). The LPC algorithm is an Nth-order model that attempts to model the current point in a time series, $x'(t_n)$, as a linear combination of the previous N points. That is

$$x'(t_n) = \sum_{i=1}^{N} a_i x(t_{n-i}) \qquad (10)$$

Third-order LPC models were developed for each column using 512-point windows with 97% overlap resulting in 480 samples of the a_i's. Over these segments of the time series the a_i's that best model the time series in a least squares sense are used as features that are assumed to be representative of the system's dynamic response during those samples. Hanning windows were applied to these data prior to the estimate of the

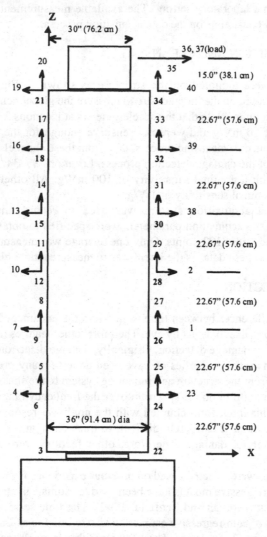

Figure 12 Accelerometer locations and coordinate system for modal testing. Accelerometers 3, 6, 9, 12, 15, 18, 21, 22, 24, 26, 28, 30, 32, and 34 are mounted in the –y direction.

coefficients. These models were developed with data from sensors 3 and 21 (Figure 12). Sensor 3 was located close to the damage, but because of the test configuration this sensor was not expected to experience large amplitude response as it primarily measures torsional motion of the structure near its fixed end. Sensor 21 was located farther from the damage and experienced some of the largest amplitude response as it primarily measured the bending response at the free end of this cantilever structure.

Over a time series, many overlapping "windows" give rise to LPC coefficient vectors, which become the multi-dimensional data samples to be analyzed in the statistical model development portion of the damage detection process. While the overlapping of windows provides a smoother estimate of the features' changes over time, samples that result from overlapping windows will not be independent.

Normalization of the data was not attempted because these tests were conducted in a laboratory environment where the input could be applied in a very controlled manner. Other considerations that led to the decision not to normalize the data included the consideration that environmental and test-to-test variability was negligible, damage was introduced in discrete increments, and it was assumed that the vibration levels were such that the physical condition of the test structures did not change during the dynamic tests.

5. 7. STATISTICAL MODEL DEVELOPMENT: FISHER'S DISCRIMINANT

Consider two data generation processes A and B, with independent multi-dimensional samples $\{x\}$ being generated by both processes. Assuming A and B have some systematic difference in the samples that they generate, Fisher's discriminant (Fisher, 1936, and Bishop, 1995) represents the optimal linear projection of the multidimensional sample space that maximally discriminates the $\{xA\}$'s from the $\{xB\}$'s. That is, it defines a linear projection $\{w\}$ such that

$$y = \{w\}^T \{x\} \tag{11}$$

produces a scalar projection, y, of the multidimensional space onto which the distribution of $\{x_A\}$'s is as distinct as possible from the distribution of $\{x_B\}$'s. Once this projection is determined from previous samples of $\{x_A\}$'s and $\{x_B\}$'s it can be used to provide the relative probability that a novel sample $\{x\}$ was generated by process A or B. Thus, the Fisher discriminant maximizes the function $F(\{w\})$, which is the distance between the means of the transformed distributions, μ_i, normalized by the total within-class covariance, s_k^2:

$$F(\{w\}) = \frac{(\mu_A - \mu_B)^2}{s_A^2 + s_B^2}, \tag{12}$$

Equation 12 can be rewritten explicitly in terms of $\{w\}$ as

$$F(\{w\}) = \frac{\{w\}^T [S_b] \{w\}}{\{w\}^T [S_w] \{w\}}, \tag{13}$$

where $[S_b]$ is the between-class covariance matrix, and $[S_w]$ is the total within-class covariance matrix.

374

Once the data have been projected down onto the scalar y dimension, the distribution of y_A and y_B points can be described by an appropriate probability density function. Since it was originally assumed that $\{x\}$ was a multi-dimensional random variable, then $y = \{w\}^T \{x\}$ is a sum of random variables and the central limit theorem is invoked to justify modeling y_A and y_B with Gaussian density functions. Novel data $\{x_{new}\}$ can be projected to get $y_{new} = \{w\}^T \{x_{new}\}$ and the likelihood, p, of y_{new} with respect to the Gaussian for class A and the Gaussian for class B can be determined.

5.8. APPLICATION OF FISHER'S DISCRIMINANT TO CONCRETE COLUMN DATA

Fisher's discriminant was defined using data from the vibration tests conducted on the undamaged columns and from the vibration tests conducted after the first level of damage corresponding to initial yielding of the steel reinforcement. Subsequent damage levels were then identified based on this same Fisher projection. As illustrated in Figure 13, when Fisher's discriminant is applied to data from both sensors on either column, there is statistically significant separation between the LPC coefficients for the undamaged cases and damage level 1 cases (solid and dashed Gaussian density functions). Also plotted as straight lines are the results of using the previously determined Fisher projection to project many samples of data from increasingly greater levels of damage into this space. While increasing damage is not necessarily related to increasing Fisher coordinate, all

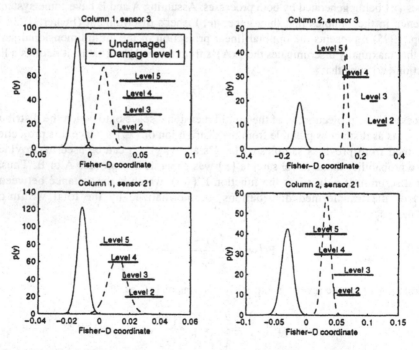

Figure 13 Distribution of LPC-generated feature vectors projected onto the Fisher coordinate. Solid horizontal lines represent widths of the distributions for higher damage levels.

damaged cases have a profile significantly different from that of the undamaged case. The discrimination between damaged and undamaged structures is obtained with both with data from both sensors. This result is significant because the response measured by Sensor 3 was of relatively low amplitude with noise contributing significantly to the measured signal . Higher-order LPC models and different size data windows produced similar results.

6. Concluding comments

Recent work in structural health monitoring and vibration-based damage detection has been briefly reviewed to show that this subject is the focus of many active research efforts and to identify some of the technical challenges in this field. A major shortcoming associated with many of these efforts is that statistical models are not applied to identify when changes in the selected features are significant. Therefore, a statistical-pattern-recognition paradigm has been proposed for the general problem of structural health monitoring. This paradigm breaks the process of structural health monitoring into the four tasks of operational evaluation, data acquisition and cleansing, feature selection, and statistical model development. Structural damage detection studies of the I-40 Bridge and of concrete columns subjected to quasi-static cyclic loading to failure were then posed in terms of this paradigm.

The application of linear damage identification methods using experimental modal data gathered from the I-40 Bridge over the Rio Grande in Albuquerque, NM was first summarized. In this study linear damage identification implies that linear dynamic models were used to model the structure both before and after damage. The nature of the damage applied to the I-40 Bridge was such that the linear damage models are applicable to these damage scenarios.

Examination of results from the experimental modal analyses verify other investigators findings that standard modal properties such as resonant frequencies and mode shapes are poor indicators of damage. The more sophisticated damage detection methods investigated herein showed improved abilities to detect and locate the damage. In general, all methods investigated in this study identified the damage location correctly for the most severe damage case; a cut through more than half the web and completely through the bottom flange. However, for several of these methods, if they had been applied blindly, it would be difficult to tell if damage had not also occurred at locations other than the actual one. The methods were inconsistent and did not clearly identify the damage location when they were applied to the less severe damage cases. The authors feel that the summary of this application to a large, *in situ* structure demonstrates the positive aspects as well as drawbacks of this emerging technology.

The I-40 Bridge tests highlight the fact that damage typically is a local phenomenon. Local response is captured by higher frequency modes whereas lower frequency modes tend to capture the global response of the structure and are less sensitive to local changes in a structure. From a testing standpoint it is more difficult to excite the higher frequency response of a structure as more energy is required to produce measurable response at these higher frequencies than at the lower frequencies. These factors coupled with the loss of information resulting from the necessary reduction of time-history measurements to modal properties add difficulties to the process of damage identification based on standard modal properties. These factors contribute to the current state where this technology is still in the research arena with only limited applications to large civil engineering infrastructure.

Based on further analysis and observations related to the I-40 bridge tests, subsequent tests on another bridge, interactions with other researchers performing similar tests, and review of the technical literature related to bridge testing, there are several things that should have been done during these tests to improve the confidence in the damage ID results. These improvements include detailed visual inspection of the bridge, performing linearity checks, performing reciprocity checks, performing false-positive studies, performing test condition sensitivity studies, and performing statistical analyses of the measured modal properties.

The results of a damage detection study applied to reinforce concrete bridge piers were then summarized. This study attempted to identify the relatively simple features of the measured data that were sensitive to damage. Criteria for selecting the features were to keep the dimension of the feature vector small and have the number of samples of the vector large. The feature vectors used were the coefficients of a third-order linear predictive coding model. A well-developed procedure for group classification, the linear discriminant operator referred to as "Fisher's Discriminant", was introduced for application this vibration-based damage detection problem. This procedure requires data to be available from both the undamaged and damaged structures.

The results of this study indicate a strong potential for using linear discriminant operators to identify the presence of damage. An attractive attribute of this statistical model is that it was applied to features obtained from response data only implying that is appropriate for structures subjected to ambient vibration from sources such as traffic or wind excitation.

The results of this study also suggest that if one or more common forms of damage occur, it may be possible to not only determine that a system is damaged but to determine which form of damage has occurred. Additional data is required to explore this possibility. Another attractive feature of the linear discriminant operator that was not fully explored during this investigation is its ability to combine data from various types of sensors. This feature will become particularly attractive when monitoring structures that experience significant variations in their dynamic response resulting from changing environmental and operating conditions. Further analyses are also required to demonstrate the ability of the linear discriminant operator to avoid false-positive indications of damage. However, multiple samples of data from the undamaged columns were not measured.

To advance the state of the art in vibration-based damage detection it is the authors' opinion that developments of non-model based pattern recognition methods will be needed to supplement the existing model-based techniques. It is anticipated that such methods will be particularly effective when analyzing a structure where the damage changes the structure from a predominantly linear system to a predominantly nonlinear system.

Data from the I-40 Bridge study and the study of the concrete columns can be downloaded from Los Alamos National Laboratory's Damage ID Web Page: http://ext.lanl.gov/projects/damage_id.

Acknowledgements

Funding for this research was provided by the Department of Energy through the Los Alamos National Laboratory's (LANL) Laboratory Directed Research and Development program, the Federal Highway Administration, LANL's University of California Interaction Office and the Department of Energy's Enhanced Surveillance Program. The work summarized here represents efforts by staff, undergraduate and graduate and post-doctoral research associates too numerous to name explicitly. With respect to the I-40 bridge project, the authors would like to acknowledge the cooperation and teamwork that was exhibited by all parties involved in these tests including engineers from Sandia National Laboratory; faculty, technicians and students from New Mexico State University; numerous people at the New Mexico State Highway and Transportation Department; and the staff at the Alliance for Transportation Research. The authors would like to express their appreciation to Mr. Tim Leary at CALTRANS and Prof. Gerard Pardoen at the University of California- Irvine who allowed us to use his test structures for the concrete column investigation..

7. References

1. Baruch, M. and Bar Itzhack, I. Y. (1978), Optimum Weighted Orthogonalization of Measured Modes, *AIAA Journal, 16(4)*, 346 –351.
2. Bendat, J. S. and Piersol, A. G. (1980), Engineering Applications of Correlation and Spectral Analysis, *John Wiley*, New York.
3. Bishop, C. M. (1995), Neural Networks for Pattern Recognition, *Oxford University Press*, Oxford, UK.
4. Doebling, S. W. and Farrar, C. R. (1998), Statistical damage identification techniques applied to the I-40 bridge over the Rio Grande, *Proc. 16th International Modal Analysis Conf.*, Santa Barbara, CA.
5. Doebling, S. W, and Farrar, C. R. (1997), Using statistical analysis to enhance modal-based damage identification, *Proc. of DAMAS 97 Conference*, Sheffield, UK.
6. Doebling, S. W., Farrar, C. R. and Goodman, R. (1997), Effects of measurement statistics on the detection of damage in the Alamosa Canyon Bridge, *Proc. 15th International Modal Analysis Conference*, Orlando, FL, February.
7. Doebling, S. W., Farrar, C. R., Prime, M B. and D W. Shevitz (1996), Damage identification and health monitoring of structural and mechanical systems from changes in their vibration characteristics: a literature review, *Los Alamos National Laboratory report LA-13070-MS*.
8. Doherty, J. E. (1987), Nondestructive Evaluation, in A. S. Kobayashi (ed.), *Handbook on Experimental Mechanics, Society for Experimental Mechanics, Chapter 12*.
9. Ewins, D. J. (1995), Modal Testing: Theory and Practice, *John Wiley and Sons, Inc.*, NY.
10. Farrar, C. R., Baker, W. E., Bell, T. M. , Cone, K. M., Darling, T. W., Duffey, T. W., Eklund, A. and Migliori, A. (1994), Dynamic characterization and damage detection in the I-40 bridge over the Rio Grande, *Los Alamos National Laboratory report LA-12767-MS*.
11. Farrar, C. R., Doebling, S. W. and Cornwell, P. J. (1998), A comparison of modal confidence interval using the Monte Carlo and bootstrap techniques, *Proc. 16th International Modal Analysis Conf.*, Santa Barbara, CA.
12. Farrar, C. R., Doebling, S. W., Cornwell, P. J. and E. G. Straser, (1997), Variability of modal parameters measured on the Alamosa Canyon Bridge, *Proc. 15th International Modal Analysis Conference*, Orlando, FL.
13. Farrar, C. R. and Jauregui, D. (1996), Damage detection algorithms applied to experimental and numerical modal data from the I-40 bridge, *Los Alamos National Laboratory report LA-13074-MS*.
14. Fisher, R.A. (1936), The use of multiple measurements in taxonomic problems, *Ann. Eugenics, Vol. 7*, Part II, 179-188.

378

15. Friswell, M. I. and Mottershead, J. E. (1995), Finite Element Modal Updating in Structural Dynamics, *Kluwer Academic Publishers*, Dordrecht, The Netherlands.
16. Guyan, R. J. (1965), Reduction of stiffness and mass matrices, *AIAA Journal, 3(2)*, 380-386.
17. Hunter, N. F. (1999), Bilinear system characterization from nonlinear time series analysis, *Proc. 17th International Modal Analysis Conf.*, Orlando, FL.
18. Maia, N. M. M., Silva, J. M. M. *et al* (1997), Theoretical and Experimental Modal Analysis, *John Wiley and Sons, Inc.*, NY.
19. Mayes, R. L. and Nusser, M. A. (1994), The Interstate-40 bridge shaker project, *Sandia National Laboratory report SAND94-0228*.
20. McConnell, K. G. (1995), Vibration Testing Theory and Practice, *John Wiley and Sons, Inc.*, NY.
21. Morgan, D. P. and Scofield, C. L. (1992), Neural Networks and Speech Pattern Processing, *Kluwer Academic Publishers*, Boston, MA.
22. Pandey, A. K. and Biswas, M. (1994), Damage detection in structures using changes in flexibility, *Journal of Sound and Vibration, 169(1)*, 3-17.
23. Pandey, A. K., Biswas, M. and Samman, M. M. (1991), Damage detection from changes in curvature mode shapes, *Journal of Sound and Vibration, 145(2)*, 321-332.
24. Prime, M. B. and Shevitz, D. W. (1996), Linear and nonlinear methods for detecting cracks in beams, *Proc. 14th International Modal Analysis Conf.*, Dearborn, MI.
25. Rabiner, L.P. and Shafer, R.W. (1978), Digital Processing of Speech Signals, *Prentice-Hall, Inc.*, Englewood Cliffs, NJ.
26. Rytter, A. (1993), Vibration based inspection of civil engineering structures, *Ph.D. Dissertation*, Dept. of Building Technology and Structural Eng., Aalborg Univ., Denmark.
27. Simmermacher, T. (1996), Damage detection and model refinement of coupled structural systems, *Ph.D. Dissertation*, University of Houston, Houston, TX.
28. Straser, E. G. (1998), A modular, wireless damage monitoring system for structures, *Ph.D. Dissertation*, Dept. of Civil Eng., Stanford Univ., Palo Alto, CA.
29. Stubbs, N., Kim, J.-T. and Farrar, C. R. (1995), Field verification of a nondestructive damage localization and severity estimation algorithm, *Proc. 13th International Modal Analysis Conference, Nashville, TN*.
30. Zhang, Z. and Aktan, A. E. (1995), The damage indices for the constructed facilities, *Proc. of the 13th International Modal Analysis Conference, Vol. 2, Nashville, TN*, 1520-1529.
31. Zimmerman, D. C. and Kaouk, M. (1994), Structural damage detection using a minimum rank update theory, *Journal of Vibration and Acoustics, 116*, 222-231.

STRUCTURAL MODIFICATION

J. HE
Victoria University of Technology
Melbourne, Australia

1. Introduction

Dynamic characteristics of a structure are usually referred to as its natural frequencies and mode shapes. The ability to alter these characteristics in order to have desired dynamic characteristics for a structure either by design, re-design or by modification has been an enduring quest by structural analysts. The need to change dynamic characteristics of a structure may come from new design requirements, solution for excessive vibration, or the necessity to control the response of the structure. It is unrealistic and unnecessary to attempt to alter all natural frequencies or mode shapes of a structure but selected changes via modification are possible. In many applications, it is often a part of the dynamic characteristics of a structure such as a particular natural frequency that needs to be changed.

Theoretically, any mass and stiffness property changes should result in changes on all natural frequencies and mode shapes. The mass and stiffness changes can be realised by changing physical parameters such as the thickness of a plate. Therefore, changes in a selected vibration mode can be achieved notionally by changes of any mass or stiffness properties. However, analysis is needed to determine where the mass or stiffness changes are most effective and by how much the changes should be made.

Structural modification is a technique that determines the relationship between the mass and stiffness changes of a structural system and its dynamic characteristic changes. Dynamic properties of a structural system are determined by the distribution of its mass, stiffness and damping properties. Therefore, it is only through the modification of these properties that can improve dynamic properties of the system to be attained. Structural modification usually constitutes two opposing approaches: (1) given prescribed dynamic characteristics such as a new natural frequency, determine the location and extent of structural changes that will best accomplish the prescribed characteristics; and (2) for suggested structural modifications, determine what dynamic characteristic changes will occur. The first is an inverse problem with no guaranteed unique solution. The second, also known as re-analysis, is a direct problem that requires mainly mathematical and numerical effort and the solution is usually unique.

Many structural modification problems in practice have a very simple objective, i.e. to overcome a problem of excessive vibration caused by resonance(s). If a natural frequency of a structure coincides with the frequency of excitation forces, excessive

379

J.M.M. Silva and N.M.M. Maia (eds.), Modal Analysis and Testing, 379–394.
© 1999 *Kluwer Academic Publishers. Printed in the Netherlands.*

vibration occurs. Changing excitation forces is not often a practical solution. This leaves modifying the structure to optimise its dynamic characteristics as the only avenue for remedying the vibration problem.

For structural modification, it is unrealistic to assume that every part of a structure can be modified and to any extent. There are practical restrictions when applying modification. Realistically, modification can only be carried out on a limited number of parameters of an existing structure. When using structural modification on design, it is possible to alter more parameters in order to meet requirements on dynamic characteristics, provided other design requirements are not violated.

Many structural modification methods were developed based on a simple mass and spring system. Such a system is physically explicit (mass and stiffness matrices) and analytically easy to solve. It permits separate mass and stiffness changes. This allows greater flexibility in implementing structural modification. Methods developed using such a simple system have to be applied to a real structure in order to fully assess their feasibility.

Structural modification based on altering the mass, stiffness or damping distributions of a system does not change its total degrees of freedom (DoFs). However, if one co-ordinate on the system is linked to the ground via a spring or two unconnected co-ordinates are linked together via a spring, the physical connectivity of the structure will have changed. In addition, structural modification can be realised by adding a substructure to the original structure. Thus, the modified structure will have increased DoFs. An example of this type of modification is the use of a mass damper on a structure. In this paper, the structural modification without altering connectivity and total DoFs will be the main subject of discussion.

2. Structural modification - Existing methods

Structural modification evolved with the development of structural dynamics. The first meaningful formulation of structural modification was given by Rayleigh [1945] who used perturbation approach to derive an approximate solution in terms of modal co-ordinates. Rayleigh found that an increase in the mass of any part of a vibrating system is attended by a prolongation of all the natural periods. His approach was later expanded by researchers to broaden the formulation and to include second order approximation. The practical application of perturbation approach was treated by Stetson and others [1976] and Sandstrom and Anderson [1982]. Their work allows specified constraints on frequencies and mode shapes and links physical parameter changes of a structure such as cross sectional area of a beam element to its modal properties. The approximation nature of perturbation approach hinders its application in large magnitude of modification.

Historical development of local structural modification in terms of dynamic absorbers can also be traced back decades ago. The work by Den Hartog and Timoshenko in this area has been frequently referred to by researchers. Up to date, many dynamic absorbers have been studied [Hunt 1979][Wager etc 1973]. A majority of these absorbers involve adding extra DoFs to a structure for dissipating energy rather than modifying its existing mass and stiffness distributions. They can be seen as a unique type of structural modification but do not fall into the scope of this paper.

A significant early study on local structural modification from the mass and stiffness characteristics of a structure was carried out by Weissenburger [1968]. He formulated the relationship between a simple lumped mass and stiffness alteration of an undamped linear dynamic system and its dynamic characteristic changes. This work was later expanded by Pomazal and Snyder [1971] who analysed the effects of adding springs and dampers to a viscously damped linear system. Their analysis included a transformation of second order equations into a set of first order equations. The complete solution of the original system is a pre-requisite of the analysis which, in today's analysis, appears to be over-demanding. This work was later extended by Hallquist [1976] who generalised and improved the outcome obtained by Pomazal and Snyder [1971]. These works often treated structural modification from the viewpoint of a direct problem. Therefore, they are more conveniently used as the re-analysis tools for a modified system.

A notable advancement on structural modification was reported by Ram and Blech [1991] who theorised the effects of modification at one degree of freedom by different mass and stiffness attachments. They proofed that the consequence of connecting a vibratory system to the ground through a simple SDoF oscillator is to increase the natural frequencies of the system which are lower than the natural frequency of the oscillator and to decrease the natural frequencies which are above that. This conclusion can be applicable to continuous systems as well as to lumped systems. This provides interesting insights to the dynamic behaviour of a structure before and after a simple modification. Nevertheless, the findings are based on a re-analysis approach. They are not applicable to structural modification.

The merit of re-analysis in structural modification is to provide efficient and reliable analytical solution of the dynamic characteristics of the modified structure after structural modification has been determined. Many re-analysis methods have been proposed. They are focused on reducing the numerical efforts such as matrix inversion or full eigenvalue solution in deriving the characteristics of the modified structure [Herbert and Kientzy 1980] [Snyder 1986] [Wang etc 1994] [Zeng etc 1998]. By nature, they are direct analysis in contrast to structural modification as an inverse analysis. Re-analysis can also be an iterative analysis when the defined structural modification is treated as a perturbation on the original system. Re-analysis is particularly useful in design optimisation process. This perhaps explains why many literature for re-analysis appear on finite element related journals and conference proceedings.

Sensitivity analysis has been a useful tool for structural modification [Wang, et al 1987] [To and Ewins 1990] [Skingle and Ewins 1988]. The sensitivity of modal properties of a structure with respect to its physical properties have been developed by many researchers such as Wilkinson [1963] Rosenbrock [1965], Rudisill [1974], Rogers [1970] and Vanhonacker [1980] [Murthy and Haftka 1988] [Adelman and Haftka 1985] [Joseph 1992]. The focus of their study was to determine the derivatives of the eigenvalues and eigenvectors of a dynamic system with respect to system changes. Since the presumption for sensitivity is small structural changes, this approach may be useful for providing a general guideline for structural modification but not useful for implementing it. Iterative sensitivity analysis used for structural modification has also been proposed but the algorithm can be tedious and inaccurate.

Structural modification as an inverse problem renders a different approach from that for re-analysis and sensitivity analysis. Inverse eigenvalue problem is a structural modification that specifies one or more natural frequencies of a structure and use modal data [Josef 1992][Ram and Braun 1991]. A separate attempt was made by Bucher and Braun [1993]. The authors formulated solutions to reallocate eigenvalues and specify eigenvectors by computing necessary mass and stiffness modifications. The solutions require only a partial set of eigensolutions that can be derived from modal testing data. The applied modifications are constrained in a way to force the selected number of eigensolutions to reside in the known subspace spanned by the original modal vectors. This circumvents the problem arising from the truncation of the modal set.

A significant progress for structural modification has recently been made [Tsuei and Yee 1987] [Yee and Tsuei 1991] in which a method was proposed to determine required mass and stiffness changes that would relocate a natural frequency of a dynamic system. This method uses frequency response function (FRF) data of modification locations rather than modal data. Thus, the amount of raw data needed in the calculation is minimum. Because it does not require an eigenvalue solution, the mass and stiffness matrices of the original system are not needed in the analysis. This method was later extended by Li et al [1994,1] [1994,2] and He and Li [1995] to study the relocation of an anti-resonance of a system and ways of optimising the properties of a system. A certain type of structural modification such as cancellation of a resonance with an anti-resonance for a selected FRF becomes possible when using this method.

3. Preliminary equations for structural modification

A linear dynamic system consists of a mass matrix [M] and a stiffness matrix [K]. Structural modification involves the modal data, FRF data and spatial model of the dynamic system. An equation describing the link between the modal data and the spatial model of the unmodified system is:

$$([K] - \omega_r^2[M])\{\phi\}_r = \{0\} \qquad r = 1, 2, ..., N \qquad (1)$$

The link between the FRF data and the spatial model of the system is defined by the following equation:

$$([K] - \omega^2[M])^{-1} = [\alpha(\omega)] \qquad (2)$$

Mathematically, each individual FRF can be derived from:

$$\alpha_{ij}(\omega) = (-1)^{i+j} \frac{\det([K]_{ij} - \omega^2[M]_{ij})}{\det([K] - \omega^2[M])} \qquad (3)$$

Here, matrix $[K]_{ij}$ is obtained by deleting the 'i'th row and 'j'th column of matrix [K], and likewise for matrix $[M]_{ij}$. Both matrices are of order of (n-1) x (n-1) and are non-symmetric unless i = j. It can be assumed that there were a dynamic system that consists of mass matrix and stiffness matrix $[K]_{ij}$.

The mass and stiffness changes can be denoted as [ΔM] for mass modification matrix and [ΔK] for stiffness modification matrix. Matrices [ΔM] and [ΔK] are of the same order as that for [M] and [K]. Usually, structural modification respects the existing physical connectivity. This means that modification only occurs on the existing physical components of a structure and no new load paths will be created. If structural modification takes the form of mass and stiffness changes as [ΔM] and [ΔK], then the modal behaviour of the modified system is governed by

$$\left([K] + [\Delta K] - \omega_r^{*2}([M] + [\Delta M])\right)\{Y\} = \{0\} \qquad r = 1,\ 2,...,N \qquad (4)$$

It is reasonable to assume that modifications are local to a structure. This means the modification matrices [ΔM] and [ΔK] are either sparse or null matrices. They can be denoted as:

$$[\Delta K] = \begin{bmatrix} [0] & [0] & [0] \\ [0] & [\Delta K^R] & [0] \\ [0] & [0] & [0] \end{bmatrix} \quad \text{and} \quad [\Delta M] = \begin{bmatrix} [0] & [0] & [0] \\ [0] & [\Delta M^R] & [0] \\ [0] & [0] & [0] \end{bmatrix} \qquad (5,6)$$

Matrix [ΔM^R] is a square submatrix reduced from [ΔM] by retaining co-ordinates for modification. Likewise can be said about matrix [ΔK^R]. For a mass-stiffness system, matrix [ΔM^R] is usually a full rank matrix but matrix [ΔK^R] can be a singular one. For instance, if structural modification is confined to a single spring connecting co-ordinates 'i' and 'j', then matrix [ΔK^R] will be a 2 by 2 matrix of rank 1. It is theoretically convenient in structural modification to describe physical elements (mass or stiffness elements) in a system is to use submatrix approach such that:

$$[M] = \sum_{r=1}^{N_m} [M]_r = \sum_{r=1}^{N_m} m_r \{e_r\} \{e_r\}^T \qquad (7)$$

$$[K] = \sum_{r=1}^{N_k} [K]_r = \sum_{r=1}^{N_k} k_r \{e_{pq}\} \{e_{pq}\}^T \qquad (8)$$

Here, matrices [M]_r and [K]_r are submatrices of the rth mass and r^th spring respectively. Vector {e_r} and {e_{pq}} are Kronecker vectors reflecting only the connectivity of a physical element. This presentation of system matrices is akin to the finite element analysis. This concept of submatrix can also be applied to modification matrices. As a result:

$$[\Delta M] = \sum_{r=1}^{u} [\Delta M]_r = \sum_{r=1}^{u} \Delta m_r \{e_r\} \{e_r\}^T \qquad (9)$$

$$[\Delta K] = \sum_{r=1}^{v} [\Delta K]_r = \sum_{r=1}^{v} \Delta k_r \{e_{pq}\} \{e_{pq}\}^T \qquad (10)$$

Here, 'u' is the number of mass elements involved in modification and 'v' that of stiffness elements in modification.

4. Anti-resonances of a dynamic structure

Like the resonance, the anti-resonance is an important manifestation of the dynamic characteristics of a structure [Mothershead 1998]. Though relatively insignificant in modal analysis, the anti-resonance is an essential feature in structural modification. Its relationship with the mass and stiffness properties of a structure needs to be well appreciated. The resonance is a global property since every FRF of a structure should share the same resonances. The anti-resonance is different. Each FRF has its own anti-resonances. Some may not have anti-resonances at all. The anti-resonances of a given receptance FRF $\alpha_{ij}(\omega)$ are the real positive roots of the following algebraic equation:

$$\det\left([K]_{ij} - \Omega^2 [M]_{ij}\right) = 0 \tag{11}$$

or the square root of the real positive eigenvalues of the following problem:

$$\left([K]_{ij} - \Omega^2 [M]_{ij}\right)\{X\} = \{0\} \tag{12}$$

Those complex or real negative eigenvalues are not for anti-resonances and their physical meanings are difficult to ascertain. Equations (11) and (12) suggests that any structural modification which occurs at co-ordinate 'i' only or co-ordinate 'j' only would not affect the anti-resonances of FRF $\alpha_{ij}(\omega)$, although it may alter the natural frequencies of the structure and anti-resonances of other FRFs. This observation precedes some structural modification methods that require participation of an anti-resonance during modification.

Modal analysis theory explains that the occurrence of an anti-resonance owes to the fact that two adjacent modal constants of an FRF have the same signs. Adjacent modal constants with different signs would produce a minimum between two resonances rather than an anti-resonance. This implies that if a structural modification method is able to prescribe a mode shape, then it is possible to create anti-resonances for an FRF at given frequencies.

5. Re-analysis for structural modification

Re-analysis serves for efficient and effective analysis of the modified structure without needing to repeat the complete eigenvalue analysis. A modified system is governed by:

$$\left[([K] + \Delta[K]) - \omega^{*2} ([M] + \Delta[M])\right]\{Y\} = \{0\} \tag{13}$$

Since the mode shape of the modified system is a linear combination of those from the unmodified one, it yields: $\{Y^*\} = [\phi]\{\eta\}$. Substituting this relationship into equation (13) will lead to the following analysis:

$$\left[\left([K]+\Delta[K]\right)-\omega*^2\left([M]+\Delta[M]\right)\right][\phi]\{\eta\}=\{0\} \tag{14}$$

$$\left([K^*]-\omega*^2[M^*]\right)\{\eta\}=\{0\} \tag{15}$$

where

$$[M^*]=[I]+[\phi]^T[\Delta M][\phi] \qquad [K^*]=[\omega_r^2]+[\phi]^T[\Delta K][\phi] \tag{16,17}$$

Equation (15) describes the relationship between the modification matrices and the modal data of the modified structure. It does not rely on the mass and stiffness matrices of the original structure. As a result, it can be conveniently used to determine the modal properties of the modified structure once modifications are determined. Modal testing can provide the modal data of the original structure. Usually, this means only m modes are measured at n co-ordinates. Both 'm' and 'n' are less than the total number of co-ordinates 'N' used for the original mass and stiffness matrices. This means the eigenvalue problem presented in equation (15) will be a reduced one with computational savings. Equation (15) only produces approximate results for the modified structure because of lack of full number of modes. The modal data from its solution will be different from that obtained from modified mass and stiffness matrices. To determine the FRF of locally modified structures, a method presented by Ozguven [1984] can be used. It derives the FRFs of a locally damped structure from their undamped counterparts. A short extension of the method can apply to the FRFs of locally modified structures.

Some methods have been proposed to iteratively determine the modal and FRF data of the modified structure from that of the original structure and modification without having to solve either full or reduced eigenvalue solution. Wang etc [1994] proposed a method based on Taylor series expansion to determine the modal data of the modified structure from that of the original structure and modifications. Its computational procedure avoided eigenvalue solution. In dealing with receptance FRFs of non-proportionally damped dynamic systems, Yang [1993] proposed an iterative method to derive the FRF from its undamped counterpart and subscribed damping matrix without solving a new eigenvalue problem.

6. Structural modification at one co-ordinate

Structural modification at one co-ordinate is a simple type of modification. Nevertheless, it can offer significant changes on the resonances and anti-resonances of a dynamic structure. Its analysis provides useful insights into how dynamic properties of a structure change with simple modifications. Possible structural modifications at one co-ordinate are simple mass modification, stiffness modification and simultaneous mass and stiffness modification. These three cases will be discussed separately.

Simple mass modification at one co-ordinate adds mass to a structure without increasing overall stiffness. As a result, in can be expected that the natural frequencies of the structure will generally be increased to varying degrees. Use a cantilever as an example with a mass added at its free end. The point FRFs of the transverse vibration at the free end before and after mass modification are shown as the thick and broken lines in

Figure 1. Since the free end is the nodal point of no modes of the cantilever, all the natural frequencies decrease and all resonances are shifted to the left.

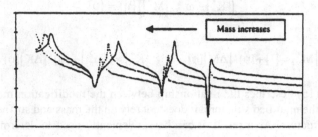

Figure 1 Change of point FRF of a cantilever at its free end due to mass modification

However, it is interesting to note that since it is a point mass modification, no anti-resonances of the point FRF are affected (see equation (12)). As resonances are changes, all anti-resonances are 'fixed'. This means every resonance (except the first one), when moving leftwards, will approach an anti-resonance. When the modification mass approach infinity, resonances will cancel with their anti-resonance partners.

A similar observation can be made when a single spring is linked between the free end of the cantilever and the ground. As the overall stiffness of the cantilever increases while the mass does not, it is expected that all the natural frequencies increase and all resonances move towards the right, as Figure 2 indicates.

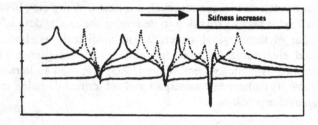

Figure 2 Change of point FRF of a cantilever at its free end due to stiffness modification

From equation (12), a single stiffness modification to ground will not change the anti-resonances of the point FRF at the free end. This means, as the resonances move while the stiffness of the spring increases, all anti-resonances are 'fixed'. This appears like that the resonances lean towards anti-resonances as the stiffness becomes significantly large. When the stiffness is infinite, these resonance and anti-resonance pairs will cancel each other.

Now consider a simple structural modification that is a combination of the single mass and stiffness modifications. A structure is modified by connecting it with a grounded SDoF system, as shown in Figure 3.

In this case, it has been found by Ram and Blech [1991] that the natural frequencies of the MDoF system which are less than the natural frequency of the SDoF modification system $\sqrt{k/m}$ will increase, while those of the system above the natural frequency of the SDoF modification system will decrease. This is shown in Figure 4 where the thin line is an FRF of the MDoF system and the thick line is the same FRF with the SDoF modification.

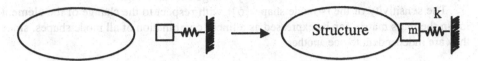

Figure 3 A MDoF structure with a SDoF modification to ground at one co-ordinate

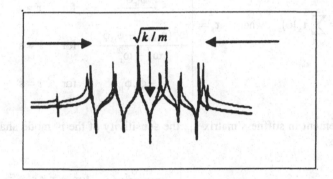

Figure 4 Shifts of natural frequencies of a MDoF system after modification

7. Sensitivity analysis for structural modification

Sensitivity analysis can usually serve for two purposes in structural modification. It provides an avenue for some iterative structural changes leading to desired dynamic characteristic changes. It can also be used for predicting the 'optimal' structural modifications with a prescribed dynamic characteristic change. Sensitivity of the modal properties of a structure with respect to its mass and stiffness changes have been used by many researchers in structural modification. The rth natural frequency square ω_r^2 with respect to the change of the element m_{ij} in the mass matrix is derived as:

$$\frac{\partial \omega_r^2}{\partial m_{ij}} = \begin{cases} -2\omega_r^2 \phi_{ir}\phi_{jr} & i \neq j \\ -\omega_r^2 \phi_{ir}^2 & i = j \end{cases} \tag{18}$$

Likewise, the sensitivity of the rth natural frequency square with respect to the change of the element k_{ij} in the stiffness matrix can be derived as:

$$\frac{\partial \omega_r^2}{\partial k_{ij}} = \begin{cases} 2\phi_{ir}\phi_{jr} & i \neq j \\ \phi_{ir}^2 & i = j \end{cases} \tag{19}$$

Equations (6.3.6) and (6.3.7) outline the sensitivity of a natural frequency with respect to a spatial parameter change. They provide a guideline for selecting the 'optimal' parameters in order to accomplish a prescribed natural frequency change. However, due to the assumption of infinitesimal changes on which sensitivity analysis is based, these equations become invalid if natural frequency change becomes large.

The sensitivity of the rth mode shape $\{\phi\}_r$ with respect to the change of the element m_{ij} in the mass matrix can be expressed as a linear combination of all mode shapes, since they are independent to one another.

$$\frac{\partial\{\phi\}_r}{\partial m_{ij}} = \sum_{s=1}^{n} \tau_s \{\phi\}_s \quad \text{where} \quad \tau_s = \begin{cases} \dfrac{\omega_r^2 \phi_{is}\phi_{ir}}{\omega_s^2 - \omega_r^2} & \text{for} \quad r \neq s \quad i = j \\[3mm] \dfrac{-\phi_{ir}\phi_{ir}}{2} & \text{for} \quad r = s \quad i = j \\[3mm] \dfrac{\omega_r^2(\phi_{js}\phi_{ir} + \phi_{is}\phi_{jr})}{\omega_s^2 - \omega_r^2} & \text{for} \quad r \neq s \quad i \neq j \\[3mm] -\phi_{ir}\phi_{jr} & \text{for} \quad r = s \quad i \neq j \end{cases} \tag{20}$$

For an element in stiffness matrix k_{ij}, the sensitivity of the rth mode shape can also be expressed as:

$$\frac{\partial\{\phi\}_r}{\partial k_{ij}} = \sum_{s=1}^{n} \gamma_s \{\phi\}_s \quad \text{where} \quad \gamma_s = \begin{cases} \dfrac{\phi_{is}\phi_{ir}}{\omega_s^2 - \omega_r^2} & \text{for} \quad r \neq s \quad i = j \\[3mm] \dfrac{\phi_{js}\phi_{ir} + \phi_{is}\phi_{jr}}{\omega_s^2 - \omega_r^2} & \text{for} \quad r \neq s \quad i \neq j \\[3mm] 0 & \text{for} \quad r = s \end{cases} \tag{21}$$

The sensitivity of an FRF of a dynamic system with respect to spatial parameter p_r of the system can be derived as:

$$\frac{\partial[\alpha(\omega)]}{\partial p_r} = [\alpha(\omega)] \frac{\partial[Z(\omega)]}{\partial p_r} [\alpha(\omega)] = [\alpha(\omega)] \left(\frac{\partial[K]}{\partial p_r} - \omega^2 \frac{\partial[M]}{\partial p_r} \right) [\alpha(\omega)] \tag{22}$$

8. Structural modification for a prescribed natural frequency

Structural modification of a MDoF undamped system can be formulated as [Tsuei and Yee 1987]:

$$\{Y\} = [\alpha(\omega^*)] (\omega^{*2}[\Delta M] - [\Delta K]) \{Y\} \tag{23}$$

This equation establishes the link between the prescribed natural frequency ω^* and the modifications $[\Delta K]$ and $[\Delta M]$. As an inverse problem, structural modification methods determine the modifications $[\Delta K]$ and $[\Delta M]$ required in order to obtain frequency ω^*. If only mass modification is made while stiffness properties are kept unchanged, then equation (23) becomes:

$$\{Y\} = \omega^{*2}[\alpha(\omega^*)][\Delta M]\{Y\} \qquad (24)$$

Assume the mass modification matrix can be expressed as $[\Delta M] = \zeta_m[\epsilon]$ where ζ_m is an unknown factor and $[\epsilon]$ is a given matrix describing the relative mass modifications among selected co-ordinates. Equation (24) will be converted to an eigenvalue problem:

$$\frac{1}{\zeta_m}\{Y\} = \omega_*^2[\alpha(\omega^*)][\epsilon]\{Y\} \qquad (25)$$

where ζ_m is the eigenvalue. A real eigenvalue that brings about realistic mass modifications to the system (not generating negative mass properties) will then create the prescribed natural frequency ω^* for the modified system. This method requires the relative mass changes at modified co-ordinates described in matrix $[\epsilon]$ to be pre-determined and it does not involve the mode shapes of the modified system. The order of the matrices and vectors in equation (25) is much smaller than that of the system mass and stiffness matrices since the equation only involves the co-ordinates of modification. The eigenvalue of this equation is $1/\zeta_m$. Once it is known, the mass modification matrix can be found as $[\Delta M] = \zeta_m[\epsilon]$ for each useful eigenvalue.

Stiffness modification can be treated likewise. Equation (23) is reduced to:

$$\{Y\} = -[\alpha(\omega^*)][\Delta K]\{Y\} \qquad (26)$$

Assume the stiffness modification matrix can be expressed as $[\Delta K] = \gamma_k[\kappa]$ where γ_k is an unknown factor and $[\kappa]$ is a given matrix describing the relative stiffness modifications among selected co-ordinates. Equation (26) will become an eigenvalue problem:

$$\frac{1}{\gamma_k}\{Y\} = -[\alpha(\omega^*)][\kappa]\{Y\} \qquad (27)$$

where γ_k is the eigenvalue. A real eigenvalue that brings about realistic stiffness modifications to the system (not generating negative stiffness properties) will then create the prescribed natural frequency ω^* for the modified system. This method requires the relative stiffness changes among modified co-ordinates described in matrix $[\kappa]$ to be pre-determined and it does not involve the mode shapes of the modified system. The order of the matrices and vectors in equation (27) is much smaller than that of the system mass and stiffness matrices since the equation only involves the co-ordinates of modification. The eigenvalue of this equation is $-1/\gamma_k$. Once it is known, the stiffness modification matrix can be found as $[\Delta K] = \gamma_k[\kappa]$ for each useful eigenvalue.

9. Structural modification for a prescribed anti-resonance of an FRF

A prescribed anti-resonance for a selected FRF may find its application when the vibration at a given location of a structure excited by a harmonic force needs to be minimised. The method to obtain a prescribed anti-resonance is identical to that for a prescribed natural frequency if matrices $[M]_{ij}$ and $[K]_{ij}$ in equation (12) are treated as the mass and stiffness matrices of an imaginary dynamic system. Therefore, the method to obtain a natural frequency can be directly applied to this imaginary dynamic system with a new eigenvalue problem:

$$\left([K]_{ij} - \Omega^2[M]_{ij}\right)\{X\} = \{0\} \tag{28}$$

The square root of the eigenvalue of this imaginary system is its 'natural frequency' which is the same as the anti-resonance of the selected FRF of the original system. The eigenvector of this imaginary system does not possess apparent physical interpretation.

10. Structural modification without eigenvalue solution

An alternative structural modification approach is to formulate the problem using linear equations [Li and He 1997]. If mass modification matrix in equation (24) is a diagonal one, then the equation can be recast as:

$$\omega^{*2}[\alpha(\omega^*)]\left[{}^{`}Y_{.}\right]\{\Delta m_r\} = \{Y\} \tag{29}$$

where diagonal matrix $\left[{}^{`}Y_{.}\right] = \text{diag}\{Y\}$. In order to solve this equation for a prescribed natural frequency, vector $\{Y\}$ as the mode shape for the modified system needs to be specified. This provides an opportunity to assign a useful mode shape as well as to prescribe a natural frequency. Equation (29) can also be transformed using $\{X\} = [\Delta M]\{Y\}$ to become:

$$[\Delta M][\alpha(\omega^*)]\{X\} = \frac{1}{\omega^{*2}}\{X\} \tag{30}$$

This leads to a set of succinct analytical solutions for mass modifications:

$$\Delta m_r = \frac{1}{\omega^{*2}} \cdot \frac{X_r}{\displaystyle\sum_{p=1}^{N_m} X_p \alpha_{rp}(\omega^*)} \qquad r = 1, 2, ..., N_m \tag{31}$$

Similarly, stiffness modification can be formulated using a set of linear equations. Using the same linear transformation, equation (26) becomes:

$$-[\alpha_{ij,kl}(\omega^*)]\left[{}^{`}Y_{pq.}\right]\{\Delta K\} = \{Y_{pq}\} \tag{32}$$

Here,

$$\alpha_{ij,kl}(\omega^*) = \alpha_{ik}(\omega^*) + \alpha_{jl}(\omega^*) - \alpha_{jk}(\omega^*) - \alpha_{il}(\omega^*) \tag{33}$$

$$\{Y_{pq}\} = \{Y_p\} - \{Y_q\} \qquad\qquad \left[{}^{`}Y_{pq.}\right] = \text{diag}\{Y_{pq}\} \tag{34,35}$$

Solution of equation yields stiffness modifications on the selected modification co-ordinates.

Alternatively, using linear transformation: $\{X_{pq}\} = \begin{bmatrix} \ddots & \Delta K & \ddots \end{bmatrix} \{Y_{pq}\}$, equation (32) can be recast as:

$$-\begin{bmatrix} \ddots & \Delta K & \ddots \end{bmatrix} [\alpha_{ij,kl}(\omega^*)]\{X_{pq}\} = \{X_{pq}\} \qquad (36)$$

This leads to a set of simple formulas for estimating stiffness modifications.

11. Cancellation of a resonance and an anti-resonance

For a given FRF, local structural modification can be sued to create a cancellation of a resonance and an anti-resonance, thus creating a nodal point at a given co-ordinate. This is sometimes referred to as pole-zero cancellation. The analysis starts from then following equation:

$$([K]+[\Delta K])\{Y\} - \omega^{*^2}([M]+[\Delta M])\{Y\} = \{0\} \qquad (37)$$

If modification matrices $[\Delta M]$ and $[\Delta K]$ satisfy:

$$([\Delta K] - \omega^{*^2}[\Delta M])\{\phi\} = \{0\} \qquad (38)$$

and ω^* and $\{\phi\}$ are a natural frequency and its mode shape of the original system, then equation (6.5.1) means that the modifications $[\Delta M]$ and $[\Delta K]$ do not have any effect on that vibration mode of the original system. Such modifications effectively fix that natural frequency and its mode shape while altering others. With this, it becomes possible via local structural modification to fix one natural frequency and shift its adjacent one, thus creating a wide frequency range free of resonances.

In order to create a pole-zero cancellation, it is possible to devise structural modifications $[\Delta M]$ and $[\Delta K]$ such that an anti-resonance of a selected FRF $\alpha_{ij}(\omega)$ is fixed. This is accomplished by designing a pair of $[\Delta M]$ and $[\Delta K]$ such that they satisfy:

$$([\Delta K]_{ij} - \Omega_*^2[\Delta M]_{ij})\{X\} = \{0\} \qquad (39)$$

for

$$([K]_{ij} + [\Delta K]_{ij})\{X\} - \Omega_*^2([M]_{ij} + [\Delta M]_{ij})\{X\} = \{0\} \qquad (40)$$

This pair of $[\Delta M]$ and $[\Delta K]$, when added to the system, will not change the anti-resonance Ω_*. The change of $[\Delta M]$ and $[\Delta K]$ without violating equation (39) will bring a resonance to this fixed anti-resonance, thus creating a pole-zero cancellation.

12. Summary

Structural modification is a technique derived from modal analysis. It studies the relationship between the mass and stiffness changes of a structure and its natural frequency

392

and mode shape changes. Structural modification by nature is an inverse problem. A common task of it is to determine necessary mass and stiffness modifications that will produce a prescribed natural frequency or mode shape.

Much of the early research in structural modification was focused on simple mass or stiffness modification. Such modification provides in depth insights into the link between spatial property and modal property changes. The results were not readily useful in many practical applications because of their unrealistic simplicity.

Re-analysis methods have been developed along with structural modification. Re-analysis provides efficient and reliable analytical solution of the dynamic characteristics of the modified structure after structural modification has been determined. There are interactive or direct re-analysis methods. Efficiency of these methods often occurs only when dynamic characteristics of only a part of the modified system are needed.

When either mass or stiffness modification takes place, it is possible to formulate structural modification as a reduced eigenvalue problem or a linear equation problem. The exact solution usually exists. This exact solution requires FRF data only from modified co-ordinates. Hence, the solution is very efficient. In real applications, the main difficulty facing structural modification is that mass and stiffness properties are interrelated. It is often impossible to change the mass of a structure without affecting its stiffness. This requires enhanced mathematical formulation for structural modification. This can lead to a higher order eigenvalue problem or a set of nonlinear equations. Numerical accuracy may suffer when solving these mathematical problems.

By carefully defining mass and stiffness modifications, it is possible to modify the modal properties of a system while keeping one of its natural frequencies constant. This opens prospects of optimising dynamic characteristics of a structure. One optimisation is to create a pole-zero cancellation, thus creating a nodal point for a given FRF. Another is to create a frequency range free of resonances.

Current methods of structural modification require minimum amount FRF or modal data for the implementation. This is an important advantage over other applications of modal analysis. However, these methods do require great accuracy from these data.

A main question that needs to be addressed satisfactorily in structural modification is to select optimal co-ordinates for structural modification. Sensitivity analysis only provides a vague direction as to which physical parameters or components should be changed in order to cause minimum changes to the structure. However, because of its mathematical origin, sensitivity analysis offers little definitive answers to structural modification.

13. References

1. Adelman, H. M. and Haftka, R. T. (1986), Sensitivity analysis of discrete structure systems, *AIAA Journal, Vol. 24, No. 5*, 823-832.
2. Baldwin, J. F. and Hutton, S. G. (1985), Natural modes of modified structure, *AIAA Journal, Vol. 23, No. 11*, 1737-1743.
3. Bucher, I. and Braun, S. (1993), The structural modification inverse problem: an exact solution, *Mechanical Systems and Signal Processing, 7(3)*, 217-238.
4. Yuan-Fang, C. and Jeng-Shyong, C. (1988), Structural dynamics modification via sensitivity analysis, *Proceedings of the 6th International Modal Analysis Conference*, 483-489.
5. Ewins, D. J. (1984), Modal Testing: Theory and Practice, *Research Studies Press*, UK.

6. Hallquist, J. O. (1976), An efficient method for determining the effects of mass modifications in damped systems, *Journal of Sound and Vibration, 44(3)*, 449-459.

7. He, J. and Li, Y. (1995), Relocation of anti-resonances of a vibratory system by local structural changes, *The International Journal of Analytical and Experimental Modal Analysis, Vol. 10, No. 4*.

8. Herbert, M. R. and Kientzy, D. W. (1980), Applications of structural dynamics modification, *SAE Paper No. 80-1125*.

9. Hunt, J B (1979), Dynamic Absorbers, Letworth, UK.

10. Joseph, K. T. (1992), Inverse eigenvalue problem in structural design, *AIAA Journal, Vol. 30, No. 12*, 2890-2896.

11. Li, Y., He, J. and Lleonart, G. (1994-1), Structural Dynamic optimisation by local structural modification, *Proceedings of the 12th International Modal Analysis Conference*, Honolulu, 127-132.

12. Li, Y., He, J. and Lleonart, G. (1994-2), Finite element implementation of structural dynamic modification, *Proceedings of International Mechanical Engineering Congress*, Perth, Australia, 157-161.

13. Li, T. and He J. (1997), Structural modification by mass and stiffness changes, *Proceedings of 15th IMAC*, Orlando, USA.

14. Mothershead, J. E. (1998), On the zeros of structural frequency response functions and their application to model assessment and updating, *Proceedings of the 16th International Modal Analysis Conference*, Santa Barbara, CA, 500-503.

15. Murthy, D. V. and Haftka, R. T. (1988), Approximation to eigenvalues and modified general matrices, *Computers and Structures, Vol. 29, No. 5*, 903-917.

16. Lord Rayleigh (1945), Theory of Sound, *Dover Publications, 2nd ed.*, New York.

17. Ozguven, H. N. (1984), Determination of receptances of locally damped structures, *Proceedings of International Conference on Recent Advances in Structural Dynamics, Vol. 2*, 887-892.

18. Pomazal, R. J. and Snyder, V. W. (1971), Local modifications of damped linear systems, *AIAA Journal*, 9, 2216-2221.

19. Ram, Y. M. and Blech, J. J. (1991), The dynamic behaviour of a vibratory system after modification, *Journal of Sound and Vibration, 150(3)*, 357-370.

20. Ram, Y. M. and Braun, S. G. (1991), An inverse problem associated with modification of incomplete dynamic systems, *Journal of Applied Mechanics, Vol. 58*, 233-237.

21. Rogers, L. C. (1970), Derivatives of eigenvalues and eigenvectors, *AIAA Journal*, 8, 943-944.

22. Rosenbrock, H. H. (1965), Sensitivity of an eigenvalue to changes in the matrix, *Electronics Letters, Vol. 1*, 278-279.

23. Sandtrom, R. E. and Anderson, W. J. (1982), Modal perturbation methods for marine structures, *Transaction of the Society of Naval Architects and Marine Engineers, Vol. 90*, 41-54.

24. Skingle, G. W. and Ewins, D. J. (1988), Sensitivity analysis using resonance and anti-resonance frequencies - a guide to structural modification, *European Forum on Aeroelasticity and Structural Dynamics*, Aachen.

25. Skingle, G. W. (1989), Structural dynamic modification using experimental data, *Ph.D thesis*, Department of Mechanical Engineering, Imperial College of Science, Technology and Medicine, London, UK.

26. Snyder, V. W. (1986), Structural modification and modal analysis – survey, *International Journal of Modal Analysis, Vol. 1 No. 11*.

27. Stetson, K. A., and Palma, G. E. (1976), Inverse of first order perturbation theory and its application to structural design, *AIAA Journal, Vol. 14*, 454-460.

28. To, W. M. and Ewins, D. J. (1990), Structural modification analysis using Rayleigh quotient iteration, *International Journal of Mechanical Sciences, Vol. 32, No. 3*, 169-179.

29. Tusei, Y. G. and Yee, E. K. L. (1987), A method to modify dynamic properties of undamped mechanical systems, *Journal of Dynamic Systems, Measurement, and Control, ASME Transactions, Vol. 111*, 403-408.

30. Vanhonacker, P. (1980), Differential and difference sensitivities of natural frequencies and mode shapes of mechanical structures, AIAA Journal, 18, 1511-1514.

31. Wagner, H. et al (1973), Dynamics of stockbridge dampers, *Journal of Sound and Vibration, 30(2)*, 207-220.

32. Wang, B. P. and Pilkey, W. D. (1980), Efficient re-analysis of locally modified structures, *Proceedings of the First Chautauqua on Finite Element Modelling*, 37-61.

33. Wang, J., Heylen, W. and Sas, P. (1987), Accuracy of structural modification techniques, *Proceedings of the 5th International Modal Analysis Conference*, 65-71.
34. Wang, Z., Lim, M. K. and Lin, R. M. (1994), Structural modification prediction using incomplete modal data, *Proceedings of the 12th International Modal Analysis Conference*, Honolulu, HI, 1736-1743.
35. Weissenburger, J. T. (1968), Effects of local modification on the vibration characteristics of linear systems, *Journal of Applied Mechanics, 35*, 327-332
36. Wilkinson, J. H. (1963), The Algebraic Eigenvalue Problem, *Oxford University Press*, London.
37. Yang, B. (1993), Exact receptances of nonproportionally damped dynamic systems, *Journal of Vibration and Acoustics, Vol. 115*, 47-52.
38. Yee, E. K. L. and Tusei, Y. G. (1991), Method of shifting natural frequencies of damped mechanical systems, *AIAA Journal, Vol.29, No. 11*, 1973-1977.
39. Zeng, X., Zhang, K. and Yang, J. (1998), A new re-analysis method for large structure modification, *Proceedings of the 16th International Modal Analysis Conference*, Santa Barbara, CA, 1598-1601.

Nomenclature

$[\cdot\ .]$	a diagonal matrix
$[I]$	unity matrix
$[K]$	stiffness matrix
$[K]_{ij}$	obtained by deleting the 'i'th row and 'j'th column of $[K]$
$[\Delta K]$	stiffness modification matrix
Δk	individual stiffness modification
$[M]$	mass matrix
$[M]_{ij}$	obtained by deleting the 'i'th row and 'j'th column of $[M]$
$[\Delta M]$	mass modification matrix
Δm	individual mass modification
N_k	number of stiffness modification elements
N_{im}	number of mass modification elements
$\{Y\}$	mode shape of modified system
$[Z(\omega)]$	dynamic stiffness matrix
$[\alpha(\omega)]$	receptance FRF matrix at frequency ω
$\alpha_{ij}(\omega)$	the ijth element in matrix $[\alpha(\omega)]$
ω	frequency
ω_r	the rth natural frequency
ω^*	prescribed natural frequency
Ω	anti-resonance frequency
Ω^*	prescribed anti-resonance
$\{\delta\}_r$	the rth Kronecker delta vector: $\{\delta\}_r$
$\{\phi\}_r$	the rth mode shape
ϕ_{jr}	the jth element of mode shape $\{\phi\}_r$

DAMPING: AN INTRODUCTION TO VISCOELASTIC MODELS

D. J. INMAN, C. H. PARK
Center for Intelligent Material Systems and Structures
Virginia Polytechnic Institute and State University
Blacksburg VA 24061-0261, USA

1. Introduction

This chapter provides an examination of viscoelastic damping normally characterized by hysteresis, a complex modulus or frequency dependent damping. Viscoelastic damping exhibited in polymeric and glassy materials as well as in some enamels. Such materials are often added to structures and devices to increase the amount of damping. Examples are rubber mounts and constrained layer damping treatments. Typically in modal analysis the simplest form of modeling damping is used. This form assumes that the damping is a linear, time invariant phenomena chosen to be viscous, or proportional to velocity, motivated by the ability to solve the equations of motion. With this as a first model, one is lead to conclude that viscoelastic behavior causes frequency dependent damping coefficients resulting in the concept of complex modulus. Thus it is not clear how to perform modal analysis of structures with viscoelastic components. Here an alternative formulation is discussed and presented for multiple degree of freedom systems that allows the treatment of hysteretic damping in dynamic finite element formulations, and hence provides a connection to modal analysis and testing.

2. Complex modulus

First we derive the complex modulus approach, connect it to material tests and define what is meant by hysteresis and frequency dependent damping. The forced response of a single degree-of-freedom system subject to a harmonic input may be written as

$$m\ddot{x} + c\dot{x} + kx = f_0 e^{j\omega t}. \tag{1}$$

Assuming that the solution will be of the form $x(t) = X e^{j\omega t}$ and substituting into (1) yields

$$m\ddot{x} + c\omega j X e^{j\omega t} + k X e^{j\omega t} = f_0 e^{j\omega t} \tag{2}$$

or

$$m\ddot{x} + (c\omega j + k) X e^{j\omega t} = f_0 e^{j\omega t} \tag{3}$$

J.M.M. Silva and N.M.M. Maia (eds.), Modal Analysis and Testing, 395–408.
© 1999 *Kluwer Academic Publishers. Printed in the Netherlands.*

The expression $X e^{j\omega t}$ can be replaced with x(t) to yield

$$m\ddot{x} + k(1+\eta j)x(t) = f_0 e^{j\omega t} \qquad (4)$$

where the term $k(1+\eta j)$ is called the *complex stiffness*, or in the case of a beam, the *complex modulus*. Here $\eta = c\omega/k$ is called the loss factor and is a function of the driving frequency ω. Note that if the system is driven at resonance, $\omega = \sqrt{k/m}$ and $\eta = 2\zeta$. Hence at resonance, the loss factor is twice the damping ratio, providing a connection to viscous damping. Details can be found in Nashif et al. (1985). Equation (4) can be used to compute the energy lost per cycle in terms of the parameter. If an experiment is performed on a viscoelastic material specimen, the energy lost per cycle can be computed as a function of the driving frequency by cycling the material in steady state at various fixed values of the driving frequency. From the material test point of view this data is then used to produce a plot of the complex modulus as a function of over a range of frequencies of interest. An example of such a plot is given in Figure 1.

It is important to note that when computing the response of a system with frequency dependent loss factor such as given in Figure 1, the response is only valid at the single driving frequency in steady state. Hence if the transient response is desired, Equation (4) cannot be used. Here we provide a method for including this type of frequency dependence by augmenting the equations of motion to provide a linear, time invariant model across a range of frequencies.

If we plot the force versus time, or stress versus strain for the system of Equation (4) for a viscoelastic material, a hysterisis loop results. Such plots correspond to constitutive stress-strain equations of the form:

$$\sigma(t) = E(t)\varepsilon(0) + \int_0^t G(t-\tau)\frac{d\varepsilon(\tau)}{d\tau}d\tau \qquad (5)$$

where $\sigma(t)$ is the stress, $\varepsilon(t)$ is the strain and $G(t)$ is the elastic modulus as developed in the theory of viscoelasticity (Christensen, 1982). The integral term gives rise to the notion of hysteresis or relaxation, in viscoelastic materials.

To see the connection between the classic hysteresis integral and the complex modulus description, take the Laplace Transform on time of Equation (5) assuming zero initial conditions. This yields $\sigma(s) = s\,G(s)\varepsilon(s)$ where s is the complex valued Laplace domain variable. If the complex variable s is restricted to lie along the imaginary axis, i.e., if $s = j\omega$ where $j = \sqrt{-1}$ and ω is the frequency, then the term $sG(s)$ becomes complex and may be written as

$$G(\omega) = G'(\omega) + jG''(\omega) = G'(\omega)[1 + j\eta(\omega)] \qquad (6)$$

where G denotes the complex modulus with real part $G'(\omega)$, called the *storage modulus*, and imaginary part $G''(\omega)$, called the *loss modulus*. Also, $\eta(\omega)$ is the loss factor in agreement with Equation (4). Furthermore, Equation (6) can be factored to reveal that

$$\eta(\omega) = \frac{G''(\omega)}{G'(\omega)} = \frac{\text{loss modulus}}{\text{storage modulus}} \qquad (7)$$

so that the loss factor is just the ratio of the complex part of G(ω) to its real part.

The loss to storage modulus given by Equation (7) is used to measure the loss factor illustrated in Figure 1. Basically a test specimen of viscoelastic material is placed between a fixed support and a known mass. The mass is excited (either sine sweep or random) with both input response F(ω) and output response X(ω) measured. These input-output measurements are manipulated in the frequency domain to produce the complex stiffness G(ω) by

$$\frac{F(\omega)}{X(\omega)} = G(\omega) = G'(\omega) + jG''(\omega) \tag{8}$$

This last expression is then used to produce Figure 1.

3. Internal variable approach

The complex modulus approach can be used to predict the steady state response of a structure with viscoelastic damping for a single harmonic disturbance at driving frequency ω. In order to compute the transient response or the response to a broad band driving force, Equation (4) must be interpreted differently. One way to account for the frequency dependent damping or hysterisis is to introduce the concept of internal variables.

Motivated by a need to produce finite element models (FEM) that are capable of predicting the dynamic response of a structure or component, Hughes and his coworkers (1985, 1993) and Lesieutre and his coworkers (1990, 1992, 1995, 1996) developed independent means of augmenting a FEM with new coordinates containing damping properties found from material loss factor curves. The Hughe's approach (called GHM) uses a second order physical coordinate system and the Lesieutre approach uses a first order state space method called the Augmenting Thermodynamic Fields (ATF) method and a generalization called Anelastic Displacement Fields (ADF). All of these methods are superior to the Modal Strain Energy (MSE) method proposed by Rogers, et. al. (1982). While MSE is substantially easier to use, ADF, AFT and GHM are more accurate and account for the possibility of complex mode shapes typical of structures with viscoelastic components. In addition, the more recent ADF method is capable of treating the temperature dependence of viscoelastic materials. These more detailed approaches are able to account for damping effects over a range of frequencies, complex mode behavior, transient responses and both time and frequency domain modeling. Inman (1989) applied the GHM approach to simple beams providing a time domain method for modeling hysterisis.

The AFT and ADF approaches focus on a first order differential equation model of the viscoelastic behavior and results in an odd order model rather then the usual second order differential equation used in dynamic finite element models. However, it does allow the treatment of temperature effects (Lesieutre, 1996). GHM on the other hand is well suited in the second order form but is of yet, unable to represent temperature effects and has numerical difficulties because of its tendency to produce nearly singular mass matrices when applied to vehicle models.

The physical motivation for seeking internal variables comes from examining the viscoelastic material at the molecular level (Ward, 1983). If a single polymer chain is stretched, the entropy change or work done may be related to stretch variables consisting of the end to end distances or principal stretch ratios. This motivates the concept of using some sort of internal variables to describe the viscoelastic material. Phenomenologically, a convenient way to mathematically describe material properties without being concerned with the molecular scale, is to hypothesize that the strain energy is a function of the principal strain invariance, which again are related to and motivate the use of some sort of internal variable to describe the energy in the viscoelastic material.

Several researchers have approached the use of internal variables to describe the complex modulus. Looking at the modulus in the Laplace domain, the loss modulus can be written as any number of different forms involving polynomials in the Laplace variable s. These are summarized in Table 1, along with a reference to their origin. In the table each method takes the approach that the loss modulus can be represented by some functional form in the Laplace domain by

$$G(s) = G_0(1 + h(s)) \tag{9}$$

where the function $h(s)$ is as defined in Table 1.

Table 1 Summary of methods for modeling viscoelastic effects using internal variables.

$h(s) = \sum \dfrac{a_i s}{s + b_i}$	Boit (1955)
$h(s) = \dfrac{E_0 + E_1 s^\beta}{1 + b s^\beta}$	Bagly and Torvik (1981)
$h(s) = \displaystyle\int_0^\infty \dfrac{\gamma(p)}{s + p} dp$	Buharivola (1982)
$h(s) = \sum_n \hat{\alpha}_n \dfrac{s^2 + 2\hat{\zeta}_n \hat{\omega}_n s}{s^2 + 2\hat{\zeta}_n \hat{\omega}_n s + \hat{\omega}_n^2}$	Hughes et al (1985)
$h(s) = 1 + \sum_i \dfrac{\Delta_i s}{s + \beta_i}$	Lesieutre (1990)
$h(s) = 1 + \sum_i \dfrac{\alpha_i \tau_i s}{\tau_i s + 1}$	Yiu (1993)

Each of the approaches listed in the table may be used to provide a damping term capable of augmenting the equations of motion to include the effects of viscoelastic components. Each has its advantages and disadvantages. Our goal, however is to use this Internal variable approach to create a damping matrix compatible with a dynamic finite element model of a structure. Hence, we will use the method of Hughes (Golla and Hughes, 1985) to develop our model because of its second order form. As is derived below, this second order form is compatible with a dynamic finite element model as it augments an existing undamped model to include a damping matrix along with the addition of new coordinates.

From the table, the complex modulus can be written in Laplace domain as

$$G^*(s) = G_0(1 + h(s)) = G_0\left(1 + \sum_{n=1}^{k} \hat{\alpha}_n \frac{s^2 + 2\hat{\zeta}_n \hat{\omega}_n s}{s^2 + 2\hat{\zeta}_n \hat{\omega}_n s + \hat{\omega}_n^2}\right) \tag{10}$$

where G_0 is the equilibrium value of the modulus, i.e., the final value of the relaxation function G(t), and s is the Laplace domain variable. The hatted terms are obtained from the curve fit to the complex modulus data for a particular viscoelastic material at a given temperature. The expansion of h(s) may be thought of as representing the material modulus as a series of mini-oscillators (second order equations) as suggested by McTavish and Hughes (1993). These terms are a representation of the internal variables motivated by molecular considerations. The number of terms kept in the expansion will be determined by the high or low frequency dependence of the complex modulus as given by curves such as the one of Figure 1. In many cases only two to four terms are necessary.

Next consider combining Equation (10) with a finite element formulation of a structure given by the usual mass and stiffness matrix. The equation of motion in the Laplace domain is

$$M\left(s^2 x(s) - s x_0 - \dot{x}_0\right) + K(s) x(s) = F(s) \tag{11}$$

where M is the mass matrix, K the complex stiffness matrix, F the forcing function, and $\dot{x}_0$ and the initial conditions. The complex stiffness matrix can be written as the summation of the contributions of the n complex moduli to the stiffness matrix such that

$$K(s) = \left(G_1^*(s)\overline{K}_1 + G_2^*(s)\overline{K}_2 + \ldots + G_n^*(s)\overline{K}_n\right) \tag{12}$$

where G_n^* refers to the nth complex modulus and $\overline{K}_n$ to the contribution of the nth modulus to the stiffness matrix. For simplicity, assume a complex modulus model with a single expansion term and zero initial conditions, so Equation (11) can be written as

$$M s^2 x(s) + G_0\left(1 + \hat{\alpha} \frac{s^2 + 2\hat{\zeta}\hat{\omega}s}{s^2 + 2\hat{\zeta}\hat{\omega}s + \hat{\omega}^2}\right)\overline{K} x(s) = F(s). \tag{13}$$

In this formulation all of the eigenvalues have dissipation modes associated with them. After some manipulation, Equation (13) can be written as

$$\begin{bmatrix} M & 0 \\ 0 & \dfrac{\hat{\alpha}}{\hat{\omega}^2} G_0 I \end{bmatrix} \begin{bmatrix} x(s) \\ \hat{z}(s) \end{bmatrix} s^2 + \begin{bmatrix} 0 & 0 \\ 0 & \dfrac{2\hat{\alpha}\hat{\zeta}}{\hat{\omega}} G_0 I \end{bmatrix} \begin{bmatrix} x(s) \\ \hat{z}(s) \end{bmatrix} s + \begin{bmatrix} (1+\hat{\alpha})G_0\overline{K} & -\hat{\alpha}G_0\overline{K} \\ -\hat{\alpha}G_0 I & \hat{\alpha}E_0 I \end{bmatrix} \begin{bmatrix} x(s) \\ \hat{z}(s) \end{bmatrix} = \begin{bmatrix} F(s) \\ 0 \end{bmatrix}$$
$$\tag{14}$$

where z(s) is the vector of dissipation coordinates. This is the final form of the Golla-Hughes-McTavish model as described by McTavish and Hughes (1993).

No theory currently exists that forms a connection between the internal stretch variables derived from molecular considerations and the internal variables z(s). Rather, the molecular model motivates the search for coordinates $z(s)$. The desire to force $z(s)$ to satisfy second order equations stems from the need to fit the final model into a dynamic finite element models and the models used in experimental modal testing.

In summary, the basic procedure of GHM is to start with plots of experimentally obtained complex modulus or loss factor data. These are plots of modulus versus frequency are available form manufactures of viscoelastic material and are curve fit to a rational polynomial as illustrated in Figure 1. This rational polynomial with coefficients reflecting the material properties of the test specimen is next used to represent the Laplace Transform of the hysteretic stress-strain relationship. The result is compared to a transfer function model of the undamped finite element model and coefficients are compared to produce a final finite element model containing expanded coordinates and a damping matrix which captures the transient decay and complex mode behavior of the structure with viscoelastic materials.

4. Experimental verification

The model of Equation (14) can be used to create a linear time invariant finite element model of a structure with viscoelastic components. This augmented finite element can then be given an impulse and the response calculated. The data from this numerical simulation can then be used to compute a transfer function. This transfer function is then compared to the experimentally obtained transfer function of the structure modeled by Equation (14) in order to verify the accuracy of the proposed modeling technique.

The experiment is performed on free-free aluminum beam, fully covered with a viscoelastic material and a constraining layer. This structure is chosen to keep the result simple enough to verify the proposed modeling approach. Other more complex verifications can be found in the literature. The dimensions of the beam are 0.381 m long, 0.038 wide and 0.003 m thick. It is suspended from the ceiling with a flexible wire which is attached 0.0254 m from either end. The beam is excited using an impact hammer, applied at the center of the beam. The accelerometer is placed on the other side of the beam also in the center. The placement of the impact and sensor is chosen to minimize excitation of the torsional modes as this are not included in the analytical model of the beam. The constraining layer is a beam of the same dimensions as the base beam, thus creating a sandwich beam. Two beams are tested, one with viscoelastic layer 0.127 mm thick and one that is 0.254 m thick made of 3M ISD 112 viscoelastic material made by 3M Corporation. This is done to insure that the analytical model accounts for the increase in damping with the increase in viscoelastic thickness as is known to be the case. Note that due to the placement of the sensor, only the odd modes will be observable and show up in the transfer function.

Figure 2 shows the acceleration bode plots for the experimental (dashed) and the analytical (solid) transfer functions. The units of the magnitude curve are g/N. As can be seen, there is a reasonable correlation in both the magnitude and phase for the first through the third mode for both values of the thickness. The correlation starts to break down for the fifth mode. This is due to fact that the accelerometer is attached to the beam using wax. The accurate data range for wax is about 0 - 2000 Hz. In addition the loss modulus was only fit in the range up to the fifth mode so that it is reasonable to expect that the accuracy of the model will fall off at higher frequencies as indicated in the figure.

The analytical model included three summations in the expansion of the material properties. By curve fitting the hatted GHM constants to the complex modulus data for ISD 112, the constants are found to be $G_0 = 5 \times 10^4$, $\hat{\alpha} = [9.6 \quad 99.1 \quad 26.2]$, $\hat{\zeta} = [73.4 \quad 1.1 \quad 3.28]$, and $\hat{\omega} = [1 \quad 2 \quad 0.5] \times 10^4$. It is necessary to include three terms to assure that the damping is modeled accurately over the full frequency range of interest. A similar curve fit is shown in Figure 1, where the points indicate the curve fit and the line indicates the manufacture's data. These simple experiments indicate that the proposed modeling technique is reasonable.

5. Modal reduction

One difficulty with the proposed approach is that it can greatly increase the order of the finite model that describes the structure with viscoelastic damping matrix. In fact for large systems it is possible that the GHM method could add more than three degrees of freedom per finite element node. Thus it is of interest to apply model back to the size the original undamped system.

Two model reduction methods have been considered. Friswell and Inman (1998) have introduced other methods specifically for GHM modeling. The standard approach to model reduction in FEM analysis is to use a condensation process or static reduction, such as Guyan's reduction (Guyan, 1965). In this approach some of the insignificant physical coordinates are removed such as rotational degrees of freedom at a node point. This leaves the reduced model in a subset of the original coordinate system but is not generally applicable to damped systems. On the other hand, the model reduction method commonly used in control theory, the internal balancing method, applies to damped systems but is cast in a first order state space formulation and it is not directly possible to express the reduced model in terms of a subset of the original states. This is solved by applying and additional coordinate transformation as introduced and applied by Yae and Inman (1993).

The internal balancing method (Moore, 1981) produces as dynamically balanced set of equations for a given input and output by considering the equation of motion in first order form. Consider the original equations of motion are taken to be

$$M\ddot{q} + D\dot{q} + Kq = f \tag{15}$$

where M, D, and K are the $n \times n$ real, symmetric, positive definite matrices. The $n \times 1$ vector q is the displacement vector. The overdots denote differentiation with respect to time. The $n \times 1$ vector f represents the external forces applied to the structure. Equation (15) is converted into the state space form such that

$$\dot{x}(t) = Ax(t) + Bu(t)$$
$$y(t) = Cx(t) \tag{16}$$

where

$$A = \begin{bmatrix} -M^{-1}D & -M^{-1}K \\ I & 0 \end{bmatrix} \quad B = \begin{bmatrix} M^{-1}B_1 \\ 0 \end{bmatrix} \quad C = \begin{bmatrix} C_1 & C_2 \end{bmatrix} \tag{17}$$

It is assumed that the system defined by A, B, C, and x is controllable, observable and asymptotically stable. The idea taken in this method is to reduce the order of a given model based on deleting those coordinates, or modes, that are the least controllable and observable. To implement this idea a measure is provided for asymptotically stable systems of the form given by Equation (16) by defining *controllability* and *observability* *grammians*, denoted by W_c and W_o, respectively and defined by

$$W_c = \int_0^\infty e^{At} BB^T e^{A^T t} dt, \qquad W_o = \int_0^\infty e^{A^T t} C^T C e^{At} dt \qquad (18)$$

where e^{At} is the state transition matrix of the open-loop system. W_c and W_o are the unique symmetric positive definite matrices which satisfy the Lyapunov matrix equations:

$$AW_c + W_c A^T = -BB^T, \qquad A^T W_o + W_o A = -C^T C \qquad (19)$$

for asymptotically stable systems. Moore (1981) has shown that there exists a coordinate system in which these two grammians are equal and diagonal. Such a system is then called *balanced*. Let the matrix P denotes a linear transformation of the system into the balanced coordinate system, which when applied to Equation (16) yields the equivalent system

$$\dot{\hat{x}}(t) = \hat{A}\hat{x}(t) + \hat{B}u(t), \quad \hat{y}(t) = \hat{C}\hat{x}(t). \qquad (20)$$

These two balanced systems are related by

$$\hat{x} = P^{-1}x, \quad \hat{A} = P^{-1}AP, \quad \hat{B} = P^{-1}B, \quad \hat{C} = CP. \qquad (21)$$

In addition, the two grammians are equal in this coordinate system:

$$\hat{W}_c = \hat{W}_o = \text{diag}[\sigma_1, \sigma_2, \cdots \sigma_{2n}] \qquad (22)$$

where $\hat{W}_c = P^{-1} W_c P$, $\hat{W}_o = P^{-1} W_o P$ and σ_i's denote the singular values of the grammians. Applying the idea of singular values as a measure of rank deficiency to the controllability and observability grammians yields a systematic model reduction method. The matrix P that transforms the original system (A, B, C) into a balanced system $(\hat{A}, \hat{B}, \hat{C})$ can be obtained using the following algorithm:

a. The reduced order model can be calculated by first calculating an intermediate transformation matrix P_1 based on the controllability grammians. Solving for W_c and find eigenvalues Λ_c and eigenvectors V_c such that $V_c^T W_c V_c = \Lambda_c$. Then define $P_1 = V_c \Lambda_c^{-1/2}$.

b. The coordinate transformation $x = P_1 \tilde{x}$ yields an intermediate system $(\tilde{A}, \tilde{B}, \tilde{C})$ calculated by $\tilde{A} = P_1^{-1} A P_1$, $\tilde{B} = P_1^{-1} B$, $\tilde{C} = CP_1$.

c. To complete the balancing algorithm, these intermediate equations are balanced with respect to $\tilde{W}_o$. Solving for $\tilde{W}_o$ and find eigenvalues $\tilde{\Lambda}_o$ and eigenvectors $\tilde{V}_o$ such that $\tilde{V}_o^T \tilde{W}_o \tilde{V}_o = \tilde{\Lambda}_o$. Let $P_2 = \tilde{V}_o \tilde{\Lambda}_o^{-1/4}$.

d. Another coordinate transformation $\tilde{x} = P_2 \hat{x}$ yields the desired balanced system $(\hat{A}, \hat{B}, \hat{C})$:

$$\hat{A} = P_2^{-1} \tilde{A} P_2 = P_2^{-1} \left(P_1^{-1} A P_1 \right) P_2 \qquad (23)$$

$$\hat{B} = P_2^{-1} \tilde{B} = P_2^{-1} P_1^{-1} B \qquad (24)$$

$$\hat{C} = \tilde{C} P_2 = C P_1 P_2 \qquad (25)$$

The transformation P is given by P_1 and P_2 as $P = P_1 P_2$. This produces the balanced system which can now be reduced by looking at the singular values of the balanced system and throwing away those coordinates which have relatively small singular values. This leaves a smaller order system with essentially the same dynamics as the full order system.

Unfortunately the coordinates left after a balanced reduction are not a subset of the finite element nodal coordinates. Thus this is not simple to relate back to the original finite element model as is the case in Guyan reduction. Yae and Inman (1993) introduced an additional coordinate transformation to produce a reduced order model in a coordinate system consisting of a subset of the original finite element coordinate system. For modal testing and finite element applications, it is desirable to provide a physical relationship between the original vector q and the reduced state vector $\hat{x}_r$. Such a relationship is found by using the fact that the balanced states are linear combinations of the original states. Symbolically this is written as:

$$\hat{x}_1 = \sum_{j=1}^{2n} c_{1j} x_j, \quad \dots, \hat{x}_{2n-k} = \sum_{j=1}^{2n} c_{(2n-k)j} x_j,$$

$$\hat{x}_{2n-(k-1)} = \sum_{j=1}^{2n} c_{(2n-k+1)j} x_j \to 0, \quad \dots, \hat{x}_{2n} = \sum_{j=1}^{2n} c_{2nj} x_j \to 0, \qquad (26)$$

where c_{ij}'s are the coefficients in the linear combinations of $\{x_1, x_2, \dots, x_{2n}\}$. Here the last k states are set to zero because they represent the least significant states in the balanced system (Moore 1981). Setting each of these summations equal to zero is equivalent to imposing k constraints on the original 2n states, which means that the modal reduction imposes dependencies on k number of the original states. In other words, one can construct a reduced order model by selecting (2n-k) states out of the original 2n states. If the (2n-k) selected states from the original system are denoted by $x_r = [x_{j1} x_{j2} \cdots x_{j2n-k}]^T$ and the corresponding (2n-k) states of the balanced system are denoted by $\hat{x}_r = [\hat{x}_1 \hat{x}_2 \cdots \hat{x}_{2n-k}]^T$, then the states in $\hat{x}_r$ are linear combinations of the states in x_r. Thus there exists a new transformation matrix P_r of order $(2n-k) \times (2n-k)$ such that $x_r = P_r \hat{x}_r$. The above constraints and the resulting transformation allow one to specify which nodes of model to be retained in the model reduction. In the following it is shown that the matrix P_r consists of certain rows and columns of the original transformation matrix P, and that there is a systematic way of constructing P_r from P.

a. Select the state variables to be retained from $\{x_1, x_2, \dots, x_{2n-k}\}$. Let the indices of those selected be $\{j_1, \dots, j_{2n-k}\}$ rows from P.

b. The transformation matrix P_r can be obtained by selecting first $2n\text{-}k$ columns and $\{j_1, \cdots, j_{2n-k}\}$ rows from P.

c. The reduced order system $\left(A_r, B_r, C_r\right)$

$$\dot{x}_r(t) = A_r\,x_r(t) + B_r\,u(t), \qquad y_r(t) = C_r\,x_r(t) \qquad (27)$$

is now expressed in terms of a subset x_r of the original state vector x, where

$$A_r = P_r\,\hat{A}_r\,P_r^{-1}, \qquad B_r = P_r\,\hat{B}_r, \qquad C_r = \hat{C}_r\,P_r^{-1} \qquad (28)$$

Thus we have provided a scheme that has the best features of the each reduction method. Here we are able to specify which coordinates to keep and provide a dynamically based reduction scheme. This allows us to remove the internal variables added to the system in order to build a damping matrix yet leaves us with a damping matrix.

A numerical example is presented in order to demonstrate the use of viscoelastic element matrices in the finite element analysis of viscoelastic beam through the reduction methods as described above. A viscoelastic beam is equally divided into four elements so that it has four node points. Each node point has six degrees of freedom, that is, one translational displacement, one rotational displacement, and four additional viscoelastic auxiliary degrees of freedom. Hence, the viscoelastic beam has twenty-four degrees of freedom in total. The time response curves of the original model and the model reduced by internal balancing are plotted together in Figure 3. Here we are able to delete the viscoelastic states, that is, GHM internal variables and maintain the elastic states. The difference between the responses of the output amplitude in original and reduced model shown by the dashed line of Figure 3 is almost zero. In this case, the difference between the full and reduced system response lies between an upper limit of (10^{-1}) and lower limit of (10^{+1}).

6. Summary

This chapter presents an approach to including the effects of viscoelastic materials in modal analysis and finite element analysis by modeling the frequency dependence with a time invariant linear system including extra dynamics to account for the hystereses effects. This produces a damping matrix that can be added to an undamped dynamics finite element model. The procedure is experimentally verified on a simple beam. One of the difficulties with the proposed method is that it greatly increases the order of the model. This problem is addressed by using a modal reduction technique. The results is a systematic procedure for adding a damping matrix to an undamped model to account for the effects of viscoelastic damping that retains the original finite element nodes (or physical coordinates) and faithfully predicts the frequency and time response of the structures. The following chapter examines other approaches to treating viscoelastic behavior.

Acknowledgments

The first author thanks his student Eric Austin, his former student Dr. Marca Lam (who produced Figure 2) and colleague Dr. M. I. Friswell for their comments and assistance. In addition, the support of the Army Research Office grant number DAAG55-98-1-0030, technically monitored by Dr. Gary Anderson is gratefully acknowledged.

7. References

1. Christensen, R. M. (1982), Theory of Viscoelasticity: An Introduction, *2nd ed., Academic Press Inc.*, New York.
2. Friswell, M. I., and Inman, D. J. (1998), Reduced order models of structures with viscoelastic elements, *in review*.
3. Golla, D. F. and Hughes, P. C. (1985), Dynamics of viscoelastic structure- a time domain, finite element formulation, *J. of Applied Mechanics, Vol. 52*, 897-906.
4. Guyan, R. J. (1965), Reduction of stiffness and mass matrices, *AIAA Journal, Vol. 3*, pp.380.
5. Inman, D. J. (1989), Vibration analysis of viscoelastic beams by separation of variables and modal analysis, *Mechanics Research Communications, Vol. 16 (4)*, 213-218.
6. Lesieutre, G. A., and Mingori, D. L. (1990), Finite element modeling of frequency-dependent material properties using augmented thermodynamic fields, *AIAA Journal of Guidance Control and Dynamics, Vol. 13*, 1040-1050.
7. Lesieutre, G. A. (1992), Finite element for dynamic modelling of uniaxial rods with frequency dependent material properties, *International Journal of Solids and Structures, Vol. 29*, 1567-1579.
8. Lesieutre, G. A., and Bianchini, E. (1995), Time domain modeling of linear viscoelasticity using anelastic displacement fields, *ASME Journal of Vibration and Acoustics, Vol. 117*, 424-430.
9. Lesieutre, G. A., and Govindswamy, K. (1996), Finite element modeling of frequency-dependent and temperature-dependent dynamic behavior of viscoelastic material in simple shear, *International Journal of Solids and Structures, Vol. 33*, 419-432.
10. McTavish, D. J. and Hughes, P.C. (1993), Modeling of linear viscoelastic space structures, *Journal of Vibration and Acoustics, Vol. 115*, 103-113.
11. Moore, B. C. (1981), Principal component analysis for linear systems: controllability, observability, and model reduction, *IEEE Trans. Automat. Contr., Vol. AC-26*, 17-32.
12. Nashif, A. D., Jones, D. and Henderson, J. P. (1985), Vibration Damping, *John Willy & Sons*, New York.
13. Rogers, L.C., Johnson, C.D., and Keinholz, D. A. (1981), The modal strain energy finite element method and its application to damped laminated beams, *The Shock and Vibration Bulletin, Vol. 51*.
14. Ward, I. M. (1983), *Mechanical Properties of Solid Polymer, 2nd ed., Wiley & Sons*, New York.
15. Yae, K. H. and Inman, D. J. (1993), Control-oriented order reduction of finite element model, *Journal of Dynamic Systems, Measurement, and Control, Vol. 115*, 708-711.

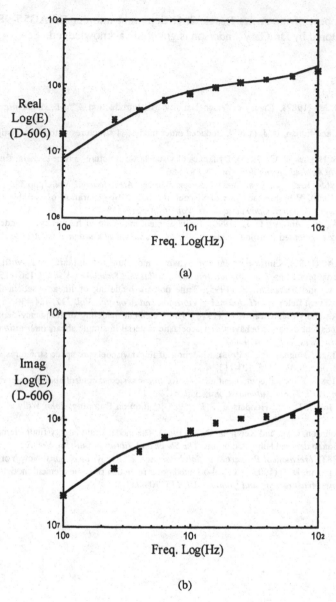

Figure 1   A sample of the real and imaginary part of the complex modulus of a viscoelastic material plotted as a function of driving frequency (*: true value, — curve-fitting).

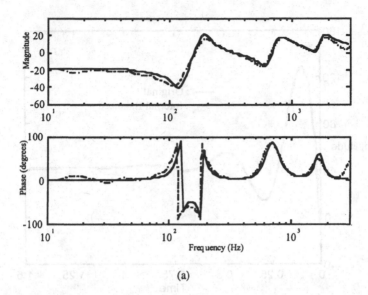

(a)

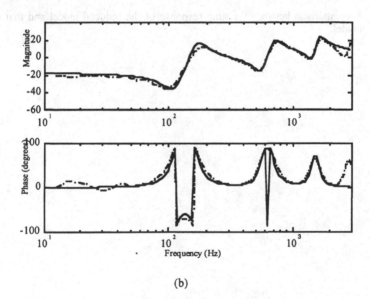

(b)

Figure 2   Both analytical and experimental functions of a beam covered with a viscoelastic layer. The dashed line is experimental and the solid line is theoretical. One set (a) is for 5 mil and the other (b) for 10 mil of ISD 112.

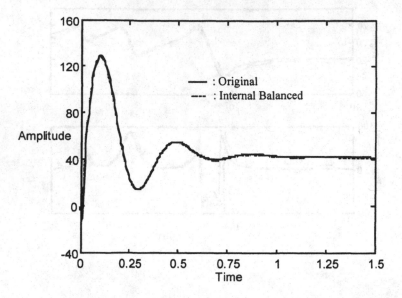

Figure 3   A comparison between the time response of the reduced model and that of the full model.

DESCRIPTION OF DAMPING AND APPLICATIONS

L. GAUL
Institute A of Mechanics, University of Stuttgart
Pfaffenwaldring 9, D-70550 Stuttgart

1. Introduction

Wave propagations and vibrations are associated with the removal of energy by dissipation or radiation. In mechanical systems damping forces causing dissipation are often small compared to restoring and inertia forces. However their influence can be great and is discussed in the present survey paper together with the transmission of energy away from the system by radiation. Viscoelastic constitutive equations with integer and fractional time derivatives for the description of stress relaxation and creep of strain as well as for the description of stress-strain damping hysteresis under cyclic oscillations are compared. Semi-analytical solutions of wave propagation and transient vibration problems are obtained by integral transformation and elastic-viscoelastic correspondence principle. The numerical solution of boundary value problems requires discretization methods. Generalized damping descriptions are incorporated in frequency and time domain formulations for the boundary element method and the finite element method.

2. Classification of Damping, Solutions of Viscoelastic Boundary Value Problems

Vibrating mechanical systems are primarily governed by cyclic transformation between potential and kinetic energy. Associated elastic and inertia properties of lumped or continuous models allow to explain natural frequencies and modes of vibrating systems or group and phase velocities of wave propagating systems. Additional mechanisms remove energy and cause the so called damping, which is responsible for the amplitude decay of free vibrations or propagating waves, for finite amplitudes and phase shifts of systems excited near resonance, for the need of external work to maintain forced vibration amplitudes and the changes in stability of non-conservative systems (Crandall 1970, Bert 1973).

Damping in metals can be caused by combinations of physical mechanisms such as thermal currents on both the micro and macro scale, grain boundary viscosity, point-defect relaxations, eddy-current effects, stress induced ordering and electronic effects (Lazan 1968).

For an important class of non-metallic materials, namely polymers and elastomers, considerable phenomenological data have been obtained. Due to the long range molecular order associated with their giant molecules, polymers exhibit rheological behavior as a combination of a reversible elastic deforming solid and a dissipative viscous liquid (Ferry 1980). The marked frequency and temperature dependence as well as relaxation and creep phenomena of such viscoelastic materials are of particular importance. Industrial applications of polymers and elastomers are widespread such as for damping layers, coatings, absorbers,resilient mounts. The purpose of developing mathematical models for rheological behavior of solids is to permit realistic response predictions from the analysis of complicated structures undergoing various conditions of loading.

TABLE I. Definition and classification of damping

Definition of Damping

    Energy dissipating property of materials and members undergoing time dependant deformations and/or displacements. Damping is primarily associated with the irreversible transition of mechanical energy into thermal energy. The energy radiaton into a surrounding domain is called radiation - or geometric - damping.

Classification of Damping

Material damping

    Energy dissipation by deformation in a medium (Irreversible intercrystal heat flux, grain boundary viscosity, etc.)

Structural Damping

    Damping in assembled structures including: Material damping of members; Frictional losses (Microslip, macroslip) at contacting surfaces (Bolted, riveted, damped, welded connections); Dissipation in medium between surfaces in relative motion (Gas pumping, squezze film damping, lubricated bearing)

Radiation damping

    Energy radiation into surrounding medium

Active/passive damping

    Damping with/without external energy and control

Internal/external damping

    Damping inside/outside defined system boundary

According to the definition in Tab. I damping is the removal of energy from a vibratory system. The energy loss is either dissipated within the system or transmitted away by radiation. The purpose of the present paper is threefold. Firstly to compare linear material damping descriptions by conventional with generalized viscoelastic constitutive equations. Replacing integer time derivatives in the conventional equations by fractional time derivatives in the generalized equations leads to improved curve-fitting properties of measured data with less parameters and assures causality (Torvik and Bagley 1987, Gaul, Klein and Kempfle 1991) which is violated by the so called constant hysteresis damping model (Crandall 1962, Gaul,

Bohlen and Kempfle 1985). Unique selection of complex roots in frequency domain of Fourier transformation is gained by mathematical and physical judgment (Gaul et al. 1991). It has been shown (Torvik and Bagley 1987) that fractional derivatives cannot be viewed merely as effective means of providing curve-fits to data, but their presence has been predicted by accepted theories for the behavior of polymer solids without cross linking for example. A comprehensive review about the application of fractional calculus to dynamic problems of linear and nonlinear hereditary mechanics of solids has been published by Rossikhin and Shitikova (Rossikhin and Shitikova 1997).

The second purpose of the paper is to derive semi-analytical solutions of one dimensional wave propagations and sdof-transient vibrations by integral transformation and adopting the elastic-viscoelastic correspondence principle (Gaul et al. 1991, Crandall 1962, Nakagawa and Kawai 1980, Beyer and Kempfle 1995, Kempfle and Gaul 1996).

The third purpose is to incorporate generalized damping descriptions in frequency and time domain formulations of discretization methods for solving 2-d and 3-d boundary value problems. This is shown for the boundary element method (BEM) in frequency domain (Gaul and Chen 1993, Gaul 1991) and in time domain (Gaul and Schanz 1994, Gaul and Schanz 1997) and the finite element method (FEM) as well (Krings 1976, Carpenter 1972, Pilkey and Pilkey 1995). Radiation damping for problems with semi-infinite or infinite exterior domains is described by surface discretization with the BEM (Pilkey and Pilkey 1995).

3. Viscoelastic Constitutive Equations

The purpose of developing viscoelastic constitutive equations is to describe the rheological stress-strain hysteresis, relaxation and creep behavior as well as the temperature influence for a homogeneously loaded volume element of material. Implementation into field equations of motion and integration over inhomogeneously loaded domains leads to response predictions of members or structures under different loadings such as sinusoidal, transient or random. Other than the microscopic approach which includes the internal structure of matter in terms of atomic and molecular interactions, the present formulation is a macroscopic approach based on phenomenological aspects of physics and the laws of thermodynamics of irreversibility. If small vibratory deflections are superimposed on static predeformations a linear description for the low-stress regime is sufficient.

Hooke's law for an elastic isotropic material relates the deviatoric state of stress and strain with shear modulus G

$$s_{ij}(t) = 2Ge_{ij}(t) \tag{1}$$

and the hydrostatic states of stress and strain with bulk modulus K

$$\sigma_{kk}(t) = 3K\varepsilon_{kk}(t) \tag{2}$$

where

$$\sigma_{ij} = s_{ij} + \frac{1}{3}\sigma_{kk}\delta_{ij}, \ \varepsilon_{ij} = e_{ij} + \frac{1}{3}\varepsilon_{kk}\delta_{ij} \tag{3}$$

The viscoelastic equation corresponding to (1) of differential operator type with time derivatives $D^k = d^k/dt^k, k \in N$ is

$$P_D s_{ij}(t) = Q_D e_{ij}(t) \tag{4}$$

with differential operators

$$P_D = \sum_{k=0}^{N} p_k D^k , \quad Q_D = \sum_{k=0}^{M} q_k D^k \tag{5}$$

or of hereditary integral types

$$s_{ij}(t) = 2 \int_{-\infty}^{t} G(t-\tau) \frac{de_{ij}}{d\tau}(\tau)d\tau$$

$$e_{ij}(t) = 2 \int_{-\infty}^{t} J(t-\tau) \frac{ds_{ij}}{d\tau}(\tau)d\tau \tag{6}$$

with the relaxation modulus $G(t)$ and the creep compliance $J(t)$ kernels describing the fading memory of the material with respect to the load history. Relaxation and creep standard tests are described by Ferry (Ferry 1980). Creep compliance and relaxation modulus are connected by the relation (Gaul 1980)

$$\frac{d}{dt} \int_{0}^{t} 2G(\tau)J(t-\tau)d\tau = 1 \tag{7}$$

The equations for the hydrostatic state are of the same structure as (5, 6). The differential operator type of (2) is

$$P_H \sigma_{kk}(t) = Q_H \epsilon_{kk} \tag{8}$$

Anisotropic constitutive equations are given by Christensen (Christensen 1971). Thermorheologically simple materials allow the introduction of the non-uniform temperature in (6) by replacing the independent variable time t by a reduced time

$$\zeta(t) = \int_{0}^{t} \Phi[T(x,\eta)]d\eta \tag{9}$$

which is based on the shift function Φ determined from experimental data (Christensen 1971, Ferry 1980).

A fruitful generalization of viscoelastic laws replaces the integer order time derivatives D^k by those of fractional order D^{α_k}, where $\alpha_0 = \beta_0 = 0, 0 < \alpha_k, \beta_k < 1$ and D^α is defined either by the nonlocal convolution operator

$$D^\alpha x(t) = \frac{1}{\Gamma(1-\alpha)} \frac{d}{dt} \int_{0}^{t} \frac{x(t-\tau)}{\tau^\alpha}d\tau \tag{10}$$

with the gamma function

$$\Gamma(1-\alpha) = \int_{0}^{\infty} e^{-x} x^{-\alpha} dx \tag{11}$$

as the inverse operation of fractional integration attributed to Riemann and Liouville (Oldham and Spanier 1974), or alternatively by a definition based on generalized finite differences according to Grünwald (Grünwald 1967)

$$D^{\alpha}x(t) = \lim_{N \to \infty} \left(\frac{t}{N}\right)^{-\alpha} \cdot \sum_{j=0}^{N-1} \frac{\Gamma(j-\alpha)}{\Gamma(-\alpha)\Gamma(j+1)} x\left[t\left(1 - \frac{j}{N}\right)\right] \qquad (12)$$

This discrete definition is more convenient in constitutive equations treated by time stepping algorithms (Padovan 1987) and can be shown to be equivalent to the definition in(10). Fractional derivatives replace the differential operators in (4) by

$$P_D = \sum_{k=0}^{N} p_k D^{\alpha_k} , \quad Q_D = \sum_{k=0}^{M} q_k D^{\beta_k} \qquad (13)$$

and change the relaxation function $G(t)$ in (6) such that sums of exponential functions for integer time derivatives are replaced by gamma functions for fractional time derivatives. A correspondence between elastic and viscoelastic equations can be found by Laplace transformation

$$\mathcal{L}\left(D^{\alpha}x(t)\right) = s^{\alpha}\mathcal{L}(x(t)) - \sum_{k=0}^{n-1} s^k D^{\alpha-1-k}x(0) , \, n-1 \leq \alpha < n \qquad (14)$$

or Fourier transformation

$$\mathcal{F}\left(D^{\alpha}x(t)\right) = (i\omega)^{\alpha}\mathcal{F}(x(t)) . \qquad (15)$$

This is why steady state time dependency such as $e_{ij}(t) = Re\left[e_{ij}^* exp(i\omega t)\right]$ or the frequency domain of (15) convert (6) and (13) to

$$s_{ij}^*(\omega) = 2G^*(\omega)e_{ij}^*(\omega) \qquad (16)$$

with

$$G^*(\omega) = i\omega\mathcal{F}\left(G(t)\right) = \sum_{k=0}^{M} \frac{q_k(i\omega)^{\beta_k}}{p_k(i\omega)^{\alpha_k}} . \qquad (17)$$

The real part of the complex shear modulus (17) is the storage modulus $G'(\omega)$, the imaginary part is the loss modulus $G''(\omega)$, their ratio is the loss factor $\eta(\omega)$

$$G^*(\omega) = G'(\omega) + iG''(\omega) = G'(\omega)\left[1 + i\eta(\omega)\right] . \qquad (18)$$

If we decompose the stress relaxation modulus $G(t) = G_{\infty} + \hat{G}(t)$ in an equilibrium modulus G_{∞} and $\hat{G}(t)$ where $\hat{G}(t \to \infty) \to 0$, the complex modulus is related to the relaxation modulus according to (17) by

$$G'(\omega) = G_{\infty} + \omega \int_0^{\infty} \hat{G}(\tau)sin(\omega\tau)d\tau , \, G''(\omega) = \omega \int_0^{\infty} \hat{G}(\tau)cos(\omega\tau)d\tau . \qquad (19)$$

An interpretation of the loss factor as energy quotient is obtained by dividing the dissipated work per cycle T, e.g. in the plane x_1, x_2,

$$D(\omega) = \int_0^T s_{12}(t)\dot{e}_{12}(t)dt = \pi G''(\omega)\hat{e}_{12}^2, \hat{e}_{12} = |e_{12}^*| \qquad (20)$$

by the 2π -fold of stored energy

$$U(\omega) = \frac{1}{2}G'(\omega)\hat{e}_{12}^2 \qquad (21)$$

$$\eta(\omega) = \frac{D(\omega)}{2\pi U(\omega)} = \frac{G''(\omega)}{G'(\omega)} = \frac{1}{Q(\omega)} = tan\delta(\omega) \qquad (22)$$

The energy ratio of (22) can as well be generalized to nonlinear stress-strain constitutive equations. Other damping measures are the quality-factor $Q(\omega)$ and the loss tangent $tan\delta(\omega)$ defined in (22). A graphical representation of equations (20), (21), (22) is given by the elliptical hysteretic loop in Fig. 1.

One advantage of the fractional derivative model of viscoelasticity are improved curve-fitting properties for relaxation and creep functions and complex moduli. This is demonstrated by an example with data taken for a high damping polymer from (Cupiał 1998) in Fig. 2. All 5 parameters of the fractional derivative model (17) with $M = N = 1$

$$G^*(\omega) = G\frac{1 + b(i\omega)^\beta}{1 + a(i\omega)^\alpha} \qquad (23)$$

with selected principal values of multi-valued function $(i\omega)^\beta$ have been found by least squares fit of the real part of (23) in the interval $1 < f/\text{Hz} < 10^4$:

$$\sum_i \left(\frac{G'(\omega_i) - G'_i}{G'_i}\right)^2 \rightarrow min. \qquad (24)$$

Proper selection of complex roots for the range of positive and negative frequencies of Fourier transform is addressed below (Gaul et al. 1991).

For comparison Fig. 2 shows results of the fitted conventional 9 parameter model $N = M = 4$ with all exponents being integers

$$G^*(\omega) = G\frac{1 + \sum_{k=0}^4 b_k(i\omega)^k}{1 + \sum_{k=0}^4 a_k(i\omega)^k}. \qquad (25)$$

Only the real part was fitted by (24) with little oscillation, whereas the loss factor was generated poorly. Thus the imaginary part had to be fitted as well and still gives rise to oscillations.

The restrictions of non-negative internal work and non-negative rate of energy dissipation (Torvik and Bagley 1987, Schanz 1994) in accordance with the second law of thermodynamics and the requirement of finite viscoelastic wave speed (Schanz 1994, Gaul and Schanz 1997) reduce (23) to a 4 parameter model

$$aD^\alpha s_{ij}(t) + s_{ij}(t) = G\left[e_{ij}(t) + bD^\beta e_{ij}(t)\right] \qquad (26)$$

if the parameters fulfill the constraints

$$0 < \alpha = \beta < 2 \qquad G \geq 0 \qquad b > a > 0 .\tag{27}$$

A powerful tool for calculating viscoelastic behavior from a known elastic structural response is the elastic-viscoelastic correspondence principle. According to this principle (Flügge 1975) the viscoelastic solution is calculated from the analytical solution by replacing the elastic moduli in the Laplace transformed domain by the transformed impact response functions of the viscoelastic material model. The viscoelastic solution is then obtained by inverse transformation.

For the above mentioned generalized 4 parameter model the elastic-viscoelastic correspondence under the assumption that the same dissipation mechanisms act in the hydrostatic and the deviatoric states is given by

$$3K \to 3K\frac{1+qs^\alpha}{1+ps^\alpha} , \qquad 2G \to 2G\frac{1+qs^\alpha}{1+ps^\alpha} ,\tag{28}$$

where K is the elastic bulk modulus and G the elastic shear modulus. In Equation (28) the transformation (14) is used for vanishing initial conditions.

An alternative formulation of the correspondence principle adopts Fourier transformation and associated complex moduli.

4. Transient Waves in Viscoelastic Solids

Transient uniaxial wave propagation in a viscoelastic rod is calculated for generalized constitutive equations by the correspondence principle before the 3-d generalization is treated numerically. The rod (Fig. 3) is fixed at $x = 0$ and loaded by the force $F(t)$ at $x = \ell$. The equation of motion

$$\frac{\partial^2 u}{\partial t^2} - c^2\frac{\partial^2 u}{\partial x^2} = 0 , \qquad c = \sqrt{\frac{E}{\rho}} ,\tag{29}$$

boundary conditions

$$u(0,t) = 0 , \qquad EA\frac{\partial u(\ell,t)}{\partial t} = -F(t)\tag{30}$$

and initial conditions

$$u(x,0) = \frac{\partial u(x,0)}{\partial t} = 0\tag{31}$$

lead to the elastic displacement field in terms of right and left propagating waves excited by the Heaviside step function $F(t) = H(t)$

$$u(x,t) = -\frac{1}{\rho c}\sum_{n=0}^{\infty}(-1)^n\left[\left(t - \frac{(2n+1)\ell - x}{c}\right)H\left(t - \frac{(2n+1)\ell - x}{c}\right)\right.$$
$$\left. -\left(t - \frac{(2n+1)\ell + x}{c}\right)H\left(t - \frac{(2n+1)\ell + x}{c}\right)\right] .\tag{32}$$

The viscoelastic solution is obtained from the Laplace transform of (32)

$$u^*(x,s) = -\frac{1}{\rho c}\frac{1}{s}\sum_{n=0}^{\infty}(-1)^n\frac{1}{s}\left[\exp\left\{-\frac{(2n+1)\ell-x}{c}s\right\}\right.$$
$$\left.-\exp\left\{-\frac{(2n+1)\ell+x}{c}s\right\}\right] \quad . \tag{33}$$

Replacing Young's modulus in the elastic wave speed c by the corresponding viscoelastic impulse response of the 4 parameter fractional derivative model (28)

$$E \rightarrow E\frac{1+qs^\alpha}{1+ps^\alpha} \; , \qquad c = \sqrt{\frac{E}{\rho}} \rightarrow c\sqrt{\frac{1+qs^\alpha}{1+ps^\alpha}} \tag{34}$$

leads to the displacement field

$$u^*(x,s) = -\frac{1}{\rho c}\frac{1}{s}\sqrt{\frac{1+ps^\alpha}{1+qs^\alpha}}\sum_{n=0}^{\infty}(-1)^n\frac{1}{s}\left[\exp\left\{-\frac{(2n+1)\ell-x}{c}\sqrt{\frac{1+ps^\alpha}{1+qs^\alpha}}s\right\}\right.$$
$$\left.-\exp\left\{-\frac{(2n+1)\ell+x}{c}\sqrt{\frac{1+ps^\alpha}{1+qs^\alpha}}s\right\}\right] \quad . \tag{35}$$

The normal stresses follow with the constitutive equation

$$\sigma^*(x,s) = \frac{1+qs^\alpha}{1+ps^\alpha}\,E\,\frac{\partial u^*(x,s)}{\partial x} \quad . \tag{36}$$

The inverse Fourier transformation is carried out numerically by an adapted version of the method of Talbot (Gaul and Schanz 1997), (Schanz 1994). Jump relations (Flügge 1975) show that the viscoelastic wave front travels with a speed calculated with the initial value $E(t=0)$ of the relaxation modulus $c_v = \sqrt{E(t=0)/\rho}$. The initial value theorem of Laplace transform governs the initial relaxation modulus derived from the 5 parameter model (26) by

$$E(t=0) = \lim_{t\to0}E(t) = \lim_{s\to\infty}sE(s) = \lim_{s\to\infty}s\left(E\frac{1+qs^\beta}{1+ps^\alpha}\frac{1}{s}\right) \quad . \tag{37}$$

Only for $\alpha = \beta$ the initial relaxation modulus $E(t=0) = E\frac{q}{p}$ exists and leads to the wave speed

$$c_v = \sqrt{\frac{E}{\rho}\frac{q}{p}} = c\sqrt{\frac{q}{p}} \quad . \tag{38}$$

This is a different justification for the reduction of the 5 parameter model (26) to a 4 parameter model.

The time dependent tip deflection of the rod in Fig. 4 shows the superposition of right and left traveling waves. After traveling the distance 2ℓ the tip displacement vanishes. Compared with the elastic solution two changes indicate the influence of material damping. Due to viscoelastic stiffening and dissipation, the wave speed increases according to (38) because the constraint (27) requires $q > p$ and the tip displacement is reduced.

While in the elastic solution the tip deflection vanishes after the wave has traveled 2ℓ (Fig. 4) this is not the case for the viscoelastic solid $\alpha \neq 1$.

In Fig. 5 the differences between the deflections corresponding to fractional derivatives of order $\alpha = 1.5, 0.5, 0.6$ and the deflections for $\alpha = 1$ are plotted.

The effect of viscoelasticity on pulse propagation is discussed next in terms of spectral analysis (Doyle 1998). The equation of motion (29) appears in spectral form

$$\frac{E^*(\omega)}{\rho} \frac{d^2 u^*}{dx^2} + \omega^2 u^*(x, \omega) = 0 \tag{39}$$

with dispersive spectrum relation from

$$u = A e^{i(k^* x \pm \omega t)}, \qquad k^* = \pm \omega \sqrt{\frac{\rho}{E^*(\omega)}}. \tag{40}$$

The nonzero imaginary part of the complex wave number k^* means there is attenuation and the effect of this is seen in a time reconstruction of a propagating pulse in Fig. 6.

If we consider the viscoelastic modulus (34)

$$E^*(\omega) = E \frac{1 + q(i\omega)^\alpha}{1 + p(i\omega)^\alpha} \tag{41}$$

it has very slow and very fast purely elastic limits

$$E^*(0) = E, \qquad E^*(\infty) = E \frac{q}{p}. \tag{42}$$

Consequently the viscoelastic dissipation occurs only in the middle range. The phase and group velocities c and c_g respectively are in general given as

$$c = \frac{\omega}{k^*} = \sqrt{\frac{E^*}{\rho}}, \qquad c_g = \frac{d\omega}{dk} = \frac{\sqrt{\frac{E^*}{\rho}}}{\left(1 - \frac{\omega}{2E} \frac{dE^*}{d\omega}\right)} \tag{43}$$

and have the same limiting values. The effect of viscosity on the propagation of a pulse is to decrease the amplitude and spread the pulse out because of the spectrum of speeds. Knauss (Knauss 1968) discusses uniaxial wave propagation in a viscoelastic material using measured material properties.

5. Transient vibration of an oscillator with different damping descriptions

The influence of different damping descriptions on the transient response of a vibratory system is discussed next. Compared are analytical approaches to incorporate the fractional derivative damping model, a limiting case of which is the Kelvin-Voigt model and the constant hysteretic model.

These models have been selected because they are frequently used in structural dynamics for steady vibrations but they require mathematical caution when transient problems are solved by integral transform. The fractional derivative model requires a unique definition of multi-valued functions (Gaul et al. 1991). The constant hysteretic model shows non causal behavior (Crandall 1962, Gaul et al. 1985) which can be avoided by properly fitted causal fractional derivative models.

The sdof oscillator in Fig. 7 is chosen as structure with a viscoelastic massless member. The transfer behavior of the member (Fig. 2) is obtained by integrating a selected constitutive equation of fractional derivative type (14) with respect to the volume for homogeneous state of stress and strain. This relates the force N to the tip displacement u in the time domain

$$N(t) = ku(t) + cD^{\alpha}u(t) \qquad 0 < \alpha < 1 \tag{44}$$

and in the frequency domain of Fourier transform

$$N^*(a_0) = k\left[1 + \xi(ia_0)^{\alpha}\right] u^*(a_0) \tag{45}$$

with stiffness $k = EA/\ell$ and viscosity $\xi = c/k$ for the limit $\alpha = 1$ of the Kelvin-Voigt model with viscous dashpot. A frequency parameter $a_0 = \omega/\omega_n$ with a scaling frequency $\omega_n = \sqrt{k/m}$ has been introduced, which leads to a non-dimensional time $\tau = \omega_n t$ according to $a_0\tau = \omega t$.

In a limited frequency band, the frequency independent, so-called constant hysteretic damping model approximates experimental results. This damping model is frequently extended over the entire range of positive frequencies to yield the force displacement relation

$$N^*(a_0) = k(1 + i\eta)\, u^*(a_0)\,. \tag{46}$$

Like the complex moduli, the complex stiffness of the members in (45),(46) can be split in storage and loss moduli

$$k^*(a_0) = \frac{N^*(a_0)}{u^*(a_0)} = k'(a_0) + ik''(a_0)\,. \tag{47}$$

But equations (45) and (46) do not as yet meet the requirements of a unique definition for the entire frequency range $-\infty < a_0 < \infty$ of the Fourier transform.

5.1. UNIQUE DEFINITION OF CONSTITUTIVE EQUATIONS IN THE FREQUENCY DOMAIN

Consider the term $(ia_0)^{\alpha}\, a_0$, $\alpha \in \mathbb{R}$ in (45) as special case of the complex expression Ω^z, where $\Omega, z \in \mathbb{C}$. A unique definition of $\Omega^z = \exp(z\mathrm{Ln}\Omega)$ is gained by restricting the arguments of $\mathrm{Ln}\Omega$ to the principal values $-\pi < \mathrm{Im}(\mathrm{Ln}\Omega) < \pi$, leading to a branch cut along the negative real axis (Gaer and Rubel 1975). This definition restricts the arguments $(ia_0)^{\alpha} = R\exp(i\phi)$ to

$$-\alpha\pi \leq \phi < \alpha\pi\,, \qquad 0 < \alpha < 1\,. \tag{48}$$

Storage and loss moduli of the member in (45)

$$k'(a_0) = k\left[1 + \xi \operatorname{Re}(ia_0)^\alpha\right], \qquad k''(a_0) = k\xi \operatorname{Im}(ia_0)^\alpha \tag{49}$$

have a unique definition with (48). A physical interpretation supports the choice of proper roots. The force $N^* \exp(ia_0\tau)$, interpreted as a rotating vector in complex plane, causes the displacement $u^* \exp(ia_0\tau)$ and has to be ahead with minimal phase shift counterclockwise for positive frequencies $a_0 > 0$, $\exp(ia_0\tau)$, and clockwise for negative frequencies $a_0 < 0$, $\exp(-i|a_0|\tau)$. Negative frequencies occur in the two-sided Fourier transform. This condition is fulfilled by (48) and requires the well-known extension of complex modulus (46) according to

$$k^* = k(1 + i\eta \operatorname{sgn} a_0). \tag{50}$$

The fractional derivative of order $\alpha = 1/2$ can be derived to be relevant for polymeric materials from molecular physics (Torvik and Bagley 1987).

According to (48) we have to choose

$$(ia_0)^{1/2} = \sqrt{|a_0|/2}\left\{ \begin{array}{l} 1 + i, \ \ a_0 \geq 0 \\ 1 - i, \ \ a_0 < 0 \end{array} \right\} \tag{51}$$

and obtain the complex modulus

$$k^*(a_0) = \left\{ \begin{array}{ll} k(1 + \xi\sqrt{a_0/2} + i\xi\sqrt{a_0/2}), & a_0 \geq 0 \\ k(1 + \xi\sqrt{|a_0|/2} + i\xi\sqrt{|a_0|/2}), & a_0 < 0 \end{array} \right. . \tag{52}$$

5.2. IMPULSE RESPONSE FUNCTION OF A DAMPED OSCILLATOR

Vibration response of the damped sdof oscillator in Fig. 8 is calculated by Fourier transform with constitutive equation (45) of the member. The FRF is given by

$$F(a_0) := \frac{ku^*(a_0)}{F^*(a_0)} = \frac{1}{1 - a_0^2 + \xi(ia_0)^\alpha}. \tag{53}$$

The real part of the FRF is plotted in Fig. 8. It shows the parameter dependence of the fractional derivative model which influences resonance frequency and damping as well. The transient response is obtained by inverse Fourier transform.

The Dirac impulse excitation $F(\tau) = \delta(\tau)$ leads to the impulse response function $u(\tau) = h(\tau)$

$$f(\tau) := m\omega_n h(\tau) = \frac{1}{2\pi} \int\limits_{-\infty}^{\infty} F(a_0)\, e^{ia_0\tau} \mathrm{d}a_0, \tag{54}$$

which governs the response for an arbitrary force $F(\tau)$ by the Duhamel convolution integral

$$u(\tau) = \frac{1}{\omega_n} \int\limits_{-\infty}^{\tau} h(\tau - \xi)F(\xi)\, \mathrm{d}\xi. \tag{55}$$

An efficient solution of the damped impulse response (55) is obtained by cutting the infinite integral in two intervals $-\infty > a_0 \geq 0$ and $0 < a_0 < \infty$ with associated unique definition of the integrand according to (51). Complex conjugate contributions lead to a real semi-infinite integral after reassembling

$$f(\tau) = \frac{1}{\pi} \int_0^\infty \frac{A(a_0)\cos(a_0\tau) + B(a_0)\sin(a_0\tau)}{A^2(a_0) + B^2(a_0)} \, da_0, \tag{56}$$

where

$$A(a_0) = 1 - a_0^2 + \xi\sqrt{a_0/2}, \quad B(a_0) = \xi\sqrt{a_0/2}. \tag{57}$$

The corresponding solution (Nashif, Jones and Henderson 1985) for the member with constant hysteretic damping in equation (50) requires substitution of (57) by

$$A(a_0) = 1 - a_0^2, \quad B(a_0) = \eta. \tag{58}$$

5.3. CAUSALITY OF IMPULSE RESPONSE FUNCTION

Causal response of the damped oscillator due to the impulse excitation requires the system to be at rest $f(\tau) = 0$ for $\tau < 0$. Causality of the transformation pair in equations (53), and (54) $f(\tau) \circ\!\!-\!\!\bullet F(a_0)$ requires (Papoulis 1962)

$$f(\tau) = \frac{2}{\pi} \int_0^\infty \operatorname{Re} F(a_0)\cos(a_0/2\tau) \, da_0 = -\frac{2}{\pi} \int_0^\infty \operatorname{Im} F(a_0)\sin(a_0/2\tau) \, da_0 \tag{59}$$

for $\tau > 0$. The abbreviations in equation (57) lead to

$$f(\tau) = \frac{2}{\pi} \int_0^\infty \frac{A(a_0)\cos(a_0\tau)}{A^2(a_0) + B^2(a_0)} \, da_0 = \frac{2}{\pi} \int_0^\infty \frac{B(a_0)\sin(a_0\tau)}{A^2(a_0) + B^2(a_0)} \, da_0 \tag{60}$$

for $\tau > 0$. The sum of the two expressions in (60) for $\tau > 0$ lead to equation (56), whereas for $\tau < 0$ equations (60) assure causality $f(\tau) = 0$ when they are inserted in equation (56). A numerical proof of the causality requirement (60) is given for (56) and (57). Criteria for existence, continuity and causality of global solutions have been presented in a rigorous mathematical framework by Beyer, Gaul, Kempfle (Kempfle and Gaul 1996, Beyer and Kempfle 1995).

The causality condition (59) is violated with equations (58) corresponding to the constant hysteretic damping model. A non-vanishing precursor $f(\tau)$ for $\tau < 0$ exists (Crandall 1962) and is bounded by $|f(\tau)| \lesssim \eta/4$ for small loss factors. Gaul et al. (Gaul et al. 1985) calculated an analytical solution of the impulse response function (56) with constant hysteretic damping (58) by theory of residues

$$f(\tau) = \frac{e^{-\lambda\tau}}{\mu^2 + \lambda^2}(\mu\sin(\mu\tau) - \lambda\cos(\mu\tau)) + I(t) \tag{61}$$

with $\mu = \sqrt{\frac{\sqrt{1+\eta^2}+1}{2}}$, $\lambda = \sqrt{\frac{\sqrt{1+\eta^2}-1}{2}}$ and a residual integral bounded by $I(t) \leq \sqrt{\frac{\sqrt{1+\eta^2}-1}{8(1+\eta^2)}}$.

5.4. NUMERICAL RESULTS OF IMPULSE RESPONSE

Fig. 9 depicts the impulse response calculated numerically from equations (56), and (57) for positive and negative time $\tau = \omega_n t$. The response proves to be causal. Parameters weighting the fractional derivative of order $\alpha = 1/2$ are chosen to be $\xi = \sqrt{2}/5$ and $\sqrt{2}/20$. Regarding equation (52), an increase of ξ does not only increase the damping but also stiffens the member according to $k(1 + \xi\sqrt{a_0/2})$. That is why stronger amplitude decay is associated with decreasing periods of zero crossing.

5.5. IMPULSE RESPONSE FOR DIFFERENT DAMPING MODELS

Fig. 10 compares the impulse response functions corresponding to three damping models: constant hysteretic model (Gaul et al. 1985); Kelvin-Voigt model with first order derivative (Crandall 1970) and with fractional derivative of order $\alpha = 1/2$.

Equivalent loss factors have been chosen at $a_0 = \omega/\omega_n = 1$ such that $\eta = a_0\xi$ (Kelvin-Voigt) $= (\sqrt{a_0/2})\xi$ (fractional derivative) $= 1/20$. This leads to nearly equal amplitude decay of all models, whereas the stiffening of the fractional derivative model decreases the periods of zero crossing.

For the generalization of the sdof oscillator description with fractional damping to mdof systems the reader is referred to contributions by Bagley et al (Bagley and Calico 1989) and Maia et al (Maia, Silva, Ribeiro and Leitão 1996) where the concepts of state space and modal analysis description are adopted.

6. Viscoelastic Solids treated by the Boundary Element Method

The simulation of 3-d dynamic response of members with relevant material damping requires discretization methods to cope for complex geometry, boundary conditions and loading. The boundary element method (BEM) provides one powerful tool (Gaul and Fiedler 1997) for the calculation of elastodynamic response in frequency and time domain. Field equations of motion and boundary conditions along with initial conditions in time domain are cast into boundary integral equations (BIE), which are solved numerically by discretization of the boundary only, thus reducing the problem dimensions by one. The boundary data are often of primary interest because they govern the transfer dynamics of members. This substructure behavior can as well be implemented in finite element models or multi-body systems (Gaul and Chen 1993). The application of BEM for modal analysis has been treated by Lanzerath (Lanzerath 1996). Another advantage of the BEM is the treatment of energy radiation into a surrounding medium of infinite or semi-infinite domains, the so called radiation or geometric damping which does not incorporate dissipation. Simple surface discretization around the radiating source provides a non-reflecting boundary because the fundamental solutions of the BIE fulfill the Sommerfeld radiation condition excluding reflections. For the material damping described by viscoelastic constitutive equations, the implementation in BEM formulations is demonstrated by the elastic-viscoelastic correspondence principle. Among numerous practical applications are the descriptions of elastomer resilient mounts or soil half-space presented herein.

6.1. ELASTIC BE-FORMULATION IN TIME DOMAIN

The Lamé field equations of a homogeneous isotropic elastic domain Ω with boundary Γ are given by

$$\left(c_1^2 - c_2^2\right) u_{i,ij} + c_2^2\, u_{j,ii} + b_j = \ddot{u}_j\left(\boldsymbol{x}, t\right) \quad \boldsymbol{x} \in \Omega \tag{62}$$

with displacement coordinates $u_j(\boldsymbol{x}, t)$ and wave speeds

$$c_1^2 = \frac{K + \frac{4}{3}G}{\varrho}, \qquad c_2^2 = \frac{G}{\varrho}. \tag{63}$$

The corresponding traction and displacement boundary conditions are

$$t_i\left(\boldsymbol{x}, t\right) = \sigma_{ik} n_k \;=\; p_i\left(\boldsymbol{x}, t\right) \qquad \boldsymbol{x} \in \Gamma_t,$$
$$u_i\left(\boldsymbol{x}, t\right) \;=\; q_i\left(\boldsymbol{x}, t\right) \qquad \boldsymbol{x} \in \Gamma_u \tag{64}$$

and the initial conditions are

$$u_i\left(\boldsymbol{x}, 0\right) \;=\; u_{0i}\left(\boldsymbol{x}\right),$$
$$\dot{u}_i\left(\boldsymbol{x}, 0\right) \;=\; v_{0i}\left(\boldsymbol{x}\right) \qquad \boldsymbol{x} \in \Omega. \tag{65}$$

The 3-d Stokes fundamental displacement tensor of the Lamé equation (62) in an unbounded space, excited by the volume force $b_j\left(\boldsymbol{x}, \boldsymbol{\xi}, t, \tau\right) = \delta\left(t - \tau\right) \delta\left(\boldsymbol{x} - \boldsymbol{\xi}\right) e_j$ is given by (e.g. (Schanz 1994))

$$\tilde{u}_{ij}\left(\boldsymbol{x}, \boldsymbol{\xi}, t, \tau\right) =$$

$$\frac{1}{4\pi\varrho} \left\{ \frac{t - \tau}{r^2} \left(\frac{3 r_i r_j}{r^3} - \frac{\delta_{ij}}{r}\right) \left[H\left(t - \tau - \frac{r}{c_1}\right) - H\left(t - \tau - \frac{r}{c_2}\right) \right] \right.$$

$$\left. + \frac{r_i r_j}{r^3} \left[\frac{1}{c_1^2} \delta\left(t - \tau - \frac{r}{c_1}\right) - \frac{1}{c_2^2} \delta\left(t - \tau - \frac{r}{c_2}\right) \right] + \frac{\delta_{ij}}{r c_2^2} \delta\left(t - \tau - \frac{r}{c_2}\right) \right\},$$

$$\tag{66}$$

where $r = \sqrt{r_i r_i}$, $r_i = x_i - \xi_i$ is the Euclidean distance between the field point $\boldsymbol{x}$ and the load point $\boldsymbol{\xi}$. The corresponding fundamental stress vector components are obtained by substituting (66) into the constitutive equation and adopting Cauchy's stress formula.

The dynamic extension of Betti's reciprocal work theorem combines two states of displacements and tractions, $(\tilde{u}_{ij}, \tilde{t}_{ij})$ and (u_i, t_i) respectively, and leads to the boundary integral equation

$$c_{ij} u_j = \int_\Gamma \left[\tilde{u}_{ij} * t_j - \tilde{t}_{ij} * u_j \right] d\Gamma + \int_\Omega \varrho \left[\tilde{u}_{ij} * b_j + \tilde{u}_{ij} v_{0j} + \dot{\tilde{u}}_{ij} u_{0j} \right] d\Omega \qquad (67)$$

where $c_{ij} = \delta_{ij}/2$ if ξ is located on a smooth boundary. The $*$ denotes the convolution with respect to time. The integral equation (67) contains boundary integrals only if the volume forces b_j and the initial conditions vanish. Discretization of the boundary integral equation in space and time leads to the boundary element formulation. Only the time discretization by n equal steps Δt is discussed here. The simplest nontrivial choice ensuring that no terms drop out in the boundary integral equation are linear shape functions for the displacements u_i and constant shape functions for the tractions t_i in time domain

$$u_i(x, \tau) = \left(U_{il}^{m-1} \frac{t_m - \tau}{\Delta t} + U_{il}^m \frac{\tau - t_{m-1}}{\Delta t} \right) \eta_l(x) \qquad (68)$$

$$t_i(x, \tau) = T_{il}^m \cdot \mu_l(x). \qquad (69)$$

The actual time step is m. The nodal values are U_{il}^m, T_{il}^m for the corresponding boundary element Γ_l at time $t_m = m\Delta t$. After substituting equations (68) and (69) the boundary integral (67) reads

$$\int_0^t \int_\Gamma \left[t_i(x, \tau) \cdot \tilde{u}_{ij}(x, \xi, t - \tau) - \tilde{t}_{ij}(x, \xi, t - \tau) \cdot u_i(x, \tau) \right] d\Gamma d\tau =$$

$$\sum_l \sum_{m=1}^n \int_{\Gamma_l} \int_{t_{m-1}}^{t_m} \left[\tilde{u}_{ij}(x, \xi, t - \tau) \mu_l(x) \cdot T_{il}^m \right.$$

$$\left. - \tilde{t}_{ij}(x, \xi, t - \tau) \cdot \eta_l(x) \cdot \left(U_{il}^{m-1} \frac{t_m - \tau}{\Delta t} + U_{il}^m \frac{\tau - t_{m-1}}{\Delta t} \right) \right] d\tau d\Gamma. \qquad (70)$$

Equation (70) can be integrated analytically (Schanz 1994). After substituting $t_m = m\Delta t$, $t_{m-1} = (m-1)\Delta t$ and $t = n\Delta t$ it can be seen that the functions depend on the difference $(n - m)$, between the observation and the excitation time only. After time and space discretization a system of algebraic equations (Schanz 1994) is obtained

$$\left[\frac{1}{2} I + \overset{(1)}{T} \right] \overset{(n)}{u} + \sum_{m=1}^{n-1} \overset{(n-m+1)}{T} \overset{(m)}{u} = \sum_{m=1}^n \overset{(n-m+1)}{U} \overset{(m)}{t}, \qquad (71)$$

where I is the identity matrix, $\overset{(m)}{T}$ and $\overset{(m)}{U}$ are the influence matrices of stresses and displacements at the time step m. The vectors $\overset{(m)}{u}$ and $\overset{(m)}{t}$ contain all nodal displacements and tractions of the time step m. Unknown boundary data are calculated in terms of known boundary data after reordering (71).

6.2. VISCOELASTIC BE-FORMULATION IN TIME DOMAIN

In order to obtain a viscoelastic boundary integral formulation from the elastic formulation (71) the elastic–viscoelastic correspondence principle (28) is applied. This requires the Laplace transformation of (71) first. The kernels of the matrices consist of the fundamental solutions of the displacements and tractions after time integration. For the sake of brevity the procedure of deducing a viscoelastic formulation is explained for the first term on the right side of (70) only.

The one sided Laplace transformation of the first term in (70) leads to

$$
\int_0^\infty \sum_{m=1}^{n} \int_{t_{m-1}}^{t_m} \tilde{u}_{ij}\left(x,\xi,t-\tau\right) \mathrm{d}\tau e^{-st}\mathrm{d}t =
$$

$$
\frac{1}{4\pi\varrho}\sum_{m=1}^{n}\Bigg\{ \int_{t_{m-1}+\frac{r}{c_1}}^{t_m+\frac{r}{c_1}} \left[f_0(r)\frac{1}{2}\left((t-t_{m-1})^2 - \left(\frac{r}{c_1}\right)^2 \right) + f_1(r) \right] e^{-st}\mathrm{d}t +
$$

$$
\int_{t_m+\frac{r}{c_1}}^{t_{m-1}+\frac{r}{c_2}} f_0\left(r\right)\left(t\, t_m - t\, t_{m-1} - \frac{t_m^2}{2} + \frac{t_{m-1}^2}{2} \right) e^{-st}\mathrm{d}t +
$$

$$
\int_{t_{m-1}+\frac{r}{c_2}}^{t_m+\frac{r}{c_2}} \left[f_0\left(r\right)\left(t\, t_m - \frac{t_m^2}{2} - \frac{t^2}{2} + \frac{1}{2}\left(\frac{r}{c_2}\right)^2 \right) + f_2\left(r\right) \right] e^{-st}\mathrm{d}t \Bigg\}, \tag{72}
$$

with f_0, f_1 and f_2 depending on spatial coordinates only. After integrating one time interval of (72), the elastic constants are replaced by the viscoelastic impact response functions. The elastic bulk and shear moduli appear only in the compression wave speed c_1 and the shear wave speed c_2. Thus the corresponding viscoelastic expressions

$$
c_{1v}^2 = \frac{1}{\varrho}\left(K\frac{1+qs^\alpha}{1+ps^\alpha} + \frac{4}{3}G\frac{1+qs^\alpha}{1+ps^\alpha} \right)
$$

$$
c_{2v}^2 = \frac{1}{\varrho}G\frac{1+qs^\alpha}{1+ps^\alpha} \tag{73}
$$

are inserted into equation (72) after integration. The method of Talbot has been selected for numerical inversion (Schanz 1994).

6.2.1. Numerical example: Waves in a 3-d viscoelastic continuum associated to the 1-d formulation in section 4

The propagation of waves in a 3-d continuum has been calculated by the present boundary element formulation in time domain. The problem geometry and the associated boundary discretization are shown in Fig. 11. Linear shape functions in space have been used. The free end is excited by a pressure jump according to a unit step function $H(t)$. The opposite end is fixed at the nodes. The time step size Δt has been chosen close to the time it takes for the viscoelastic compression wave to travel across one element.

The viscoelastic material data of a corning glass at $550^{o}C$ (Tab. II) are used in this example.

TABLE II.   Material data for corning glass

E	$=$	$2.075 \cdot 10^{9} \frac{N}{m^2}$	p_K	$=$	p_G	$= \quad 3.5\,s^{0.635}$
ν	$=$	0.25	α_K	$=$	α_G	$= \quad 0.635$
ϱ	$=$	$1000\frac{kg}{m^3}$	q_K	$=$	q_G	$= \quad q\,s^{0.635}$

Fig. 12 shows the longitudinal displacement in the center of the free end cross section versus time for several values of the damping parameter q in the constitutive equation (63). Wave reflections at the fixed and the free end show up. The viscoelastic wave speed of the compressional wave front is given by

$$c_{1v}^{2} = \frac{1}{\varrho}\left(K\frac{q_K}{p_K} + \frac{4}{3}G\frac{q_G}{p_G}\right) \tag{74}$$

Increasing damping parameters q stiffen the solid and increase the wave speed. Similar to the 1-d solution in Fig. 4 shorter travel times show up in Fig. 12 and the stiffening leads to smaller deflections.

6.2.2.   Numerical example with radiation damping: Waves in semi-infinite soil

The propagation of waves in an elastic concrete foundation slab ($E = 3 \cdot 10^{8} N/m^2$, $\rho = 2000\,kg/m^3$, $\nu = 0.2$) bonded on a viscoelastic soil half-space ($\nu = 0.35, E = 1.38 \cdot 10^{8}\,N/m^2, \rho = 1966\,kg/m^3, \alpha = 1.3$, $p = 1s^{-1.3}$) has been calculated by the presented BEM in time domain. Both domains are coupled by a substructure technique based on displacement- and traction-continuity at the interface. The assumption of welded contact does not allow the nonlinear effect of partial uplifting.

The problem geometry and the associated boundary discretization are shown in Fig. 13. The soil discretization is truncated after a distance of the foundation length. The surface of the foundation slab is exited by a positive and negative pressure jump according to Fig. 13. Linear spatial shape functions have been used. Similar to the Courant criteria, the time step size Δt has been chosen close to the time needed by the viscoelastic compression wave to travel across the largest element.

In Fig. 14 vertical surface displacement at point A is plotted versus time for different values of the constitutive parameter q. Obviously the wave speeds increase for higher values of q, because of a stiffening of the material with growing influence of viscoelasticity. This is associated with a significant displacement reduction. Propagation of the wavefronts across the boundary of the discretized area shows the capability of the BEM to describe radiation damping without reflection (Schanz 1994).

6.3. VISCOELASTIC BE-FORMULATION IN FREQUENCY DOMAIN

An integral representation for steady-state elastodynamics is obtained after Fourier transformation of (10) with transformed variables

$$U_i(\boldsymbol{x}, \omega) = \mathcal{F}[u_i(\boldsymbol{x}, t)] \tag{75}$$

$$
c_{ij} U_j(\boldsymbol{\xi}) = \int_{\Gamma} \left[U_{ij}^*(\boldsymbol{x}, \boldsymbol{\xi}) \, t_j(\boldsymbol{x}) - T_{ij}^*(\boldsymbol{x}, \boldsymbol{\xi}) \, U_j(\boldsymbol{x}) \right] \mathrm{d}\Gamma +
$$
$$
+ \int_{\Omega} U_{ij}^*(\boldsymbol{x}, \boldsymbol{\xi}) \left(B_i/\rho \right) \mathrm{d}\Omega \tag{76}
$$

where $\boldsymbol{x}$ are the boundary points to which the integral extends. $U_i(\boldsymbol{x})$ and $t_i(\boldsymbol{x})$ are displacement and traction components at $\boldsymbol{x}$, $U_{ij}^*(\boldsymbol{x}, \boldsymbol{\xi}, \omega)$ and $T_{ij}^*(\boldsymbol{x}, \boldsymbol{\xi}, \omega)$ represent displacement and traction components of the fundamental solution at the field point $\boldsymbol{x}$ when a unit point load is applied at the load point $\boldsymbol{\xi}$ following the i-direction, and c_{ij} is a coefficient that depends on the geometry of the boundary at $\boldsymbol{\xi}$ ($c_{ij} = \delta_{ij}/2$ for smooth boundary). One can solve the integral equation for a sufficiently large number of ω values and numerically invert $U_i(\boldsymbol{x}, \omega)$ to obtain the time-dependent displacement. Viscoelastic material properties enter the Fourier transformed fundamental solution (Gaul and Fiedler 1993)

$$U_{ij}^* = \frac{1}{4\pi G^*} \left[\psi(r, c_1^*, c_2^*) \, \delta_{ij} - \chi(r, c_1^*, c_2^*) \, r_{1i} r_{1j} \right] \tag{77}$$

after having replaced the elastic moduli G, K by complex moduli in the wave speeds $c_1^{*2} = \frac{K^* + \frac{4}{3}G^*}{\rho}, c_2^{*2} = \frac{G^*}{\rho}$ in the functions ψ and χ which depend on the distance $r = (r_i r_i)^{1/2}$, $r_i = x_i - \xi_i$.

Only boundary integrals remain in (76) if the body forces B_i vanish. The integrals in (76) are discretized into the sum of integrals extending over the boundary elements in Fig. 15. This yields for a boundary node α

$$c_{ij}^{(\alpha)} U_j^\alpha = U_{ij}^{\alpha\beta} t_j^\beta - T_{ij}^{\alpha\beta} U_j^\beta \tag{78}$$

where

$$U_{ij}^{\alpha\beta} = \int_{\Gamma_{el}} U_{ij}^{*\alpha} \Omega^\beta \mathrm{d}\Gamma, \qquad T_{ij}^{\alpha\beta} = \int_{\Gamma_{el}} T_{ij}^{*\alpha} \Omega^\beta \mathrm{d}\Gamma, \tag{79}$$

contain the shape functions $\Omega^\beta(\boldsymbol{x})$ for the boundary elements [$\Omega^\beta(\boldsymbol{x}) = 1$ for constant elements], U_j^β and t_j^β are the displacement and traction nodal values. After numerical integration, the matrix formulation of (78) for all nodes and smooth boundary yields

$$\hat{\boldsymbol{T}} \boldsymbol{u} = \boldsymbol{U} \boldsymbol{t}, \qquad \hat{\boldsymbol{T}} = \frac{1}{2} \boldsymbol{E} + \boldsymbol{T}, \tag{80}$$

where

$$
\begin{aligned}
\boldsymbol{u} &= \{U^1 U^2 \cdots U^\alpha \cdots U^n\}^{\mathrm{T}}, \qquad U^\alpha = \{U_1^\alpha U_2^\alpha U_3^\alpha\}^{\mathrm{T}} \\
\boldsymbol{t} &= \{t^1 t^2 \cdots t^\alpha \cdots t^n\}^{\mathrm{T}}, \qquad t^\alpha = \{t_1^\alpha t_2^\alpha t_3^\alpha\}^{\mathrm{T}} \quad .
\end{aligned} \tag{81}
$$

Unknown boundary data are solved in terms of known boundary data after reordering (80).

6.3.1. *Numerical example: Substructure dynamics of elastomer isolators*

The present approach improves the modeling of elastomer supports mounts Fig. 15 by describing the 3-d elastodynamics with boundary elements instead of using static spring characteristics. The mobility matrix of the mount is calculated from the mixed boundary value problem of the elastomer domain Fig. 15 with two vulcanized steel plates. In the plate interfaces I and II rigid body displacement fields are required while the tractions at the free surface (a) vanish. The aim of the following analysis is to relate the 12 stress resultants, namely forces $F_i^{I,II}$ and moments $M_i^{I,II}$ to the 12 degrees of freedom namely displacements $u_i^{I,II}$ and the angles of rotation $\phi_i^{I,II}$, where $i = 1, 2, 3$ in a dynamic frequency dependent stiffness matrix.

By taking the stiffness matrix of a cylindrical elastomer support mount into account, the isolation factor of a single stage mounting system in Fig. 16 has been calculated. The 3-d BE approach describes the measured lower second longitudinal resonance frequency properly whereas the 1-d rod theory fails to do so. It includes lateral vibrations which create an anti-resonance at 1100 Hz. Proper viscoelastic constitutive description leads to a good prediction of experimental results in a broad frequency range.

7. Damping Description for the Finite Element Method

The finite element method (FEM) is a well established domain discretization approach. It has found increasing applications in damping design of complex structures. Several approaches exist for damping modeling in the context of mdof structures discretized by FEM (Pilkey and Pilkey 1995, Garibaldi and Onah 1996) such as

- viscoelastic material behavior
- hysteretic damping with assumed frequency dependence
- local damping mechanisms .

A special case of viscoelastic behavior is the viscous damping assumption. This model extends the mass and stiffness matrix in the FE equation of motion by a viscous damping matrix. Various forms of constructing this matrix exist. Several of these such as the so called proportional damping (Rayleigh damping), the Caughey series (general Rayleigh damping), the direct procedure which constructs the damping matrix with real modes and modal damping ratios and the hysteretic damping allow to decouple the equations of motion by modal transformation (Garibaldi and Onah 1996). The application of fractional derivatives to modal analysis is discussed in (Bagley and Calico 1989, Maia et al. 1996). A variety of solution procedures is based on real modes in time or frequency domain. If the damping mechanisms of the structures do not validate the decoupling assumption, complex modes exist. Decoupling can be obtained by rewriting the equations of motion in first order state space notation and transformation with right and left eigenvectors (Krämer 1984). Radiation damping can be simulated in finite element models by means of infinite elements (Bettes 1992). The near field between the radiating surface and an artificial boundary has to be discretized. On the artificial boundary one can prescribe boundary conditions which incorporate (exactly or approximately) the far field behavior into the FEM model (Givoli 1992). Infinite elements use for example in frequency domain analytical solutions in radial direction and shape functions in circumferential direction.

7.1. FINITE ELEMENT FORMULATION OF VISCOELASTIC SOLIDS

The implementation of viscoelastic constitutive equations of conventional differential operator type (85) or generalized type (92) as well as associated hereditary integral formulations (86) into the FE formulation for an elastic structure is discussed next. Spatial discretization reduces the elastodynamics of a continuous structure to the FE equation of motion

$$[M]\{\ddot{u}\} + [K]\{u\} = \{F\} \tag{82}$$

with n dofs in the displacement vector $\{u\} = [u_1\, u_2\, \ldots\, u_n]^T$ with mass and stiffness matrix $[M]$ and $[K]$ respectively and external force vector $\{F\}$. The stiffness matrix is assembled of element stiffness matrices $[K]_e$ which relate nodal dofs and nodal forces

$$[K]_e\{u\}_e = \{F\}_e . \tag{83}$$

The principles of mechanics lead to the stiffness matrix

$$[K]_e = \int_V [B]^T[E][B]\, dV \tag{84}$$

by volume integration over the element domain. The elasticity matrix $[E]$ relates the stress state with the strain state

$$\{\sigma\} = [E]\{\varepsilon\} . \tag{85}$$

In displacement theory, the element displacements $\{v\}$ are expressed in terms of the nodal displacements as

$$\{v\} = [N]\{u\}_e \tag{86}$$

by the matrix of shape functions $[N]$. The strain displacement relation with the differential operator matrix $[D]$

$$\{\varepsilon\} = [D]\{v\} = [D][N]\{u\}_e = [B]\{u\}_e \tag{87}$$

defines the element strain displacement matrix $[B]$. The elastic-viscoelastic correspondence principle is adopted to implement viscoelastic constitutive equations. A matrix notation of Hooke's law (82), (83), (84) and (85) is given by

$$\{\sigma\} = (2G[E]_G + 3K[E]_K)\{\varepsilon\} \tag{88}$$

with the vectors

$$\{\sigma\} = [\sigma_{xy} \, \sigma_{yz} \, \sigma_{zx} \, \sigma_{xx} \, \sigma_{yy} \, \sigma_{zz}]$$
$$\{\varepsilon\} = [\varepsilon_{xy} \, \varepsilon_{yz} \, \varepsilon_{zx} \, \varepsilon_{xx} \, \varepsilon_{yy} \, \varepsilon_{zz}] \tag{89}$$

and the matrices

$$[E]_G = \begin{bmatrix} 1 & 0 & 0 & 0 & 0 & 0 \\ & 1 & 0 & 0 & 0 & 0 \\ & & 1 & 0 & 0 & 0 \\ & & & \frac{2}{3} & -\frac{1}{3} & -\frac{1}{3} \\ \text{sym.} & & & & \frac{2}{3} & -\frac{1}{3} \\ & & & & & \frac{2}{3} \end{bmatrix}, \quad [E]_K = \begin{bmatrix} 0 & 0 & 0 & 0 & 0 & 0 \\ & 0 & 0 & 0 & 0 & 0 \\ & & 0 & 0 & 0 & 0 \\ & & & \frac{1}{3} & \frac{1}{3} & \frac{1}{3} \\ \text{sym.} & & & & \frac{1}{3} & \frac{1}{3} \\ & & & & & \frac{1}{3} \end{bmatrix}. \tag{90}$$

7.2. DIFFERENTIAL OPERATOR FORMULATION

The corresponding viscoelastic constitutive equation in differential operator notation is obtained by replacing the bulk modulus K by $\frac{1}{3}(Q_H/P_H)$ and the shear modulus G by $\frac{1}{2}(Q_D/P_D)$. As division with differential operators is not defined, (88) is rewritten as

$$P_D P_H \{\sigma\} = ([E]_G Q_D P_H + [E]_K Q_H P_D) \{\varepsilon\} . \tag{91}$$

The elastic constitutive equation (85) is now replaced by the viscoelastic one (91). The obtained equation of motion of one finite element with the mass matrix $[M]_e$

$$[M]_e P_D P_H \{\ddot{u}\}_e + ([K]_G Q_D P_H + [K]_K Q_H P_D) \{u\}_e = P_D P_H \{F\}_e \tag{92}$$

includes the volume integrals

$$[K]_G = \int_V [B]^T [E]_G [B] \, dV , \quad [K]_K = \int_V [B]^T [E]_K [B] \, dV . \tag{93}$$

The equation of motion of the structure associated to (82) is obtained by assembling the finite element equations of motion (92).

7.3. HEREDITARY INTEGRAL FORMULATION

The equivalent hereditary integral formulation of the constitutive equation is the integrated differential operator formulation (91)

$$\{\sigma\} = (2G_0[E]_G + 3K_0[E]_K) \{\varepsilon\} + \int_0^\infty \left(2\tilde{G}(\tau)[E]_G + 3\tilde{K}(\tau)[E]_K \right) \{\dot{\varepsilon}(t - \tau)\} \, d\tau \tag{94}$$

with relaxation modulus for shear $G(t) = G_0 + \tilde{G}(t)$ and compression $K(t) = K_0 + \tilde{K}(t)$. The associated finite element equation of motion is an integro-differential-equation

$$[M]_e \{\ddot{u}\}_e + (2G_0[K]_G + 3K_0[K]_K) \{u\}_e +$$
$$+ \int_0^\infty \left(2\tilde{G}(\tau)[K]_G + 3\tilde{K}(\tau)[K]_K \right) \{\dot{u}(t - \tau)\}_e \, d\tau = \{F\}_e . \tag{95}$$

Both formulations, the differential operator type and the hereditary integral type, have in common that the element matrices $[M]_e$, $[K]_G$ and $[K]_K$ in (92), (93) and (95) can be generated by commercial FEM software with elastic constitutive equations.

7.4. SOLUTION APPROACHES FOR FE EQUATIONS

Available approaches for solving the structure equation of motion (Carpenter 1972, Krings 1976, Padovan 1987) often prefer the differential operator formulation 13 for numerical reasons. Such approaches for calculating the displacement response $\{u\}$ are:

- Numerical integration (Padovan (Padovan 1987) uses the Grünwald definition of fractional derivatives in (12) for time-stepping algorithms)
- Finite time elements
- Integral transforms (Laplace, Fourier)
- Modal analysis (A summary of approaches is presented in (Pilkey and Pilkey 1995, Garibaldi and Onah 1996, Maia et al. 1996, Bagley and Calico 1989))
- Matrix functions

After the displacement response is known, differential equations for the stresses have to be integrated (Carpenter 1972, Krings 1976).

7.4.1. *Example: Viscoelastic finite rod element*

The equation of motion of a finite rod element in Fig. 17 shall be adapted to a viscoelastic conventional 3 parameter model with uniaxial constitutive equation

$$p_1 \, \dot{\sigma}_{xx} + \sigma_{xx} = q_0 \, \varepsilon_{xx} + q_1 \, \dot{\varepsilon}_{xx} \,. \tag{96}$$

One corresponding rheological model consists of a spring E_0 in parallel with spring E_1 and a viscous dashpot R_1 in series such that the parameters are related to (96) by

$$p_1 = \frac{R_1}{E_1} \qquad q_0 = E_0 \qquad q_1 = R_1 (1 + \frac{E_0}{E_1}) \,. \tag{97}$$

For the stress and strain state of the slender rod the correspondences

$$E \rightarrow \frac{Q}{P} = \frac{q_0 + q_1 \frac{d}{dt}}{1 + p_1 \frac{d}{dt}} \quad 2G \rightarrow \frac{Q_D}{P_D} = \frac{1}{1+\nu} \frac{Q}{P} \quad 3K \rightarrow \frac{Q_D}{P_D} = \frac{1}{1-2\nu} \frac{Q}{P} \tag{98}$$

with real Poisson's ratio ν hold.

Linear shape functions lead to the consistent mass matrix $[M]_e$ and elastic stiffness matrix $[K]_e$

$$[M]_e = \frac{\varrho A \ell}{6} \begin{bmatrix} 2 & 1 \\ 1 & 2 \end{bmatrix} \qquad [K]_e = \frac{E_0 A}{\ell} \begin{bmatrix} 1 & -1 \\ -1 & 1 \end{bmatrix} \tag{99}$$

with modal displacement and force vectors

$$\{F\}_e = \begin{bmatrix} F_i & F_j \end{bmatrix}^{\mathrm{T}} \qquad \{u\}_e = \begin{bmatrix} u_i & u_j \end{bmatrix}^{\mathrm{T}} \tag{100}$$

according to Fig. 17.

The differential operator type FE equation of motion (92) leads to

$$p_1 [M]_e \{\ddot{u}\}_e + [M]_e \{\ddot{u}\}_e + \frac{q_1}{E_0} [K]_e \{\dot{u}\}_e + [K]_e \{u\}_e = \{F\}_e + p_1 \{\dot{F}\}_e . \qquad (101)$$

With the relaxation modulus

$$E(t) = E_0 + E_1\, e^{-\frac{t}{p_1}} \qquad (102)$$

and relaxation moduli

$$2G(t) = \frac{E(t)}{1 + \nu} \qquad\qquad 3K(t) = \frac{E(t)}{1 - 2\nu} \qquad (103)$$

the hereditary integral type FE equation of motion (94) leads to

$$[M]_e \{\ddot{u}\}_e + [K]_e \{u\}_e + \int_0^\infty \frac{E_1}{E_0} e^{-\frac{\tau}{p_1}} [K]_e \{\dot{u}(t - \tau)\}_e \, \mathrm{d}\tau = \{F\}_e . \qquad (104)$$

The Kelvin-Voigt model of spring E_0 and dash-pot R_1 in parallel is obtained from (101) and (104) by taking the limit $E_1 \to \infty$

$$[M]_e \{\ddot{u}\}_e + \frac{R_1}{E_0} [K]_e \{\dot{u}\}_e + [K]_e \{u\}_e = \{F\}_e . \qquad (105)$$

The differential operator type FE equation for the constitutive equation (96) but with fractional order time derivatives $\alpha = 1/2$ has been treated by Bagley (Bagley and Torvik 1985).

8. Summary

The paper provides a unified approach for conventional and generalized linear models of viscoelastic constitutive behavior. Creep, relaxation and hysteresis effects of materials and structures are described consistently. Advantages of the fractional derivative concept are outlined. Mathematical consequences resulting from operator non-locality in time domain and uniqueness questions arising in frequency domain are addressed. The elastic-viscoelastic correspondence principle serves as a tool to obtain as well analytical and numerical BEM and FEM solutions of wave propagation and vibration problems by transform methods. Characteristics of viscoelastic waves and vibrations are discussed. The paper is focussed on material damping but includes aspects of radiation damping description by discretization methods as well.

Important aspects of damping description are beyond the scope of selected topics of this survey paper. This is why additional reading is recommended on the following subjects: Thermo-viscoelasticity and nonlinear viscoelasticity (Christensen 1971, Ferry 1980, Lazan 1968), determination of mechanical properties by experimental methods (Ferry 1980, Garibaldi and Onah 1996, Mahrenholtz and Gaul 1997, Bert 1973), damping devices and surface damping treatment (Nashif et al. 1985, Cremer, Heckl and Ungar 1973, Garibaldi and Onah 1996), material damping data (Lazan 1968, Nashif et al. 1985) and structural damping (Ruzicka 1960) including the nonlinear dissipation in mechanical joints such as bolted or riveted connections (Gaul and Lenz 1997, Gaul and Sachau 1997, Ottl 1985). A list which is by far not complete.

References

Bagley, R. L. and Calico, R. A.: 1989, The fractional order state equations for the control of viscoelastically damped structures, *AIAA Journal* **89-1213**, 487–496.

Bagley, R. L. and Torvik, P. J.: 1985, Fractional calculus in the transient analysis of viscoelastically damped structures, *AIAA Journal* **23**(6), 918–925.

Bert, C. W.: 1973, Material damping: An introductory review of mathematical models measures and experimental techniques, *J. Sound Vibration* **29**(2), 129–153.

Bettes, P.: 1992, *Infinite elements*, Penshaw Press, Cleadon.

Beyer, H. and Kempfle, S.: 1995, Definition of physically consistent damping laws with fractional derivatives, *ZAMM* **74**(8), 623–635.

Carpenter, W. M.: 1972, Viscoelastic stress analysis, *Int. J. Num. Meth. Engg.* **4**, 357–366.

Christensen, R. M.: 1971, *Theory of viscoelasticity*, Academic Press, New York.

Crandall, S. H.: 1962, Dynamic response of systems with structural damping, *in* S. Lees (ed.), *Air, Space and Instruments, Draper Anniversary Volume*, McGraw-Hill, New York, pp. 183–193.

Crandall, S. H.: 1970, The role of damping in vibration theory, *J. Sound Vibration* **11**(1), 3–18.

Cremer, L., Heckl, M. and Ungar, E. E.: 1973, *Structure borne sound*, Springer, Berlin.

Cupiał, P.: 1998, Some approaches to the analysis of nonproportional damped viscoelastic structures, *in* D. Besdo (ed.), *International Symposium an Dynamics of Continua*, Shaker Verlag, Aachen.

Doyle, J. F.: 1998, *Wave propagation in structures*, Springer, New York.

Ferry, J. D.: 1980, *Viscoelastic properties of polymers*, John Wiley & Sons, New York.

Flügge, W.: 1975, *Viscoelasticity*, Springer, Heidelberg.

Gaer, M. C. and Rubel, L. A.: 1975, Fractional calculus and its applications, *in* A. Dold and A. Eckmann (eds), *The Fractional Derivative and Entire Functions*, Vol. 457 of *Lecture Notes in Mathematics*, Springer, Berlin, pp. 171–206.

Garibaldi, L. and Onah, H. N.: 1996, *Viscoelastic material damping technology*, Becchis Osivide, Torino.

Gaul, L.: 1980, *Dynamics of soil structure interaction*, Habilitation thesis, University of Hanover. (in german).

Gaul, L.: 1991, Dynamic transfer behaviour of elastomer isolators; boundary element calculation and experiment, *Mechanical Systems and Signal processing* **5**(1), 13–24.

Gaul, L. and Chen, C. M.: 1993, Modelling of viscoelastic elastomer mounts in multibody systems, *in* W. Schielen (ed.), *Advanced Multibody System Dynamics*, Kluwer Academic Publ., Dordrecht, pp. 257–276.

Gaul, L. and Fiedler, C.: 1993, Improved calculation of field variables in the domain based on BEM, *Engineering Analysis with Boundary Elements* **11**, 257–264.

Gaul, L. and Fiedler, C.: 1997, *Boundary element method in statics and dynamics*, Vieweg Verlag, Braunschweig.

Gaul, L. and Lenz, J.: 1997, Nonlinear dynamics of structures assembled by bolted joints, *Acta Mechanica* **125**, 169–181.

Gaul, L. and Sachau, D.: 1997, Nonlinear active damping of adaptive space structures, *in* F.-K. Chang (ed.), *Structural Health Monitoring*, Technomic Publishing Co., Lancaster, pp. 208–219.

Gaul, L. and Schanz, M.: 1994, Dynamics of viscoelastic solids treated by boundary element approaches in time domain, *European Journal of Mechanics A/Solids* **13**, 43–59.

Gaul, L. and Schanz, M.: 1997, Boundary element calculation of transient response of viscoelastic solids based on inverse transformation, *Meccanica* **32**, 171–178.

Gaul, L., Bohlen, S. and Kempfle, S.: 1985, Transient and forced oscillations of systems with constant hysteretic damping, *Mechanics Research Communications* **12**(4), 187–201.

Gaul, L., Klein, P. and Kempfle, S.: 1991, Damping description involving fractional operators, *Mechanical systems and signal processing* 5(2), 81–88.

Givoli, D.: 1992, *Numerical methods for problems with infinite domains*, Elsevier, Amsterdam.

Grünwald, A. K.: 1967, über begrenzte derivationen und deren anwendungen, *Zeitschrift für Mathematik und Physik* 12, 441–480.

Kempfle, S. and Gaul, L.: 1996, Global solutions of fractional linear differential equations, *Proc. of ICIAM95, ZAMM 76 suppl. 2* pp. 246–251.

Knauss, W. G.: 1968, Uniaxial wave propagation in a viscoelastic material using measured material properties, *J. Applied Mechanics* 35, 449–453.

Krämer, E.: 1984, *Machine dynamics*, Springer, Berlin. (in german).

Krings, W.: 1976, Contribution to FEM for linear, viscoelastic material, *Mitteilungen aus dem Institut für Mechanik 3*, Ruhr-Universität Bochum. (in german).

Lanzerath, H.: 1996, The application of BEM for modal analysis, *Mitteilungen aus dem Institut für Mechanik 103*, Ruhr-Universität Bochum.

Lazan, B. J.: 1968, *Damping of materials and members in structural mechanics*, Pergamon Press, Oxford.

Mahrenholtz, O. and Gaul, L.: 1997, Damping questions, VDI-Bildungswerk BW 2950, 1-33. (in german).

Maia, N. M. M., Silva, J. M. M., Ribeiro, A. M. R. and Leitão, J. J. A. A.: 1996, On the possible application of fractional derivatives to modal analysis, *14th Int. Modal Analysis Conference (IMAC XIV)*, Vol. 1, Dearborn Detroit, pp. 172–177.

Nakagawa, N. and Kawai, R.: 1980, Transient wave propagation in viscoelastic bars, *Memoirs of the Faculty of Engineering, Kobe University* 26, 1–17.

Nashif, A. D., Jones, D. I. G. and Henderson, J. P.: 1985, *Vibration damping*, John Wiley & Sons, New York.

Oldham, K. B. and Spanier, J.: 1974, *The fractional calculus*, Academic Press, San Diego.

Ottl, D.: 1985, *Nonlinear damping in space structures*, number 73 in *VDI-Fortschrittsberichte Reihe 11*, VDI Verlag. (in german).

Padovan, J.: 1987, Computational algorithms for FE formulations involving fractional operators, *Computational Mechanics* 2, 271–287.

Papoulis, A.: 1962, *The fourier integral and its applications*, McGraw-Hill, New York.

Pilkey, W. and Pilkey, B. (eds): 1995, *Shock and vibration computer programs*, The Shock and Vibration Information Analysis Center. Booz, Allen & Hamilton, Arlington.

Rossikhin, Y. A. and Shitikova, M. V.: 1997, Applications of fractional calculus to dynamic problems of linear and nonlinear hereditary mechanics of solids, *Applied Mechanics Reviews* 50(1), 15–67.

Ruzicka, J. E. (ed.): 1960, *Structural damping*, Pergamon Press, Oxford.

Schanz, M.: 1994, A boundary element formulation in time domain for generalized viscoelastic constitutive equations, *Bericht aus dem Institut A für Mechanik*, Universität Stuttgart. (in german).

Torvik, P. J. and Bagley, D. L.: 1987, Fractional derivatives in the description of damping materials and phenomena, *ASME DE-5: The role of damping in Vibration and Noise Control*, pp. 125–135.

434

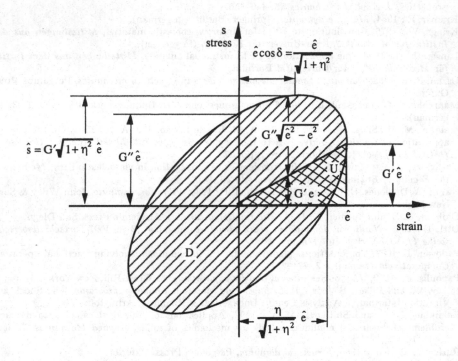

Figure 1   Elliptical hysteresis loop characteristic of linear damping.

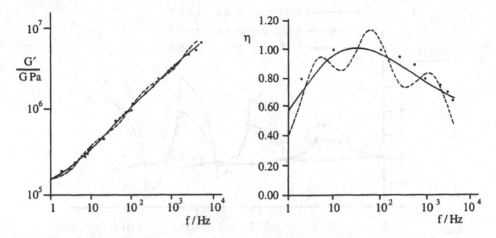

Figure 2   Storage modulus G´(ω) and loss factor η(ω); ..... Experimental data; ——— 5 parameter fractional derivative model (G = 0.87 x 10⁵ GPa, a = 0.039, α = 0.39, β = 0.64); - - - - Conventional 9 parameter model

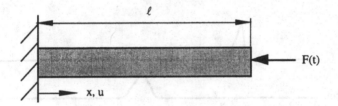

Figure 3   Fixed-free thin rod with coordinate x, displacement u

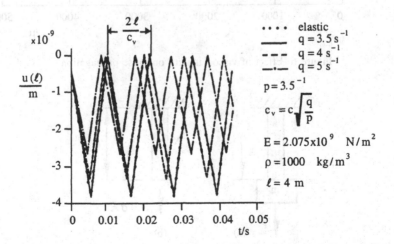

Figure 4   Tip deflection due to wave propagation in elastic and viscoelastic solid

436

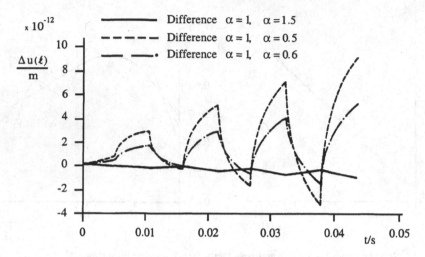

Figure 5 Tip deflection difference

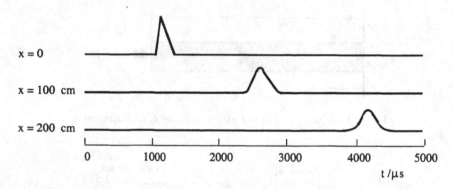

Figure 6 Effect of viscoelasticity on pulse propagation

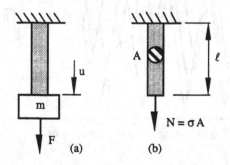

Figure 7 (a) sdof oscillator; (b) massless member

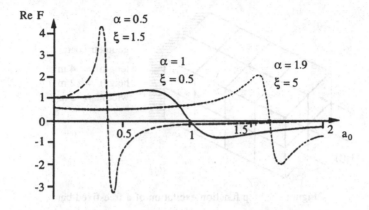

Figure 8 FRF of sdof oscillator with fractional derivative damping model

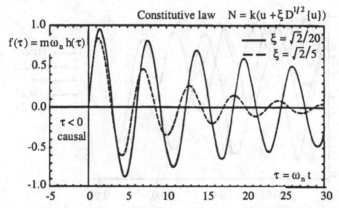

Figure 9 Impulse response of oscillator with fractional derivative damping model

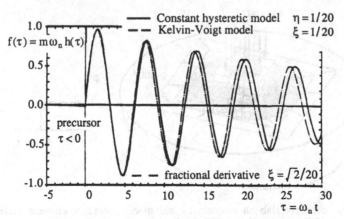

Figure 10 Impulse response for different damping models

438

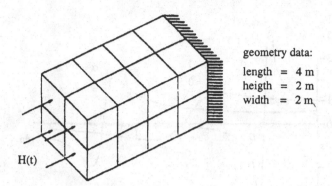

geometry data:

length = 4 m
heigth = 2 m
width = 2 m

Figure 11 Step function excitation of a free-fixed bar

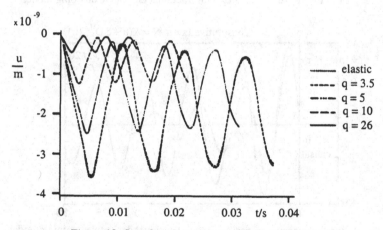

Figure 12 Step function response of a free-fixed bar

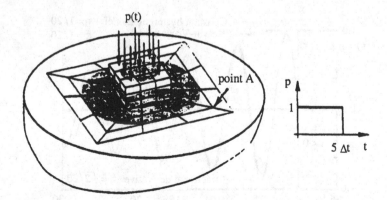

Figure 13 Elastic concrete slab on viscoelastic halfspace: boundary element discretization, surface displacement wave fronts, losding function

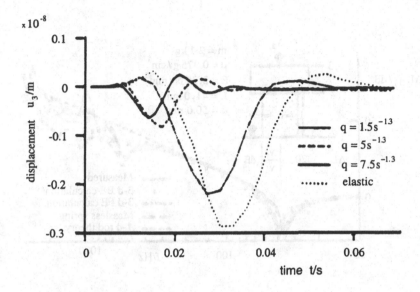

Figure 14 Displacement u3 perpendicular to surface on point A for different values q

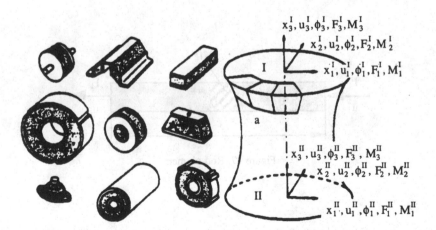

Figure 15 Elastomer mounts and BE discretization

440

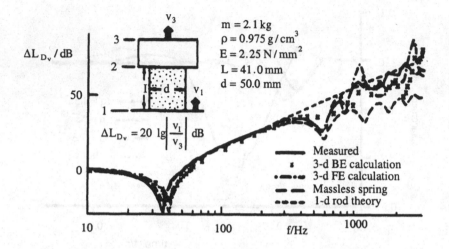

Figure 16 Elastomer isolator

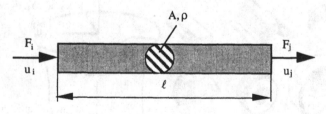

Figure 17 Rod element

EXISTENCE AND NORMALIZATION OF COMPLEX MODES FOR POST EXPERIMENTAL USE IN MODAL ANALYSIS

S. R. IBRAHIM
Professor of Mechanical Engineering
Old Dominion University
Norfolk, VA, U.S.A.

Summary

The eigen characteristics of a dynamical system offer a vector sub-space suitable for performing canonical transformations on the system of equations of the structural dynamics problems being considered. While the concept is mathematically fully developed for both damped and undamped systems, practitioners -- at all levels -- tend to indiscriminately use the system's normal modes as a basis for applications containing nonproportional damping. Such a practice in most cases is a reasonable approximation and results in small, if not infinitesimal, errors. However, with the increase of sophistication and accuracy requirements in certain applications of modal analysis, these approximations must be fully analyzed and understood.

In this paper, emphasis is directed to some specific applications for which it is generally a common practice to use normal, or undamped, modes as a vector subspace for use with nonproportionally damped system. These are:

a) Expansion of measured mode shapes for use in model updating procedures.

b) Computation of normal modes from identified complex mode shapes.

c) Response calculations and modal filtering.

Detailed background on complex modes is presented. Commonly used transformations are examined and error models are derived and quantified.

Nomenclature

A, B	state space system's matrices	Z	real mode matrix
M, K, C	mass, stiffness, viscous damping matrices	Ψ	eigenvector matrix
		Λ	spectral matrix
E, E'	error matrices	ψ_i	i^{th} eigenvector
S	matrix of singular values	e	error vector in Eq. (22)
T	transformation matrix	q(t)	complex modal coordinates
U, V	matrices of left and right singular vectors	x(t)	system response
		$\eta(t)$	principal coordinates

J.M.M. Silva and N.M.M. Maia (eds.), Modal Analysis and Testing, 441–452.

m	number of measured modes
n	number of measured dofs
δ	damping factor
$\varepsilon, \varepsilon_1$	relative errors
λ	complex eigenvalue

1. Introduction

Dissipative self adjoint structures, in many cases, possess highly complex modes due to the existence of nonproportional damping mechanisms. Such situations are becoming more common, rather than exceptional, due to the recent trends of using additional passive damping in modern structures to reduce noise and vibration.

Unlike the classical phase resonance method, which is based on measuring an approximation to the normal modes of structures under test, the majority of the modern modal identification algorithms readily identify the complex mode shapes. Even though techniques have the option of identifying the normal modes of a non-proportionally damped structures, such options remain to be based on some inherent assumptions and approximations in their underlying theories.

On the other hand, and due to a lagging state of the art in damping modeling and analysis, the normal mode theory remains to be the basis for the majority of modal analysis applications such as modeling, structural modification, component mode synthesis, sensitivity analysis, modeling error localization and model updating; among many others.

The concepts and usefulness of normal modes are basic common knowledge. Those involved with mechanical vibrations and modal analysis quite frequently deal with complex modes and understand how they can be used as an orthogonal basis able to uncouple the equations of motion, in presence of nonproportional damping. However, since the original contribution of Caughey and O'Kelly [1], many papers have been written on complex modes, indicating that this subject is not yet fully understood.

Recently, motivated by problems arising in advanced applications of modal analysis, some authors reanalyzed many aspects of complex modes. Mitchell [2] presented a clear review on the subject, Lang [3] tried to explain when and how modal analysis leads to complex modes and what is the effect of their representation. Liang et al. [4] posed and analyzed very interesting problems: whether the concept of normal modes is equivalent to the concept of proportional damping, whether the existence of complex modes is an indicator of non-proportional damping and how is a mode influenced by damping.

The need to use the measured or identified modes, which are usually complex, in techniques or algorithms which assume normal mode, presented a definite dilemma for researchers for the past several decades. The past practices of simply ignoring the complexity of the measured modes by dealing only with their magnitudes, can produce needlessly erroneous results in a field which is striving for higher accuracy. Thus, research to normalize complex modes or to compute a corresponding set of normal modes have increased in recent years. Majority of the contributions in such an area assume that measured mode complexity is solely due to nonproportional damping.

Deblauwe and Allemang [5] however, and correctly, pointed out that there could be other reasons for the complexity of identified modal vectors. These could be:

1. Aliasing
2. Leakage
3. Mass loading effects
4. Measurements noise
5. Nonlinearities
6. High modal density
7. dentification errors.

Since it is difficult, if not impossible, to simultaneously consider all such factors in normalizing complex modes, it will be also assumed here that the complexity is caused only by the nonproportional damping. Any procedure however is hoped to indirectly deal with the other factors, which cause complexity of measured modes, by means of smoothing or filtering out their effects or contribution to mode complexity.

The major obstacle in computing normal modes from measured complex modes is the fact that the set of measured modes is incomplete. The number of modal coordinates (number of identified modes) is usually much smaller than the number of physical coordinates (measurements).

Leuridan [6] suggested using the measured data to directly estimate the mass, stiffness and damping matrices after which the damping is set to zero and normal modes are calculated. Ibrahim [7], [8] used oversized identification models with noise added to complete the response rank deficient matrices as well as adding some assumed modes to complete the measured set. Niedbal [9] solved the problem by applying the physical characteristics of the real normal formulation. The imaginary parts of the complex modes are minimized by the least square solution and the orthogonality conditions of the generalized mass and stiffness matrices are maintained by the eigenvalue calculation. Zhang and Lallement [10] proposed a method which is based on using orthogonality constraints. Natke and Roberts [11] studied computing normal modes via the improvement of a mathematical model, using the identified modal data, and later setting damping to zero.

Zhang and Lallement [12], Wei, Allemang and Brown [13] and Wei, Zhang, Allemang and Brown [14] studied the use of a transformation matrix to transform the incomplete system's physical coordinates to the modal coordinates, subspace coordinates or principal coordinates and thus deal with a reduced complete system. These approaches however are iterative since they are based on assuming an initial transformation matrix which is to be updated till convergence is achieved. Later work [15-18] addressed the same issues.

2. Theoretical background on complex modes

Any real structure is dissipative. The linear dissipation mechanisms can be related to different state variables. For structural damping the proportionality linear function is with displacement, for viscous damping dissipation is linearly related to velocity; in other cases a nonlinear relationship may exist and dissipation can depend on higher order terms of displacements and/or velocities.

444

The linear models generally used in structural dynamics are the viscous damping and the structural damping. Both of these models can be represented either as proportional or non-proportional. Here the non-proportional viscous model is considered.

When using a viscous model, the real modes of vibration Φ of the associated conservative system diagonalize the mass and stiffness matrices, but not the damping matrix. As a matter of fact those modes are not energetically independent when damping is present, i.e. there is an energy dissipation in a given mode, due to the effect of adjacent mode. This effect is quantitatively defined by the off-diagonal terms of the generalized damping matrix $\Phi^T C \Phi$, resulting in a modal coupling which depends on the value of damping and the modal density of the system. In order to determine a set of eigenvectors capable of uncoupling the equation of motion for the non conservative system, a new eigenvalue problem must be formulated for the non conservative system. In fact, it is known that the eigenvectors that solve the homogeneous equation of the viscous damped system do not uncouple the equation of motion:

$$M\ddot{x} + C\dot{x} + Kx = f \qquad (1)$$

with M, C and K of order m×m. In order to determine the required set of eigenvectors that diagonalize the above three matrices, the original homogeneous problem must be transformed into the normal form,

$$A\dot{y} = By \qquad (2)$$

with

$$y = \begin{bmatrix} x \\ \dot{x} \end{bmatrix} \qquad A = \begin{bmatrix} C & M \\ M & 0 \end{bmatrix} \qquad B = \begin{bmatrix} -K & 0 \\ 0 & M \end{bmatrix}$$

This transformation preserves the matrices symmetry. The eigenvalue problem:

$$[-B + \Lambda A]\Phi = 0$$

yields real or complex conjugate eigenvalues λ, and real or complex conjugate eigenvectors. Therefore, it is possible to build up an extended modal matrix Φ_E and an external spectral matrix Λ_E (both usually complex conjugates for underdamped engineering structures) which have the form:

$$\Phi_E = \begin{bmatrix} \Psi & \Psi* \\ \Psi\Lambda & \Psi*\Lambda* \end{bmatrix} \qquad \Lambda_E = \begin{bmatrix} \Lambda & 0 \\ 0 & \Lambda* \end{bmatrix}$$

where Ψ is the eigenvector matrix of the original problem

2.1 ENERGETICAL CONSIDERATIONS

The eigenvectors Φ_E:

- are defined within an arbitrary (complex) scale constant;

- are orthogonal with respect to the matrices A and B previously defined;
- represent a complete set of 2m orthogonal eigenvectors which uncouple the system written in normal form, i.e.:

$$\Phi_E^T A \Phi_E = \overline{A}_E \qquad \Phi_E^T B \Phi_E = \overline{B}_E$$

with $\overline{A}_E$ are $\overline{B}_E$ diagonal matrices.

It is worthwhile to point out that Φ_E is the set of modes which satisfies the previous properties, but not Ψ (the eigenvectors of the original system). However, in order to understand the energetic properties related to the eigenvectors Ψ we can rewrite the previous equations in terms of Ψ, i.e.:

$$\Psi^T M \Psi \Lambda + \Lambda \Psi^T M \Psi + \Psi^T C \Psi = \text{diag}[a_i] = \overline{A} \tag{3}$$

$$\Psi^{*T} M \Psi \Lambda + \Lambda^* \Psi^{*T} M \Psi + \Psi^{*T} C \Psi = 0 \tag{4}$$

$$\Lambda \Psi^T M \Psi \Lambda - \Psi^T K \Psi = \text{diag}[b_i] = \overline{B} \tag{5}$$

$$\Lambda^* \Psi^{*T} M \Psi \Lambda - \Psi^{*T} K \Psi = 0 \tag{6}$$

Analogous relationships hold for the complex conjugate pair. These relations show that the Ψ matrix alone does not diagonalize the physical matrices M, C, K. Therefore, the eigenvectors Ψ_i are not anymore energetically independent. Due to damping there is coupling and the energy stored and dissipated by a single mode depends on the other modes. In particular it can be shown that the orthogonality conditions state a global transfer energy balance among interacting modes. The energy, dissipated by one mode, coupled with another mode, is equal to the energy that is transferred to it by the inertial interaction with the second mode. Analogous relations holds for the elastic and inertial interaction.

2.2 SCALING OF COMPLEX MODES

It is known that the scaling process does not affect the shape of real modes, i.e. their geometrical appearance. This situation is no more valid for complex modes. As the real modes, complex modes can be arbitrarily scaled, but the way in which scaling is performed changes the appearance of complex modes, because there is a transfer of information between the real and imaginary parts of modes. Any choice is possible, but the scaling commonly used in modal analysis are:

(i) largest component (in modulus) is real and equal to one

(ii) $$\Psi^T M \Psi \Lambda + \Lambda \Psi^T M \Psi + \Psi^T C \Psi = I \tag{7}$$

and

$$\Lambda \Psi^T M \Psi \Lambda - \Psi^T K \Psi = \Lambda$$

(iii) $$\Psi^T M \Psi \Lambda + \Lambda \Psi^T M \Psi + \Psi^T C \Psi = 2j\Lambda_I \qquad (8)$$

and $$\Lambda \Psi^T M \Psi \Lambda - \Psi^T K \Psi = 2j\Lambda_I \Lambda$$

in which $\Lambda = \Lambda_R + j\Lambda_I$.

If the (iii) normalization is used, the imaginary parts of the modes are minimized and the real parts are maximized. Thus the real part holds most of the information. Moreover this normalization is equivalent to unit mass normalization used for real modes, because it yields a modal mass ($\Psi_i^T M \Psi_i$) closet to the identity matrix.

The problem of complex modes scaling is particularly important in some applications on complex modes, e.g. in the identification of real modes associated with complex modes.

3. Investigation of the use of some common equations

3.1 EXPANSION OF COMPLEX MODES IN THE SUBSPACE OF REAL MODES

In several investigations on complex mode normalization a common assumption is that the complex eigenvectors Ψ (let us call them modes) can be expanded into the subspace of real modes, namely:

$$\Psi = Z T \qquad (9)$$

in which Ψ is the complex modal matrix, Z the real modal matrix and T a complex matrix. The above expression represents a linear transformation of the complex modes into the real ones. Ψ is assumed to be a maximum rank. Both T, $\Psi \in \mathbb{C}$ such that:

$$\Psi = \Psi_R + j\Psi_I \qquad T = T_R + jT_I$$

To investigate the above assumption, a system in which the number of measured d.o.f.s n equals the whole set of measured modes is considered. For the sake of simplicity we call it a complete system whereas for incomplete system we mean a system in which the number of measurements is larger than the number of identified modes. Note however that this definition is not correct because a complete system can be strictly defined only for a discrete system and is meaningless for continuous systems. For a complete system we have:

$$\Psi_R + j\Psi_I = Z(T_R + jT_I)$$

i.e.:

$$\Psi_R = Z T_R \qquad \Psi_I = Z T_I$$

Z can be determined by an inversion operation to give:

$$Z = \Psi_R T_R^{-1} \qquad Z = \Psi_I T_I^{-1} \tag{10}$$

from which:

$$\Psi_R T_R^{-1} = \Psi_I T_I^{-1}$$

i.e.

$$T_I = T_R \Psi_R^{-1} \Psi_I = T_R L \tag{11}$$

with $L \in \mathbb{R}$ ($m \times m$), which links the imaginary part of T to its real part, through a new real matrix $\Psi_R^{-1} \Psi_I$, as well the imaginary part of Ψ with its real part, i.e:

$$\Psi_I = Z T_I = Z T_R L = \Psi_R L \tag{12}$$

Consider now an incomplete system. Is equation (9) still valid for the incomplete system where, as it is usual in experimental modal testing, the number of measurements n is larger than the number of identified modes m? In more precise terms the question can be posed as follows: is it possible to find a matrix T such that Eq.(9) still holds? If it exists, then:

$$\Psi_{n \times m} = Z_{n \times m} T_{m \times m}$$

But, if the above is true, then, for the complete system, we could write, with different matrix dimensions:

$$\Psi_{n \times n} = Z_{n \times n} T_{n \times n}$$

and, after matrix partitioning:

$$\Psi_{n \times m} = Z_{n \times m} T_{m \times m} + Z_{n \times (n-m)} T_{(n-m) \times m} \tag{13}$$

If equation (13) must be satisfied, still being valid equation (9) for the complete system, the matrix T should be uncoupled, namely $T_{(n-m) \times m} = T_{m \times (m-n)}$, such that the higher real modes, not accounted for in the incomplete system, would not have influence on the lower complex modes. This is presumably true under two conditions:

(i) complex modes are almost real, such that there is an almost direct correspondence between every complex and real mode. In such a case $T_I \approx 0$, and T_R is not only banded but almost diagonal with elements approximately equal to 1.

(ii) damping is proportional, i.e only real modes exist. In this case $T_I = 0$, and T_R is strictly a unit matrix, i.e. $\Psi_I = 0$ and $\Psi_R = Z$.

Excluding the previous conditions, it can be stated that in any other case it is not possible for an incomplete system to express T_I as a linear combination of T_R and thus it is not correct to expand the complex modes into the linear subspace of real modes.

Another way to show that equation (9) does not generally hold is through the use of the singular value decomposition (SVD) of the Ψ matrix. Let us decompose the complex matrix $\Psi_{n \times m}$ into its singular values, i.e.:

$$\Psi_{nxm} = [X + jY] = U_{x_{nxm}} S_{x_{mxm}} V_{x_{mxm}}^T$$
$$+ jU_{y_{nxm}} S_{y_{mxm}} V_{y_{mxm}}^T$$

with $X = \Psi_R$ and $Y = \Psi_I$. Since we wish to check whether Ψ can be written as the product of a real matrix (the real mode matrix Z) and a complex matrix, let us express U_x in terms of Z and U_y in terms of U_x as follows:

$$U_x = ZC$$

$$U_{y_{nxm}} = U_{x_{nxm}} Q_{mxm} \qquad (14)$$

where C and Q are suitable transformation matrices. With the above positions we can write:

$$\Psi = Z \left[C S_x V_x^T + j C Q S_y V_y^T \right]$$

In this relation the right hand side in square brackets is a complex matrix, say $\hat{T}$, i.e.:

$$\Psi = Z \hat{T} \qquad (15)$$

and it can be easily shown that $\hat{T} \equiv T$ defined in Eq.(9).

Now, if $n = m$, the above formulation is correct because we can always determine a matrix Q such that relation (14) holds, provided that U_x is not singular, which here is excluded because of the nature of our problem. In fact, it can be written that:

$$Q = U_x^{-1} U_y$$

On the contrary, when $n > m$, the matrix Q does not generally exist. Therefore, in this case equation (14) can be rewritten but with the introduce a matrix of errors E:

$$U_y = U_x Q + E \qquad (16)$$

that is equation (14) is not strictly valid, which means that the complex eigenvectors are not generally expandable into the real ones.

It is worthwhile to point out that the error is zero in the following two circumstances:

- $n = m$, i.e. the number of measurements is equal to the number of measured modes. Note however, that this does not generally mean that the identified real modes are equal to those determined from the correspondent conservative system;

- the system is undamped or exhibits a proportional damping mechanism. In both these cases Y = 0 provided that the eigenvectors are normalized to unit mass or such that the real parts are maximized (Eq.8). For any other normalization condition the above conditions yield a direct proportionality between X and Y, i.e. between U_x and U_y. In this case the rank r of U_y ($r \le \min(n, m)$) is equal to the rank of the augmented matrix $(U_x | U_y)$ ($r' = \text{rank}\ (U_x | U_y)$) and the basic results of matrix theory show that the equations are consistent, i.e. a unique solution exists if m = r. Therefore, the normalization operation usually introduces a rotation in the complex plane and, consequently, a fictitious imaginary part in the eigenvectors.

3.2   RELATION BETWEEN PHYSICAL AND PRINCIPAL COORDINATES

The second equation that is frequently assumed in many modal analysis applications such as transformation complex modes into real ones, response computation, control system design, etc., is:

$$x(t)_{nx1} = Z_{nxm}\ \eta(t)_{mx1} \qquad (17)$$

in which x(t) is the response of equation (1) (namely damped response), Z the real modal matrix and η the principal coordinates. This equation would mean that the damped physical coordinates can be related to the principal coordinates through the real modes.

For viscous damping the physical coordinates x may be expanded into the principal ones as:

$$\begin{bmatrix} x(t) \\ \dot{x}(t) \end{bmatrix}_{2nx1} = \begin{bmatrix} \Psi & \Psi^* \\ \Psi\Lambda & \Psi^*\Lambda^* \end{bmatrix}_{2nx2m} \begin{bmatrix} q(t) \\ q^*(t) \end{bmatrix}_{2mx1}$$

where q(t) is necessarily complex. Rewriting the previous equation for x only:

$$x(t)_{nx1} = \Psi_{nxm}\ q(t)_{mx1} + \Psi^*_{nxm}\ q^*(t)_{mx1}$$

i.e.

$$x(t)_{nx1} = \hat{\Psi}_{nx2m}\ \hat{q}(t)_{2mx1} \qquad (18)$$

Eq.(18) is straightforward but seems to contradict Eq.(17), because 2m modes are needed instead of m. Can they both be valid?

Since Ψ and q(t) are complex, we can write:

$$\Psi = \Psi_R + j\Psi_I \qquad q(t) = q_R(t) + jq_I(t)$$

so that:

$$x(t) = \Psi q(t) + \Psi^* q^*(t) = 2\Psi_R q_R(t) - 2\Psi_I q_I(t)$$

(and in fact x must be real), i.e.:

$$x(t)_{nx1} = 2\begin{bmatrix} \Psi_R & -\Psi_I \end{bmatrix}_{nx2m} \begin{bmatrix} q_R(t) \\ q_I(t) \end{bmatrix}_{2mx1} \tag{19}$$

Therefore, for non proportional damping, it is not possible to strictly define a real matrix of order $n \times m$ (real modes) on which expand the physical coordinates. Of course, if Ψ is almost real, then: $\Psi_R + j\Psi_I \approx \Psi_R = Z$, and then Eq.(17) can hold.

It was already shown that, if the system is complete, real and imaginary parts of eigenvectors can be related through a linear transformation L as $\Psi_I = \Psi_R L$, in which $L = \Psi_R^{-1}\Psi_I$. In this case Eq.(19) writes:

$$x(t) = 2\begin{bmatrix} \Psi_R & -\Psi_R L \end{bmatrix} \begin{bmatrix} q_R(t) \\ q_I(t) \end{bmatrix}$$

or

$$x(t)_{nx1} = 2\left\{ \Psi_R q_R(t) - \Psi_R L q_I(t) \right\} \tag{20}$$

Reminding that:

$$\Psi_R = Z T_R$$

$$\Psi_I = Z T_I = Z T_R L$$

$$x(t) = 2\left\{ Z T_R q_R(t) - Z T_R L q_I(t) \right\}$$

i.e.

$$x(t) = Z_{nxn}\left\{ a_1(t) - a_2(t) \right\}_{nx1} = x(t) = Z \eta(t)$$

where $a_1 = Z T_R q_R$, $a_2 = Z T_R L q_I$, and $\eta = a_1 - a_2$. However, this solution is valid only for a complete system, and for an incomplete system in general we have:

$$\Psi_I = \Psi_R L + E' \tag{21}$$

and consequently:

$$x(t) = Z \eta(t) + e = Z \eta(t) - 2 E' q_I \tag{22}$$

where E' and e are an error matrix and an error vector, respectively. The error is increasingly important as damping and incompleteness become significant. In fact, L can be only determined in a least-square sense (no solution exists) and the error in Eq.(21) is related to this lack of solution.

4. Concluding remarks

From the foregoing analyses, it is concluded that some of the frequently used transformations which utilize normal modes as a basis, to expand complex modes and responses of nonproportionally damped systems, are approximate if not totally invalid. They become acceptable approximations when the applications, for which the results are intended to be used in, are relatively insensitive to introduced errors. In modal analysis such applications are numerous.

However, there exist a few applications in which the use of such above approximations can lead to erroneous and misleading results. One of these applications is the use of measured mode shapes to update analytical models. Such update is based on the direct amount of derivation of experimental "normal" mode shapes from those of the analytical model. Using normal modes obtained through the discussed transformations can lead to the erroneous updating of the analytical model to match the erroneous experimental modes.

The other application of interest which is as sensitive to errors in experimental data is the use of experimentally measured mode shapes for structural integrity monitoring. If monitoring is based on deviations of normal mode shapes, the transformations discussed in this paper should be avoided.

5. References

1. Caughey, T. K., O'Kelly, M. M. J. (1965), Classical normal modes in damped linear dynamic systems, *Trans. ASME, J. Applied Mech. 32*, 583-588.
2. Mitchell, L. (1990), Complex modes: a review, *Proc. 8th IMAC*, Kissimmee, FL.
3. Lang, G. F. (1989), Demystifying complex modes, *Sound and Vibration, 23*, 36-40.
4. Liang, Z., Tong, M., Lee, G. C. (1992), Complex modes in damped linear dynamic systems, *Int. J. of Analytical and Experimental Modal Analysis, 7*, 1-20.
5. Deblauwe, F., Allemang, R. J. (1986), A possible origin of complex modal vectors, *Proceedings of the 11th International Seminar on Modal Analysis, paper A2-3*, Katholic University of Leuven.
6. Leuridan, J., Brown, D. and Allemang, R. J. (1982), Direct system parameter identification of mechanical structures with application to modal analysis, *AIAA paper 82-0767*, 548-556.
7. Ibrahim, S. R. (1982), Dynamic modeling of structures from measured complex modes, *AIAA Journal, 21*, 898-901.
8. Ibrahim, S. R. (1983), Computation of normal mode from identified complex modes, *AIAA Journal, 21*, 446-451.
9. Niedbal, N. (1984), Analytical determination of real normal modes from measured complex responses, *SAE Paper No. 84-0095*, Palm Springs, CA.
10. Zhang, Q, Lallement (1985), New method of determining the eigensolutions of the associated conservative structure from the identified eigensolutions, *Proceedings of the 3rd International Modal Analysis Conference*, 322-328.
11. Natke, H. G. and Robert, D. (1985), Determination of normal modes from identified complex modes, *Z. Flugwiss. Weltraumforsch, g.*, 82-88.
12. Zhang, Q., Lallement, G. (1987), Comparison of normal eigenmodes calculation methods based on identified complex eigenmodes, *J. of Spacecraft*, 69-73.
13. Wei, M. L., Allemang, R. J., Brown, D. L. (1987), Real-normalization of measured complex modes, *Proceedings of the 5th International Modal Analysis Conference*, London, 708-712.
14. Wei, M. L., Zhang, Q., Allemang, J. and Brown, D. L. (1987), A time domain subspace iteration method for the normalization of measured complex modes, *Proc. 12th International Seminar on Modal Analysis, paper No. S2-3*, Katholic University of Leuven, Belgium.
15. Ibrahim, S. R., Fullekrug, U. (1990), Investigation into exact normalization of incomplete complex modes by decomposition transformation, *Proc. 8th IMAC*, Las Vegas.
16. Placidi, F., Poggi, F., Sestieri, A. (1991), Real modes computation from identified modal parameters with estimation of generalized damping, *Proc. 9th IMAC*, Florence, 1991.

17. Imregun, M. and Ewins, D. J. (1993), Realization of Complex Modes, *Proc. 11th IMAC*, 1303-1309.
18. Sestieri, A. and Ibrahim, S. R. (1994), Analysis of errors and approximations in the use of modal coordinates, *Journal of Sound and Vibrations, 177 (2)*, 145-157.

ACTIVE CONTROL OF STRUCTURES

L. GAUL, U. STÖBENER

Institute A of Mechanics, University of Stuttgart
Pfaffenwaldring 9, D-70550 Stuttgart

1. Introduction

Modern design and tailored materials have lead to structures with significant reduction of weight going along with an improvement of stiffness properties. But as the properties of such conventional passive systems can not be varied time dependent they have reached their limit of performance in a design state.

Growing requirements even for the off design state created a new class of technical systems, the so called active systems. In the present article the term active is devoted to active mechanical structures which are equipped with integrated sensors and actuators such that they can react against environmental disturbances governed by controller and associated algorithm.

In many contributions to active systems (book references (Beards 1995, Smith, Peters and Owen 1996, Shetty and Kolk 1997, Preumont 1997, Fuller, Elliott and Nelson 1996, Nelson and Elliott 1995)) new word creations for such systems or parts of them, like 'smart structures', 'intelligent materials', 'adaptive structures' and 'adaptronic', 'mechatronic' and 'structronic', 'adaptive control' etc. can be found. They all stand in a context to active systems but focus to different aspects of this topic which are not identified here. It can be noticed, that many fields of engineering contribute to the developing of active systems. New classes of materials have been created with adaptive properties due to electrical, magnetic and thermal stimulation. The manufacturing and structural integration of such materials require new scills. The development of constitutive equations, models and simulation tools are required along with new implemented control strategies for the design of adaptive structures.

The interdisciplinary aspects of active structures are discussed below by their fundamentals. More details are presented for a few selected ideas and applications only.

J.M.M. Silva and N.M.M. Maia (eds.), Modal Analysis and Testing, 453–486.

2. Materials for Active Structures

Materials for active structures are 'intelligent' or 'smart' materials. 'Smart' indicates some features of this materials, which are comparable to functions of biological muscles and nerves. The most important materials of this group are shape memory alloys (SMA), piezoelectric and magneto restrictive materials, electrorheological fluids (ERF) and optical fibers. In recent research reports hydro polymer gel with pseudo muscle behavior is added.

2.1. SHAPE MEMORY ALLOYS

Shape memory alloys (SMA) are metals which after deformation deform back to their original shape above a certain temperature revert. Best known is the family of nickel-titanium alloys. They were created by Bühler et al. in 1962 at the Naval Ordonance Laboratory (now the Naval Surface Warfare Center) and therefore sometimes they are called Nitinol (Ni for nickel, Ti for titanium and NOL for Naval Ordonance Laboratory). Already since the early seventies NiTi-alloys were industrially used for different applications like tube connectors or medical pliers. Other shape memory alloys are copper-aluminium-nickel (CuAlNi) and copper-zinc-nickel (CuZnNi).

There are several properties which describe the macroscopical behavior of SMA. A specimen of SMA is undergoing a test procedure such, that traction is applied until a plastic strain occurs, it is unloaded and then heated up to a certain temperature. After unloading the specimen is longer, than it was before. This, of course, would happens to every metal by exceeding the yield stress and is known as plastic deformation. In the case of an SMA the temperature rise to a certain value leads to the recovering of the former shape. This behavior is typical for SMA and is called 'quasi plastic material behavior' or 'shape memory effect'. The maximum recovering strain e.g. for NiTi is about 8 percent and for the copper alloys about 4 to 5 percent. An illustration of the test procedure and the stress-strain diagram is given in Fig. (1). By repeating this procedure but with a higher temperature from the beginning of the traction load it is observed, that the plastic deformation occures but vanishes instantaneously after unloading the specimen. This is called the 'pseudo elastic material behavior' which is presented in Fig. (1) as well.

The stress-strain diagrams of Fig. (1) shows, that for some values of stress two states of strain are possible. Which of them is actually present is determined by the history of load. This effect is called the 'two way behavior of SMA' and can be used for mechanical switch functions.

The reason for the above described phenomena is a phase transition between martensit and austenit in the micro structure of the SMA. This phase transition is motivated by the temperature change. The martensit start temperature M_s fixes the temperature point where the growth of martensit begins and at the martensit finish temperature M_f the growth of martensit ends. The corresponding temperatures for austenit are A_s and A_f. For most SMA $M_f < M_s < A_s < A_f$ is valid.

The material behavior of SMA is thus highly nonlinear. Therefore nonlinear constitutive equations are necessary. Although SMA's have been known and used for decades, there is a lack of a constitutive law which on one hand explains the physical

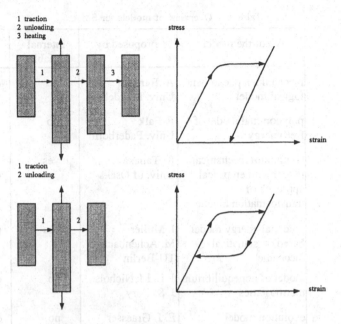

Figure 1. Macroscopical material behavior of SMA

effects in the microstructure of the material and on the other hand is compact and efficient enough to be able to be implemented in a discretization method like FEM. In the table (I) and in the references ((Bertram 1982),(Falk 1980),(Cahn and Larche 1984),(Tanaka and Nagaki 1982),(Tanaka and Iwasaki 1985),(Tanaka, Sato and Kobayashi 1985), (Tanaka 1986),(Tanaka and Sato 1988),(Liang and Rogers 1990),(Brinson 1993),(Brinson and Lammering 1993),(Müller and Achenbach 1983), (Müller 1986),(Müller 1989),(Müller 1992),(Achenbach 1989),(Fu, Huo and Müller 1993),(McNichols 1987), (Warlimont, Delaey and Krishnan 1974),(Graesser and Cozzarelli 1994)) a collection of different proposed models for the description of SMA is given. "Physical" in the column 'orientation' indicates, that the model explains the internal effects of phase transitions and 'engineer' that the model gives an explicit relation between stress and strain which can be easily implemented in a calculation code.

In the following we will present the free energy model by F. Falk (Falk 1980) (also known as Ginzburg-Landau model) because of its simplicity. The key to his model is the describtion of the free energy per volume as a polynominal of the strain ε and the temperature θ:

$$F(\varepsilon, \theta) = \alpha\varepsilon^6 - \beta\varepsilon^4 + (\delta\theta - \gamma)\varepsilon^2 + F_0(\theta), \tag{1}$$

where α, β, γ and δ are positive material constants to be identified by experiments. According to Falk this polynominal is the simplest function which satisfies the stress-strain diagrams shown in Fig. (1). A scaled nondimensional form is obtained

Table I. Overview of models for SMA

typ of the model	proposed by	internal variables	orientation
mechanical, phenomeno-logical model	A. Bertram Univ. Magdeburg	no	engineer
polynominal model of free energy	F. Falk Univ. Paderborn	no	physical/ engineer
continuum mechanical model with empirical approach of transformation kinetic	K. Tanaka Univ. of Osaka	yes	physical/ engineer
potential energy model based on statistical mechanic	I. Müller M. Achenbach TU-Berlin	yes	physical
model of nonequilibrium thermostatics	J. L. McNichols J. S. Cory	no	engineer
evolution model	E. J. Graesser Naval Surface Warfare Center F. A. Cozzarelli Univ. New York	no	engineer

by introducing

$$f = \frac{\alpha^2}{\beta^3} F, \quad e = \sqrt{\frac{\alpha}{\beta}} \varepsilon, \quad \vartheta = \frac{\alpha\delta}{\beta^2}\theta - \frac{\gamma\alpha}{\beta^2} - \frac{1}{4} \tag{2}$$

and

$$f(e, \vartheta) = e^6 - e^4 + (\vartheta + \frac{1}{4})e^2 + f_0(\vartheta). \tag{3}$$

By this scaling the free energy is equal for all kinds of SMA. Figure (2) shows the free energy as function of the strain for different temperatures. The doted line sections marks regions, where the cristal structure is not stable. For high temperatures ($\vartheta > 1/12$) the function f has one minimum for vanishing e. This means in a stress free state only austenit can exist in a stable manner. For $-1/4 < \vartheta < 1/12$ we find three minima. Both minima on the right and left represent the martensit twins and the minimum in the center indicates the austenit phase. For temperatures below -1/4 only two minima belonging to the martensit twins exist. In this temperature range no austenit exists even in a stress free state. To change the phase it is necessary to overcome the energy barrier. This can be done by mechanical work or by heating.

The stress strain relation is found as derivative of function F with respect to the strain

$$\sigma(\varepsilon, \theta) = \frac{\partial F(\varepsilon, \theta)}{\partial \varepsilon}, \tag{4}$$

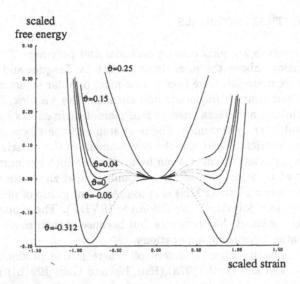

Figure 2. Helmholtz free energy

which results in,

$$\sigma(e,\vartheta) = \frac{\partial f(e,\vartheta)}{\partial e} = 6e^5 - 4e^3 + 2(\vartheta + \frac{1}{4})e. \tag{5}$$

This relation between stress an strain can be implemented e.g. into a finite element formulation as usefull calculation tool for SMA. In the references ((Brinson and Lammering 1993),(Savi, Braga, Alves and Almeida 1998)) finite element formulations for SMA are proposed.

After this brief introduction about SMA one modern application is discussed. Whereas traditional applications (see reference (Wayman 1980)) use the features of SMA in a simple way, like e.g. clamps, modern applications (references (Mooi 1992), (Ikegami, Wilson, Anderson and Julien 1990),(Baz, Iman and McCoy 1990), (MacLean, Patterson and Misra 1991b),(MacLean, Draper and Misra 1991a)) provide SMA elements the function of actuators controled by computers. One such application is the active eigenfrequency tuning. Wires of SMA embedded in a host structure are connected with an electric circuit. By applying voltage the electric resistance of the wires causes an increase of heat. As described above this procedure leads to a phase transition and would normaly change the lenghts of the wires. In this case the wires are constrained by the host structure. The result is the generation of an internal stress which changes the stiffness of the structure. Because eigenfrequencies depend on stiffness and inertia properties the spectrum can thus be tuned.

2.2. PIEZOELECTRIC MATERIALS

Other smart materials are piezoelectric ceramics and polymers. Since the first scientific investigations about the piezoelectric effect by Jacques and Pierre Curie in 1880, piezoelectric materials have become the most popular smart materials. In the first decades of our century physicists and chemists like Vasalek, Busch, Scherrer, Zunhauer, Goldmann and Tisza tried to find piezoelectric effects by different types of potassium and later by ceramics. The next step was the discovery of the fundamentals of piezoelectricity. Next was the development of the polarization procedure to reinforce the piezoelectric effect, even by materials which are normaly non piezoelectric. The most common piezoelectric ceramic is lead zirconate titanate (PZT). Also lead magnesium niobate (PMN) is in use. A recent group of piezo materials are special polymers, like polyvinylidene diflouride (PVDF). These polymers exhibit a smaller piezoelectric effect than ceramics, but because of their mechanical flexibility they have advantages for some applications.

In the field of active structural control, where is one of the authors project (references (Hsu, Lin and Gaul 1997a),(Hsu, Lin and Gaul 1997b)) PVDF films are bonded to thin plates and used as sensors and actuators. Because of their flexibility PVDF films can be cut in desired shapes, e.g. modal shapes, and bonded even to arced surfaces. Contributions to active structural control for plates can be found in ((Fuller 1990),(Dimitriadis and Fuller 1991)).

The industrial interest in piezoelectric materials increased in the sixties. A lot of technical applications were developed in this time, like microfones and force transducers. Technical progress in manufacturing allowed new design concepts and at the beginning of the nineties the first piezoelectric actuators were commercially offered. Nowadays position control systems in an accuracy range of microns are nothing exceptional.

There are good reasons to use piezo electric elements in sciences and industry. The accuracy of the position was named above. The deformation of piezoelectric actuators has only small losses and wear. Piezoelectric elements have a long durability. The efficiency of transforming electrical work to mechanical work and reverse is very good. To hold a selected position no further energy is needed. High forces can be generated due to the piezoelectric effect. Actuators with a maximum force of 50,000 N are available. The speed of the actuators is high, thus piezoelectric elements are especially suited for generation of ultrasound.

The physical background of the piezoelectric effect is a non-symmetry of the charge distribution within the crystal structure. Deformation of the crystal causes a polarization. That means one side of the crystal has an excess of negative ions and the opposite side has a lack of negative ionics. Different circumstances can result in a loss of the piezoelectric features by destroying the non-symmetry of the charge distribution. At the Curie-temperature the piezoelectric material becomes para electric. Above this temperature the direction of polarization disappears and even after cooling under the Curie-temperatur there is no longer a piezoelectric effect.The same happens if the applied electric field exceeds a certain limit and has an invers orientation to the original polarization. This limit is called coercitivity field strength and occures by 10 to 20 percent of the operating voltage.

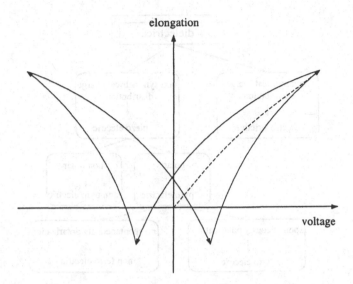

Figure 3. Hysteresis of piezoelectric material

Piezoelectric materials exhibit a nonlinear behavior. By measuring the extension of a piezoelectric element during the variation of applied voltage, we get the curve shown in Fig. (3). The initial conditions are zero extension and zero voltage. Then the voltage increases and the piezo element elongates. At the end of the initial curve the voltage is reduced. Instead of returning on the initial curve back to the origin, a new curve appears. At zero voltage a remanent extension remains. The reason for this elongation is, that a part of the dipols generated by the voltage are stable. By applying a negative voltage we can compensate the remanent elongation and at the end of our experiment we find a closed hysteresis cycle. There are more nonlinear physical effects, like drifting and temperature influence, which make the material behavior of piezoelectric elements complicate. To describe these phenomena is bejond the scope of this paper. In chapter 3 the constitutive equations and a finite element formulation are treated.

2.3. ELECTROSTRICTIVE MATERIALS

Electrostrictive materials have a close relationship to piezoelectric materials. Fig. (4) shows the relations between the phenomena of dielectric materials. Whereas all piezoelectric materials have a non symmetry in their charge distribution, electrostrictive materials are symmetrical.

In every dielectric material an electric field leads to a deformation. The direction of this deformation does not depend on the orientation of the electric field, because of the quadratic influence of the polarization. This statement is valid for piezoelectric materials, too. But opposite to electrostrictive materials, the piezoelectric effect, with it's dependence on the orientation of the electric field, is dominante for them.

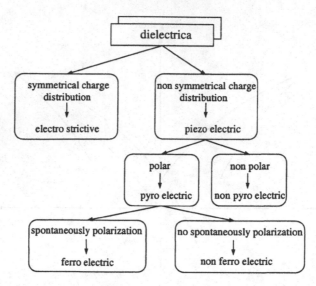

Figure 4. Family of dielectrica

2.4. MAGNETOSTRICTIVE MATERIALS

The magnetostrictive effect describes length change of ferro magnetic materials by applying a magnetic field. The physical background of the magnetostrictive effect is a rotation of magnetic domains (Weisssche Bezirke) until they line up with the external magentic field. This effect was described by Weiss in 1842. Fundamental scientific investigations with rare earth alloys were carried out by Clark. Since 1963 the magnetostrictive effect was measuret with terbium (Tb) and dysprosium (Dy) for low temperatures, with 1 percent of elongation, scientifics were searching for materials which exhibit the effect for higher temperatures. Today TERFENOL-D is the most common magnetostrictive material. Actuators made of TERFENOL-D are being investigated for use in active vibrationdamping.

Like piezoelectrics magnetostrictive materials have a nonlinear behavior with hysteresis. The expansion of magnetostrictive material depends strongly on the mechanical pre stress. For small strains the linear constitutive 1-d equations can be used

$$\varepsilon = s\sigma + \kappa H, \tag{6}$$

$$B = \kappa\sigma + \mu H. \tag{7}$$

Here ε is the strain, s is the compliance, σ is the stress, κ is the magnetostrictive coefficient, H is the magnetic field, B is the magnetic displacement and μ is the magnetic field constant. Besides the magnetostrictive effect there exists other physical phenomena in this material group, e.g. the magneto calorical effect and vortex streams.

2.5. ELECTRORHEOLOGICAL FLUIDS

Electrorheological fluids (ERF) were first investigated and proposed by W. Winslow in 1949. ERF are fluids with the capability of changing their viscosity due to an applied electric field. The transformation from the liquid to the solid state happens within milliseconds.

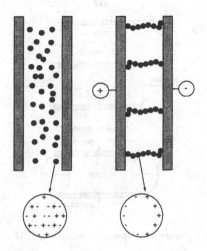

Figure 5. Chain forming by ERF

A fluid with this feature should enjoy a variety of technical applications but in fact there are only few commercial ERF devices (see reference (Wolff and Wendt 1994)). The reasons for this lack are some problems associated with these fluids. There are firstly the high abrasive wear due to dispersed solids, secondly the insufficient reproducibility of ERF effect in manufacture and thirdly ERF can only be produced in small quantities, but not in industrial scale.

Of course there are various activities to find solutions for these problems and it seems that there are ways to overcome them. The first generation of ERF were mixed with water to create dipoles. BAYER AG developed an ERF based on soft, crosslinked polyurethane particles in silicone oil (see reference (AG 1994)). This water-free ERF guarantees constant properties even in high temperature applications. The polyurethane particles contain dissolved metal ions which are necessary for fast polarization and the electrorheological effect. As modification of the first ERF generation with hard inorganic particles this new generation of ERF is less abrasive and can be produced in an industrial scale.

Fig. (5) shows an illustration of the physical phenomenon in ERF. Without an electric field the particles have no order and there are no dipoles. This situation changes instantaniously by applying an electric field. The dipoles in the particles appear and the particles form chains in the direction of the electric field.

Depending on the application different operation modes of ERF can occure. Couplings and breaks use the shear-mode, hydraulic systems and dampers work in the flow-mode and small-amplitude-dampers are based on the squeeze-mode. In

Fig. (6) all modes are listed. Shear stresses up to 2000 Pa can be reached by electric field strenght of 3 kV/mm.

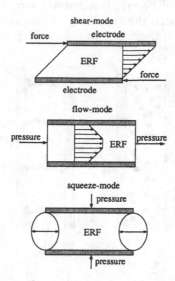

Figure 6. Operation modes of ERF

Besides well tested applications such as dampers with adjustable damping coefficient, shock absorbers and hydraulic valves without mechanical abrasion, there is a new research need for active ERF acoustic absorbers. The mode of operation of these absorbers can be based on an adjustable impedance (see reference (Wicker, Eberius and Guicking 1996)). Incoming sound pressure waves impact to a membrane which is supported by an ERF inside the absorber. Due to the variable viscosity the combination of membrane an ERF changes their impedance and the absorber can be adapted to the actual incoming wave to maximize the absorbtion.

2.6. OPTICAL FIBERS

Different to the materials discussed above optical fibers work as sensors only. They are circular dielectric waveguides which transport optical energy and information. They have a central core surrounded by a concentric cladding with slightly lower refractive index. Fibers are typically made of silica with index modifying dopants such as GeO_2. A protective coating of one or two layers of cushioning material (such as acrylate) is used to reduce the crosstalk between adjacent fibers and the loss increasing microbending that occurs when fibers are pressed against rough surfaces. For a greater environmental protection, fibers are commonly incorporated into cables. Typical cables have a polyethylene sheat that encases the fiber within a strenght member such as steel or kevlar strands. In Fig. (7) shows a schematic of an optical fiber.

Optical fibers embedded in a smart material can provide data by several ways. First they can simply provide a steady light signal to a sensor. Breaks in the ligth

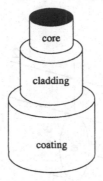

Figure 7. Structure of an optical fiber

beam indicate a structural flaw that has snapped the fiber. The second more subtle approach involves looking at the key characteristics of the light, intensity, phase, polarization or a similar feature. The US National Aeronautics and Space Administration and other research centers have used such a fiber-optic system to measure the strain in composite materials. Fiber-optic sensors can also measure magnetic fields, deformations and accelerations. Resistance to adverse environments and immunity to electrical or magnetic noise are among the advantages of optical fibers.

3. Finite Element Formulation for Piezoelectrics

Active structures with embedded smart materials have to be calculated for design and simulation of operation. One of the most established and powerful tools for structure calculations are finite element programs. Often commercial finite element software offer the possibility to implement own user defined elements and constitutive equations. For the calculation of active structures finite element programs can be extended by adding such elements. Some of the commercial finite element programs offer material laws for smart materials (see reference (Hib 1989)).

In the nexts chapters the derivation of a finite element formulation for a piezoelectric actuator in stacked design is presented. This kind of 1-d actuator has been selected because it is one of the most frequently used. An adapted finite element formulation for a 1-d rod is needed.

3.1. FINITE ROD ELEMENT

Fig. (8) represents a 3-node finite element. The rod length is L and its cross section is A. Nodal forces are $\mathbf{N}^i$, $\mathbf{N}^j$, $\mathbf{N}^k$, the nodal displacements are $\mathbf{u}^i$, $\mathbf{u}^j$, $\mathbf{u}^k$, the nodal electric charges are q^i, q^j, q^k and the nodal electric potentials are φ^i, φ^j, φ^j. The superscripts i,j,k denote node numbers. The forces and displacements are vectors with components in cartesian coordinates (e.g. $\mathbf{u}^i = \mathbf{u}^i_x + \mathbf{u}^i_y + \mathbf{u}^i_z$) and the charges and potentials are scalar quantities.

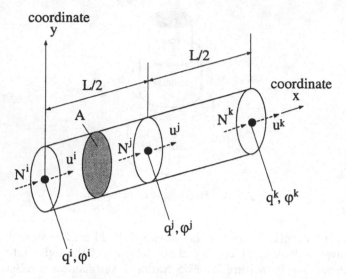

Figure 8. Piezoelectric rod

3.2. VIRTUAL WORK

The equation of motion is derived from D'Alembert's generalization of principle of virtual displacements

$$\int_V (\rho \delta u_i \ddot{u}_i + \delta \varepsilon_{ij} \sigma_{ij}) \, dV = \int_V \delta u_i b_i \, dV + \int_A \delta u_i t_i \, dA, \qquad (8)$$

with volume V and cross section A of the rod, ρ is the mass density, δu_i is the virtual displacement, $\ddot{u}_i$ is the acceleration, $\delta \varepsilon_{ij}$ is the virtual strain, σ_{ij} is the stress, b_i is the volume force and t_i is the traction. For the electric flux conservation holds the principle

$$\int_V \delta E_i D_i \, dV + \int_V \delta \varphi q_V \, dV + \int_A \delta \varphi q_A \, dA = 0. \qquad (9)$$

D_i is the electric flux or electric displacement, δE_i is the virtual electric field, q_V is the charge per volume, q_A is the charge per area and $\delta \varphi$ is the virtual electric potential. Superposition of these equations leads to

$$\int_V (\rho \delta u_i \ddot{u}_i + \delta \varepsilon_{ij} \sigma_{ij}) \, dV - \int_V \delta E_i D_i \, dV =$$
$$\int_V \delta u_i b_i \, dV + \int_A \delta u_i t_i \, dA + \int_V \delta \varphi q_V \, dV + \int_A \delta \varphi q_A \, dA. \qquad (10)$$

3.3. CONSTITUTIVE EQUATIONS

As next step the constitutive equations for piezoelectric material are introduced. The nonlinear material behavior of piezoelectrics has been discussed before. With small electric field strenght and small mechanical stresses the effects of hysteresis can be neglected. Under this conditions the following linear set of equations is used

$$\sigma_{ij} = C_{ijkl}^E \varepsilon_{kl} - e_{ijm} E_m, \tag{11}$$

$$D_m = e_{mij} \varepsilon_{ij} + \xi_{mn}^\sigma E_n. \tag{12}$$

In this equations σ_{ij} are the stress coordinates, C_{ijkl}^E are the elasticity constants measured for constant electric field, ε_{kl} are the strain coordinates, E_m and E_n are the electric field coordinates, e_{ijm} are the piezoelectric stress coefficients, D_m is the electric flux and ξ_{mn}^σ is the dielectric coefficient measured for constant stress. Inserting the constitutive Eqn. (10) we get

$$\int_V \rho \delta u_i \ddot{u}_i dV + \int_V \delta\varepsilon_{ij} C_{ijkl}^E \varepsilon_{kl} dV - \int_V \delta\varepsilon_{ij} e_{ijm} E_m dV$$

$$- \int_V \delta E_m e_{mij} \varepsilon_{ij} dV - \int_V \delta E_m \xi_{mn}^\sigma E_n dV =$$

$$\int_V \delta u_i b_i dV + \int_A \delta u_i t_i dA + \int_V \delta\varphi q_V dV + \int_A \delta\varphi q_A dA. \tag{13}$$

Coupled mechanical and electrical field equations can be derived from (13).

3.4. FINITE ELEMENT APPROXIMATION

The finite element method approximate the displacement field by nodal weighted shape functions. A common group are the Hermitian polynomals. They are formulated in natural coordinates, according to Fig. (9). For a three node rod element quadratic polynomals and their derivatives with respect to the natural coordinates given by

$$H^1 = \frac{1}{2}\left(r^2 - r\right), \quad B^1 = r - \frac{1}{2}, \tag{14}$$

$$H^2 = 1 - r^2, \quad B^2 = -2r, \tag{15}$$

$$H^3 = \frac{1}{2}\left(r^2 + r\right), \quad B^3 = r + \frac{1}{2}. \tag{16}$$

A matrix formulation is introduced as follows

$$\mathbf{H} = \begin{bmatrix} \mathbf{H}_m & 0 \\ 0 & \mathbf{H}_{el} \end{bmatrix} = \begin{bmatrix} H^1 & H^2 & H^3 & 0 & 0 & 0 \\ 0 & 0 & 0 & H^1 & H^2 & H^3 \end{bmatrix}, \tag{17}$$

$$\mathbf{B} = \begin{bmatrix} \mathbf{B}_m & 0 \\ 0 & -\mathbf{B}_{el} \end{bmatrix} = \begin{bmatrix} B^1 & B^2 & B^3 & 0 & 0 & 0 \\ 0 & 0 & 0 & -B^1 & -B^2 & -B^3 \end{bmatrix}. \tag{18}$$

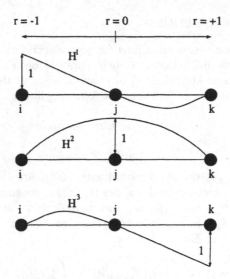

Figure 9. Hermitian polynomals

Now the fields of displacements and of electric potentials are approximated by

$$\mathbf{p} = \begin{bmatrix} \mathbf{u} \\ \varphi \end{bmatrix} = \begin{bmatrix} \mathbf{H}_m & 0 \\ 0 & \mathbf{H}_{el} \end{bmatrix} \begin{bmatrix} \mathbf{u}_e \\ \varphi_e \end{bmatrix} = \mathbf{H}\mathbf{p}_e. \tag{19}$$

In (19) $\mathbf{p}$ contains the continuous displacement and potential field and $\mathbf{p}_e$ contains nodal values. The coordinates in the element state vector $\mathbf{p}_e$ are given by

$$\mathbf{p}_e = \begin{bmatrix} \mathbf{u}_e \\ \varphi_e \end{bmatrix} = \begin{bmatrix} u^i \\ u^j \\ u^k \\ \varphi^i \\ \varphi^j \\ \varphi^k \end{bmatrix}. \tag{20}$$

Strains and the electric field are approximated by

$$\begin{bmatrix} \varepsilon_m \\ \mathbf{E}_{el} \end{bmatrix} = \begin{bmatrix} \mathbf{B}_m & 0 \\ 0 & -\mathbf{B}_{el} \end{bmatrix} \begin{bmatrix} \mathbf{u}_e \\ \varphi_e \end{bmatrix}. \tag{21}$$

3.5. EQUATION OF MOTION

The equation of motion is obtained by implementing the finite element approximation in the coupled mechanical and electrical field Eqn. (13). This yields

$$\begin{bmatrix} \delta\mathbf{u}_e \\ \delta\varphi_e \end{bmatrix}^T \int\limits_{V_e} \rho \begin{bmatrix} \mathbf{H}_m^T\mathbf{H}_m & 0 \\ 0 & 0 \end{bmatrix} dV \begin{bmatrix} \ddot{\mathbf{u}}_e \\ \ddot{\varphi}_e \end{bmatrix}$$

$$\begin{bmatrix} \delta \mathbf{u}_e \\ \delta \varphi_e \end{bmatrix}^T \int_{V_e} \begin{bmatrix} \mathbf{B}_m^T E \mathbf{B}_m & \mathbf{B}_m^T e \mathbf{B}_{el} \\ \mathbf{B}_{el}^T e \mathbf{B}_m & -\mathbf{B}_{el}^T \xi \mathbf{B}_{el} \end{bmatrix} dV \begin{bmatrix} \mathbf{u}_e \\ \varphi_e \end{bmatrix}$$

$$= \begin{bmatrix} \delta \mathbf{u}_e \\ \delta \varphi_e \end{bmatrix}^T \int_{A_e} \begin{bmatrix} \mathbf{H}_m^T \\ \mathbf{H}_{el}^T \end{bmatrix} \begin{bmatrix} \mathbf{t} \\ q_s \end{bmatrix} dA. \tag{22}$$

Equation (22) can be simplified as follows. The elasticity matrix is reduced Young's modulus E for the 1-dimensional case. No additional volume forces and volume charges are present. This means that normal forces act and charges are located on the front and rear faces which is consistent with the operational mode of the real actuator. The boldface symbols are vectors or matrices. The rest are scalar quantities. In a compact form we can write the virtual work equation as

$$\delta \mathbf{p}_e^T \left(\mathbf{M}_e \ddot{\mathbf{p}}_e + \mathbf{K}_e \mathbf{p}_e - \mathbf{F}_e \right) = 0, \tag{23}$$

and conclude the equation of motion as

$$\mathbf{M}_e \ddot{\mathbf{p}}_e + \mathbf{K}_e \mathbf{p}_e = \mathbf{F}_e. \tag{24}$$

By comparising Eqn. (22) with Eqn. (23) the physical meaning of the integrals is obvious. The first integral is the consistent mass matrix $\mathbf{M}_e$, the second is the stiffness matrix $\mathbf{K}_e$ and the third is the load vector $\mathbf{F}_e$. The shape functions (14,15,16) lead to the matrices

$$\mathbf{M}_e = \frac{\rho A L}{30} \begin{bmatrix} 4 & 2 & -1 & 0 & 0 & 0 \\ 2 & 16 & 2 & 0 & 0 & 0 \\ -1 & 2 & 4 & 0 & 0 & 0 \\ 0 & 0 & 0 & 0 & 0 & 0 \\ 0 & 0 & 0 & 0 & 0 & 0 \\ 0 & 0 & 0 & 0 & 0 & 0 \end{bmatrix}, \tag{25}$$

$$\mathbf{K}_e = \frac{A}{12L} \begin{bmatrix} 7E & -8E & 1E & 7e & -8e & 1e \\ -8E & 16E & -8E & -8e & 16e & -8e \\ 1E & -8E & 7E & 1e & -8e & 7e \\ 7e & -8e & 1e & -7\xi & 8\xi & -1\xi \\ -8e & 16e & -8e & 8\xi & -16\xi & 8\xi \\ 1e & -8e & 7e & -1\xi & 8\xi & -7\xi \end{bmatrix}, \tag{26}$$

$$\mathbf{F}_e = \begin{bmatrix} N^i \\ N^j \\ N^k \\ Q^i \\ Q^j \\ Q^k \end{bmatrix}. \tag{27}$$

3.6. ACTIVE VIBRATION ISOLATION

In this chapter a single-degrees-of-freedom oscillator is treated for example, which can be controlled by one piezoelectric stacked actuator. The oscillator is excited by a harmonic force disturbance (see Fig.(10)). The vibration force is transmitted by a rod to the receiving structure. This system can serve as a simple model for a real world system like an engine with resilient mount. The passive mount is replaced by a stacked actuator and a small quartz force transducer. This combination makes it possible to sense the transmitted force and to control this force such that an active bearing is obtained.

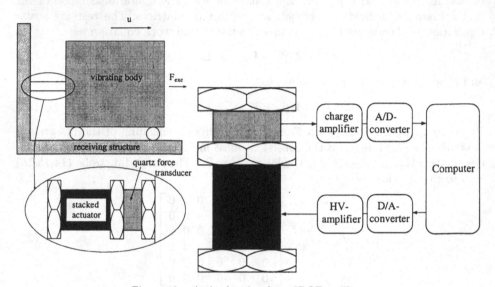

Figure 10.   Active bearing for a SDOF oscillator

The control circuit is very simple. The equation of motion is given as a special case of Eqn. (24) with neglected mass of the actuator

$$M\ddot{u} + K_{act}u + K_{act}d_{33}U_{act} = \hat{F}_{exc}e^{i\omega t}, \tag{28}$$

where M is the mass of the oscillator, K_{act} is the stiffness of the actuator, d_{33} is the piezoelectric charge coefficient, U_{act} is the voltage applied to the actuator, $\hat{F}_{exc}$ is the excitation force amplitude and ω is the excitation frequency. The displacement u of the oscillator vanishes if the voltage is

$$\hat{U}_{act} = \frac{1}{K_{act}d_{33}}\hat{F}_{exc}. \tag{29}$$

In the control circuit shown in Fig.(10) the transmitted force, which equals the excitation force, is measured. Equation (29) is solved by the computer. The result governs the actuator voltage via the high voltage amplifier. The vanishing displacement is a kind of active motion control (AMC). Another control goal could be the force cancelation at the receiving structure which is called active vibration isolation (AVI). AVI can be achieved by compensating the actuator stiffness by the actuator force. The actuator voltage is thus

$$\hat{U}_{act} = \frac{1}{\omega^2 M d_{33}} \hat{F}_{exc}. \tag{30}$$

In Fig. (11) the data for a benchmark example are listed. The graph included in Fig. (11) shows the actuator voltage, which is needed for AVI, as a function of the frequency.

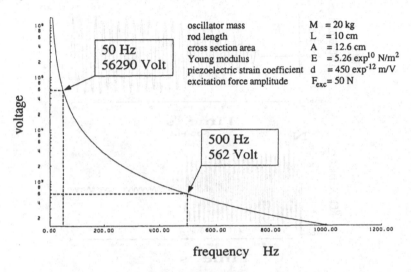

Figure 11. Benchmark example

With the values of the benchmark example a finite element simulation is performed by using the finite rod element derivated above. The time response of the force and the displacement are calculated. Initially the actuator voltage is zero. Without control the excitation force is equal to the reaction force at the receiving structure. Setting the actuator voltage to the precalculated value (see Fig. (11)) we obtain a much smaller force. The associated displacement increases.

4. Control Strategies

The broad field of control strategies is only briefly adressed. Several characteristics distinguish and classify the control strategies, such as linear and nonlinear control, feedback and feedforward control, model based and non model based control and adaptive and non adaptive control. Here the focus is on feedback and feedforward control as state of the art in active control. Nevertheless, so called intelligent control is a subject of rapid development and intensive research. Such control strategies include genetic algorithms, neural networks, fuzzy logic and expert systems. A short

470

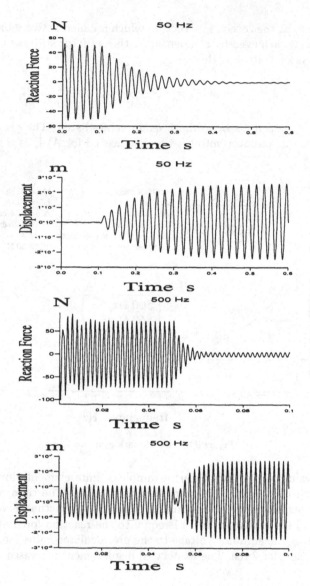

Figure 12. Results of FE-calculation

overview about this topic is given in reference ((Lu 1996),(Noor and Jorgensen 1996)). Because of their similarity in the mode of operation to biological intelligence these control strategies are very universal control tools with an enormous wide range of applications. With this performance characteristics they might become the control strategies of future decades.

4.1. PRINCIPLES OF FEEDBACK CONTROL

The main feature of feedback control is the closed control loop. The response of a mechanical system which is excited by a disturbance is measured with one or more sensors. Then the sensor signal is used by the controller to determine an actuator signal. This actuator signal drives the actuator to operate on the mechanical system as a secondary source. In Fig. (13) two closed loops are shown.

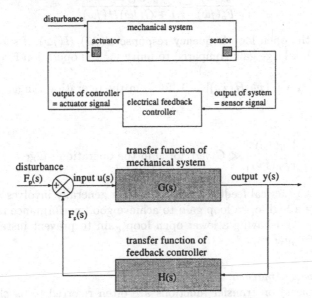

Figure 13. Closed feedback control loop

The upper loop is related to the physical components and in the other loop of Fig. (13) the mechanical system and the controller are replaced by their transfer functions in the Laplace domain. This loops are especially adopted to the task of structural control, such that the value of the output should be zero. This typ of depiction is called the block diagram. The transfer function of the mechanical system G(s) is defined as the ratio of the Laplace transform of the response y(s) to the Laplace transform of the net excitation $F_p(s) - F_s(s)$ such we can write

$$y(s) = G(s)\left[F_p(s) - F_s(s)\right]. \tag{31}$$

$F_p(s)$ is the primary disturbance and $F_s(s)$ is the secondary actuator excitation. The transfer function of the controller H(s) is defined by the ratio of the secondary actuator excitation $F_s(s)$ and the Laplace transform of the response y(s). With this definition we obtain

$$F_s(s) = H(s)y(s). \tag{32}$$

Combination and some simple algebraic manipulations lead to the closed loop transfer function

$$\frac{y(s)}{F_p(s)} = \frac{G(s)}{1 + G(s)H(s)}. \tag{33}$$

Several classical controller design methods that can be used to determine a desired transfer function behavior, e.g. root locus method, Hurwitz method and Nyquist method. The control aim can be a minimization of disturbance response, which is often the objective in active control of structures. This means that the transfer function must be minimized with the transfer function in frequency domain. Minimization of

$$\frac{y(j\omega)}{F_p(j\omega)} = \frac{G(j\omega)}{1 + G(j\omega)H(j\omega)}, \tag{34}$$

is achieved in the open loop frequency response $G(j\omega)$ $H(j\omega)$, if small phase shift goes along with a large gain compared to unity for the operation frequency.

$$|1 + G(j\omega)H(j\omega)| \gg 1 \text{ for } \omega \text{ in the operation range}, \tag{35}$$

which means

$$\left|\frac{y(j\omega)}{F_p(j\omega)}\right| \ll G(j\omega) \text{ for } \omega \text{ in the operation range}. \tag{36}$$

The design of a practical feedback control system generally involves a compromise between having a high open loop gain to achieve good performance in the working frequency range and having a lower open loop gain to prevent instability outside the operation frequency.

4.1.1. State Space Approach

The methods based on transfer functions are often referred to as classical methods. They lead to respectable results in the design of single-input-single-output (SISO) control systems, but they are difficult to apply for multi-input-multi-output (MIMO) control systems. A more modern approach is the state space method. The state space method uses the possibility of transforming a higher order differential equation in a set of first order differential equations. The advantage of this procedure is that any higher order differential equation, that describes the behavior of a mechanical system, can be rewritten in a standard form such that standard controller design methods can be applied. Due to the transformation, a set of internal state variables occurs.

There are two equations in the state space formulation. The first

$$\dot{\mathbf{x}} = \mathbf{A}\mathbf{x} + \mathbf{B}\mathbf{u} + \mathbf{E}\mathbf{w}_1 \tag{37}$$

is called the system equation, which consists of a set of first order differential equations, and the second

$$\mathbf{y} = \mathbf{C}\mathbf{x} + \mathbf{D}\mathbf{u} + \mathbf{w}_2 \tag{38}$$

is called the output equation, which describes the system response as a function of the internal variables, the input and the measurement noise. The symbols in Eqn. (37) and (38) denote

$\mathbf{x}$ state variable vector,	$\mathbf{A}$ system matrix,
$\mathbf{u}$ input vector,	$\mathbf{B}$ input matrix,
$\mathbf{y}$ output vector,	$\mathbf{C}$ output matrix,
w_1 system noise,	$\mathbf{D}$ feedthrough matrix,
w_2 measurement noise,	$\mathbf{E}$ system noise input matrix.

The system noise w_1 is a result of environmental loads, modelling errors, not simulated dynamics, nonlinearities and the noise inherent in the input vector. The measurement noise exists because of sensor noise and modelling errors. The feedthrough matrix D has to be considered if there is a feedthrough component in the system. Such a component may appear for some types of sensors or sensor locations or as a result of modal truncation.

In Fig. (14) the classical transfer function is associated to the state space approach. The system and the measurement noise are neglected in Fig. (14).

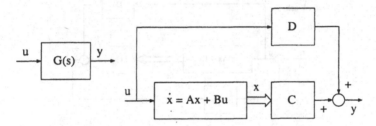

Figure 14. Transfer function and state space approach

From Fig. (14) it is obvious that the state space approach itself includes no controller. It is just a kind of formalism to describe the system dynamics. There are two possibilities to create the necessary controller. If the state variables are directly measurable a direct state feedback control can be applied (see Fig. (15). Otherwise a state observer or state estimator is necessary to calculate the state variables from the output. In both cases the state variables are the inputs for the controller gain matrix. This is in contrast to the output feedback where the system output is directly used as input for the controller gain matrix (see Fig. (15). Of course often some components of the output vector are equal to some state variables such that the difference between state feedback and output feedback becomes undistinguishable in practice.

4.1.2. *Optimal Control*
Linear optimal control, as used for linear quadratic regulators LQR (see reference (Siouris 1996)), is a design method for the gain matrix G by using state feedback control. Objective of the optimal control is the minimization of a cost function or a performance index. Asymptotic stability implies that $\mathbf{x} \to \mathbf{0}$ as $t \to \infty$ such that

474

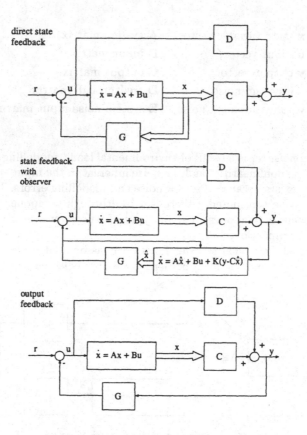

Figure 15. Types of feedback

the quadratic cost function written as a time integral

$$J = \int\limits_{t=0}^{\infty} \left(\mathbf{x}^T \mathbf{Q} \mathbf{x} + \mathbf{u}^T \mathbf{R} \mathbf{u} \right) dt \qquad (39)$$

is minimized. The quadratic weighting matrices $\mathbf{Q}$ and $\mathbf{R}$ often contain only diagonal elements. Under this condition element entries of high values lead to a fast reduction of state variables and input vector elements. The $\mathbf{Q}$ term with the state vector accounts for the system response and the $\mathbf{R}$ term with the actuator vector weights the control effort, respectively. The choise of $\mathbf{Q}$ and $\mathbf{R}$ are not unique and therefore crucial for the controller design. It can be shown (see e.g. (Preumont 1997),(Fuller et al. 1996)) that the cost function is related to the solution $\mathbf{P}$ of algebraic Riccati equation

$$\mathbf{PA} + \mathbf{A}^T\mathbf{P} + \mathbf{Q} - \mathbf{PBR}^{-1}\mathbf{B}^T\mathbf{P} = \mathbf{0}. \qquad (40)$$

Therefore the LQR controller is also called Riccati controller. Finally the gain matrix $\mathbf{G}$ is obtained as

$$\mathbf{G} = \mathbf{R}^{-1}\mathbf{B}^T\mathbf{P} \qquad (41)$$

4.1.3. *Modal control*

Another very useful and simple design method for state feedback controllers is the modal control. The concept of modal control was proposed by H.H. Rosenbrock in 1962. The basic idea of modal control is to transform the state variables in a special form such that they are independent from each other. The advantage of independent state variables is the selective controlability of system dynamics. Independent state variables can physically be interpreted as modal coordinates, providing the name of this method.

Modal coordinates are introduced by transformation with the modal matrix Φ consisting of the eigenvectors of the system matrix $\mathbf{A}$

$$\mathbf{x} = \Phi\mathbf{z}. \tag{42}$$

The associated eigenvalue problem leads to the representations with the spectral matrix of eigenvalues Λ which is diagonal

$$\mathbf{A} = \Phi\Lambda\Phi^{-1}, \ \Lambda = \Phi^{-1}\mathbf{A}\Phi, \tag{43}$$

which changes the system Eqn. (37) to

$$\dot{\mathbf{z}} = \Phi^{-1}\mathbf{A}\Phi\mathbf{z} + \Phi^{-1}\mathbf{B}\mathbf{u}, \tag{44}$$

by neglecting the system and measurement noise. The product $\Phi^{-1}\mathbf{A}\Phi$ can be replaced by the diagonal matrix Λ, such that Eqn. (44) change to

$$\dot{\mathbf{z}} = \Lambda\mathbf{z} + \Phi^{-1}\mathbf{B}\mathbf{u}. \tag{45}$$

Equation (45) represents the system state in decoupled modal form and is the basis for modal controllers which allows independent control of the natural eigenfrequencies and thus damping of these modes without influencing their form. Now, a way has to be selected how to transform the output vector in the modal space. The output Eqn. (38) is used by neglecting the feedthrough and the noise component

$$\mathbf{y} = \mathbf{C}\mathbf{x} = \mathbf{C}\Phi\mathbf{z}. \tag{46}$$

Equation (46) can be transformed to

$$\mathbf{z} = \Phi^{-1}\mathbf{C}^{-1}\mathbf{y} = \tilde{\mathbf{C}}\mathbf{y}. \tag{47}$$

Applying Eqn. (47) to the output of the system we obtain a mode analyser (see Fig. (16)). The regulator consists of the gain matrix $\mathbf{G_z}$ and is coupled to the system input via the modal synthesizer $\mathbf{B}^{-1}\Phi$. This construction leads to

$$\mathbf{u} = \mathbf{r} - [\mathbf{B}^{-1}\Phi]\,\mathbf{G_z}\mathbf{z} = \mathbf{r} - \tilde{\mathbf{B}}\mathbf{G_z}\mathbf{z}, \tag{48}$$

where the vector $\mathbf{r}$ is the disturbance as depicted in Fig. (15) and the closed loop system matrix can be rewritten as

$$\dot{\mathbf{z}} = [\Lambda - \mathbf{G_z}]\,\mathbf{z} + \Phi^{-1}\mathbf{B}\mathbf{r}. \tag{49}$$

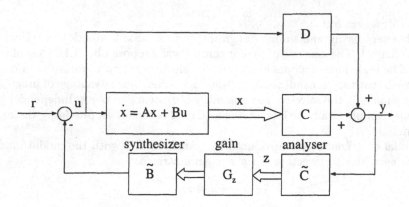

Figure 16. Modal state feedback controller

If G_z is of diagonal forrm the objective of independent change of eigenvalues without changing the eigenvectors is achieved. There are disadvantages which have not been discussed here, e.g. that there must be as many inputs as state variables.

4.2. PRINCIPLES OF FEEDFORWARD CONTROL

The basic idea of feedforward control is to build a second transmission path which is parallel to the primary source path to cancelate the disturbance. This principle is depicted in Fig. (17). It is obvious that for this concept the disturbance signal must be known because it has to be used as a reference signal for the input of the feedforward controller. If the reference signal is directly available, e.g. due to the powersupply, and the feedforward controller is not adaptive, feedforward control needs no sensor. But often the reference signal has to be measured by a sensor and often feedforward controllers are adaptive such that they need an error signal, also measured by a sensor (see Fig. (18)). In contrast to feedback control the error signal only has the task to accomodate the controller by adjusting its weighting factors to timevariant changes of the primary transmission path and the behavior of the mechanical system. The error signal does not act like a variable which passed through the controller to be used as the actuator signal.

The block diagram of Fig. (17) is a simplification and does not show the real complexity of a feedforward controller. One problem e.g. is the feedback effect of the secondary actuator. Because of the bidirectional nature of the primary transmission path the influence of the actuator changes the excitation signal. This feedback effect can be compensate by adding a filter to the controller that has the same response equal to the feedback path but with opposite sign.

Assuming that there is no noise in the reference signal, the desired cancelation of primary and secondary forces would be perfect for

$$H(s) = P(s) \quad \rightarrow \quad e(s) = 0. \tag{50}$$

The equality of the transfer functions $H(s)$ and $P(s)$ means that the magnitude and phase response of the controller has to be equal for all frequencies to those

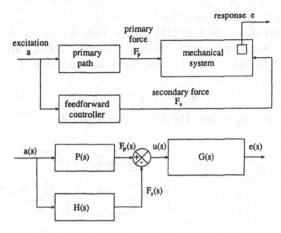

Figure 17. Principle of feedforward control

of the primary path. This statement is also valid for the impulse response of the controller and makes the method of feedforward control to a problem of electrical filter design. Nowadays filtering is usually done in a digital manner because of the problems in designing complicate analogue filters. Using digital filters in an analogue environment means that it is necessary to convert the signals, what can be done by A/D- and D/A-converters. To prevent aliasing the converters are equipped with analogue low-pass-filters which causes a phase shifting. The phase shifting and the processing time of the controller results in an inevitable delay. Thus the initial part of the impuls of the primary path can't modeled by a digital filter.

It may seem as if digital filters loose their attractiveness because of their inherent delay, but there is an enormous adavantage besides them. Often the characteristics of the primary transmission path and the mechanical systems change slowly with respect to time. To fulfill its tasks the controller has to accomodate his transfer behavior to this change. Digital filters allows to accomodate their transfer behavior simply by adjust their weighting factors (see Fig. (18)).

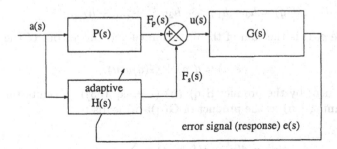

Figure 18. Adaptive feedforward controller

4.2.1. *Digital Feedforward Controllers*

The block diagram shown in Fig. (17) can be redrawn in the form shown in Fig. (19) as was described in reference ((Fuller et al. 1996)). The Laplace variable s has to replaced by the sample number n and the delay operator q. The delay operator is defined by

$$q^{-1}a(n) = a(n-1), \tag{51}$$

what means that q^{-1} represents the delay of one sample.

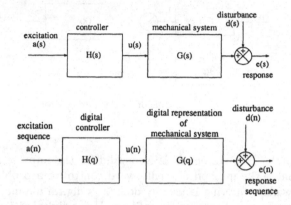

Figure 19.   Block diagram of the digital feedforward controller

In the following discussion we focus our interest to finite impuls response (FIR) filters, whose names indicates that the duration of their impuls response has a finite lenght. FIR filters produce an output that is created as the sum of weighted previous inputs expressed by

$$u(n) = \sum_{i=1}^{I-1} h_i a(n-1) \tag{52}$$

or in form of the delay operator

$$u(n) = H(q)a(n) \tag{53}$$

with

$$H(q) = h_0 + h_1 q^{-1} + h_2 q^{-2} + \dots + h_{I-1}q^{I-1}. \tag{54}$$

The response e(n) is the sum of the disturbance and the output of the mechanical system.

$$e(n) = d(n) + G(q)u(n). \tag{55}$$

Substituting u(n) by the product H(q)a(n) (see. eq. (53)), permute the multiplication and defining r(n) as the product of G(q)a(n) leads to

$$e(n) = d(n) + H(q)r(n) = \sum_{i=1}^{I-1} h_i r(n-1). \tag{56}$$

From Eqn. (56) it can be seen that the response e(n) depends on each weighting factor h_i of the controller. The ultimate task is now to find the best factors. This task can be solved by the minimization of a cost function. It can be shown (see e.g. (Nelson and Elliott 1995)) that the so called filtered-x LMS algorithm

$$h_i(n+1) = h_i(n) - \alpha e(n)r(n-1) \tag{57}$$

satisfy this task. The coefficient α indicates the convergence. The filtered-x LMS algorithm was first proposed by Morgan in 1980 and independently for feedforward control by Burgess in 1981.

5. Active Damping in Structures with Bolted Joints

Lightweight flexible structrures e.g. for deployable space systems often consist of truss structures. The truss elements are connected by bolted joints. Microslip and macroslip in the joints contact surfaces are the dominating dissipation mechanisms as compared to material damping and environmental damping if no additional damping measures are applied. So far only position control by actuators in truss elements is realized. The present approach aims to control the nonlinear transfer behavior of joints by adapting the control pressure. This combines the avtive damping and stiffness changes.

5.1. ACTIVE SPACE STRUCTURE

Leightweight space structures have to perform precise manoeuvres. To achieve this goal, vibrations of the structure due to internal and environmental disturbances have to be damped out quickly. This can be done by adaptive structures, where sensors and actuators adapt the system behavior.

5.1.1. *Simulation of Space Truss Structure*
Vibrations of a spacecraft with flexible antenna boom are simulated. The antenna with a mass of 35 kg is attached to the spacecraft by a truss structure of lenght 20,3m and a mass of 95 kg as shown in Fig. (20).

For the simulation of large reference or rigid body motions with superimposed small deflections a combination of the multibody system approach (MBS) and the finite element method (FEM) has been proposed (see reference (Gaul, Lenz and Sachau 1998)). The equations of motions are consistently linearized such that the so called geometric stiffening terms are preserved by a pertubation technique which splitts low and high frequency modal contents. The flexibility is treated in the sence of a Ritz approximation by superposition of structural modes calculated with FEM (see reference (Lenz 1997)).

480

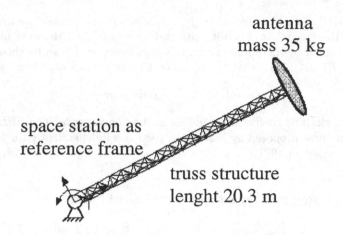

antenna
mass 35 kg

space station as
reference frame

truss structure
lenght 20.3 m

Figure 20.   Model of a space vehicle with a flexible truss

5.2. NONLINEAR TRANSFER BEHAVIOR OF BOLTED JOINTS

The dominant dissipation in vibratory response of the truss structure in Figure (20) is due to the nonlinear transfer behavior of the joints. Such transfer behavior of isolated joints has been measured for bolted lap joints as part of suspended resonators for longitudinal or torsional forced excitation (see reference (Gaul and Lenz 1997)).

The Valanis model known from endochronic plasticity turned out to be well suited for the transient as well as steady state generalized force and deformation description (see reference (Lenz 1997)). A proper form for simulation is the ODE

$$\dot{F}(u, \dot{u}, F) = \frac{E_0 \dot{u} \left[1 + \frac{\lambda}{E_0} sgn\dot{u} \left(E_t u - F\right)\right]}{1 + \kappa \frac{\lambda}{E_0} sgn\dot{u} \left(E_t u - F\right)}, \tag{58}$$

which relates the rate of generalized force $\dot{F} = dF/dt$ to the force itself, the generalized relative deformation u and the velocity $\dot{u}$. The four parameters E_0, E_t, σ_0 and κ in Eqn. (58) can be identified from measured hysteresis (see reference (Gaul and Lenz 1997)). They are depicted in a closed hysteresis in Fig. (21).

5.3. ACTIVE BOLTED JOINTS IN SPACE TRUSS STRUCTURE

The passive space structure of Fig. (20) is changed to a semi active system by replacing the passive bolted joint between the connecting rod and the spacecraft by an active joint depicted in Fig. (22).

According to a patent submitted by the first author (Gaul 1997) a piezoelectric stack is used to modify the joint normal pressure P by voltage control. Significant variation of the hysteresis with different contact pressure are shown in Fig. (23) for transmitted torque versus relative angle of rotation.

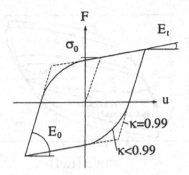

Figure 21. Closed hysteresis of the Valanis model of joint

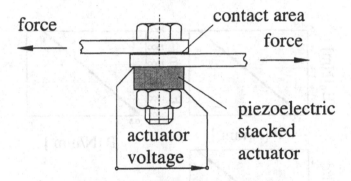

Figure 22. Active bolted joint connection

The dependency of the parameters E_0, E_T and σ_0 from the normal pressure P can be approximated by linear functions, plotted as lines in Fig. (24). The selected parameter κ is approximately 0.01 for all cases.

With a single semi active joint a single input single output control system (SISO) of low complexity is proposed to control the structure. The antenna displacement is measured directly with a sensor, e.g. an optical device. Figure (25) shows the control system block diagram. The controller calculates the adapted variable P for the bolted joint connection depending on the control deviation $u_A = u_A^d - u_A^a$. The control algorithm most widely used in standard controller design features is a PID feedback. In the present case the integrator (I-element) is not used.

For simulation purposes the controller has to be described in time domain. The state space form is given with the varaible x as

$$\dot{x} = -N_0 x + u_A, \quad x(0) = 0$$
$$P = (Z_0 - Z_1 N_0) x + Z_1 u_a. \tag{59}$$

Herewith all portions of the controled space structure are expressed in the form of nonlinear ODE, ready for time integration within the multibody simulation. Due to the nonlinear behavior of the bolted joint connection, a linearization of the equations

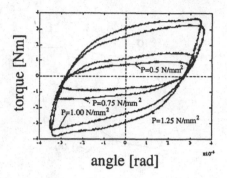

Figure 23. Hysteresis plot of lap joint with varying normal pressure

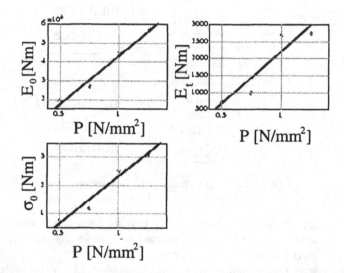

Figure 24. Dependency of the Valanis parameters on the joint normal pressure P

of motion is inadvisable. Therefore the dynamic behavior of the system has to be investigated in time domain via simulation.

The space station is considered to be inertial fixed. The system reaction to an oscillating force in y-direction at the antenna is investigated. The excitation acts with an amplitude of 10 N. The system behavior near the lowest eigenfrequency (1.36 Hz) of the truss structure with fixed bolted joint connection is investigated. Here the frequency band from 0.5 Hz up to 2.5 Hz is considered.

For discrete excitation frequencies in this range, the maximal antenna deflection has been calculated and plotted versus the excitation frequency in Fig. (26) for different constant joint normal pressure P in the range of 50 up to 500 N/mm^2. The best passive system behavior which means the smallest antenna excursion is achieved by choosing $P_{\mathrm{I}} = 150 N/mm^2$.

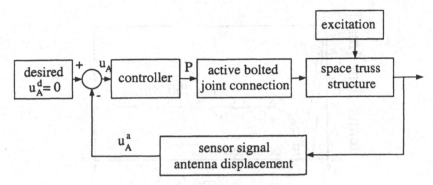

Figure 25. Control system block diagram

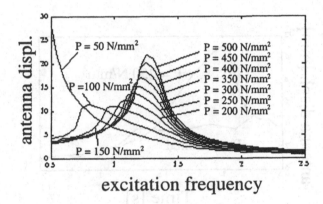

excitation frequency

Figure 26. Maximal absolute antenna displacement for different constant joints normal pressure P

Now the PD-controller is used. Suitable controller parameters Z_0 and Z_1 are found from systematic parameter variation where the parameters are varied from 0 up to 10^6 in steps of 10^5 $N/(m^3 s)$ respectively N/m^3. The parameters N_0 is fixed to $N_0 = 100$ $1/s$. The large number of simulations is performed with the MBS program SIMPACK. Figure (27) shows the achieved enhancement of the system behavior, after selecting the control parameters $Z_0 = 2 \cdot 10^5$ $N/(m^3 s)$ and $Z_1 = 7 \cdot 10^5$ N/m^3. Especially the amplitudes in the frequency range from 0.9 Hz up to 1.5 Hz are much smaller compared with the uncontroled passive system.

The achieved improvement of the system behavior can as well be seen by the free vibration response, Fig. (28). An initial antenna tip displacement of 60 mm attenuates after a few oscillations down to 7 mm for the passive system with pressure $P_1 = 150$ N/mm^2. Due to microslip the amplitude reduction is small during the following oscillations. Much more efficient is the controled system behavior. Here the deflection reduces quickly to a much lower level of about 2 mm.

484

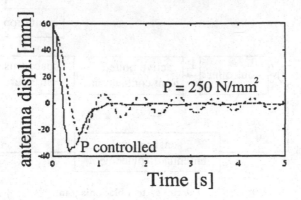

Figure 27. Maximal absolute antenna displacement a) controled with $Z_0 = 2 \cdot 10^5 \ N/(m^3 s)$, $Z_1 = 7 \cdot 10^5 \ N/m^3$ b) constant $P = 150 \ N/mm^2$

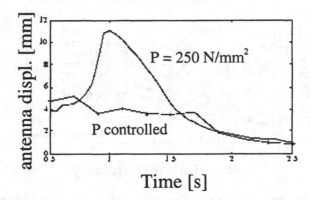

Figure 28. Antenna deflection during free vibration

5.4. ACTIVE DAMPING MAXIMIZATION

An alternative control concept proposed in (Lenz 1997) aims at the maximization of energy dissipation in the semi active joint by microslip under the constraint, that macroslip in each joint is avoided with a preselected safety factor. According to this concept, the transmitted force in the joint has to be measured e.g. by strain gauge as depicted in Fig. (29). The control unit directly varies the control contact pressure in the joint by the applied voltage on the piezoelectric element.

As the contact pressure determines the sticking force, it is controled in each time step depending on the measured transmitted force as low as possible to maximize macroslip such that gross slip, the so called macroslip, is excluded with a safety factor.

Simulation results for an active truss structure (see reference (Lenz 1997)) compared to the passiv response prove the dissipation efficiency of this concept which does not require preinformation about structural loading.

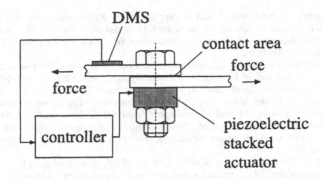

Figure 29. Active bolted joint with strain gauge

References

Achenbach, M.: 1989, A model for an alloy with shape memory, *International Journal of Plasticity* pp. 397–395.

AG, B.: 1994, Provisional productinformation rheobay, Bayer AG, Germany.

Baz, A., Iman, K. and McCoy, J.: 1990, The dynamics of helical shape memory actuators, *Journal of Intelligent Material Systems and Structures* 1, 105–133.

Beards, C.: 1995, *Engineering Vibration Analysis with Application to Control Systems*, Edward Arnold.

Bertram, A.: 1982, Thermo-mechanical constitutive equations for the description of shape memory effects in alloys, *Nuclear Engineering and Design* pp. 173–182.

Brinson, L.: 1993, One dimensional constitutive behaviour of shape memory alloys: Thermo-mechanical derivation with non-constant material functions and redefined martensit internal variable, *Journal of Intelligent Material Systems and Structures* 4, 229–242.

Brinson, L. and Lammering, R.: 1993, Finite element analysis of the behavior of shape memory alloys and their applications, *International Journal of Solids and Structures* 30, 3261–3280.

Cahn, J. and Larche, F.: 1984, A simple model for coherent equilibrium, *Acta Metallurgica* 32, 1915–1923.

Dimitriadis, E. and Fuller, C.: 1991, Active control of sound transmission through elastic plates using piezoelectric actuators, *AIAA Journal* 29, 1771–1777.

Falk, F.: 1980, Model free energy, mechanics and thermodynamics of shape memory alloys, *Acta Metallurgica* 28, 1773–1780.

Fu, S., Huo, Y. and Müller, I.: 1993, Thermodynamics of pseudoelasticity - an analytical approach, *Acta Mechanica* 99, 1–19.

Fuller, C.: 1990, Actice control of sound transmission/radiation from elastic plates by vibration inputs, *Journal of Sound and Vibration* 136, 1–15.

Fuller, C., Elliott, S. and Nelson, P.: 1996, *Active Control of Vibration*, Academic Press.

Gaul, L.: 1997, Active control of joints in mechanical members and structures, Deutsches Patentamt, Offenlegungsschrift DE 197 02 518.8.

Gaul, L. and Lenz, J.: 1997, Nonlinear dynamics of structures assembled by bolted joints, *Acta Mechanica* 125, 169–181.

Gaul, L., Lenz, J. and Sachau, D.: 1998, Active damping of space structures by contact pressure control in joints, *International Journal Mechanics of Structures and Machines* 26, 81–100.

Graesser, E. and Cozzarelli, F.: 1994, A proposed three dimensional constitutive model for shape memory alloys, *Journal of Intelligent Material Systems and Structures* 5, 78–89.

Hib: 1989, *ABAQUS Program Documentation*.

Hsu, C., Lin, C. and Gaul, L.: 1997a, Shape control of composite plates by bonded actuators with high performance configuration, *Journal of Reinforced Plastics and Composites* 16, 1692–1710.

Hsu, C., Lin, C. and Gaul, L. (eds): 1997b, *Suppression of Distributed Dynamical Deflections by Piezoelectric Materials for Harmonic External Loads*, Smart Structures and Materials, SPIE.

Ikegami, R., Wilson, D., Anderson, J. and Julien, G.: 1990, Active vibration control using nitinol and piezoelectric ceramics, *Journal of Intelligent Material Systems and Structures* 1, 189–206.

Lenz, J.: 1997, *The Influence of Micro- and Macroslip in Joints on Structural Dynamics*, PhD thesis, University of Stuttgart.

Liang, C. and Rogers, C.: 1990, One dimensional thermomechanical constitutive relations for shape memory materials, *Journal of Intelligent Material Systems and Structures* 1, 207–234.

Lu, Y.: 1996, *Industrial Intelligent Control*, Wiley Interscience Publication.

MacLean, B., Draper, J. and Misra, M.: 1991a, Development of a shape memory actuator for adaptive truss applications, *Journal of Intelligent Material Systems and Structures* 2, 261–280.

MacLean, B., Patterson, G. and Misra, M.: 1991b, Modeling a shape memory integrated actuator for vibration control for large space structures, *Journal of Intelligent Material Systems and Structures* 2, 72–94.

McNichols, J.: 1987, Thermodynamics of nitinol, *Journal of Applied Physics* 61, 972–984.

Mooi, H.: 1992, *Active control of structural parameters of a composite strip using embedded shape memory alloy wires*, Master's thesis, Deutsche Forschungsanstalt für Luft- und Raumfahrt.

Müller, I.: 1989, On the size of hysteresis in pseudoelasticity, *Continuum Mechanics and Thermodynamics* 1, 125–142.

Müller, I.: 1992, Thermoelastic properties of shape memory alloys, *European Journal of Mechanics* 11, 173–184.

Müller, I. and Achenbach, M.: 1983, Creep and yield in martensitic transformations, *Ingenieur Archiv* 53, 73–83.

Müller, I. (ed.): 1986, *Pseudoelasticity in Shape Memory Alloys. An Extreme Case of Thermoelasticity*, Accademia Nazionale dei Lincei.

Nelson, P. and Elliott, S.: 1995, *Active Control of Sound*, Academic Press.

Noor, A. and Jorgensen, C.: 1996, A hard look at soft computing, *Aerospace America* pp. 35–39.

Preumont, A.: 1997, *Vibration Control of Active Structures*, Kluwer Academic Publishers.

Savi, M., Braga, A., Alves, J. and Almeida, C. (eds): 1998, *Finite Element Model For Trusses with Shape Memory Alloy Actuators*, EUROMECH 373 Colloquium.

Shetty, D. and Kolk, R.: 1997, *Mechatronics System Design*, PWS Publishing Company.

Siouris, G.: 1996, *Optimal Control and Estimation Theory*, Wiley Interscience Publication.

Smith, B., Peters, R. and Owen, S.: 1996, *Acoustics and Noise Control*, Addison Wesley Longman Limited.

Tanaka, K.: 1986, A thermomechanical sketch of shape memory effect: One dimensional tensile behaviour, *Res Mechanica* 18, 251–263.

Tanaka, K. and Iwasaki, R.: 1985, A phenomenological theory of transformation superplasticity, *Engineering Fracture Mechanics* 21, 709–721.

Tanaka, K. and Nagaki, S.: 1982, A thermomechanical description of materials with internal variables in the process of phase transitions, *Ingenieur Archiv* 51, 287–299.

Tanaka, K. and Sato, Y.: 1988, Estimation of energy dissipation in alloys due to stress-induced martensitic transformation, *Res Mechanica* 23, 381–393.

Tanaka, K., Sato, Y. and Kobayashi, S.: 1985, Pseudoelasticity and shape memory effect associated with stress induced martensitic transformation: A thermomechanical approach, *Transactions of Japan Society of Aero and Space Science* 28, 150–160.

Warlimont, H., Delaey, L. and Krishnan, R.: 1974, Thermoplasticity, pseudoelasticity and the memory effect associated with martensic transformations, *Journal of Material Science* 9, 1536–1555.

Wayman, C.: 1980, Some applications of shape memory alloys, *Journal of Metals* pp. 129–137.

Wicker, K., Eberius, C. and Guicking, D. (eds): 1996, *Acoustic Properties of Electro Rheological Fluids*, ACTUATOR, Bremen.

Wolff, C. and Wendt, E. (eds): 1994, *Application of Electrorheological Fluids in Hydraulic*, ACTUATOR, Bremen.

ACOUSTIC MODAL ANALYSIS

P. SAS
KULeuven, Div. PMA
B-3001 Heverlee, Belgium

F. AUGUSZTINOVICZ
Technical University of Budapest
H-Budapesti Müssaki Egyeten, Hongary

1. Introduction

The phenomena, related to the existence of acoustic modes, were already known in the ancient world and our ancestors, though instinctively, have even exploited some of the acoustic effects [1]. The first treatments of scientific character of the field date back to the 19th century [2,3] while the basics of the modal theory of room acoustics were developed in the first half of this century [4-7]. Nevertheless, a revival of the acoustic modal theory and its experimental aspects seems to be worthwhile for a couple of reasons.

In the first place, the ever increasing demand for lower noise levels and personal comfort necessitates to attack also those problems, for which the traditional armoury of noise control engineering is no longer sufficient. The booming noise in cars or the propeller noise in small aircrafts are typical examples of a low frequency noise problem, where the consideration of the single acoustic modes is inevitable.

In the second place, a wide range of modern computational, experimental and noise control techniques are based on or closely related to the acoustic modes. We can cite here both the Finite Element and the Boundary Element method, active noise and/or vibration control of sound field in closed spaces and others.

A further and very important reason of the renewed interest in acoustical modal analysis (AMA) is that the rapid development in the experimental techniques of structural dynamics has just recently enabled us to render the AMA method from a pure theoretical calculation procedure, burdened with serious application limitations, to an experimental engineering routine. This transition is however not without dangers, since the analyst can obtain misleading results if the structural methods are used for acoustic applications without due foresight.

The aim of these courses notes is to summarise the basic notions of the acoustical modal theory and the inter-relations thereof, to shed light on the existence and limitations of the analogy between the modal behaviour of mechanical and acoustic systems and to give some hints for those who are interested in the practical details in the experimental modal analysis in acoustics and vibro-acoustics.

2. Physics and mathematics of modes in one-dimensional acoustic systems

2.1. ANALYSIS OF A WAVEGUIDE, TERMINATED BY GENERAL ACOUSTIC IMPEDANCES

J.M.M. Silva and N.M.M. Maia (eds.), Modal Analysis and Testing, 487–506.

2.1.1. *Mathematical Analysis*

We will now repeat the analysis of a one-dimensional acoustic waveguide on the basis of a general acoustic formulation: the acoustic wave equation [21].

Consider a finite circular tube of length L and of diameter d so that $d \ll \lambda$. Let the air in the tube be at rest, with uniform density and temperature. Then the wave motion in the tube can be described by the one-dimensional acoustic wave equation

$$\frac{\partial^2 p}{\partial x^2} = \frac{1}{c^2} \frac{\partial^2 p}{\partial t^2} \tag{1 a}$$

and similarly for the particle velocity

$$\frac{\partial^2 v}{\partial x^2} = \frac{1}{c^2} \frac{\partial^2 v}{\partial t^2} \tag{1 b}$$

Seeking for a solution to Eq. (6 a) in the form

$$p(x,t) = p_x(x) \, p_t(t)$$

the spatial and the temporal variables can be separated:

$$\frac{d^2 p_x(x)}{dx^2} + k^2 p_x(x) = 0 \tag{2 a}$$

$$\frac{1}{c^2} \frac{d^2 p_t(t)}{dt^2} + k^2 p_t(t) = 0 \tag{2 b}$$

Eq. (2 a) is the familiar homogeneous Helmholtz-equation (here in one-dimensional form), describing the spatial variation of the pressure along the tube while Eq. (2 b) can be used to determine the temporal variation. The solution $p_x(x)$ has to satisfy both the equation and the boundary conditions at the end of the tube. We face here a classical eigenvalue problem: one has to determine, for which k values does a non-trivial solution exist and what is the sound field associated with these eigenvalues.

The solution is assumed to be in the usual complex form:

$$p(x,t) = p_0 \, e^{j\omega t} \left(a \, e^{-jkx} + b \, e^{jkx} \right) \equiv \psi(x) e^{j\omega t} \tag{3 a}$$

The particle velocity can again be obtained from the Euler-relation:

$$v(x,t) = -\frac{1}{j\omega\rho} \frac{\partial p}{\partial x} = \frac{p_0}{\rho c} e^{j\omega t} \left(a \, e^{-jkx} - b \, e^{jkx} \right) \tag{3 b}$$

To obtain the modes of the investigated system, Eq. (2 a) has to be solved with appropriate boundary conditions at the ends of the tube. Assume that the tube is closed at one end ($x = 0$) by a specific acoustic impedance Z_1:

$$\frac{p_x(x)}{v_x(x)}\bigg|_{x=0} = -Z_1 \tag{4}$$

and similarly at $x = L$:

$$\frac{p_x(x)}{v_x(x)}\bigg|_{x=L} = Z_2 \tag{5}$$

(Note that the negative sign in Eq. (4) is caused by the fact that the outward normal of the enclosing surface is opposite to the direction of the x-axis.) Substituting Eqs. (3 a) and (3 b) in Eqs. (4) and (5) we get the characteristic equation of the system:

$$\frac{Z_1+\rho c}{Z_1-\rho c}\frac{Z_2+\rho c}{Z_2-\rho c}e^{j2kL} = 1 \tag{6}$$

Fortunately, it can be solved in closed form by using the complex logarithm function and then one obtains for the nth modal frequency a complex value:

$$\omega = \omega_n + j\delta$$

where the real value of ω,

$$\omega_n \equiv \frac{nc\pi}{L} - \frac{c\phi\{\zeta_{12}\}}{2L} \tag{7}$$

represents the frequency of the eigenvibration while the imaginary value stands for the damping, caused by the energy losses in the system:

$$\delta \equiv \frac{c\ln|\zeta_{12}|}{2L} \tag{8}$$

and

$$\zeta_{12} \equiv \frac{Z_1+\rho c}{Z_1-\rho c}\frac{Z_2+\rho c}{Z_2-\rho c} \tag{9}$$

Calculating now the wavenumber

$$k = \frac{\omega}{c} = k_n + j\kappa$$

and substituting back to the homogeneous Helmholtz-equation, the mode shape can eventually be calculated:

$$\psi(x) = \frac{\cosh\kappa x\left(Z_1\cos k_n x + j\rho c\sin k_n x\right)}{Z_1 - \rho c} - \frac{\sinh\kappa x\left(\rho c\cos k_n x + jZ_1\sin k_n x\right)}{Z_1 - \rho c} \tag{10}$$

490

2.1.2. *Qualitative discussion of some special cases*

To discuss the behaviour of our system quantitatively, it is instructive to calculate the complex modal frequencies and the mode shapes for various simple parameter combinations.

First assume that the closing impedances are purely imaginary: $Z = jX$. The modal frequency then becomes purely real, but different from the closed tube resonances ($\omega_n = nc\pi/L$)

$$\omega_n = \frac{nc\pi}{L} - \frac{c \tan^{-1}(\rho c/X)}{L} \tag{11}$$

Depending on the sign of the closing reactances, the impedances at the end of the tube make the tube seemingly longer or shorter than the closed tube. Depicting the mode shape, Eq. (10), in the Nyquist plot, one gets a straight line having a slope which depends on the ratio of $X/\rho c$ as depicted in Figure 1. These kind of modes are usually referred to in structural dynamics as *collinear modes*, in spite of the fact that the mode shape function is actually a complex function.

If the terminating (identical) impedances are purely real and greater than the specific acoustic impedance of the fluid (ρc), the damped modal frequency remains unchanged with respect to the rigid-rigid termination but one gets finite damping:

$$\delta = \ln[(R + \rho c)/(R - \rho c)]/L \tag{12}$$

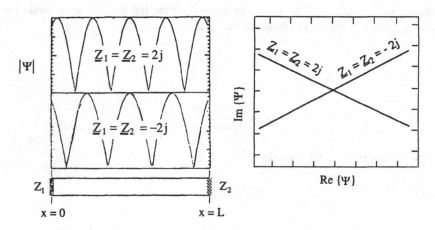

Figure 1   Calculated mode shapes of a one-dimensional waveguide, terminated by pure imaginary impedances

The mode shape is accordingly no longer collinear but becomes a truly complex one as shown in Figure 2.

In the general case both effects take place: the mode will be damped and the modal frequency is shifted, and the mode shape will be complex, too.

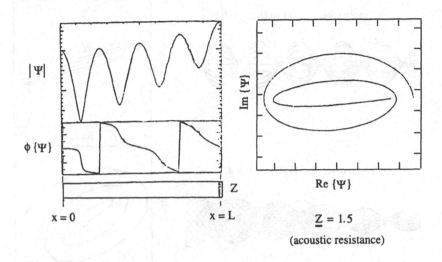

Figure 2   Calculated mode shape of a one-dimensional waveguide, terminated by pure real
impedance

2.1.3. *Verification experiments*

The aim of our experiments was to conduct an acoustical modal analysis test for the
verification of the calculations discussed above. The apparatus used for the experiments
was a standard acoustic impedance tube of Type Brüel & Kjær 4002. This instrument is
normally used for measurements of the specific acoustic impedance and absorption
coefficient of circular cut samples of acoustic materials. The design and the dimensions of
the tube ensure that only one-dimensional wave propagation can exist in the tube within
the specified frequency range.

Unlike in the course of the standard procedure, the excitation of the tube was ensured
by means of a random noise generator and the excitation signal was parallel fed to the
input of an experimental modal analysis system. The sound field was sampled by means
of a small electret condenser microphone Type AKG CK-67/3, traversed along the axis of
the tube by means of the original microphone carriage system. The first modal analysis
test was carried out without any sound absorbing material. Then a 5 cm thick
polyurethane foam disc was inserted in the sample holder and the experiment was
repeated.

As anticipated, the first experiment resulted in a number of lightly damped
eigenfrequencies of the tube. It is interesting to see, however, that the frequencies are not
strictly harmonic; the slight deviations are thought to be caused by the finite acoustic
impedance of the driving loudspeaker. The residuals of the experimental AMA test, which
are proportional to the mode shape values, are shown in Figure 3 a), depicted both as a
spatial function along the tube and in a Nyquist plot. In agreement with our findings
above, the locus is nearly collinear and the mode shape corresponds to a clear standing
wave. In case of the absorber at the end of the tube, see Figure 3 b), the locus of the
residuals is truly complex.

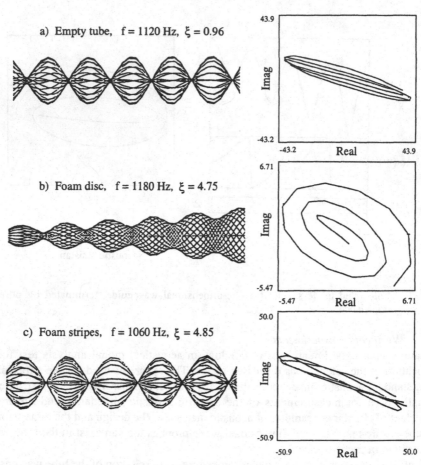

a) Empty tube,   f = 1120 Hz, ξ = 0.96

b) Foam disc, f = 1180 Hz, ξ = 4.75

c) Foam stripes, f = 1060 Hz, ξ = 4.85

Figure 3   Measured mode shapes and locii of residuals of a one-dimensional waveguide.
a) Tube rigidly terminated, b) tube terminated by a 50mm thick foam absorber, c)
tube provided with absorber stripes along the whole tube

2.2.   ANALYSIS WITH DISTRIBUTED DAMPING

The above analysis represents a rather special case, with non-rigid boundary conditions at
the ends of a one-dimensional vibrating system only. It is more realistic to assume that
the damping is distributed along the length of the tube. The mathematical treatment of
this problem is feasible but too complex to repeat here in full details; the reader is rather
referred to the relevant literature [9,10]. Nevertheless, it is instructive to show the results
of a simple experiment: the modal analysis of the tube discussed in paragraph 2.1.3 with
stripes of sound absorbing material along the whole tube. The measuring system and the
analysis technique is just the same as it was before. As one can see in Figure 3 c), the
modal frequency is again different from the case with rigid termination but the mode
shape corresponds very well to the undamped situation, in spite of the fact that the modal
damping of the test is higher than it was with the sound absorbing termination.

This observation can be explained by means of the analogy between mechanical and acoustical systems, to be evolved later on in paragraph 4.1. As it is well known from structural dynamics, it is not the damping itself which is responsible for the complexity of the mode shape but rather the local distribution of it. (Of course, if damping is present in a system, the modal frequencies must always be complex.) Local damping is an inherent characteristics of the overwhelming majority of acoustical systems since the viscous or thermal conductivity losses of the acoustic fluid - which are distributed damping sources - are negligible with respect to damping caused by the dissipative boundaries of the fluid in the relevant frequency range of modal analysis. As a consequence, in most of the cases one encounters complex modes, the interpretation of which can cause various problems.

3. Modes and forced waves in three-dimensional systems

3.1. MODES IN UNDAMPED SYSTEMS

If we want to extend our analysis for a general, three-dimensional, bounded space, the one-dimensional wave equation, Eq. (1 a) has to be written in the more general form of

$$\nabla^2 p = \frac{1}{c^2} \frac{\partial^2 p}{\partial t^2} \tag{13}$$

The space-dependent Helmholtz-equation then becomes

$$\left(\nabla^2 + k^2\right)p = 0 \tag{14}$$

This equation is not easy to solve in closed form, unless one uses a coordinate system in which the variables are separable and the boundary conditions are simple enough. A classical, simple, yet important case is a rectangular room as shown in Figure 4. Assume that the boundaries are perfectly rigid; then we have the boundary conditions

$$v_x = 0 \quad \text{at} \quad x = 0 \quad \text{and} \quad x = L_x, \tag{15 a}$$
$$v_y = 0 \quad \text{at} \quad y = 0 \quad \text{and} \quad y = L_y, \tag{15 b}$$
$$v_z = 0 \quad \text{at} \quad z = 0 \quad \text{and} \quad z = L_z, \tag{15 c}$$

Since the individual boundary conditions depend on only one of the coordinates, they can be fulfilled independently. Let us compose the solution by the product of three functions, each dependent only on x, y and z, respectively:

$$p(x,y,z) = p_x(x)\, p_y(y)\, p_z(z) \tag{16}$$

It is rather plausible that all these terms differ from each other in the independent variables only and the spatial variation in one of the directions should be the same as in the one-dimensional case [7]. Thus we obtain:

494

$$p(x,y,z) = \cos \frac{n_x \pi x}{L_x} \, \cos \frac{n_y \pi x}{L_y} \, \cos \frac{n_z \pi x}{L_z} \qquad (17\,a)$$

where the numbers n_x, n_y and n_z are non-negative integers. The corresponding eigenfrequency is then :

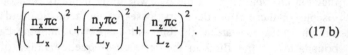

$$\sqrt{\left(\frac{n_x \pi c}{L_x}\right)^2 + \left(\frac{n_y \pi c}{L_y}\right)^2 + \left(\frac{n_z \pi c}{L_z}\right)^2} \, . \qquad (17\,b)$$

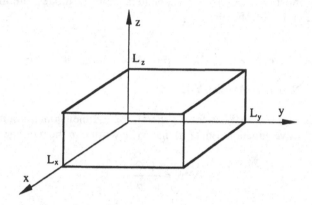

Figure 4 Sketch of a simple three-dimensional room

If one of the coefficients n_x, n_y, n_z is different from zero, the modes are similar to the one dimensional case and are called 'axial modes'. In cases where two coefficients are different from zero, we speak of 'tangential modes' and in the general case of 'obligue modes'.

The eigenfrequencies of a three dimensional rectangular room can be visualised in several ways. For axial or tangential modes the neutral or nodal lines can be shown as depicted in Figure 5 a). For tangential modes the wavefronts can also be drawn, see Figure 5 b). The true three-dimen-sional "modal model" can be depicted as distorted wireframes, by colors etc. Nevertheless, the schematic representation of the modal frequencies by means of a three-dimensional lattice does not only represent the modes in a very concise way, but it can help to understand the limitations of the AMA method as well. Let us take a three-dimensional spatial mesh in a Cartesian coordinate system, with unit mesh widths of $c\pi/L_x$, $c\pi/L_y$ and $c\pi/L_z$, respectively, parallel to the axes as shown in Figure 5 c). Any vector from the origin to a nodal point of this lattice then corresponds to a particular eigen-frequency, because the length of this vector is equal to (17 b). Obviously, as the mode numbers are increasing, more and more combinations can result in almost identical vector lengths and thus closely spaced modal frequencies. If the number of modes per unit frequency is too high, the modes can no longer be distinguished in a proper way by experimental methods. Along with the matter of mode complexity as discussed above, these features of the acoustic eigenmodes can pose serious practical limitations to the feasibility of meaningful acoustic modal analyses.

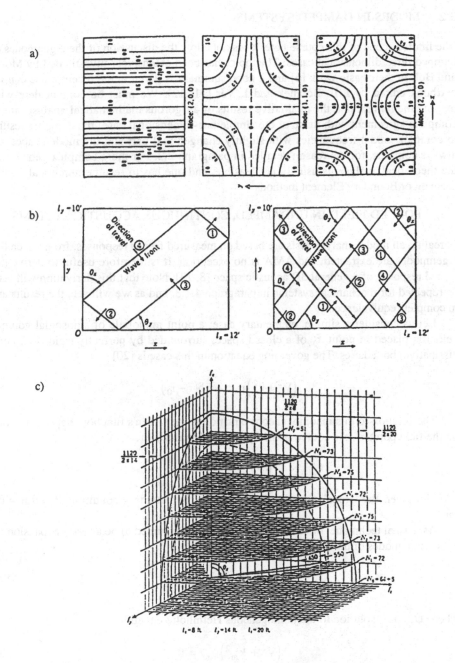

Figure 5   Possibilities of representation of acoustic modes in three-dimensional cavities
a) 2D demonstration of nodal lines, b) 2D demonstration of wavefronts, c) 3D
demonstration of modal frequencies [15]

3.2. MODES IN DAMPED SYSTEMS

The limited scope of this course notes does not allow the discussion of the eigenmodes of damped three-dimensional cases. The topics has been extensively investigated by Morse and Bolt in their basic work [6] and a good summary is given in the recent noise control textbook edited by L.L. Beranek and I. Vér [11, 12]. We must be content here with making the remark that the qualitative results of a rigorous mathematical analysis are in complete agreement with those of the simple one-dimensional case: the damping results in complex modal frequencies with slightly changed real part and the mode shapes are always complex, for the reason detailed in paragraph 2.2. For more complex geometries the theoretical analysis is usually not feasible, and one has to resume numerical, Finite Element or Boundary Element methods.

3.3. FORCED FIELD IN UNDAMPED, CONTINUOUS ACOUSTIC SYSTEMS

In reality, all modal analysis tests are based on measured forced responses, from which the eigenmodes are extracted, and AMA is no exception. It is therefore useful to derive the forced response of a general 3D acoustic space [8, 13]. Note that this derivation will also be repeated for mechanical systems in paragraph 4.1.3. and as we will see, the results are in complete equivalence.

Let us consider a simple, elementary case: a point monopole of sinusoidal volume velocity, placed in point $\vec{r}_0$ of a closed space, surrounded by perfectly rigid (i.e., non-dissipative) boundaries. The governing equation in this case is [20]

$$\left(\nabla^2 + k^2\right)p = -\dot{q}\,\delta(\vec{r} - \vec{r}_0) \tag{18}$$

The solution to this inhomogeneous equation is the Green's function, depending both on the field point $\vec{r}$, the source point $\vec{r}_0$ and the frequency ω:

$$p(\vec{r}) = g(\vec{r}, \vec{r}_0, \omega)$$

Moreover, the Green's function has to satisfy the boundary condition $\partial g / \partial\,\bar{n} = 0$, too.

As a usual technique, the Green's function can be assumed to be a series expansion of the normal modes of the closed space such as

$$g(\vec{r}, \vec{r}_0, \omega) = \sum_m a_m\,\psi_m \tag{19}$$

where ψ_m is a solution to the homogeneous Helmholtz equation

$$\left(\nabla^2 + k_m^2\right)\psi_m = 0 \tag{20 a}$$

Let ψ_n be another solution

$$\left(\nabla^2 + k_n^2\right)\psi_n = 0 \tag{20 b}$$

Multiplying Eq. (20 a) by ψ_n and Eq. (20 b) by ψ_m and summing we obtain:

$$\psi_m\left(\nabla^2+k_n{}^2\right)\psi_n - \psi_n\left(\nabla^2+k_m{}^2\right)\psi_m = 0$$

But $\psi_m\nabla^2\psi_n - \psi_n\nabla^2\psi_m$ is the divergence of $\psi_m\nabla\psi_n - \psi_n\nabla\psi_m$. Thus, integration over the volume and application of Gauss' theorem yields :

$$\oiint_S (\psi_m\nabla\psi_n - \psi_n\nabla\psi_m).dS + (k_n^2 - k_m^2)\oiiint_V \psi_n\psi_m.dV = 0.$$

As $\nabla\psi_n$ and $\nabla\psi_m$ equal zero on the boundary surface S, one get :

$$(k_n^2 - k_m^2)\oiiint_V \psi_n\psi_m.dV = 0.$$

As $k_n \neq k_n$, this means that

$$\oiiint_V \psi_n\psi_m.dV = 0, \tag{21}$$

what can be interpreted as the orthogonality relation of acoustic eigenmodes.

The series of Eq. (19) is now substituted to Eq. (18). Making use of Eq. (20 b) we have

$$\nabla^2 g + k^2 g = \sum_m a_m\left(k^2 - k_m^2\right)\psi_m = -\dot{q}\,\delta(\vec{r} - \vec{r}_0).$$

Multiplying both sides by ψ_n and integrating over the volume, one get :

$$\sum_m a_m(k^2 - k_m^2)\oiiint_V \psi_n\psi_m.dV = \oiiint_V -\dot{q}\psi_n\delta(\vec{r} - \vec{r}_0).dV.$$

Using the orthogonality of the modes and the definition formula of the Dirac function, we obtain

$$a_n(k^2 - k_n^2)\Lambda_n = -\dot{q}\psi_n(\vec{r}_0), \text{ whereby } \Lambda_n = \oiiint_V \psi_n^2.dV..$$

As a consequence, the weighting factors a_m equal

$$a_m = \frac{\dot{q}\psi_m(\vec{r}_0)}{\Lambda_m(k_m^2 - k^2)},$$

resulting in

$$g(\vec{r},\vec{r}_0,\omega) = \sum_m \dot{q}\,\frac{\psi_m(\vec{r})\,\psi_m(\vec{r}_0)}{\Lambda_m\left(k_m^2 - k^2\right)} \tag{22}$$

This equation is important for later developments: the forced response of an acoustic system is expressed by its normal modes. Note that the equation is symmetric in the source and response position, caused by and, at the same time, representing the reciprocity of acoustic systems.

4. Methods and tools of experimental modal analysis

Before we tackle the practical aspects of AMA, it is worth to survey the correspondence of mechanical and acoustic systems in analytical terms. This analogy is known for a long time [14-17], but nowadays is again of increased importance due to the availability and increasing use of structural analysis systems and methods for acoustic problems.

4.1. ANALOGIES BETWEEN ACOUSTIC, MECHANICAL AND ELECTRICAL SYSTEMS

4.1.1. *Lumped parameter acoustic elements*

Our analysis, presented in paragraph 2.1.1., is based on the assumption that the dimensions of the considered acoustic system is negligible with respect to the wavelength in all but one direction. This restriction can be fully extended in all of the directions, and this way one can come to the notion of the *lumped* or *concentrated parameter acoustic elements*.

Without going into much details of the elaboration (a good summary can be found e.g. in [15]), one can obtain the definition equations as follows.

The *acoustic mass* is the ratio of the pressure and the rate of change of the volume velocity, that is, the volume acceleration caused by the pressure:

$$m_a \equiv \frac{p}{\dot{q}} = \frac{p}{\alpha} = \frac{\rho l}{A} \tag{23}$$

It is associated with a mass of air accelerated by a net force which acts to displace the gas without appreciably compressing it. In structural terms, an acoustic mass has one degree of freedom, namely, its velocity displacement.

The *acoustic capacity* is described by the *acoustic compliance*: the ratio between the volume displacement and the pressure:

$$c_a \equiv \frac{\int q\, dt}{p} = \frac{\xi}{p} = \frac{V}{\rho c^2} \tag{24}$$

It is associated with a volume of fluid that is compressed by a net force without appreciable average displacement of the centre of gravity of the volume.

Eventually, the *acoustic resistance* is the ratio of the pressure and the volume velocity, caused by the pressure:

$$r_a \equiv \frac{p}{q} \tag{25}$$

Comparing these basic equations with their mechanical and electrical counterparts, a complete analogy can be established. Making use of this equivalence, real-life acoustic systems can be modelled by electric or mechanical models and the acoustic problem can readily be solved by well established methods of these fields.

4.1.2. *Matrix description of acoustical and mechanical systems*
Continuing the analogy, just like the Kirchoff equations in electricity theory and the equations of motion in mechanics, the acoustic equations can also be summarised in a matrix form:

$$[m_a]\{\ddot{\xi}\} + [r_a]\{\dot{\xi}\} + \left[\frac{1}{c_a}\right]\{\xi\} = \{p\} \tag{26}$$

where, for the sake of consistency, we have used the volume displacement ξ instead of the volume velocity q.

Another kind of acoustic analogy can also be established, in which the volume velocity corresponds to force and pressure corresponds to velocity. Beranek refers to this equivalence as a "mobility type" analogy as opposed to the earlier one which he calls an "impedance type". Other authors use the terms "direct" and "inverse" instead of impedance and mobility analogy. As the names show, the first one is more straightforward and it is traditionally used for electroacoustical applications while the latter one is more useful in vibroacoustic applications.

In order to obtain the second formulation, recall the inhomogeneous form of the wave equation:

$$\nabla^2 p - \frac{1}{c^2}\ddot{p} = -\rho\dot{q} \tag{27}$$

Using the methods of discrete system theory and introducing damping terms, several researchers [see e.g. 17, 18, 19] have shown that Eq. (27) can be converted into a matrix form like

$$[A]\{\ddot{p}\} + [B]\{\dot{p}\} + [C]\{p\} = \{\dot{q}\} \tag{28}$$

Comparing Eqs. (26) and (28) with the usual form of the mechanical equations of motion [22]

$$[M]\{\ddot{x}\} + [C]\{\dot{x}\} + [K]\{x\} = \{f\} \tag{29}$$

the formal analogy between Eq. (26) or (28) and Eq. (29) becomes obvious.

In principle there is nothing to prevent us from using either Eq. (26) or Eq. (28) as a basis to perform an AMA test, by using the methods originally developed to solve Eq. (29). However, there are some essential difference between them. The matrix description in Eq. (26) is based on the assumption that the system can be appropriately described by interconnecting individual lumped parameters, whereas Eq. (28) is free from this serious limitation. Moreover, the lumped parameter description as given in Eq. (26) needs the volume velocity to be measured as the response parameter and the system has to be excited by an ideal pressure source, which isn't easy to realise in practice. Eq. (28) is therefore much more appropriate for experimental purposes. One also have to note, nevertheless, that no direct physical meaning can be attributed to the matrix terms [A], [B] and [C] while the use of Eq. (26) itself often gives a good physical insight into the nature of the problem. In summary, Eq. (26) is more appropriate for simple, quick-look calculations while the AMA experiments are based on Eq. (28).

4.1.3. *Forced response of discrete mechanical systems*

The derivation of Eq. (33), in other words, the discretisation of the continuous acoustic wave equation tends towards the use of the finite element formulation. However, this transition step can be avoided and the acoustic-mechanical analogy can be illustrated from a different aspect, if we proceed in opposite direction. Below we show that the forced response of a discrete mechanical system can be described in the same form as the forced response of a continuous acoustic system.

Assume an undamped multiple degree of freedom mechanical system, excited by sinusoidal forces. This system will be described by the matrix equation

$$\left[[K] - \omega^2 [M] \right] \{x\} = \{f\} \tag{30}$$

The response vector $\{x\}$ can be calculated by inverting the matrix term on the left side, usually referred to as *system matrix* and denoted by [B]:

$$\{x\} = [B]^{-1} \{f\} \tag{31}$$

The inverse of the system matrix can be easier calculated, if we introduce the so called *principle* or *modal coordinates* by means of the transformation

$$\{x\} = [\Psi]\{q\} \tag{32}$$

where $[\Psi]$ is the *modal matrix:* a matrix whose columns are the modal vectors of the original system, and q stands throughout this paragraph for the new, modal coordinates [22]. The advantage of this coordinate transformation is that the original system of equation, consisting of general matrices, is decoupled and the new matrix equation contains diagonal matrices only:

$$\left[\langle K \rangle - \omega^2 \langle M \rangle \right] \{q\} = [\Psi]^T \{f\} \tag{33}$$

Introducing the notation $\langle B \rangle = \langle K \rangle - \omega^2 \langle M \rangle$, the response of the system expressed in modal coordinates can be calculated, assuming that the frequency of excitation is different from any resonance frequency and thereby the diagonalized system matrix can be inverted:

$$\{q\} = \langle B \rangle^{-1} [\Psi]^T \{f\} \tag{34}$$

One can show [23] that the inverse of $\langle B \rangle$ will also be diagonal, containing the elements

$$\beta_i = \frac{1}{m_i \left(\omega_i^2 - \omega^2\right)} \tag{35}$$

Combining Eqs. (31), (32), (34) and (35) we eventually get

$$[B]^{-1} = [\Psi] \langle B \rangle^{-1} [\Psi]^T = \sum_{i=1}^{n} \frac{\left\{\Psi_r^{(i)}\right\}\left\{\Psi_e^{(i)}\right\}}{m_i \left(\omega_i^2 - \omega^2\right)} \tag{36}$$

One element, b_{re} of this matrix represents the response of the system in the rth DOF, excited by a unity force in the eth DOF

$$b_{re} = \sum_{i=1}^{n} \frac{\Psi_r^{(i)} \Psi_e^{(i)}}{m_i \left(\omega_i^2 - \omega^2\right)} \tag{37}$$

The expression Eq. (37) may be compared with that given in Eq. (22). This proves the accuracy of the lumped parameter acoustic elements method.

4.2. EXPERIMENTAL ACOUSTIC MODAL ANALYSIS

The first experimental modal analysis test, to the author's knowledge, has been reported in 1972, that time still without detailed theoretical background [24]. The first attempt to give a solid justification for the use of structural methods for acoustical problems stems from Nieter and Singh [25]. Their original technique, first applied for one-dimensional cases only, has later been extended to three-dimensional systems and an outline of the derivation of Eq. (28) was given [26]. A number of further publications about the topics are cited in the references [30-35].

As already mentioned shortly above, a wider spread of experimental AMA methods is impeded by a few principal and practical difficulties. One of these is that unlike in structural mechanics where almost all conceivable transfer functions can be readily measured in practice, in acoustics only the sound pressure can be determined reliably. Another difficulty, partly related to the previous one is the problem of how to provide appropriate excitation and reference signals therefrom. Below we overview the methods of experimental AMA and give some hints which may be able to overcome the difficulties encountered.

4.2.1. *Basics and methods of the analysis*
Consider a three-dimensional closed acoustic system with rigid or finite impedance but non-vibrating boundaries (the modal analysis of vibro-acoustic systems goes beyond the scope of this paper). The governing equation of the system can be written in the form

$$\nabla^2 p(\vec{r},t) - \frac{1}{c^2} \ddot{p}(\vec{r},t) = -\rho \dot{q} \, \delta(\vec{r} - \vec{r}_0) \qquad (= \text{Eq. (27)})$$

Assume now that a number of point monopoles of known volume velocity are placed in the space and the sound pressure across the volume is sampled at an appropriate number of points. The continuous wave equation can then be substituted by its discrete equivalent

$$[A]\{\ddot{p}\} + [B]\{\dot{p}\} + [C]\{p\} = \{\dot{q}\} \qquad (= \text{Eq. (28)})$$

Taking the Laplace-transform and assuming zero initial conditions we get

$$\left[s^2[A] + s[B] + [C]\right]\{p(s)\} = s\{q(s)\} \tag{38}$$

As usual in structural dynamics, the inverse of the matrix term can be substituted by the frequency response matrix H(s):

$$p(s) = \left[H(s)\right] s\{q(s)\} \tag{39}$$

The frequency response matrix can be expressed as a partial fraction expansion of modal parameters

$$\left[H(s)\right] = \sum_{i=1}^{n} \frac{Q_r \{\Psi\}_i \{\Psi\}_i^T}{s - \lambda_r} + \frac{Q_r^* \{\Psi\}_i^* \{\Psi\}_i^{*T}}{s - \lambda_r^*} \tag{40}$$

Substituting now s by $j\omega$ and using Eq. (39) it becomes obvious, that the modal parameters of the system can be gained from the FRF measurements where the sound pressures across the volume are referenced to the volume velocities of the sources. In acoustic terms, the transfer impedances of the field have to be measured:

$$Z_{re}(\omega) = \frac{p(\omega)}{q(\omega)} = j\omega \sum_{i=1}^{n} \frac{(r_{re})_i}{j\omega - \lambda_i} + \frac{(r_{re})_i^*}{j\omega - \lambda_i^*} \tag{41}$$

The expressions (38) to (41) are in complete analogy, up to the constant $j\omega$, with those being used in structural dynamics [22], therefore the usual structural methods and software packages can be used without modification.

4.2.2. Equipment Requirements and Simplification Possibilities

In principle, no correct experimental AMA can be conducted without using a well-controlled volume velocity source. Unfortunately, such actuators are commercially not available. A few experimental systems have been reported on in the literature [25-28], out of which the converted acoustic driver method seems and actually has been found to be the most practicable [29].

Imagine an electrodynamic loudspeaker which is provided with a closed, sealed housing behind the diaphragm. The most obvious realisation could be to use a horn driver. Unfortunately, these loudspeakers are generally designed for high frequency sound reproduction and sometimes cannot radiate sufficient acoustic power in the frequency range relevant for AMA applications. A good quality medium-range loudspeaker with closed housing or, in case of even lower frequencies, a closed box loudspeaker unit may be helpful. (Note that the use of any bass-reflex boxes should be avoided, whatever attractive their low-frequency characteristics would be. The reflex opening of these units acts as another, unwanted and uncontrolled local radiator around resonance frequency and the resonance of the system can cause interpretation problems in the course of the analysis if not damped out sufficiently.)

If the back cavity's dimensions are considerably smaller than the wavelength, one can assume that the pressure is constant everywhere in the cavity. Then we have an acoustic capacity excited by the backward radiation of the diaphragm, causing a pressure in the cavity which can be calculated by means of Eq. (24):

$$p = q_{back} \frac{\rho c^2}{j\omega V} \tag{42}$$

By measuring this pressure a good reference signal can be derived. In order to calibrate the whole system, another cavity of known volume can be connected to the loudspeaker and using the same formula the volume velocity, radiated forward can be related to the pressure measured in the back cavity.

This method has one single practical drawback, namely, that the pressure in the back cavity is very often too high, amenable to measurement. Another possibility is to measure the displacement of the diaphragm, implicitly assuming of course that the whole diaphragm moves with the same amplitude and phase. Substituting $q = Av$ in Eq. (42) it is easy to show that the pressure in the back cavity is proportional to the displacement.

To demonstrate the applicability of the technique, the cross section of an instrumented medium range speaker (Type Philips AD 50 060) is shown in Figure 6. along with the measured frequency response functions of the back pressure and the diaphragm displacement, measured by means of a Bentley proximity probe; both referenced to the input voltage of the loudspeaker. The similarity of the two FRF's supports that the diaphragm can indeed be considered as a rigid piston in the used frequency range.

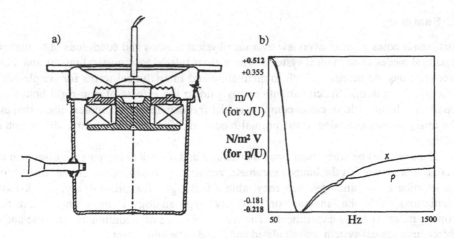

Figure 6   Controlled volume velocity source. a) Cross section of the loudspeaker, b) measured p/U and x/U frequency responses

504

If the analyst is interested in the modal frequencies and the mode shapes of the system only and a correct modal model is of no importance, the volume velocity source can be substituted by a simple loudspeaker. Then the reference signal can be taken directly from the input clamps of the speaker. (Needless to say, that the reference signal cannot be derived by using a microphone in the close vicinity of the source. The sound pressure measured in any point of the volume is a response rather than an excitation signal. If one aims this way just to detect the modal frequencies only, it can happen that even strong normal modes will be missed if the microphone comes to a local maximum.) One has to be aware of the fact that in this case the loudspeaker itself becomes an element of the system to be investigated, and any possible resonances of the exciter appear in the analysis as supplementary modes. These false modes are not easy to distinguish from the actual modes of the system in case of strong damping. The same holds for the microphones. The solution is that before the actual test the frequency response function of the actuators and sensors have to be carefully controlled.

As far as the sensors for the measurement of the responses are concerned, the difficulties are much smaller but a bit care is appropriate here, too. The acoustic field has to be sampled by using microphone positions which are closer than $\lambda/6$ (a general rule of thumb used for discrete acoustic methods). In case of large dimensions the number of microphone positions can run high and if one does not use a large number of parallel channels, the total measurement time can be long, enabling the loudspeaker to heat up the air in the volume, thereby changing the sound speed in air which in turn can cause the variation of the modal frequencies. In case of low damping, i.e., for sharp peaks in the FRF's, even a slight frequency shift can cause serious problems during postprocessing of the data. The problem can be overcome by using microphone arrays. Since, however, the costs of such arrays can be prohibitive if good quality measuring microphones are to be used, new types of low-cost microphones have been developed and commercialised, especially for acoustic modal analysis application. We have found good results by using electret studio microphones, too.

5. Summary

This course notes gave an overview over the physical reasons and conditions of formation of natural modes in acoustical systems. The notions related to acoustical modes and their inter-relations are discussed both in qualitative and in analytical terms for simple one-dimensional systems. Special attention was paid to the effects of non-rigid boundary conditions, being able to cause complex modal frequencies and complex mode shapes. The analysis was extended to the normal modes and forced field of three-dimensional systems.

The analogies between acoustical, mechanical and electrical systems are discussed in details. It is shown that the lumped parameter acoustic approach, based on the "direct" or "impedance type" analogy, has inevitable advantages for preliminary, quick-look calculations while the "inverse" or "mobility" type analogy is more appropriate for acoustic modal analysis experiments. The forced response of a continuous acoustic and a discrete mechanical system are calculated and found to be equivalent.

The outcome of theoretical considerations supports and justifies the engineering practice that those methods, originally developed for problems of structural mechanics, can also be applied in acoustics. However, high values and strong unproportionality of damping and the overlapping of modes at higher frequencies make the interpretation of results more difficult in acoustics than it is usual in structural dynamics. A further inconvenience is caused by the fact that the graphical interfaces of the presently available modal analysis systems are designed to animate spatial structures rather then scalar field variables

Eventually, some practical hints are given, how the equipment requirement of acoustic modal analysis can be met by simple methods.

6. References

1. See e.g. the application of the so called *Vitruvius vases* to improve the acoustics of ancient Greek theatres and Turkish minarets. Referenced by Tarnóczy: Acoustical design. Müszaki Könyvkiadó, Budapest, 1966. (In Hungarian)
2. Duhamel, J.M.C. (1849), On the vibration of a gas in cylindrical, conical etc., tubes, *J. Math. Pures Appl., 14,* 49-110.
3. Rayleigh, J.W.S. (1945), The Theory of Sound, *Vol. 2, 2nd ed., Dover,* New York, .
4. Schuster, K. and Wetzmann, E. (1929), On reverberation in closed spaces, *Ann. Phys., 1 (5),* 671-695.
5. Strutt, M.J.O. (1929), On the acoustics of large rooms, *Phil. Mag., 8 (7),* 236-250.
6. Morse, Ph.M. and Bolt, R.H. (1944), Sound waves in rooms, *Rev. Modern Physics, 16 (2),* 69-147.
7. Cremer, L. (1950), Die wissenschaftlichen Grundlagen der Raumakustik., *Vol.2, Wellentheoretische Raumakustik.,* S. Hirzel, Stuttgart.
8. Pierce, A.D. (1981), Acoustics. An introduction to its physical principles and applications, *McGraw-Hill,* New York.
9. Singh, R., Lyons, W. M. and Prater, G. (1989), Complex eigensolution from longitudinally vibrating bars with a viscously damped boundary, *J. Sound Vib., 133 (2),* 364-367.
10. Prater, G. and Singh, R. (1991), Complex modal analyis of non-proportionally damped continuous rods, *The Int. Journ. Analytical and Experimental Modal Analysis, 6 (1),* 13-24.
11. Beranek, L.L. and I., Vér (Eds.) (1992), Noise And Vibration Control Engineering, *Chap. 6, McGraw-Hill,* New York, .
12. Nefske, D.J. and Sung, Sh.H. (1990), Sound in small enclosures, *General Motors Res. Lab. Research Publication, GMR-7069,* Warren.
13. Skudrzyk, E. (1971), The Foundations Of Acoustics. Basic Mathematics And Basic Acoustics, *Springer,* Wien.
14. Firestone, F.A. (1938), The mobility method of computing the vibrations of linear mechanical and acoustical systems: mechanical-electrical analogies, *J. Appl. Phys., 9,* 373-387.
15. Beranek, L.L. (1954), Acoustics, *McGraw-Hill,* New York, .
16. Barát Z. (1975), Technical Acoustics, *Course notes,* University of Technology, Budapest, . (Manuscript; in Hungarian)
17. Everstine, G.C. (1981), Structural analogies for scalar field problems, *Int. Journ. Numerical Methods in Engineering, 17,* 471-476.
18. M. Petyt (1983), Finite element techniques for acoustics, in P. Filippi (ed.), *ICMS Courses and Lectures No. 277, Theoretical acoustics and numerical techniques , Springer,* Wien, .
19. Göransson, P. (1992), Acoustic finite elements, *Proc. 17th Int. Seminar on Modal Analysis, Course on advanced techniques in applied acoustics, Part VI,* Leuven, .
20. Fahy, F. (1985), Sound And Structural Vibration. Radiation, Transmission And Response. *Academic Press,* London, .
21. Augusztinovicz F., Sas, P., Otte, D. and Larsson, P.O. (1992), Analytical and experimental study of complex modes in acoustical systems, *Proc. 10th Int. Modal Analysis Conference, Vol. I,* San Diego, 110-116.
22. Formenti, D., Allemang, R., Rost, R. et al. (1992), Analytical and experimental modal analysis, *Proc. 17th Int. Seminar on Modal Analysis, Basic course on experimental modal analysis,* Leuven, .

506

23. Bishop, R.E.D., Gladwell, G.M.L. and Michaelson, S. (1979), The Matrix Analysis of Vibration, *Cambridge University Press*, Cambridge, UK.
24. Smith, D.L. (1976), Experimental techniques for acoustic modal analysis of cavities, *Proc. Inter-Noise 76*, Washington, 129-132.
25. Nieter, J.J. and Singh, R. (1982), Acoustic modal analysis experiment, *J. Acoust. Soc. Amer. 72 (2)*, 319-326.
26. Kung, Ch.H. and Singh, R. (1985), Experimental modal analysis technique for three-dimensional acoustic cavities, *J. Acoust. Soc. Amer., 77 (2)*, 731-738.
27. Singh, R. and Schary, M. (1978), Acoustic impedance measurement using sine sweep excitation and known volume velocity technique, *J. Acoust. Soc. Amer. 64 (4)*, 995-1005.
28. Salava, T. (1974), Sources of constant volume velocity and their use for acoustic measurements, *J. Audio Eng. Soc., 22*, 145-153.
29. Peter van der Linden, *Personal communication*.
30. Sas, P., Augusztinovicz, F., Van de Peer, J. and Desmet, W. (1993), Modelling the vibro-acooustic behaviour of a double wall structure, *Proc. of the 4th Int. Seminar on Applied Acoustics*, Leuven.
31. Brown, D.L. and Allemang, R.J. (1978), Modal analysis techniques applicable to acoustic problem solution, *Proc. Inter-Noise 78*, 909-914.
32. Cafeo, J.A. and Trethewey, M.W. (1984), An experimental acoustical modal analysis technique for the evaluation of cavity characteristics, *Proc. Inter-Noise 84*, Honolulu, 1367-1370.
33. Fyfe, K.R. and Ismail, F. (1987), Application of modal analysis to the acoustic modelling of vibrating cylinders, *Proc. 5th IMAC*, 336-342.
34. Okubo, N. and Masuda, K. (1990), Acoustic sensitivity analysis based on the results of acoustic modal testing, *Proc. 8th IMAC*, 270-274.
35. Fyfe, K.R., Cremers, L., Sas, P. and Creemers, G. (1991), The use of acoustic streamlines and reciprocity methods in automotive design sensitivity studies, *Mech. Systems Signal Proc., 5 (5)*, 431-441.

NEURAL NETWORKS FOR MODAL ANALYSIS

N. A. J. LIEVEN
Department of Aerospace Engineering
University Walk, University of Bristol

1. Definition of Artificial Neural Networks

A generic definition of an Artificial Neural Network - which is more commonly referred to simply as a Neural Network - has been stated as [1]:

A structure (network) composed of a number of interconnected units (artificial neurons). Each unit has an input/output (I/O) characteristic and implements a local computation or function. The output of any unit is determined by its I/O characteristic, its interconnection with other units, and (possibly) external inputs. Although "hand crafting" of the network is possible, the network usually develops and overall functionality through one or more forms of training.

As such a Neural Network provides a means of mapping a set of input variables onto a set of outputs by means of a non-linear function. These networks are directly analogous to many functions of the brain and have formed the motivation for the development of new algorithms. The characteristics that are associated with the functions of the brain and which are also desirable properties of neural networks are:

(i)   the brain integrates and stores experiences;

(ii)   the brain is able to make accurate predictions about new situations on the basis of previously self-organised experiences;

(iii) the brain considers new experiences in the context of stored experiences;

(iv) the brain represents a fault tolerant architecture;

(v)   the brain does not require perfect information;

(vi) the brain seems to have available - perhaps unused - neurons ready for use;

vii) the brain does not provide, through microscopic or macroscopic examination of its activity, much information concerning its operation at a high level; and

(viii) the brain tends to lead to behaviour which errs towards a state of equilibrium.

2. Introduction

Neural Networks (NN) have become widely accepted in a large range of applications such as image processing, signal processing, medicine, financial systems and control. Many of these areas can be divided into three topics:

507

J.M.M. Silva and N.M.M. Maia (eds.), Modal Analysis and Testing, 507–528.
© 1999 *Kluwer Academic Publishers. Printed in the Netherlands.*

(i) *classification* problems, in which an input is assigned to one of a number of distinct classes or categories;

(ii) *regression*, in which there is a continuous mapping from inputs to an output or outputs; and

(iii) *novelty detection*, in which abnormal or unusual features are detected.

Examples of classification include character recognition, speaker identification and object recognition. For these cases the NN has one or more output nodes that are discrete and commonly binary valued. In circumstances where regression is required, the NN serves as a nonlinear functional map between the input nodes and outputs, the outputs being continuously valued nodes. Examples of this type of application would be risk insurance in underwriting, adaptive control for robotics and prediction of currency exchange rates. The final role of NNs is in novelty detection, where a system is trained on normal data and applied to detect divergence from the standard conditions. Recent applications have been in fault detection in machinery, the detection of abnormal cells in cancer research and flight control monitoring systems.

The first application included in this paper involves modelling of time domain behaviour. This is perhaps the most common application of NNs in the control area. The technique is enjoying more widespread use as it has the ability to track abrupt non-linearities. Although these are not normally present in structural dynamics problems, they frequently occur in coupled modal analysis/control/forced response applications, such as flexible aircraft response. The other applications of NNs discussed in this paper are areas which are of particular interest to the modal analysis community: damping classification, model updating and damage detection. They are included as they neatly fall into the categories of classification, regression and novelty detection and so give an indication of the scope of NNs for the vibration/modal analysis community.

3. Neural Network theory

The function of a Neural Network (NN) is to map a set of input vectors **P**, onto a corresponding set of target output vectors usually denoted by **T**. All networks consist of neurons which modify their input(s) to an output. By forming layers of neurons and interconnecting them the input to the network is transformed to an output which has the ability to model non-linear functions. Before applying NN it is worth considering the following:

(i) are NNs suitable for the problem?

(ii) which useful NN architectures can readily be applied or modified to the problem?

(iii) are there any practical pointers which would indicate the complexity of the NN?

Even allowing for the above points, it is worth remembering that NNs can easily be applied to form the perfect "garbage-in, garbage-out" black box - you have been warned! Having made the decision to use an NN the following design steps should be pursued to increase chances of success:

(i)   select the parameters to be quantified;

(ii)   determine the availability of the measurement data;

(iii)   assess the adequacy of the computational resources;

(iv)   consider the availability and quality of the training and test data;

(v)   select the type of NN architecture;

(vi)   develop an NN simulation;

(vii)   train the NN;

(viii) simulate the NN system performance using unseen test data; and

(ix)   be prepared to alter any of the above and iterate!

4. Types of Neural Networks

In the field of Modal Analysis we are mainly concerned with classification and regression applications of NNs. Typically, Back Propagation and Radial Basis Function (RBF) networks are a common choice for these types of problems. There are significant differences between these two approaches which will be discussed separately below.

4.1.   BACK PROPAGATION

The strategy for creating back propagating NNs is twofold: construction and training. Back propagation techniques are normally applied to Multi-Layer Perceptron (MLP) networks. The construction phase is based on defining layers of neurons which pass transformed information from one layer to the next. Thus a complete pass of information will map the inputs onto the outputs. The building bricks of any network are its neurons, these can be linear or non-linear functions ϕ which act on a weighted sum of its inputs plus an offset bias. An example is shown schematically in Figure 1.

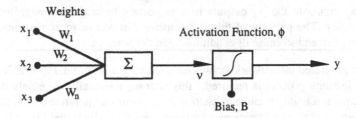

Figure 1 Multiple Input Neuron

Thus, the output of the neuron is given by

$$y = \phi\left(B + \sum_{i=1}^{n} W_i x_i \right) \qquad (1)$$

510

Where activation functions are normally set to be of log sigmoid form, $\phi(v) = 1/(1 + e^{-v})$, or of hyperbolic tan-sigmoid form $\phi(v) = \tanh(v)$. Note that a linear neuron simply scales and offsets the input, whereas the tansigmoid function constrains the output to ±1 through a non-linear distribution. In order to achieve flexibility to perform a nonlinear mapping, it is necessary to construct multi layered networks, which take the output from one layer and use this as an input to the next. Thus, having the basic building bricks we can construct layers of neuron which transform the x_i inputs to the y_j outputs as shown in Figure 2.

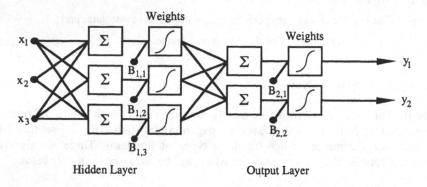

Figure 2 Multi-Layer Perceptron Network

The rationale for having multi layered networks is that each layer can increase the complexity and therefore the ability to achieve nonlinear mapping. For the applications of interest here, we will be using a two layered network. The first layer will consist of tansigmoid neurons, to scale the input vectors within the range ±1 to stabilise conditioning. The second layer will consist of linear neurons, to scale the outputs to their true values, note these are not bounded.

Practical Tip:

For each x_i inputs to the y_j outputs it is important to scale the magnitude of the vectors to unity. The purpose of this is to ensure that one or more neurons does not become saturated and so cause ill-conditioning in the network.

Having generated the network architecture, training is required. For MLPs an entirely supervised learning process is required - this is computationally extremely demanding. The NN output is a highly nonlinear function and requires an iterative process to form the correct weights and biases. The training phase adjusts the weights and biases in each layer so that an input in the Euclidean range of the training data can be transformed to the desired output. The approach to training a back propagating network, is to minimise the error between the output and the target vectors.

Thus for every possible selection of weights and biases an error surface can be constructed. The training mechanism seeks the minimum point on this surface. Although several strategies have been suggested [2], the minimisation of the sum-of-squares error is found to be ideal for most training problems:

$$E = \sum_{n=1}^{N} \left(y^U - f\left(x^U\right)\right)^2 \qquad (2)$$

Having defined the error function, the next stage is to minimise this objective function in parameter space. Two training strategies are commonly used:

(i) the Widrow-Hoff rule: this minimises the sum squared error by taking its derivative with respect to the weights. The selection of weights and biases is calculated by moving in the direction of the maximum negative gradient thereby reducing the neurons' errors. Two modifications are made to this method to increase the speed of convergence: an adaptive learning rate and momentum. By incorporating an adaptive learning rate the learning step is kept as large as possible whilst maintaining stability. Momentum uses information from the previous presentation of training data to the network — usually referred to as an epoch — to allow the network to pass through local minima on the error surface. The inclusion of momentum and an adaptive learning rate increases both the speed and reliability of the network.

(ii) the Levenberg-Marquardt approximation: this is an approximate Newton method which is based on the calculation of the Jacobian matrix of each error to each weight. Thus the error surface is calculated over a larger range of parameters than those used for the gradient descent method described above. This allows far more rapid convergence, however, the method is limited by its intensive memory requirements.

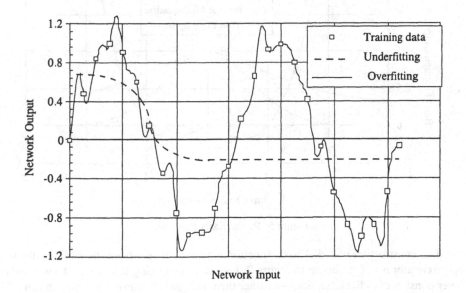

Figure 3 Overfitting and Underfitting using NNs

512

4.2.   RADIAL BASIS FUNCTIONS

Whilst back propagation provides an effective and reliable method of system modelling, the user is always left with the dilemma of deciding how many neurons should be used. If too few are selected the output will underfit the target data, similarly an extravagant choice will lead to over fitting, an example of which is shown in Figure 3. This is akin to selecting the wrong order polynomial to fit unknown data. Thus a self generating network which minimises errors through an optimal number of neurons would be desirable. One such method is the radial basis approach to NNs.

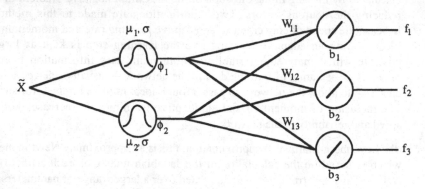

Figure 4 Schematic Diagram of an RBF Network

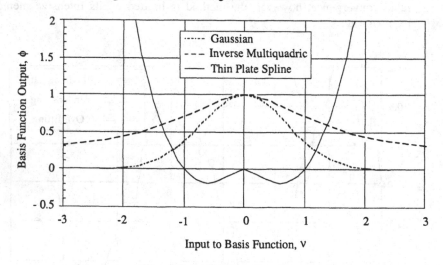

Figure 5 Radial Basis Function

The construction of radial basis function networks differs from the general function approximator network shown in Figure 2, the difference being that the first - and only - layer consists of radial basis neurons rather than tansigmoid neurons. Usually the an RBF network will only have a single nonlinear layer to provide the mapping as shown in Figure 4.

The behaviour of various radial basis neurons is shown in Figure 5. Figure 4 shows that the radial basis function has a maximum output of unity when the input is zero. Thus if the input vector is calculated to be the difference between the weight **W** and the network input **P**, then the neuron will act as a detector which outputs unity when these vectors coincide. The width of these functions is determined by a spread factor which allows each neuron to cover a vector space of the input vector. The network is then trained by adding individual neurons until there is a sufficient density to cover the vector space. Typically, more radial basis function neurons are required than using standard back propagation tansigmoid neurons, however, the solution will always converge to within the required tolerance.

The training procedure for RBFs is generally significantly quicker than training methods for multi-layer perceptron networks. Training is a two-stage approach: the location and width of the basis functions are determined using unsupervised techniques while the final layer weights are found by a solution of a linear problem by singular value decomposition.

The section of centres for the basis functions is not obvious and is difficult to achieve in an optimal way. The simplest method is to set them equal to a random set of the input vectors. However, this is not very effective and more sophisticated methods are normally relied upon. The two methods that are most commonly employed are K-means clustering [3] and the orthogonal least-squares approach [4]. K-means clustering attempts to find a set of centres that reflects the distribution of data. The data points are divided into K subsets S_j, containing N_j data points so as to minimise the sum-of-squares clustering function:

$$J = \sum_{j=1}^{K} \sum_{n \in S_j} \left\| x^n - \mu_j \right\|^2 \qquad (3)$$

where μ_j is the mean of the data set in S_j and is given by:

$$\mu_j = \frac{1}{N_j} \sum_{n \in S_j} x^n \qquad (4)$$

The primary drawback of K-clustering is the requirement to specify the number of clusters K, before training. The optimum number of clusters is not known in advance, so a modified version of orthogonal least-squares is usually adopted.

It is normally not clear how much each basis vector contributes to the overall error of the system, because there is generally correlation between each vector. However, if a set of orthogonal vectors is constructed in the space spanned by the vectors of hidden activations, then the orthogonality makes it possible to determine the amount of error due to each centre. This procedure enables automatic selection of the next basis function centre at the data point that results in the largest decrease in sum-squared error. This algorithm is repeated until a sufficient number of basis functions has been selected to reduce the overall error to an acceptable level.

5. Network optimisation of MLPs

Although RBFs generate sufficient neurons to map the inputs to the outputs, they have a tendency to model noise very effectively. It is difficult to define and control the sum-squared error so that they model the structural characteristics of interest and filters the noise. An approach adopted for MLPs, where the network architecture is predefined, is to use constructive algorithms. A major issue in the development of feedforward NNs is the selection of the network architecture. Too few neurons will lead to underfitting whereas too many leads to a high order simulation, prone to susceptibility to noise. The method of "Pruning" has been developed specifically to address this problem. The approach starts with a relatively large network of neurons and connections and iteratively removes them until an optimal criterion is reached. Clearly, the most important aspect of this approach is which weights are to be removed. It is often not obvious which weights will have the least significance and so can be removed with impunity. Thus we need to quantify the measure of relative importance, or *saliency*, of different weights.

The simplest measure of importance of a weight is its magnitude $|w|$. However, this indicates which weights are larger than others and so has little relevance as a measure of importance. A more justifiable approach is to define saliency in terms of the error in output of the network caused by a change in weight. Thus, if we consider an error function due to a small change in the values of the weights so that w_i is changed to $w_i + \delta w_i$ then the corresponding change in the error function is given by:

$$\delta E = \sum_i \frac{\delta E}{\delta w_i} \delta w_i + \tfrac{1}{2} \sum_i \sum_j H_{ij} \delta w_i \, \delta w_j + \vartheta(\delta w^3) \tag{5}$$

where H_{ij} are the elements of the elements of the Hessian matrix

$$H_{ij} = \frac{\partial^2 E}{\partial w_i \partial w_j} \tag{6}$$

Having assumed that the training process has converged allows us to discard the first term. Le Cun [2] proposes that the Hessian can be approximated by omitting the non-diagonal terms. Neglecting higher order terms in the expansion in equation 6 leads to the form:

$$\delta E = \frac{1}{2} \sum_i H_{ii} \delta w_i^2 \tag{7}$$

Thus, if a weight has an initial value of w_i and it is set to zero by pruning, then the increase in error is given by $\delta w_i = w_i$ and the saliency value for each weight is approximately $H_{ii} w_i^2 / 2$. Procedurally the practical implementation would consist of the following steps:

(i)   chose a relatively large network architecture;

(ii)   train the network until the normal stopping criterion is reached;

(iii) compute the second derivative for each of the weights H_{ij} and hence evaluate the saliencies $H_{ii} w_i^2 / 2$;

(iv) rank the weights by saliency and delete the lowest saliency weights; and

(v) go to (ii) and repeat until an overall stopping criterion is reached.

This method is referred to as *Optimal Brain Damage* [2]. A further refinement which includes the effect of the off diagonal terms of the Hessian matrix is referred to as *Optimal Brain Surgery*. Not surprisingly, the latter technique is favoured.

6. Neural Networks for time domain prediction

Although not of direct application in Modal Analysis, the prediction of non-linear behaviour in the time domain is of considerable interest in a wide range of engineering disciplines. The approach used in Neural Networks is to include recursion around the network. This involves using the output from the network at time "t" as an input to predict the behaviour at time "t+1". A typical example for an aircraft model is shown in Figure 6. In this example the output of the network is the coefficient of lift "cl' at time t with the inputs t network being angle of attack - "alpha" - and cl at time t-1 and t-2. Thus previous time step predictions of the coefficient of lift are used to predict the next step. Note that in this application the network encapsulates the combined nonlinear behaviour of both the structural dynamics and aerodynamics.

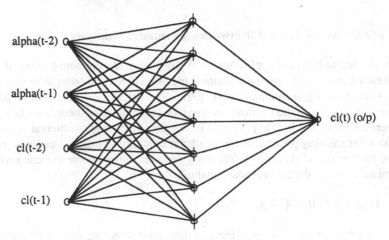

Figure 6 Time Domain Regressive Neural Network (NNARX) model.

Figure 7 demonstrates the time domain predictive capabilities of the simple recursive network shown in Figure 6 when presented with unseen data. Note that the method has anticipated the response of this highly nonlinear system very well. The sudden change in behaviour after 600 sample points is caused by vortex shedding in the aerodynamic flow and so is a discrete nonlinear characteristic. Even in the presence of this disturbance the prediction follows the trend in a stable manner after a one sample lag.

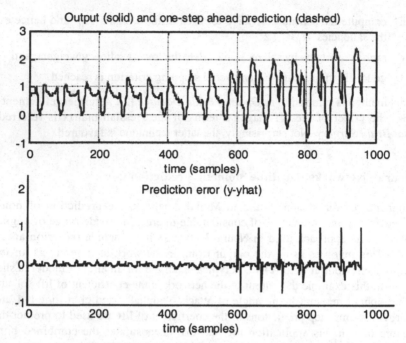

Figure 7 NN simulated time domain response for unseen data

7. Applications of Neural Networks in modal analysis

As with all Neural Network applications it is worth considering which areas of research have already been solved by more traditional means. There is no point in re-inventing the wheel when existing techniques work perfectly well. For direct modal parameter extraction it has been shown there is little benefit in using neural techniques [5]. However, for more intractable problems in the presence of experimental noise, neural networks offer exciting possibilities, which from evidence in other disciplines [6] should prove to be robust and stable. The following sections outline some current applications of neural networks in the area of modal analysis.

7.1.   DAMPING IDENTIFICATION

Recent work has shown that the effects of nonlinear damping can be identified using MLPs [7]. This is particularly useful for time domain simulations as has been mentioned above (see Figure 7). Another useful application of NNs has been the identification of different types of damping within a single system [5]. In this instance the network is being asked to solve two problem simultaneously: classification and regression. This is particularly demanding as the network must not only identify the type, but also its magnitude. Conventional methods [8] - such as the carpet plot of responses before and after resonance and Dobson's line fitting method [9] - provide only an indication of damping type. A curved damping plot as shown in Figure 8 is typical of a combined hysteretic and viscous damped mode. However, the right hand figure shows the effect of noise contamination on the measured data, obscuring the visible trends. Similarly, Dobson's method will indicate purely viscous damping by a horizontal imaginary part of the inverse receptance, whereas hysteretic damping causes the line to be of constant slope. Both methods can be severely inhibited by the presence of measurement noise, and are difficult to interpret in a quantifiable sense.

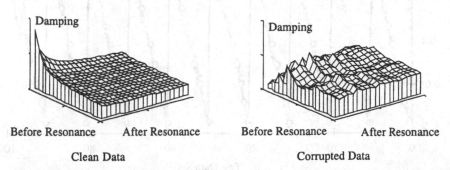

Figure 8   Carpet Plots of a Viscously-Damped Oscillator, with and without Noise Corruption

The process adopted by Ratcliffe [5] is to create the training data such that the modal stiffness is normalised out by rescaling the receptance data. The normalisation process follows three steps:

(i)   fit a circle to the receptance data and calculate the radius;

(ii) divide the receptance data by the radius, so that the radius becomes unity, irrespective of the damping or modal stiffness; and

(iii) if desired, calculate the inverse receptance after the normalisation process - thus allowing presentation of data in a form akin to Dobson's method.

For the results presented here, the inverse method data were used to train the network as this has been shown to be more reliable for this application. The effectiveness of the trained RBF network for classification of combined viscous and hysteretic damping is shown in Figure 9. Two types of basis function have been used, the Gaussian distribution and the Thin Plate Spline Function (TPSF - see Figure 5) Note, the circle fit is not an exact procedure, particularly for large amounts of viscous damping, where the locus of the receptance data is not longer circular. This approximation clearly affects the quantification of the damping parameters at high levels. However, it is pleasing to observe that the correct classification has been achieved throughout the range of interest.

A consideration in all modal analysis applications is the influence of noise on the analysis process. A perceived advantage of NNs is that they are relatively robust in the presence of noise on the test data, compared with analytical methods [10]. Figure 10 shows the classification performance for 100 unseen data sets of both Gaussian and TPSF basis function with increasing noise contamination. Again the TPSF demonstrates its superiority, which reflects its more common usage in the field.

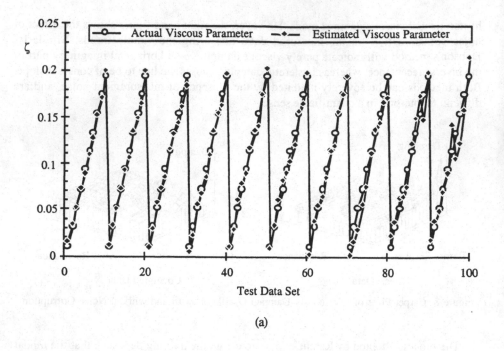

(a)

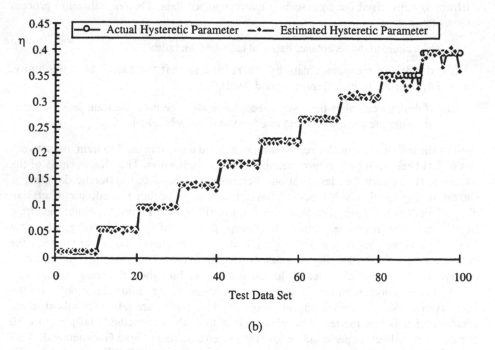

(b)

Figure 9 Damping Estimates using Inverse Presentation - (a), (b): Gaussian RBF Network

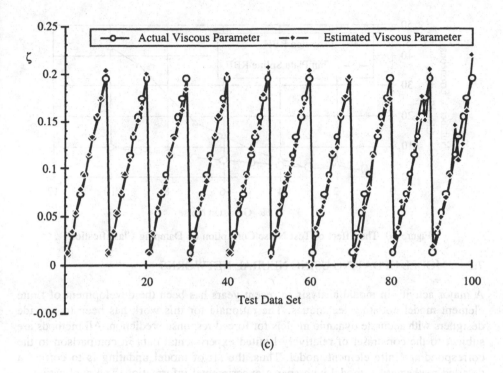

(c)

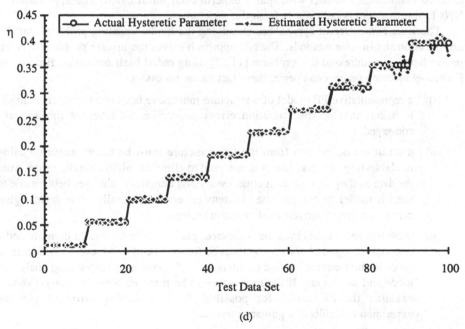

(d)

Figure 9 (*Cont.*) Damping Estimates Using Inverse Presentation - (c), (d): TPSF RBF Network

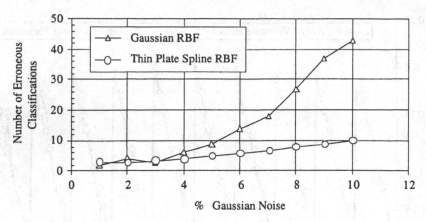

Figure 10 The Effect of Test Noise Corruption on Damping Classification

7.2. MODEL UPDATING USING NEURAL NETWORKS

A major activity in modal analysis in recent years has been the development of finite element model updating techniques. The rationale for this work has been to provide designers with accurate dynamic models for forced response prediction. All methods are subject to the constraint of relatively limited experimental data in comparison to the corresponding finite element model. Thus, the art of model updating is to correct a detailed mathematical model with sparse experimental information. The application of NNs to this problem is not exempt from this stricture.

There are two different approaches to solving the model updating problem, namely optimisation and inverse methods. The NN approach solves the inverse problem. Several researchers have addressed this problem [11,12] using radial basis networks. Before any FE model updating process can begin, three factors are necessary:

(i)   a representative FE model of a structure must have been produced; care should be taken that no discretisation errors occur, i.e. all relevant modes have converged;

(ii)   a set of measurements from the test structure must be taken, acquired using modal testing best practice. It is essential to eliminate all systematic errors from the data as they will result in unnecessary and unrealistic changes being made to the FE model to compensate. Systematic errors generally arise from signal processing errors and test configuration problems.

(iii)   these two sets of data must be compared, preferably in the modal domain, and a reasonable degree of correlation must be found. Specifically, every experimental mode in the frequency range of interest should have a corresponding analytical mode and *vice-versa*. If the modes cannot be matched then the analyst should examine the FE model for possible gross modelling errors, using the experimental results as a guidance tool.

If all three of these criteria are present, then it may be feasible to update the FE model.

7.3. INVERSE PROBLEM FORMULATION

If a set of alterations to an FE model is presented, it is a simple matter to generate the set of modal data that results from the updated model. This is done by solving the following eigenproblem:

$$\left([K_A]+[\Delta K_A^U]\right)\{\phi_A^U\}_r = \left(\lambda_A^U\right)_r\left([M_A]+[\Delta M_A^U]\right)\{\phi_A^U\}_r \tag{8}$$

where $[K_A]$ is the original analytical global stiffness matrix

$[M_A]$ is the original analytical global mass matrix

$[\Delta K_A^U]$ is the stiffness change due to updating

$[\Delta M_A^U]$ is the mass change due to updating

$\left(\lambda_A^U\right)_r$ is the rth updated analytical eigenvalue

$\{\phi_A^U\}_r$ is the rth updated analytical eigenvector, or mode shape

The updating approach considered here examines the inverse of this problem – given a set of (experimental) modal data, what set of alterations to the FE model would result in these data? An exact analytical solution to this inverse problem is only possible using linear algebra techniques when full data are available, i.e. when n modes have been acquired, each with every DOF measured, for an n-DOF model. These circumstances never arise in practical model updating scenarios, so the inverse problem must be solved by another, nonlinear, method.

Before tackling the inverse problem, it is first necessary to consider what possible alterations to the FE model are allowable. Usually, elemental parameters such as density, Young's Modulus, cross-sectional area, etc. are valid. Thus, it seems reasonable to pick a set of updating parameters from the possible sets of all of the individual elemental parameters.

The particular choice of parameters to update is exceedingly important to all parametric model updating methods. If too few parameters are chosen then the analytical and experimental models cannot be reconciled. If too many or the wrong parameters are chosen then the updating results will not be physically realistic. For the example shown here, it is assumed that an appropriate set of parameters has been obtained.

Given the set of updating parameters, the inverse problem is solved using the following approach:

(i) Generate a set of observations (the training data) from the FE model. The observations consist of pairs of input and target vectors $(\{x_i\},\{y_i\})_{i=1}^{p}$, where the ith target vector $\{y_i\}$ contains the changes made to the updating parameters, and the ith input vector $\{x_i\}$ contains data obtained from the updated model. Each observation requires equation (8) to be solved.

(ii) Train an RBF neural network using this training data and the OLS algorithm. The trained net should generalise, such that it will produce the correct updating parameter values when shown data from an adjusted FE model, despite having not encountered the data before.

(iii) Show the experimental data to the trained net. This will result in the net suggesting some updating parameter values. These updating parameters are then applied to the FE model.

This process results in a new analytical FE model. Consequently, the process can be re-applied to this new model to obtain a further improved model, i.e. an iterative strategy is used. Stages (i) (ii) and (iii) are repeatedly executed until the FE model converges on the final updated FE model.

There are still many details to be filled in before this algorithm can be used in practice. The most notable omissions concern stage (i) where two questions remain, namely:

(i) 'What strategy of adjusting the updating parameters should be used; i.e. how many observations should be generated in total, and what changes to the parameters should be made for each observation?', and

(ii) 'For each observation made, what modal (or frequency) data should be stored in the vector $\{x_i\}$?'

These are both significant questions and in attempting to answer them, the problem of 'where to begin?' is encountered. For example, strategies for generating training data cannot be tested without first deciding both the choice of updating parameters and the neural network type. Yet it is difficult to fix on a network methodology without detailed knowledge of the type of training data that will be encountered. Since it is necessary to begin somewhere, the approach taken here is to choose an initial data generation strategy and updating parameter set, and to test the updating strategy on two case studies. The modal domain will be used for reasons of training data size economy. The strategy for generating training data is create training sets which span the errors likely to present in the finite element model. Thus for a given number of parameters, or p-values, a range of values should be selected either individually or randomly distributed across the parameter space. The latter method is favoured [13] as it minimises the size of the training set. The outcome of this process is a training set consisting of modeshapes and their corresponding eigenvalues for each parameter variation.

For demonstration purposes, the T-shaped mild steel plate of 3.5 mm thickness shown in Figure 11 is used here. The simulated test case scaled the stiffnesses of all ten elements to be in error according to the p-values shown in table 1.

The ideal training data generation strategy would create a comprehensive sub-sample of all the possible updating parameter settings. However, it is apparent that this approach is infeasible unless the number of updating parameters is trivial. Consequently, we have to select a representative subset of the possible variations.

Let $\{p\}$ be the set of N_p p-values, each one assigned to an elemental parameter of the FE model. Each p-value setting will produce an observation from the FE model, by solving equation (8). The resulting modal data $(\lambda_A^U)_r$ and $(\phi_A^U)_r$ can be assembled into an experimental vector $\{E\}$. Thus, $\{E\}$ is a function of $\{p\}$. The input vector / target vector pair that is added to the training data set is $(\{E\}, \{p\})$.

Each observation from the inverse problem generates a set of modal data. The complete data set is available, i.e. every DOF of the model has accurate mode shape information, and a large number of modes is available. It would be unrealistic to use all of these data for simulated model updating purposes, as no modal test could hope to acquire the corresponding experimental results. For the purposes of a simple case study, only the first five eigenvectors and eigenvalues are included in each experimental training vector $\{E_{tr}\}$. Additionally, all rotational DOFs are removed from the mode shapes, as they are presently difficult to measure accurately in vibration tests.

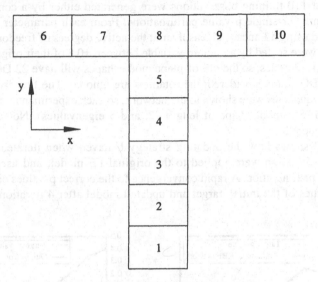

Figure 11 T-shaped free-free plate model

Table 1 Target p-values for simulated experimental T - shaped plate model

p-value	Setting	p-value	Setting
P_1	0.40	P_6	-010
P_2	0.30	P_7	-020
P_3	0.20	P_8	-030
P_4	0.10	P_9	-0.40
P_5	0.00	P_{10}	-0.50

It is important that $\{E_{tr}\}$ is not dominated by a few large terms such as the eigenvalues, or the remaining information will be ignored by the neural networks. To prevent this, the eigenvalues $(\lambda_A^U)_r$ will be scaled by the initial FE model eigenvalues $(\lambda_A)_r$. It is also important that the experimental and analytical modes correspond with each other. Different observations from the FE model may have different mode orders. Consequently, the analytical modes have to be matched for correlation against the (simulated) experimental modal data. This can be achieved by picking the Correlated Mode Pairs (CMPs), using the highest terms of the MAC matrix.

Each eigenvector is mass normalised. However, mass normalisation does not completely determine the eigenvector, because a mode shape can be multiplied by a factor of −1 without affecting the normalisation. It is important that each analytical mode shape is scaled to have the same 'algebraic sign' as its corresponding experimental mode shape, to prevent distortion of the training data. This scaling can be achieved by calculating the Modal Scale Factor (MSF) between the two modes. If the MSF is negative, then the analytical mode shape should be negated before it is stored in $\{E_{tr}\}$.

A total of 110 training observations were generated either by a combination of p-value variations, or single p-value perturbations. From each parameter set five modes were extracted using a Lanczos eigensolver at the active degrees of freedom of the model. The p-values were scaled by a random variable between ±0.8 of their original magnitude. The model has 22 nodes, so the out-of-plane mode shapes will have 22 DOFs as only the translational DOFs are considered; the rotations are ignored. The first five mode shapes and natural frequencies were shown to the network, so each experimental input vector was of length 115 (5 x mode shape of length 22, and 5 eigenvalues). Note: all rigid-body modes were ignored.

Figure 12 shows how the updating strategy behaved when iterated, i.e. when the results of one iteration were applied to the original FE model, and used as the initial model for the next iteration. A rapid convergence to the correct p-values can be seen. The modal properties of the initial, target and updated model after 4 iterations are shown in Table 2.

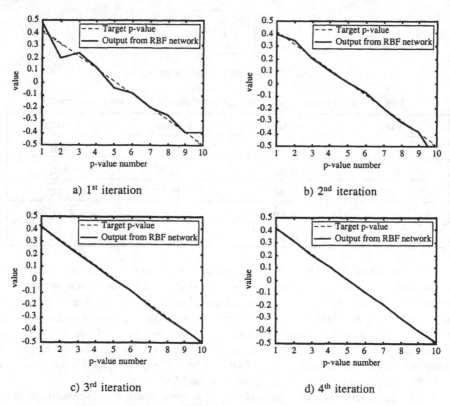

a) 1st iteration

b) 2nd iteration

c) 3rd iteration

d) 4th iteration

Figure 12 Results of iterative approach for T-shaped plate

Table 2 modal properties of T-shaped plate, before and after updating

	Natural Frequencies, Hz			MAC Values of CMPs	
	Initial	Updated	Target	Initial	Updated
1st Mode	43.1	45.0	44.9	0.9809	1.0000
2nd Mode	77.9	66.6	66.5	0.9803	1.0000
3rd Mode	105.2	103.6	103.5	0.9872	1.0000
4th Mode	126.7	130.7	130.5	0.9762	1.0000
5th Mode	214.4	182.0	181.9	0.8070	1.0000

Both the physical properties and the modal description of the updated model show a high level of convergence. Note that although MAC values have converged to unity, the predicted natural frequencies are not perfectly matched to the target frequencies. The reasons for this are twofold: firstly, the MAC is a relatively insensitive measure, which can mask slight inaccuracies; and secondly, the training data for the network were dominated by modeshapes thus giving more weight to the eigenvectors than the natural frequencies. Further work has been undertaken to address this problem [13], which optimises the choice of training data and parameter selection. There is still considerable work to be undertaken in this field, but NNs provides a promising approach which bypasses the problems of coordinate incompleteness normally encountered in other methods.

7.4. TIME DOMAIN DAMAGE DETECTION

An area which has justifiable attracted widespread attention in the field of modal analysis is that of damage detection. The scope of this activity ranges from interstitial delamination in composites to crack detection on highway bridges. Although the physical scale of these problems is considerably different the technical challenges they face are identical. The two major difficulties which must be addressed if damage detection is to be realisable are:

- Spatially incomplete measurements
- Limited frequency range of measured data

These give rise to the issue of sensitivity. If the measurement data is not sensitive to the location of the damage then no method will detect it. All too frequently techniques to detect damage have been hampered by the adequacy of the test data – neural networks are no exception. If the measure data only encompasses the lower order modes, these will generally exhibit global deformation. Herein lies a problem for modal analysis, to detect damage, or indeed to carry out localised model updating, higher order modes are required. These are difficult to extract due to the increased modal density. The approach adopted here, is use time domain data, thus avoiding the problems of modal extraction. By keeping the data I this domain, both lower and higher order characteristics are preserved.

The example used to demonstrate the approach to damage detection using NNs, or in this case impact detection, using a simple free-free beam with a high frequency accelerometer situated at one end as shown in see Figure 13. The training data are generated by impacting the beam at known locations along its length. The architecture of the network in this case was two layer back-propagating network with seven tan-sigmoid neurons. Typical responses are shown in Figure 14. It should be noted that the beam is undamaged by the impact, so its structural properties remain consistent throughout the training period. Clearly this would not be the case if true damage were to occur. This is an issue still to be fully addressed in this application.

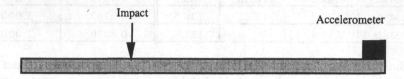

Figure 13 Impact detection on free-free beam

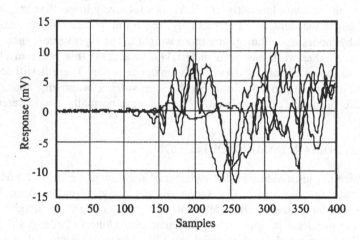

Figure 14 Time domain response after impacts

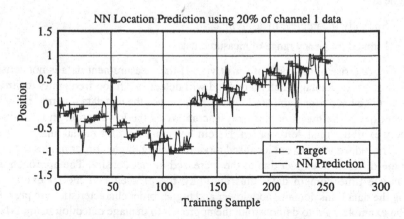

Figure 15 Impact prediction using Recursive NN

The prediction of impact detection from time domain signals presented to trained network are shown in Figure 15. The network is able to predict 95% of the unseen impacts to within 5% accuracy. This result exhibits considerable promise, not only for the application of NNs to this problem but also the use of time domain data. As can be seen from Figure 14, the characteristic response alters significantly upon varying the impact location. This is in large part caused by the dispersion characteristics, or frequency dependent wave velocity within the material — see Figure 16. Although these modes are difficult to extract, the measurement in time domain allows the response to reveal the superposition of all of the excited modes, whether dispersive or not. This is a distinct advantage for NNs as they act on a pattern recognition, consequently the more idiosyncratic the response the more likely the network is to be able to identify the impact location – regardless of structural complexity.

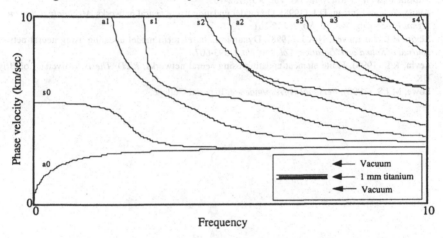

Figure 16 Dispersion Characteristics for Titanium [14]

8. Concluding remarks

The use of NNs in Modal analysis offers the prospect of solving some previously intractable problems in the field. In particular, the fields of adaptive control, fault detection and model updating seem well suited to the NN approach. A word of caution though, it is all too tempting to use NNs when they are not necessary. If an existing method is adequate then use it. The training and development of NNs presents a set of problems unique to their own solution process, which will absorb as much time as you have available. However, if the modal analysis problem is non-linear, time dependent, coupled and suffers from coordinate incompleteness - or any combination of these - then NNs provide a promising way forward.

9. References

1.   Schalkoff, R.J. (1997), Artificial Neural Networks, *McGraw-Hill, ISBN 0-07-057118-X.*
2.   Bishop, C.M. (1995), Neural Networks for Pattern Definition, *OUP, ISBN 0-19-853864-2.*

528

3. Moody, J. and Darken, C.J. (1989), Fast-learning in neural networks of locally-tuned processing units, *Neural Computation, 1*, 281-294.
4. Chen, S. and Billings, S.A. (1992), Neural networks for nonlinear dynamic system modelling and identification, *International Journal of Control, Vol. 56(2)*, 319-346.
5. Ratcliffe, M.J. (1997), Identification and application of measured frequency domain data for structural dynamics, *PhD Thesis*, University of Bristol, UK.
6. Lisboa, P.G.J. (1992), Neural Networks: Current Applications *Chapman and Hall, ISBN 0-412-42790-7.*
7. Worden, K. and Tomlinson, G. (1998), Frequency domain analysis of NARX neural networks, *Journal of Sound and Vibration, in press.*
8. Maia, N.M.M. *et al* (1997), Theoretical and Experimental Modal Analysis, *Research Studies Press, ISBN 0-86380-208-7.*
9. Dobson, B.J. (1987), A straight line techniques for extracting modal properties from FRF data, *Mechanical Systems and Signal Processing, Vol.1(1)*, 57-59.
10. Jean, J.S.N. and Wang, J. (1994), Weight smoothing to improve network generalisation, *IEEE Transactions on Neural Networks, Vol. 5(5)*, 752-763.
11. Atalla, M.J. and Inman, D.J. (1998), On model updating using neural networks, *Mechanical Systems and Signal Processing, Vol. 12, No. 1*, 135-161.
12. Levin, R.I. and Lieven, N.A.J. (1998), Dynamic finite element model updating using neural networks, *Journal of Sound and Vibration, Vol. 250, No.5*, 593-607.
13. Levin, R.I. (1998), Finite element updating using neural networks, *Ph.D. Thesis*, University of Bristol, UK.
14. Lowe, M.J.S. (1998), *Personal correspondence*, Imperial College.

ADVANCED OPTIMISATION METHODS FOR MODEL UPDATING

N. A. J. LIEVEN
Department of Aerospace Engineering
University Walk, University of Bristol

1. Introduction

The challenge of model updating is to refine a finite element (FE) description of a structure so that it exhibits the same dynamic behaviour as its experimental counterpart. In this sense, the FE model is optimised with respect to the measurements. The following sections discuss the fundamentals of optimisation and then describes two advanced techniques – genetic algorithms and simulated annealing. These techniques enable minimisation of high order, complex functions of which can be defined in model updating terms by declaring a difference between the measured characteristics and the FE model's representation. Genetic Algorithms have deservedly received particular attention from the optimisation community in a variety of applications. The technique uses natural selection and genetics to form search algorithms. Simulated Annealing is also based on a physical phenomena, the state of a metal at a particular temperature during cooling. As such these methods differ significantly in form from standard optimisation techniques. So in addition to describing genetic algorithms and simulated annealing along with their applications in modal analysis, this paper will provide a brief introduction to the field of optimisation and the way in which some of the approaches differ. As with many techniques in modal analysis, several may seem appropriate at the outset. The field of optimisation is no exception. However, the introduction of new techniques offers new opportunities which have the potential to exploit the increased use of parallel architecture in computing.

2. Optimisation

Optimisation provides a means of finding the best solution to a problem. In mathematical terms, this involves seeking the minimum or maximum value of a function, $f(x_1, \ldots \ldots x_n)$, which has n variables. This function may be optimised as it is in an unconstrained way, or constrained by additional functions which usually relate to some physical quantity associated with the system. In structural dynamic optimisation for instance, the total mass of the system or symmetry in the structural matrices may be appropriate constraints. Thus, optimisation for dynamicists is a tool to aid the decision making process to achieve desired structural properties. Note that it does not use statistical means to arrive at a decision, but solves a mathematical model which represents one or more characteristics of the system.

J.M.M. Silva and N.M.M. Maia (eds.), Modal Analysis and Testing, 529–548.

Although it is important to understand the methods used in optimisation, it is imperative to understand the derivation of the function to be minimised. In dynamics we usually have a choice between the modal and frequency domain information to form the basis of the problem. Often it is convenient to minimise the difference between two set of modal data rather than the corresponding FRFs. However, modal data may exclude significant residual effects from out of range modes or nonlinearities removed by the analysis process. On the other hand the FRF data does not include explicit information relating to mode numbering - as would be provided by a MAC calculation - and has variable and usually significant levels of noise distributed unevenly throughout the frequency range. All this goes to show that it is often not obvious what choice of optimisation function should be made, but in all cases an appreciation of what form the solution should take will help assist in the formulation of the problem.

3. Classical optimisation

Before discussing some of the applications of optimisation in modal analysis, it is perhaps worthwhile to outline some of the procedures commonly used. The first "school" of approaches is unconstrained optimisation. These methods minimise - or maximise - a function which is not subject to any constraints. The variables of the function $f(x)$ can either be discrete or continuous. Thus to find the minimum of a function with n variable we simply have to satisfy:

$$\frac{\partial f}{\partial x_i} (x_1,........x_n) = 0 \qquad (i = 1,......,n) \qquad (1)$$

Unfortunately a closed form solution to this set of equations is rarely available as the functions are usually couple, nonlinear and complex. If these conditions exist then iterative means are usually sought. The approach which is modified according to the type of problem, normally adopts the following strategy:

(i)   select a trial solution to $f(x)$, say x_k;

(ii) find a direction from this trial solution in which $f(x)$ decreases;

(iii) find a point x_{k+1} in this direction so that $f(x_{k+1}) < f(x_k)$;

(iv) repeat the process from the new trial solution.

Perhaps the simplest application of this procedure is the *one variable at a time method*. This technique starts with a trial solution and alters each variable in turn, so that the function $f(x)$ is reduced for fixed values of the other variables. This method is both computationally inefficient and also prone to failure when two or more of the equations are orthogonal. A more robust and efficient technique is to use gradient descent, which uses the first and/or second derivatives of the objective function to indicate the next position in the search for the minimum point.

The earliest of the gradient descent methods is credited to Cauchy [1] in 1847. Since then his method has been modified and improved, though the basic principle is still applied. A simple application of the method would be as follows. Determine the normalised gradient vector at the current (i[th])point:

$$d_i = \frac{\left(\dfrac{\partial f}{\partial x_1}, \dfrac{\partial f}{\partial x_2}, \dots\dots\dots, \dfrac{\partial f}{\partial x_n}\right)}{\sqrt{\displaystyle\sum_{j=1}^{n}\left(\dfrac{\partial f}{\partial x_j}\right)^2}} \qquad (2)$$

Then calculate the next point a distance h_i way from the first trial point by:

$$x_{i+1} = x_i + h_i d_i \qquad (3)$$

Typically, this process is repeated until the value of the function no longer decreases. At this point the step length is reduced and the process is repeated until a new plateau is reached. The procedure is shown diagrammatically in Figure 1 for a single step length over a two dimensional arbitrary contour.

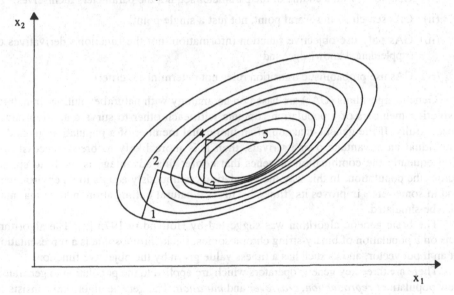

Figure 1 Equal Step Method of Gradient Descent

Various improvements have been suggested to this method, which involve the use of the Hessian (second derivative of the function) and nonlinear step lengths [2]. Although these methods are efficient in terms of computation, the storage requirements for the Jacobian and Hessian matrices for high order systems can be prohibitive. Indeed when considering high order optimisation, it is worth noting that the associated functions will usually have several minima. Although the goal of all optimisation methods is to find the global minimum, a common failing of gradient descent methods is to settle on a local minimum which may well be some considerable distance from the optimum solution. Modifications to these methods which include momentum and adaptive step length help to resolve this difficulty by including information from previous points, to take the solution process past a small local minimum and onto the next trough. However, they are not robust and a computational penalty is incurred.

4. Genetic Algorithms

From this background, the approach to optimisation offered by Genetic Algorithms seems quite attractive. The method provides the facility to find a global optimum solution to complicated - previously intractable - optimisation problems. The method does not require the objective function to be differentiable and so can be applied to discrete as well as continuous functions. Also Genetic Algorithms are not dependent on the starting position - as can be the case for gradient descent methods - and so are highly robust. To summarise, Genetic Algorithms are different from conventional optimisation techniques in the following respects:

(i) GAs work with a coding of the parameter set, not the parameters themselves;

(ii) GAs search from several point, not just a single point;

(iii) GAs only use objective function information, not the function's derivatives or supplemental knowledge; and

(iv) GAs use probabilistic transition rules not deterministic criteria.

Genetic algorithms (GAs) are based on an analogy with natural evolution. In natural evolution members of a population compete with each other to survive and reproduce successfully. If the genetic makeup of an individual member of a population gives that individual an advantage over its rivals, then it is more likely to breed successfully. Consequently the combination of genes that confer this advantage is likely to spread across the population. In this way the population continuously adapts to its environment and in some sense improves its 'fitness'. This is a natural optimisation method that may also be simulated.

The basic genetic algorithm was suggested by Holland in 1975 [3]. The algorithm acts on a population of binary-string chromosomes. Each chromosome is a representation of an input vector, and as such has a fitness value given by the objective function.

There are three key genetic operators which are applied to the population to generate a new population: *reproduction*, *crossover* and *mutation*. The genetic algorithm consists of generating a new population of chromosomes from the old population using these three operators. A flowchart outlining the genetic algorithm is given in Figure 2.

The reproduction operator assigns each chromosome a relative probability of being reproduced according to the fitness of the chromosomes. The fittest chromosomes may typically be reproduced two or three times, the least fit chromosomes may not be reproduced at all. The reproduction probabilities may be assigned according to the objective function values directly, or according to a pre-defined table based on the ranking of the chromosomes. The tabular option has two advantages over the direct probabilities option: firstly the tabular option reduces the risk of early convergence of the population on an initially fit chromosome, and secondly it encourages fine-tuning towards the end of the GA run, when the fitnesses of the population are similar. Consequently the tabular reproduction method is the option chosen for the examples shown here.

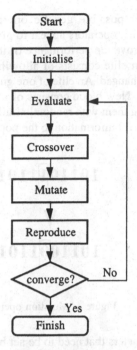

Figure 2 Genetic Algorithm flowchart

The crossover operator mixes genetic information amongst the population. The chromosomes are randomly paired together. A crossover point is then randomly selected along the length of each pair of chromosomes. Each chromosome is cut at this point, and re-joined with the corresponding cut section of its pair, i.e. genetic information is swapped between chromosomes, see Figure 3. The crossover operator is applied to each pair of chromosomes with probability p_c (typically 0.6). The crossover operator has the potential to join successful genetic fragments together to form fitter individuals.

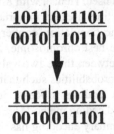

Figure 3 Crossover operator

The mutation operator has the potential to re-introduce genetic information that has been lost from the population. After crossover has occurred, each binary digit of each chromosome has a small probability p_m (typically 0.01 to 0.001) of mutating. A binary digit that mutates is simply inverted, see Figure 4.

534

There are many other possible genetic operators that may be applied to the population, but few additional operators appear to give any real benefit to the algorithm. Two operators that do improve the performance of the algorithm are the introduction of an *elite* and *new blood*. An elite consists of allowing the best few chromosomes from each generation to pass unchanged. An elite of one guarantees that the best member of the population will not be lost. New blood consists of removing the worst few members of the population and replacing them with randomly initialised chromosomes. This can have the effect of introducing useful information to the population.

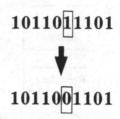

Figure 4 Mutation operator

There are several parameters that need to be set before the genetic algorithm may be used, namely the population size N_p the crossover probability p_c, the mutation probability p_m, the reproduction table probabilities, the elite size N_e and the new blood size N_b. The most important parameter is the population size, which depends on the length of the chromosomes, which in turn depends on the level of discretisation acceptable. A population that is too large results in much wasted effort, whereas too small a population results in failure to locate the global minimum. A rule of the thumb commonly used is to have N_p a factor of k times bigger than the chromosome length, where k can vary between two and twenty. Higher values of k produce more robust algorithms at the expense of slower execution times . In the model updating example shown later the value k = 4 will be used. Hence, with 8-bit discretisation, $N_p = 32d$.

The reproduction table probabilities are also important parameters; these probabilities control how quickly or slowly the population converges. Typical values would be a relative probability of two for the best chromosome, zero for the worst and the other probabilities calculated linearly between these two values. Actual probabilities may be determined by scaling the relative probabilities such that their sum is one.

It is necessary to define a coding process between the input vectors and chromosomes. Two commonly used coding methods are the standard binary encoding and Gray encodings [4]. The standard binary encoding has the disadvantage that certain close numbers may be very distant in terms of their encoding. For example, 3 encodes as 011 and 4 as 100, requiring every bit to be inverted. This can lead to 'walls' that the GA finds hard to pass. Gray codes avoid this wall - every number can reach its neighbours via a single flip of a bit. See Figure 5. Recent publications in the field of modal analysis have exploited different codings to improve the performance of GAs [5, 6]. Although coding is well advised it appears to make only a marginal difference to the overall performance of the procedures. Before an input vector is encoded it is necessary to decide the degree of precision for each parameter, e.g. 8 bit precision gives 256 possible settings for the parameter, spread linearly about its range.

Instructions:
Note: a high bit is set to 1, a low bit to 0.

1. the ith high bit represents a value of 2^i - 1
   (i counts from *right* to *left*)
2. the most significant high bit adds its value,
   the next high bit subtracts its value, etc.
   (the sign alternates for each high bit)

Example: (3-bit encoding)

```
[100] -> 7    [010] -> 3
[101] -> 6    [011] -> 2
[111] -> 5    [001] -> 1
[110] -> 4    [000] -> 0
```

Figure 5 Gray coding

5. An example of GAs applied to a aimple problem.

To understand fully how a particular algorithm works it is useful to develop a simple worked example. The example given by Goldberg [7] and shown here, is to maximise the function $f(x) = x^2$ over the range of x between 0 and 31. Although this is a trivial example, it will serve the purpose of demonstrating some of the fundamental steps required to use genetic algorithms effectively.

The first stage is to select the length of the chromosome to represent our function. In this case we will select a binary code of length 5 digits. This effectively discretises the problem into 31 levels between 0 (00000) and 31 (11111). Although a convenient choice for this problem, it can be easily seen than a longer or shorter chromosome will give greater or less accuracy as is required by the problem.

Having selected the length of the chromosome, we are then required to choose a random population to act as initial estimates. In this case we will choose 4. For each of these 4 chromosomes a random selection of 5 binary digits is made to give an initial representation of the problem. Thus, 20 random binary digits are generated. This gives each chromosome a value corresponding to x and in turn a fitness value f(x). This is shown in Table 1.

Having generated a first estimate of the representation of the function, the next stage is to select the most appropriate string which satisfies our optimisation goal. In other words which of the strings has the highest values and so maximises f(x). The selection of the most appropriate string is achieved on a probabilistic basis, with those most likely to satisfy the objective being selected more often than those which are not. The selection is made by giving a proportional weight to those with the maximum likelihood, thus penalising those which are a long way from the optimum solution. This is achieved and shown in the tabular form in Table 2.

Table 1 Initial Genetic Representation of f(x)

String Number	Initial population	X	fitness value, $f(x)=x^2$
1	0 1 1 0 1	13	169
2	1 1 0 0 0	24	576
3	0 1 0 0 0	8	64
4	1 0 0 1 1	19	361

Table 2 Probabilities of Reproduction after initial calculation of fitness

String Number	Initial Population	x	Fitness Value $f(x)=x^2$	Probability of selection $f_1 / \sum f$	Expected Sample f_1/f	Rounded actual count
1	0 1 1 0 1	13	169	0.14	0.58	1
2	1 1 0 0 0	24	576	0.49	1.97	2
3	0 1 0 0 0	8	64	0.06	0.22	0
4	1 0 0 1 1	19	361	0.31	1.23	1
Sum			1170	1.00	4.00	4
Average			293	0.25	1.00	1
Maximum			576	0.49	1.97	2

From these calculations, string number two, by virtue of its high fitness value is clearly the most likely candidate for the optimum solution. By calculating the probability of selection in proportion to the sum of the fitness values, we are left with indicators for the number of samples to be selected for reproduction for the next stage of the genetic algorithm. With only four chromosomes to carry forward to the next stage, the rounded counts indicate that string number two should be carried forward twice, along with strings one and four being represented once, thereby eliminating the third string. This is not surprising, given sample three's low fitness value.

Having achieved reproduction, the next stage of the genetic algorithm is to introduce crossover and mutation. For crossover, the four reproduced strings are randomly paired into "mates" and at a random crossover site the binary digits are swapped. This creates four new strings based on the fittest chromosomes.

Mutation - the last step in the process - is achieved by inverting a single bit within a string. It is normal to assume a low mutation rate, in order to avoid undoing the good work of the reproduction stage. In this instance the probability of mutation is chosen to be 0.0001. Therefore for the 20 bits within the population there is a likelihood of $20*0.0001 = 0.02$ mutations. Thus for this relatively small population no mutations are indicated. Accordingly no bits are inverted and crossover remains the last operator in this generation. Having completed the new generation, a corresponding fitness value is calculated for each string. The process of mating and crossover is shown in Table 3.

Having calculated the new fitness values the first iteration of the genetic algorithm is complete. By comparing the average fitnesses before and after the first generation, we can see it has improved from 293 to 439. This indicates that the strings are now converging towards the maximum for the function. Although it is unwise to draw conclusions from one "loop" of a probabilistic procedure, from this point it is easy to visualise how the next generations would tend to the optimum solution.

Table 3 Strings after Reproduction and Crossover

Mating pool after Reproduction (Cross site shown by |)	Mate (Randomly Selected)	Crossover Site (Randomly Selected)	New Population	x Value	$f(x) = x^2$
0 1 1 0 | 1	2	4	0 1 1 0 0	12	144
1 1 0 0 | 0	1	4	1 1 0 0 1	25	625
1 1 | 0 0 0	4	2	1 1 0 1 1	27	729
1 0 | 0 1 1	3	2	1 0 0 0 0	16	256
Sum					1754
Average					439
Maximum					729

6. Simulated Annealing – Theory

Simulated annealing is derived from an analogy with the annealing process of material physics. It is well known that certain materials have multiple stable states, and that these states have differing molecular distributions and energy levels. For example many materials have one crystalline low-energy state and multiple glassy high-energy states. Annealing is a method of finding the lowest energy state from all of the possible stable states.

The annealing process consists of heating the substance until it is molten, then slowly and discretely lowering the temperature. The substance is allowed to reach *thermal equilibrium* at each temperature. Eventually the temperature is lowered until the material freezes. If the temperature is lowered sufficiently slowly then the annealing process will always pick out the global minimum energy state from the almost infinite number of other possible states. Annealing is a natural optimisation process, and it is this process that is simulated.

To simulate annealing it is necessary to consider the underlying thermodynamics behind the process. In 1877 Boltzmann described how fluids in thermal equilibrium behave. The state or configuration of the system is described by the spatial position of each component of the system, i.e. the position of every atom. There is an exceedingly large number of possible states for even a small number of atoms. Each state s has an energy E(s) associated with it (the precise nature of the energy function E is unimportant here). If the system is in thermal equilibrium at a temperature T then the probability $\pi_T(s)$ of the system being in a given state s is given by the Boltzmann distribution:

$$\pi_T(s) = \frac{e^{\frac{-E(s)}{kT}}}{\sum\limits_{w \in S} e^{\frac{-E(w)}{kT}}} \qquad (4)$$

where k is the Boltzmann constant and S is the set of all possible states.

However, the Boltzmann distribution contains no information about how a fluid reaches thermal equilibrium at a given temperature. In 1953 Metropolis *et al* [6] developed an algorithm that simulates this process. The *Metropolis algorithm* consists of repeating the following procedure: if the system is in state s_{old} with energy $E(s_{old})$, a randomly chosen atom is perturbed resulting in a state s_{new} with energy $E(s_{new})$. This new state is either accepted or rejected according to the *Metropolis criterion*: if $E(s_{new}) < E(s_{old})$ then the new state is automatically accepted, but if $E(s_{new}) \geq E(s_{old})$ then the probability of accepting the new state is given by:

$$P(\text{accept } s_{new}) = e^{\left\{ \frac{-(E(s_{old}) - E(s_{new}))}{kT} \right\}} \qquad (5)$$

When this process of generating new states and either accepting or rejecting them is continued, it can be shown that the probability of the system being in a given state tends to the Boltzmann distribution (4), i.e. the system eventually reaches thermal equilibrium. N.B. The Metropolis criterion may be stated as: accept the new state if and only if

$$E(s_{new}) \leq E(s_{old}) - T \log(U) \qquad (6)$$

where U is a uniformly distributed random variable between 0 and 1 and ln is the natural logarithm.

The Metropolis algorithm models how a system reaches equilibrium at a single temperature T. In 1983 Kirkpatrick derived a simulated annealing (SA) algorithm based on the Metropolis algorithm. The analogy between optimisation and thermodynamics was defined as follows: the state of the system becomes the choice of input parameter values, the energy function becomes the objective (or cost) function, the temperature becomes a control parameter and finding the lowest energy state becomes finding the global minimum. For simulated annealing a new state is generated from an old state by a problem-dependent *neighbourhood function* that returns a random new state in the neighbourhood of the old state.

Kirkpatrick introduced the concept of an *annealing schedule* that describes how and when the temperature is lowered throughout the annealing process. Under the simulated annealing algorithm, the Metropolis algorithm is executed at each temperature on the annealing schedule in turn until thermal equilibrium is reached. A flowchart outlining the simulated annealing algorithm is shown in Figure 6. The annealing schedule starts at a high temperature, T_0, and discretely lowers the temperature until the system is *frozen* (i.e. no transitions are accepted), hopefully at the global minimum.

7. Simulated Annealing - Implementation details

The above formulation of SA has left many of the implementation details undecided, such as the precise nature of the annealing schedule and the choice of neighbourhood function. Various workers have suggested differing implementations of simulated annealing. This section will present three SA algorithms.

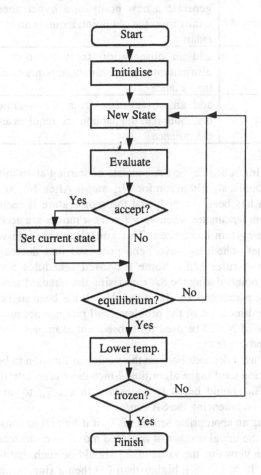

Figure 6 Simulated Annealing flowchart

To define a particular SA algorithm fully it is necessary to define a cooling schedule and neighbourhood function. A selection of neighbourhood functions suggested in the literature is shown in Table 4. Many more neighbourhood functions have been proposed, generally based on different statistical distributions. In practice these other functions offer little improvement over the functions outlined in Table 4.

Normal adjustment and Cauchy adjustment are not generally adopted since the results tend to be similar to results obtained for fixed radius adjustment.

Table 4 Simulated Annealing Neighbourhood Functions

Function name	Description
line adjustment	pick one parameter at random, and then change its value to a new random value chosen uniformly within the parameter's valid range
Fixed radius adjustment	generate a new point on a hypersphere that is a fixed radius from the old point; requires an extra parameter - the radius
Normal adjustment	add an observation from the n-dimensional normal distribution to the parameters; requires an extra parameter - the variance
Cauchy adjustment	add an observation from the n-dimensional Cauchy distribution to the parameters; requires an extra parameter - the variance

A standard cooling schedule for SA consists of starting at an initial temperature T_0 and letting the Metropolis algorithm run for N_{sa} steps. After N_{sa} steps it is hoped that thermal equilibrium has been reached, and the temperature is reduced by a factor ρ, $0<\rho<1$. The algorithm terminates when very few new moves are accepted at a particular temperature, i.e. the system has frozen. This annealing schedule will be used in this paper, although other schedules have been proposed that decrease the temperature according to different rules [7,8]. Some proposed schedules actually increase the temperature at some points during the SA run. Using the standard annealing schedule, the only problem specific parameters are T_0, ρ and N_{sa}. It has been suggested that using $N = 100d$, where d is the dimension of the problem, will produce acceptable results on most problems. This value of N will be used for subsequent examples. Two values of ρ have been used, $\rho = 0.9$ and $\rho = 0.95$.

The value of T_0 used depends both on the particular function to be optimised and the neighbourhood function used in the algorithm. Since there is no intuitive way of deciding what the value of T_0 should be, it is necessary to set T_0 to an appropriate value automatically before commencing the SA run.

To aid in deciding an appropriate setting of T_0 it is helpful to consider the acceptance ratio, ϕ_0, defined as the initial number of accepted moves over the total number of moves made. It is a common view that the value of T_0 should be such that the value of ϕ_0 lies between 0.5 and 0.9. If ϕ_0 is much higher than 0.9 then a significant percentage of the SA run will be spent in a molten state, wasting effort on little more than a random walk. If ϕ_0 is much lower than 0.5 then the chances of getting trapped in a local minimum increase. Hence an appropriate T_0 may be estimated by measuring ϕ_0 and adjusting T_0 if ϕ_0 is out of range. ϕ_0 is typically measured over 20 transitions.

Three variations of SA that are readily applicable to model updating are:

1. Using the standard cooling schedule with the line adjustment neighbourhood function. This method tends to work well when the parameters of the objective function are relatively independent.

2. Using the standard cooling schedule with the fixed radius adjustment neighbourhood function. This method works better when the parameters of the objective function are coupled, but if the radius is set too small then the algorithm often fails to escape from local minima.

3. Combining the previous two algorithms, i.e. using both the line adjustment and fixed radius adjustment. This is a new variation on the SA algorithm, presented for the first time, which should hopefully combine the strengths of both of the neighbourhood functions. It is necessary to have two temperatures for this algorithm, one for each neighbourhood function. These temperatures may be obtained separately using the method outlined above. After N_{sa} steps at each temperature, both temperatures are reduced by a factor of . The neighbourhood functions are used for alternate SA steps at each temperature. This algorithm is termed Blended Simulated Annealing (BSA).

It is the third of these which has been shown to be most appropriate for model updating [11] and so will be adopted as the approach for the test case shown below.

8. Genetic Algorithms and Simulated Annealing for model updating.

Having demonstrated the genetic algorithm applied to a simple problem and outlined the approach for Simulated Annealing, it is now appropriate to highlight some of their relative strengths and shortcomings on a typical problem in the field of modal analysis. Finite Element model updating, is considered to be one of the most demanding, generating more papers - and a book devoted to the subject [12] - than perhaps any other current issue in modal analysis. Therefore, it is appropriate to apply optimisation here in this context [5].To consider the effects of using the genetic or SA algorithms for dynamic FE model updating it is necessary to have a simple test model. The model used for this section is an undamped ten element two-dimensional cantilevered beam shown in Figure 7.

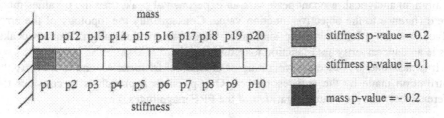

Figure 7 Target p-values

Each element of this model has two updating parameters (p-values), mass and stiffness, hence the model has twenty p-values to update. It is necessary to decide the domain in which the updating will occur, i.e. the modal or frequency domain. It is well known that modal analysis of frequency domain data can introduce systematic errors to the data, so the frequency domain will be used to avoid this problem.

Table 5 first ten natural frequencies of a ten element beam

Natural Frequency / Hz	Description
64.3	1st bending
402.8	2nd bending
1127.0	3rd bending
2206.7	4th bending
3646.7	5th bending
4314.6	1st extensional
5450.5	6th bending
7626.3	7th bending
10184.3	8th bending
13050.4	2nd extensional

The first ten natural frequencies of the model are shown in Table 5. The simulated experimental data consist of ten FRFs. Each FRF has 801 frequency points distributed evenly between 0 and 4000Hz, hence including the first five bending modes. The target p-values are shown in Figure 7. All ten Y-DOF FRFs will be considered, initially with no noise added. Note that the rotational and extensional FRFs are not included.

It is necessary to decide the form of the objective function. An obvious first choice is the sum-squared difference between the magnitude of the experimental and analytical FRFs, summed over every available frequency point and FRF. Note that the phase is excluded because the model is undamped.

$$f_1(\{p\}) = \sum_j \sum_\omega \left(\left\| {}_X\alpha_{jk}(\omega) \right\|_2 - \left\| {}_A^U\alpha_{jk}(\omega) \right\|_2 \right)^2 \qquad (7)$$

This function performs badly in practice. This failure is due to the function being dominated by the contributions made at the FRF resonant peaks. If a set of p-values does not align an analytical resonant peak with an experimental peak, then the p-values make little difference to the objective function value. Consequently the topology of the error surface is mostly flat, with sharp sink-holes where the p-values align resonant peaks. This is an unnecessarily hard function to optimise.

It is desirable to reduce the weighting of the natural frequencies, and to increase the contribution made by the anti-resonances. One possible approach is to consider the difference between the natural logarithm of the FRF magnitudes, i.e.

$$f_2(\{p\}) = \sum_j \sum_\omega \left(\log\left(\left\| {}_X\alpha_{jk}(\omega) \right\|_2 \right) - \log\left(\left\| {}_A^U\alpha_{jk}(\omega) \right\|_2 \right) \right)^2 \qquad (8)$$

This objective function produces a smoother topology. A more refined approach to objective function definition can be considered through application of regularisation methods [12]. The p-values obtained from using this function and the SA and Genetic Algorithm are shown in Figures 8 and 9 respectively.

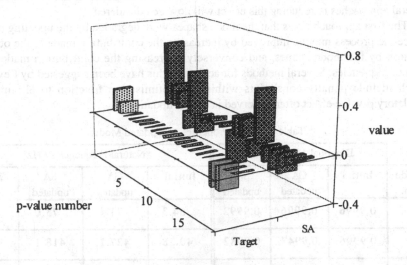

Figure 8 Updated and target p-values using SA

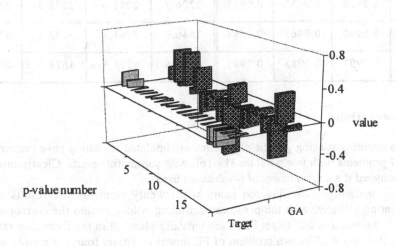

Figure 9 Updated and target p-values using GA

It is evident that the resulting p-values are significantly different from the target p-values, oscillating around the correct values. It is clear that both Simulated Annealing and Genetic Algorithm have not been successful, since they have degraded the MAC values without significantly improving the natural frequencies (Table 6).

It is clear that the responses of both the updated and target FE models are similar in the frequency range of interest, despite having significantly different p-values. It is not until the frequency range becomes much higher that the models start to diverge. This is a problem that many FE model updating methods have to cope with; oscillatory mass and stiffness p-values can produce almost exactly the same response as the correct p-values. Several approaches to reducing this effect will now be considered.

The first approach notes that the mode shapes were degraded by the updating process. Hence the process may be improved by increasing the contribution made to the objective function by the mode shapes, and conversely decreasing the contribution made by the natural frequencies. Several methods for achieving this have been suggested by Levin [11] which include penalty constraints within the optimisation function to eliminate the oscillatory p-value effect often observed in model updating.

Table 6 Modal Properties of Updated Model

Mode No.	Diagonal MAC values			Natural Frequencies / Hz			
	Initial	GA updated	SA updated	Initial	GA updated	SA updated	Target
1	0.9996	0.9996	0.9992	64.3	71.1	73.0	69.1
2	0.9996	0.9947	0.9982	402.8	427.1	418.1	417.3
3	0.9997	0.9933	0.9974	1127.0	1203.8	1182.0	1180.7
4	0.9988	0.9955	0.9948	2206.7	2291.6	2278.2	2278.2
5	0.9990	0.9465	0.9964	3646.7	3761.4	3772.3	3772.7
6	0.9997	0.9983	0.9993	4314.6	4715.3	4678.3	4574.7

9. Observations

Previous attempts at using genetic algorithms or simulated annealing have concentrated on small problems with few p-values [13-16], with successful results. Clearly problems are encountered if a larger number of p-values are used.

When updating a cantilevered beam with twenty elements the results appear disappointing initially, with the p-values oscillating wildly around the correct values. However, the updated and target FRFs are virtually identical in the frequency range of interest. This is a well-known problem of FE model updating, found in many updating methods. It stems from the nature of the dynamic behaviour of structures; the 'oscillating' p-value structure actually does behave very similarly to the correct smoothly p-valued structure at low and medium frequencies. This suggests that higher frequency ranges are desirable to update FE models, which is widely known. Unfortunately it is often impractical to measure the high frequency responses of structures.

The sensitivity of the objective function f_2 can be seen by considering the error surface produced by the function (Figure 10) when varying two of the p-values (the stiffnesses at the tip). The error surface contains a nearly flat trough which holds the global minimum point. A small change to another p-value can slightly change the gradient of the trough, drastically changing the position of the global minimum. This is why the function is so hard to minimise, even with one of the best optimisation algorithms currently available.

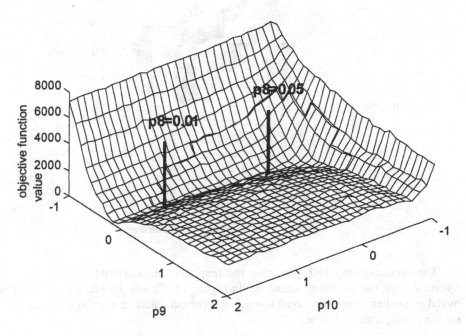

Figure 10 Error surface of objective function, with minimum positions highlighted (all other p-values set to zero)

Clearly, model updating has unique problems associated with ill-conditioning and sensitivity which have been dealt with extensively elsewhere [17, 12]. The poorly defined minimum shown in Figure 10 presents a particular difficulty for the genetic algorithm. The reason for the technique not finding the global minimum of the objective functions is that the genetic algorithm is a discrete operator. Thus, if the global minimum lies between two adjacent binary states of the genetic string, the global minimum may well be missed. For discrete application of genetic algorithms, such as fault detection [4], this is not so much of a problem. However, for model updating a two stage approach is advisable: apply the genetic algorithm to find a point on the error surface near to the global minimum, and then apply an efficient local optimisation technique - such as the nonlinear Simplex method [18] - to iterate to the true solution.

By way of an example, a final application of the Blended Simulated Annealing approach followed by the nonlinear Simplex Algorithm is demonstrated. This is used in conjunction with parameter substructuring in order to improve the conditioning of the updating problem. Therefore, by minimising the number of variables to be updated using a optimised topology objective function the conditioning of the problem is placed in the best possible light. The effect of this is shown in Figure 11.

546

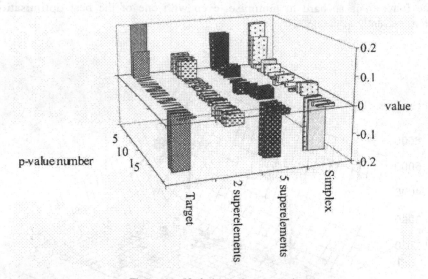

Figure 11 Updating using superelements

The corresponding FRFs showing the response characteristics before and after updating using this procedure are shown in Figure 12. Clearly the FRFs – and resulting modal properties - have converged towards the correct solution, even in the presence of significant experimental noise.

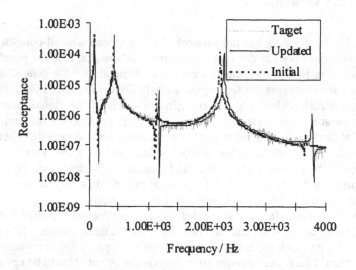

Figure 12 Updating results using BSA, with imposed 10% Gaussian noise

10. Concluding remarks

Simulated Annealing and Genetic Algorithms provide powerful and efficient means of finding the global minimum of an objective function. The techniques are not distracted by the local minima en route to their final destination. In cases where the objective function is shallow in the region of the global solution, a continuous optimisation method should be used. Simulated Annealing offers a promising way forward in this respect. In applications where discrete solutions are required [4], then Genetic Algorithms perhaps provide the most effective means of optimisation currently available. However, it should be borne in mind that even most effective the optimisation methods will not compensate for a poorly posed - and therefore ill-conditioned - updating problem. Before embarking on any exercise in this area the physics of the problem should be well understood and carried through the problem by minimising the number of variables and improving the topology of the objective function.

11. References

1. Cauchy, A. (1847), *C.r. hebd, Seanc. Academie Sci, Vol 25*, Paris, pp. 536.
2. Davidon, W.C., Variable metric method for minimisation, *AEC Research and Development, Report ANL 5990 (Rev. TID 4500, 14th Ed.)*.
3. Holland, J. (1975), Adaptation in natural and artificial systems, *University of Michigan Press, ISBN 0-472-08460-7*.
4. Dunn, S.A., The use of genetic algorithms and stochastic hill-climbing in dynamic finite-element model identification, *Computers and Structures, in press*.
5. Levin, R.I. and Lieven, N.A.J. (1998), Dynamic finite element model updating using simulated annealing and genetic algorithms, *Mechanical Systems and Signal Processing, Vol. 12(1)*, 91-120.
6. Yap, K.C. and Zimmerman, D.C. (1998), The effect of coding in genetic algorithm based structural damage detection, *Proc. of the 16th International Modal Analysis Conference*, Sta. Barbara, USA.
7. Goldberg, D.E. (1989), Genetic algorithms in search, optimisation and machine learning, *Addison-Wesley, ISBN 0-201-15767-5*.
8. Metropolis, N., Rosenbluth, A.W., Rosenbluth M.N, Teller A.H. (1953), Equations of state calculations by fast computing machines, *Journal of Chemical Physics, Vol. 21, No. 6*, 1087-1092.
9. Kirkpatrick, S., Gelatt, C.D., Vecchi, M.P. (1983), Optimization by simulated annealing, *Science, Vol. 220, No. 4598*, 671-680.
10. Van Laarhoven, P.J.M, Aarts, E.H.L. (1987), Simulated Annealing: Theory and Applications, *Kluwer Academic Publishers, ISBN 90-277-2513-6*.
11. Levin, R.I. (1998), Finite element updating using neural networks, *Ph.D. Thesis*, University of Bristol, UK.
12. Friswell, M.I. and Mottershead, J.E. (1995), Finite Element Model Updating in Structural Dynamics, *Kluwer Academic Publishers, ISBN 0-7923-3431-0*.
13. Larson, C.B., Zimmerman, D.C. (1993), Structural model refinement using a genetic algorithm approach, *Proc. of the 11th International Modal Analysis Conference*, Florence, 1095-1101.
14. Rajeev, S., Krishnamoorthy, C.S. (1992), Discrete optimization on structures using genetic algorithms, *Journal of Structural Engineering A.S.C.E., Vol. 118, No. 5*, 1233-1249.
15. Friswell, M.I., Penny, J.E.T., Garvey ,S.D. (1996), A combined genetic and eigensensitivity algorithm for the location of damage in structures, *Proc. Int. Conf. on Identification in Engineering Systems*, 357-367.
16. Mares, C., Surace, C. (1996), Finite element model updating using a genetic algorithm, *DTA NAFEMS 2nd International Conference on Structural Dynamics Modelling*, 41-52.

548

17. Waters, T.P. (1995), Finite element model updating using frequency response functions, *Ph.D. Thesis*, Department of Aerospace Engineering, University of Bristol, UK..
18. Nelder, J.A., Mead, R. A (1965), Simplex method for function minimization, *The Computer Journal, Vol. 7*, 308-313.

MODAL ANALYSIS FOR ROTATING MACHINERY

D. J. EWINS
Imperial College of Science, Technology & Medicine
London, UK

1. Introduction

1.1. OBJECTIVES AND STRUCTURE OF THE PAPER

The primary objective of this paper is to lay the foundations for the application of modal analysis and testing technology to rotating machinery. Although this technology is already used to some extent by the manufacturers and users of rotating machines, such use is generally limited to stationary parts of the machine, or to rotating components in a stationary state. Relatively little use of modal testing has been possible until quite recently for a number of reasons: mostly to do with the practical difficulties involved but also because of complications which arise in the he underlying theory, not all of which issues are fully understood by all who need them. In this paper, we shall try to dispel these confusions and show the ways in which modal testing may become more widely used in this important area.

The paper will commence with a review of the problem areas in rotating machinery dynamics in which modal testing can provide valuable assistance, and will summarise some of the special vibration properties exhibited by rotating machinery components, even when they are not rotating, as these constitute a major element in determining the vibration properties under rotating conditions. Then, the basis of modal theory for rotating structures is presented, followed by discussion of practical issues of excitation and response measurement under rotating conditions. The modified analysis procedures which must be followed to conduct a modal analysis of data measured on rotating structures is then explained.

1.2. THE FIELD OF ROTATING MACHINERY DYNAMICS

The areas in which vibration has a significant impact on the performance or reliability of rotating machines are numerous and include many problems which manifest themselves generally in 3 areas, depending upon whether their primary area of influence lies in:

(i)   the main casing or external body of the machine and its supports;

(ii)   the rotors (shafts) and their bearings (so-called 'rotor dynamics' problems)

(iii) the rotating components of blades, discs, wheels, impellers, gears,...

J.M.M. Silva and N.M.M. Maia (eds.), Modal Analysis and Testing, 549–568.
© 1999 *Kluwer Academic Publishers. Printed in the Netherlands.*

550

There are a number of specific problems which involve more than one of these groups of components, and important amongst these are a number of quite serious rotor/stator interaction problems, any of which can turn into instabilities with disastrous consequences. Some rotor dynamics problems come into this category, but there have been a small number of serious and destructive such vibration phenomena encountered in high-speed rotating machines in recent years. The mechanisms of such interactions stem either from metal-to-metal contact in rubs, of rotors on bearings, of rotating and stationary parts of seals or of blades on casings, or through fluid coupling brought about in seals and other close-tolerance gaps between rotating and stationary parts. Most other vibration problems are associated with fatigue failures brought about by excessive long-term vibration, usually of the rotating components themselves which are also stressed by the steady centrifugal loads resulting from the rotation.

In all these problem areas, the solution is to anticipate and to avoid by good design analysis but in many cases there are so many potential problems that prediction of all is very demanding and often not achieved: dependence on measurements is still high, either in tests undertaken as part of the design/development cycle (i.e. modal tests to validate theoretical models), or in tests undertaken on prototype or in-service machines to establish exactly what vibrations are experienced under operating conditions. These latter tests are necessary not only to confirm the designer's ability to analyse the dynamics of the machine, or to predict its modes of vibration correctly, but primarily to determine the nature and levels of excitation forces which are developed under operating conditions as these are often much more difficult to predict than the structural dynamic properties.

We are concerned here essentially with the modal tests which are carried out to validate theoretical models of machines or their structural components as part of the above processes which seek to develop vibration-free and reliable machines. There are many examples and opportunities to use conventional modal testing on the stationary parts of these machines, and these will not be discussed here save for a brief resume of the most important modal properties of the typical components found in these machines. Most of the ensuing contents will be concerned with the extension of this technology to cases where the test structure, or part of it, is rotating.

1.3. SUMMARY OF MAIN VIBRATION PROPERTIES OF ROTATING MACHINE STRUCTURES

It is found that many of the most critical (i.e. vibration-prone) components in rotating machines fall into a class of structure which is most generally described as "quasi-axisymmetric". The various types of structure which are:

(i)   axisymmetric

(ii) cyclically periodic

(iii) slightly asymmetric, or aperiodic (which means slightly imperfect structures of the first and second types), referred to as "quasi-periodic"

are those which concern us here. We shall not describe the vibration properties of these structures in detail, but simply report here that the essential modal properties are as follows:

- axisymmetric structures (plain discs, wheels, etc.) have modes whose shapes are described in the circumferential direction, θ, by variations of the form:

$$\phi(\theta) = \cos(n\theta + \alpha_n)$$

and therefore known as "n-nodal diameter" or nND modes;

- cyclically-periodic structures (such as bladed discs, impellers, gear wheels) have modes whose shapes are described in the circumferential direction, θ, by expressions of the form:

$$\phi(\theta) = \{A_n \cos(n\theta + \alpha_n) + A_{N-n} \cos((N-n)\theta + \alpha_{N-n})$$
$$+ A_{N+n} \cos((N+n)\theta + \alpha_{N+n}) + \ldots\}$$

where N is the number of blades, vanes, teeth etc. These are also referred to as "n-nodal diameter modes" although this description is less precise in these cases. (It should be noted that if such mode shapes are defined by determining the amplitude ratios only at the N discrete points around the rim which carry the blades/vanes/etc., then the discrete Fourier description which results will be incapable of discriminating above the first term in this series – hence the classification simply as 'n-nodal diameter' modes.);

- quasi-periodic structures, in which the loss of symmetry is small, have mode shapes which are described (in the circumferential direction) by: $\phi(\theta) = \sum \cos(r\theta + \alpha_r)$, where r = 1, 2, 3, ..., but is generally – although not always - dominated by n, N-n, N+n, etc When these modes are so dominated by a single, or few, terms of this type, the nodal diameter label is still used, although it is much less precise than for the other cases.

A second feature of the modes of these quasi-axisymmetric structures concerns their natural frequencies. In the case of type (i) and (ii) structures, most of the modes exist in pairs of 'double' modes: two modes with identical natural frequencies and mode shapes which differ only in the angular orientation of the nodal lines, i.e. in αn. As is the case generally, when there are two or more modes with identical natural frequencies, any combination of the individual two mode shapes is also a mode shape. This can lead to some unexpected features in the case of these axisymmetric structures where, for example, a valid mode shape can be produced by a combination of $1 \times \cos(n\theta + \alpha_n)$ plus $i \times \sin(n\theta + \alpha_n)$, the result of which is a $\cos n\theta$ mode shape rotating around the structure in a travelling wave motion.

One consequence of these features is that the components which display such modal properties are much more difficult to test than are structures with single modes. As a result, measured modal properties of axisymmetric, or quasi-axisymmetric, structures are often in error, sometimes through ignorance on the part of the analyst and sometime because of the inherent difficulties in making the measurements.

1.4. PROBLEMS OF APPLYING MODAL TESTING TO ROTATING MACHINES

Here we shall simply list the areas of additional difficulty which are encountered when seeking to apply modal testing methodologies to structures which are rotating. It is the case that most such structures fall into one of the above-listed three categories and so carry the additional complications borne by all such structures but here we shall be concentrating on four areas of particular concern:

(i)   the underlying modal theory which is applicable to structures which are rotating;

(ii) the problems of exciting a rotating component, and of measuring the excitation force(s);

(iii) the problems of measuring the response of the rotating components; and

(iv) the selection and analysis of the measured response functions in order to extract the required modal properties of the rotating components.

These areas will be dealt with in turn in the he following sections of the paper.

2.   Modal analysis theory for rotating structures

2.1.   ESSENTIAL FEATURES OF ROTATING SYSTEMS

The essential features of rotating structures which make them different from stationary structures, are (i) the existence of gyroscopic and centrifugal effects and (ii) the different axes which may be used to describe or observe the resulting vibration behaviour. The latter feature can be found to be responsible for Doppler-like effects in which the vibration frequencies on rotating components are different from those of the excitation which is applied in a stationary frame of reference. There are also particular effects which result from electromagnetic, aerodynamic and hydrodynamic effects forces which are generated between rotating and stationary components during the normal operation of the machine. In addition, there is usually some out-of-balance effects which generate vibration. These generally constitute excitation forces and so do not appear in the equations until the forced response is sought.

2.2.   MODAL ANALYSIS OF THE ESSENTIAL ROTATING STRUCTURE

2.2.1.   *Coordinate axes*

In order to illustrate the essential features of the modal properties of rotating structures in general, we shall use a simple example which contains many of the features found in much more general cases. This example is illustrated in Figure 1 by the 2DOF system shown which has a rigid disc carried by a rigid shaft in flexible bearings. The various parameters of interest (I_O, L, k_x, k_y, J, Ω_z) are defined in the figure.

It is first necessary to determine which coordinate set is to be used for writing the equations of motion which will describe the vibration of this system. There are essentially two choices: (ii) those fixed in space: x_S, y_S, z_S (usually abbreviated to: x, y and z) and those rotating with the shaft: x_R, y_R, z_R. The relationship between these two coordinate sets is simple to define, and is as follows:

$$x_R = x_S \cos\Omega_z t + y_S \sin\Omega_z t$$

$$y_R = -x_S \sin\Omega_z t + y_S \cos\Omega_z t$$

(1)

2.2.2. *Equations of motion in fixed coordinates and solution for free vibration*

Using first the fixed axes set of coordinates, we can write the equations of motion for free vibration of the system shown in terms of x, y, and z, as follows:

$$
\begin{bmatrix} \dfrac{I_0}{L^2} & 0 \\[2ex] 0 & \dfrac{I_0}{L^2} \end{bmatrix} \begin{Bmatrix} \ddot{x} \\ \ddot{y} \end{Bmatrix} + \begin{bmatrix} c & \dfrac{J\Omega_z}{L^2} \\[2ex] -\dfrac{J\Omega_z}{L^2} & c \end{bmatrix} \begin{Bmatrix} \dot{x} \\ \dot{y} \end{Bmatrix} + \begin{bmatrix} k_x & 0 \\ 0 & k_y \end{bmatrix} \begin{Bmatrix} x \\ y \end{Bmatrix} = \begin{Bmatrix} 0 \\ 0 \end{Bmatrix}
\tag{2}
$$

It should be noted that if the rotation is zero ($\Omega_z = 0$), then the two equations revert to those of the very simple 2DOF system in which the two modes of vibration are (a) pure horizontal motion of the disc centre, (b) pure vertical motion of the disc centre, with natural frequencies

$$
\omega_a = \sqrt{(k_x L^2 / I_0)} \qquad \omega_b = \sqrt{(k_y L^2 / I_0)}
$$

If the bearings are symmetric – a special case, but one of practical importance – then these two modes have identical natural frequencies with the concomitant properties exhibited by the mode shapes, along the lines already summarised for axisymmetric structures in general. If the rotation of the disc is not zero, then the solution to the general form of these equations of motion yields the following eigenproperties:

$$
\begin{bmatrix} 1 & -i \\ -i & 1 \end{bmatrix}; \quad \begin{bmatrix} \omega_1^2 & 0 \\ 0 & \omega_2^2 \end{bmatrix}; \quad \begin{bmatrix} 1 & i \\ i & 1 \end{bmatrix}
\tag{3}
$$

$$
\omega_{1,2} = \omega_\Omega \mp (\gamma \Omega_z / 2); \quad \omega_\Omega^2 = \omega_0^2 + (\gamma \Omega_z / 2)^2; \quad \gamma = J / I_0; \quad \omega_0^2 = k L^2 / I_0
$$

These show that the skew-symmetric velocity-dependent terms have caused there to be two sets of eigenvectors – the left vectors and the right vectors - and for these eigenvectors to become complex but not the eigenvalues, whose values are real. Thus we have a conservative system with two distinct natural frequencies, with no damping terms, but with mode shapes which are complex. The precise form of these two mode shapes is of considerable interest: if the system is symmetric in the sense that the horizontal and vertical stiffnesses are identical, then it should be noted that the two natural frequencies are not identical (as was the case in the non-rotating system) but that the two mode shapes represent orbits, one (that corresponding to the lowest natural frequency) travelling in the reverse sense to the spin of the rotor and the other, corresponding to the higher of the natural frequencies, travelling in the same direction as the spin. In the more general case where the two stiffnesses are not identical, then the two modes constitute backward and forward travelling elliptical orbits.

These are the essential modal properties of a rotating system which is prone to the gyroscopic forces which derive from the Coriolis accelerations that result when a system is rotated simultaneously about two perpendicular axes, as happens when a spinning disc is caused to vibrate about a diameter at the same time as spinning about its principal axis. Graphical displays of the variation of the two natural frequencies are shown in Figure 2.

2.2.3. *Equations of motion in rotating coordinates and solution for free vibration*

If we transform the equations of motion previously written in terms of the fixed-axes coordinate set to those for the rotating-axes set, using the usual transformation which, in this case, is:

$$\begin{Bmatrix} x \\ y \end{Bmatrix} = \begin{bmatrix} c & -s \\ s & c \end{bmatrix} \begin{Bmatrix} x_R \\ y_R \end{Bmatrix} \quad \text{and} \quad \begin{Bmatrix} x_R \\ y_R \end{Bmatrix} = \begin{bmatrix} c & s \\ -s & c \end{bmatrix} \begin{Bmatrix} x \\ y \end{Bmatrix} \tag{4}$$

where $c = \cos(\Omega t)$ and $s = \sin(\Omega t)$, we shall obtain an alternative set of equations of motion, as follows:

$$\begin{bmatrix} \dfrac{I_0}{L^2} & 0 \\ 0 & \dfrac{I_0}{L^2} \end{bmatrix} \begin{Bmatrix} \ddot{x}_R \\ \ddot{y}_R \end{Bmatrix} + \begin{bmatrix} 0 & -\dfrac{2\Omega_z I_0}{L^2} + \dfrac{J\Omega_z}{L^2} \\ \dfrac{2\Omega_z I_0}{L^2} - \dfrac{J\Omega_z}{L^2} & 0 \end{bmatrix} \begin{Bmatrix} \dot{x}_R \\ \dot{y}_R \end{Bmatrix}$$

$$+ \begin{bmatrix} -\dfrac{I_0\Omega_z^2}{L^2} + \dfrac{J\Omega_z^2}{L^2} + (c^2 k_x + s^2 k_y) & sc(k_y - k_x) \\ sc(k_y - k_x) & -\dfrac{I_0\Omega_z^2}{L^2} + \dfrac{J\Omega_z^2}{L^2} + (s^2 k_x + c^2 k_y) \end{bmatrix} \begin{Bmatrix} x_R \\ y_R \end{Bmatrix} = \begin{Bmatrix} 0 \\ 0 \end{Bmatrix} \tag{5}$$

It can be seen that, in this general case where the two stiffnesses, k_x and k_y are different, these equations of motion present something of a problem for anyone seeking to solve them. The main problem is that the coefficients in the three matrices are no longer all constant terms, those relating to the stiffnesses now having periodically time-varying values. The solution of this class of eigenproblem is a non-standard form, and will be discussed later. For the moment, it is of direct interest to revert to the special case where the two stiffnesses are identical as in this case the equations of motion in the rotating coordinates reduce to a set of 'standard' equations in which all the coefficients are constants:

$$\begin{bmatrix} \dfrac{I_0}{L^2} & 0 \\ 0 & \dfrac{I_0}{L^2} \end{bmatrix} \begin{Bmatrix} \ddot{x}_R \\ \ddot{y}_R \end{Bmatrix} + \begin{bmatrix} 0 & -\dfrac{2\Omega_z I_0}{L^2} + \dfrac{J\Omega_z}{L^2} \\ \dfrac{2\Omega_z I_0}{L^2} - \dfrac{J\Omega_z}{L^2} & 0 \end{bmatrix} \begin{Bmatrix} \dot{x}_R \\ \dot{y}_R \end{Bmatrix}$$

$$+ \begin{bmatrix} -\dfrac{I_0\Omega_z^2}{L^2} + \dfrac{J\Omega_z^2}{L^2} + k & 0 \\ 0 & -\dfrac{I_0\Omega_z^2}{L^2} + \dfrac{J\Omega_z^2}{L^2} + k \end{bmatrix} \begin{Bmatrix} x_R \\ y_R \end{Bmatrix} = \begin{Bmatrix} 0 \\ 0 \end{Bmatrix} \tag{6}$$

Solution of these equations for free vibration describes just the same behaviour as the earlier solution, based on the fixed-axes coordinate set, although the eigenvalues and eigenvectors are different in this case.

2.2.2. *Equations of motion for non-axisymmetric rotor*

In all of the preceding cases, the rotor has been assumed to be axisymmetric and this is indeed the case for a great many of the cases actually encountered in practice. However, there will be special circumstances in which this symmetry is not present and it is appropriate to include comment here concerning the implications for the relevant equations of motion and free vibration (modal) solution to these equations. It can be seen that two cases are possible: (i) where the stator is symmetric (i.e. where $k_x = k_y$) and (ii) where the stator is not symmetric. It can also be shown that in both of these two cases the equations of motion written in terms of fixed-axes coordinates will have the periodically time-varying coefficients which were encountered above (and which present a non-standard eigenproblem). It is also found that the latter case, (ii), also exhibits this same property when the equations are written in terms of the rotating-axes coordinates. However, when the former of these last two cases are described by equations written in terms of the rotating-axes coordinates, then these equations adopt the simpler form of constant coefficients and are thus amenable to a conventional modal analysis solution. So, in summary, we can show the pattern of equations and axis sets as follows:

Rotor	Stator	EOM in fixed-axes	EOM in rotating-axes
SYMM	SYMM	LTI	LTI
SYMM	NONSYMM	LTI	Periodic
NONSYMM	SYMM	Periodic	LTI
NONSYMM	NONSYMM	Periodic	Periodic

where LTI = Linear, time-invariant (i.e. constant coefficients)

2.3. FORCED VIBRATION ANALYSIS (AND FRFs) OF THE ESSENTIAL ROTATING STRUCTURE

In this section we shall extend the above free vibration analysis to the case of forced vibration, for the particular cases of excitations which might be considered for use in modal testing: i.e. single excitations at selected points; multiple phased excitations at selected points; excitations caused by out-of-balance.

2.3.1. *Single-point excitation*

Here we consider the case where a single harmonic force is applied at a fixed point in space, $f_x(t) = F_0 e^{i\omega t}$, as this would be the most likely excitation used in a modal test. The equations of motion in fixed-axes are extended by adding some viscous dampers in parallel with the springs and the externally-applied force, $f_x(t)$:

$$\begin{bmatrix} \dfrac{I_0}{L^2} & 0 \\ 0 & \dfrac{I_0}{L^2} \end{bmatrix} \begin{Bmatrix} \ddot{x} \\ \ddot{y} \end{Bmatrix} + \begin{bmatrix} c & \dfrac{J\Omega_z}{L^2} \\ -\dfrac{J\Omega_z}{L^2} & c \end{bmatrix} \begin{Bmatrix} \dot{x} \\ \dot{y} \end{Bmatrix} + \begin{bmatrix} k & 0 \\ 0 & k \end{bmatrix} \begin{Bmatrix} x \\ y \end{Bmatrix} = \begin{Bmatrix} f_x \\ f_y \end{Bmatrix} = \begin{Bmatrix} F_0 \\ 0 \end{Bmatrix} e^{i\omega t} \qquad (7)$$

The solution of these equations leads directly to the following expressions for the FRFs between the harmonic responses, X and Y, and the excitation force F_0:

$H_{xx}(\omega) =$

$$X/F_0 = \frac{(kL^2 + i\omega cL^2 - \omega^2 I_0)}{(k^2L^2 + 2i\omega ckL^2 - \omega^2(c^2L^2 + 2I_0k + (J\Omega_z/L)^2) - 2i\omega^3 cI_0 + \omega^4(I_0/L)^2)}$$

(8)

$H_{yx}(\omega) =$

$$Y/F_0 = \frac{i\omega J\Omega_z}{(k^2L^2 + 2i\omega ckL^2 - \omega^2(c^2L^2 + 2I_0k + (J\Omega_z/L)^2) - 2i\omega^3 cI_0 + \omega^4(I_0/L)^2)}$$

Similar expressions can be obtained for the other 2 FRFs in the he 2x2 matrix for this system. It should be noted that the transfer FRFs (between x and y directions) are zero when there is no rotation but that these are decidedly non-zero, and have a magnitude of the same order as the point FRFs, once there is some rotation of the system. It should also be noted that the two transfer FRFs are not identical, as would be expected in a system which exhibited reciprocity but are complex conjugates of each other, a feature which can be traced directly to the skew symmetry in the velocity-dependent system matrix.

2.3.2. *Synchronous out-of-balance excitation*

Another excitation of considerable interest is that which can be generated 'internally' in the rotating structure by virtue of its own out-of-balance, whether inherent or added deliberately. The effect of this type of excitation force can be seen by introducing the relevant forces to the equations of motion. Take, as an example, the equations of motion for the symmetric rotor, symmetric stator case (from equation (2)) and introduce the forces from an out-of-balance of magnitude (mr) which are:

$$F_x = (mr)\Omega_z^2 \cos\Omega_z t \qquad F_y = (mr)\Omega_z^2 \sin\Omega_z t \qquad (9)$$

or

$$\left\{\begin{matrix} F_x \\ F_y \end{matrix}\right\} = F_{OOB}\left\{\begin{matrix} 1 \\ -i \end{matrix}\right\}e^{i\Omega_z t} \qquad (10)$$

The essential features of the response of our simple rotating structure to this excitation can be illustrated first using the symmetric-symmetric example described above. In that case, the full equations of motion are:

$$\begin{bmatrix} \dfrac{I_0}{L^2} & 0 \\ 0 & \dfrac{I_0}{L^2} \end{bmatrix}\left\{\begin{matrix} \ddot{x} \\ \ddot{y} \end{matrix}\right\} + \begin{bmatrix} 0 & \dfrac{J\Omega_z}{L^2} \\ -\dfrac{J\Omega_z}{L^2} & 0 \end{bmatrix}\left\{\begin{matrix} \dot{x} \\ \dot{y} \end{matrix}\right\} + \begin{bmatrix} k & 0 \\ 0 & k \end{bmatrix}\left\{\begin{matrix} x \\ y \end{matrix}\right\} = \left\{\begin{matrix} f_x \\ f_y \end{matrix}\right\} = \left\{\begin{matrix} 1 \\ -i \end{matrix}\right\}F_{OOB}\,e^{i\Omega_z t} \quad (11)$$

and their solution, in terms of the fixed-axes coordinates, is as follows:

$$\left\{\begin{matrix} X \\ Y \end{matrix}\right\}e^{i\Omega_z t} = \left\{\begin{matrix} A \\ -iA \end{matrix}\right\}F_{OOB}\,e^{i\Omega_z t} \qquad (12)$$

where $A = \dfrac{L^2}{I_0(\omega_0^2 - (1-\gamma)\Omega_z^2)}$.

It can be seen from these expressions that only a single resonance is encountered as the rotor is run up from zero to a rotation speed which is higher than both the two natural frequencies, in place of the two resonances which might be expected as a result of the two natural frequencies possessed by the system. The explanation for this effect is quite simple: the excitation generated by an out-of-balance rotating with the structure has a shape as well as a frequency and that shape is orthogonal to one of the two mode shapes: in fact, this synchronous out-of-balance excitation will only excite the higher of the two modes of vibration – the one whose mode shape is a forward-whirling circular orbit. The other mode – the backward travelling circular orbit – cannot be excited by this type of excitation.

If we next consider the case of a symmetric rotor with an asymmetric stator, then in this case both modes can be excited because neither mode shape is purely orthogonal to the excitation pattern and as the asymmetry gets bigger, the extent to which the backward travelling mode is excited itself gets greater.

2.3.3. Nonsynchronous out-of-balance excitation

The next 'standard' excitation to be considered is that of a non-synchronous out-of-balance and this can be generated by a concentric rotor with an out-of-balance being added to the test rotor and spun at a speed which is different to and independent of the test rotor speed. (One of the limitations of the synchronous out-of-balance excitation is that it is not possible to measure a conventional response function in which the frequency of excitation is varied through a range of interest while the structure properties are kept constant.) With such non-synchronous excitation it is possible to excite the structure in a way which reflects the travelling nature of many in-service excitation sources but without attracting the frequency-speed constraint which applies with synchronous out-of-balance.

Thus, for the standard case of the symmetric-symmetric system, we can write the equations of motion (defining the excitation frequency as $(\beta\Omega)$ as:

$$\begin{bmatrix} \dfrac{I_0}{L^2} & 0 \\ 0 & \dfrac{I_0}{L^2} \end{bmatrix} \begin{Bmatrix} \ddot{x} \\ \ddot{y} \end{Bmatrix} + \begin{bmatrix} 0 & \dfrac{J\Omega_z}{L^2} \\ -\dfrac{J\Omega_z}{L^2} & 0 \end{bmatrix} \begin{Bmatrix} \dot{x} \\ \dot{y} \end{Bmatrix} + \begin{bmatrix} k & 0 \\ 0 & k \end{bmatrix} \begin{Bmatrix} x \\ y \end{Bmatrix} = \begin{Bmatrix} f_x \\ f_y \end{Bmatrix} = \begin{Bmatrix} 1 \\ -i \end{Bmatrix} F_{OOB} e^{i\beta\Omega_z t} \quad (13)$$

and the solution is then found to be:

$$\begin{Bmatrix} X \\ Y \end{Bmatrix} e^{i\beta\Omega_z t} = \begin{Bmatrix} A \\ -iA \end{Bmatrix} F_{OOB} e^{i\beta\Omega_z t}$$

$$(14)$$

$$\text{where } A = \dfrac{L^2}{I_0(\omega_0^2 - \beta(\beta-\gamma)\Omega_z^2)}$$

2.3.4. Multi-point excitations

The last of the 'standard' excitations to be considered is that which is created when 2 or more point excitations are applied simultaneously. This is a standard technique in modal testing of stationary structures and has specific and very direct applications to the cases of rotating structures as well. The main application is to the simulation of the rotating excitations such as those just discussed caused by out-of-balance effects. Such a simulation is possible in this case simply by applying fixed sinusoidal point excitations in the x and y directions respectively, when the two forces are phased 90 degrees apart relative to each other (in time) so that they are expressed by:

$$f_x = F_x \cos \omega t \qquad f_y = F_y \sin \omega t \qquad (15)$$

Of course, such an excitation pattern is exactly the same as that caused by the general out-of-balance case above, save for the fact that the two forces can be controlled independently of each other (it is not automatically the case that $F_x = F_y$). However, this is a technique which is amenable to further developments in more complicated cases where the test structure is more flexible than the rotor system we have been studying so far. For the time being, we can simply note this option which, in some cases, is referred to as "complex" excitation.

2.4   FORCED VIBRATION OF ROTATING FLEXIBLE COMPONENTS

2.4.1   *Vibration modes of rotating flexible components*
In this section we shall consider briefly some of the issues concerning the vibration analysis of flexible structures such as discs and bladed wheels which are carried on the rotating shaft.

It will be noted at this stage that the vibration modes of these components can be influenced by the effects of rotation, such as those from centrifugal/centripetal effects, but it will be considered here that such effects are either small or of no great consequence to the concerns of modal testing. The problems which are encountered are more concerned with the frequency differences which occur between vibration response of the rotating structure and those of excitation forces which are usually derived from phenomena which are fixed in space. Thus we shall focus our attention on this latter aspect, and will not concern ourselves with details of the other effects which cause the natural frequencies of various modes to change from the values they have at rest: the effect of rotation on the mode shapes is generally much less than we have experienced so far in respect of the rotating shaft itself, where gyroscopic effects play an important part. These (gyroscopic) effects do not usually play a significant role in the great majority of the vibration properties of the discs and wheels that are carried by the shaft.

Consider now that we are concerned with the flexural vibration of the disc which is carried by the shaft previously shown in Figure 1. The flexural (axial) vibration modes of the disc will be dominated by modes whose mode shapes are essentially of the form:

$$\phi(\theta) = \cos(n\theta + \alpha_n)$$

or, in the case of a cyclically-symmetric structure such as a bladed disc:

$$\phi(\theta) = \{A_n \cos(n\theta + \alpha_n) + A_{N-n} \cos((N-n)\theta + \alpha_{N-n}) + A_{N+n} \cos((N+n)\theta + \alpha_{N+n}) + \ldots \},$$

or, in the most general case of a non-symmetric structure:

$$\phi(\theta) = \sum \cos(r\theta + \alpha_r).$$

It should be noted at this stage, that in order for modes such as these to be excited, the excitation forces must contain a component in their circumferential distribution of the essential $\cos n\theta$ form (or the secondary terms in the series) and this property will play a significant role in the excitation/response properties which we are seeking.

2.4.2. *Point excitation of flexible axisymmetric components*
If we consider first a single point excitation, $F(t) = F_0 \cos \omega t$, located at some point (α_0) around the rim of the disc (which, for the moment is stationary), then it is possible to express this excitation as follows:

$$F(\theta, t) = (F_0 / \pi) \sum_n \cos(n\theta + \alpha_0) \cos \omega t$$

From this expression, it is possible to see that such an excitation will excite all modes in the nND series by virtue of the circumferential decomposition of a single-point force. A typical FRF that might be found from such a structure is shown in Figure 3.

Now let us consider the same excitation being applied – still fixed in space – to a disc which is rotating past the excitation point with speed, Ω. In order to determine what will be the response of the rotating disc to this excitation, it is necessary to transform the excitation force, defined above in fixed-axes coordinates, to the rotating-axes frame, and this can be done using the standard transformation between θ_S, and θ_R:

$$\theta_R = \theta_S - \Omega t \tag{16}$$

This transformation results in the rotating structure experiencing an excitation which is described by the following expression:

$$F(\theta_R, t) = (F_0 / \pi) \sum_n \cos((n\theta_R + \Omega t) + \alpha_0) \cos \omega t \tag{17}$$

which, in turn, becomes :

$$F(\theta_R, t) = (0.5 F_0 / \pi) \sum_n \left[\cos(n\theta_R + \alpha_0)(\cos(\omega - n\Omega)t + \cos(\omega + n\Omega)t) \right.$$

$$\left. + \sin(n\theta_R + \alpha_0)(\sin(\omega - n\Omega)t - \sin(\omega + n\Omega)t) \right] \tag{18}$$

From this expression, it can be seen that a single-frequency, single-point excitation applied at frequency, ω, appears to the rotating disc as a myriad of excitations, each with a particular shape (and therefore capable of exciting particular modes) and each with a particular frequency. The list of such frequencies at which the disc is excited is:

ω, $(\omega - \Omega)$, $(\omega + \Omega)$, $(\omega - 2\Omega)$, $(\omega + 2\Omega)$, $(\omega - 3\Omega)$, $(\omega + 3\Omega)$,... $(\omega - n\Omega)$, $(\omega + n\Omega)$, ...

Thus, the apparently simple excitation provided by a single point harmonic force will generate an extremely complex vibration response on the rotating component, making the definition of frequency responses (FRFs) much more difficult than is the case for a non-rotating structure. This is simply the same effect that was encountered in the simpler case of a rotating shaft whenever the coordinates used for the response and the excitation were not in the same axes set. Indeed, it is possible to relate those earlier cases (of the rotating shaft) to these more general ones here by noting that vibration of a rigid disc is confined to the two families of modes identified as 0- and 1-nodal diameter modes and that a shaft whose section does not deform is, in effect, similar to a rigid disc. Thus we can consider the previous cases of rotor dynamics as a special case of the more general flexible structure, such as a disc, or bladed wheel.

It should be noted that the magnitudes of the components in each harmonic (here uniformly $(0.5 F_0/\pi)$) can be adjusted by the application of more than one point force. Two such forces, strategically placed around the circumference of the machine and suitably phased in their temporal excitation signals, can be used to excite selectively certain orders (certain values of n) at the expense of suppressing excitation and response in other orders. Such a technique has long been used to simulate the excitation experienced by a rotating structure without actually rotating either the test structure or the excitation source. It is, in effect, a special version of the classical multi-point excitation method known as 'appropriation' or 'normal mode' testing.

3. Modal testing of rotating structures

3.1. BASIS OF MODAL TESTING FOR ROTATING STRUCTURES

So far, most of our attention has been directed towards a better understanding of the underlying theory upon which any modal testing of rotating structures will have to be based. Now it is appropriate to consider some of the practical issues of performing a modal test on a structure which is rotating as well as vibrating.

The main problems to be resolved, in addition to all those which apply to modal testing of structures in general, are:

(i)   the difficulty of applying and measuring suitable excitations;

(ii)   the difficulty of measuring responses on a rotating structure;

(iii) the absence of reciprocity and similar features which are often used to check the quality of measured data in modal testing;

(iv) the need for the measurement of a greater number of FRFs in order to construct a mathematical model from identified modal parameters;

(v)   the difficulty of extracting genuine modal parameters from measured response functions, in view of the fact that many such responses are not true FRFs, and

(vi) the likely incidence of non-linearities, and other effects which can limit the quality of data measured in the relatively hostile environment of a running rotating machine.

In this section, we shall focus our attention on two of these items: (i) and (ii) as these are more fundamental and crucial to being able to embark on a modal test of a rotating structure. Many of the other issues are more matters of quantity and quality of the measured data, although (iv) and (v) will be discussed in the next section.

3.2. EXCITATION DEVICES SUITABLE FOR ROTATING STRUCTURES

There are essentially two requirements of excitation devices to be used in modal testing of rotating machines. The first is of the devices which are needed to excite vibration of the rotor(s) themselves, and the second relates to those which will be used to excite flexural vibration in the rotating components such as discs and wheels.

3.2.1. *Excitation of rotors*

Considering first the methods which can be used to excite a rotating shaft: there are essentially three approaches possible:

(i)   by exciting with a conventional shaker applied radially on the outer race of an auxiliary free bearing whose inner race rotates with the shaft;

(ii) by using an out-of-balance device which rotates with the shaft, either at its own rotation speed or at a non-synchronous speed; or

(iii) by using active magnetic bearings (AMBs).

The first of these approaches has been used successfully for some years and can be implemented with just one exciter or with two or more, if it is desired to create an excitation in the rotating-axes coordinate frame (desirable if response measurements are to be made on the rotating components, rather than with transducers fixed in space). An example of its application is shown in Figure 4 with a typical response function which is obtained using it. This approach has the advantage that it is relatively easy to measure the applied force, but one disadvantage is the need to intrude into the rotating machine itself in order to install the auxiliary bearing. In many practical machines, such intrusion would be impractical.

The second approach is used in some sectors of the rotating machinery industry involves adding an independent rotor onto the test shaft, usually at the end, in such a way that it can be driven at a rotation speed which is independent of that of the test shaft itself. The added rotor has an adjustable out-of-balance device which is used to generate a rotating excitation force at any desired frequency while the rotor is spinning at an operating speed of interest for the measurement. These excitation devices, while providing a more representative excitation source than the single-point drive, also require a significant modification to the test machine and will not always be a viable option.

The third option represents a recent development of new technology and one which offers distinct possibilities for the future. The use of active magnetic bearings (AMBs) in high-speed rotating machinery is still at a relatively early stage of development but these devices offer the prospect of using the same bearing to support the rotor and to inject a controlled excitation into it at the same time, thereby permitting the measurement of response functions under running conditions without adding any additional components to the machine. An example of a test rig developed around this concept is shown in Figure

5, together with an FRF measured using this type of excitation. In this case, it is relatively straightforward to generate a range of different excitation patterns, from the single-point radial excitation delivered by the first type above, to the combined x and y direction forces which can be phased to introduce a rotating excitation patter to the rotor at any effective rotation speed. The applied forces can be determined at source using Hall sensors to measure the magnetic flux and these can be calibrated to indicate the instantaneous forces being transmitted to the rotor. These devices can provide significant control of excitation suitable for modal testing.

3.2.2. *Excitation of rotating components*

The second class of excitation device required for modal testing of rotating machinery structures is that necessary to excite vibration in the flexible discs or wheels which are carried by the rotor. This can often be more difficult than the excitation of the rotor itself because it is virtually impossible to make a direct physical connection to the test object itself. The most likely excitation sources are (i) air jet(s) impinging on the surface of the rotating structure or (ii) magnet(s) placed very close to the moving surface (assuming that to be magnetic material).

In both cases there are problems of control and of accurate measurement of the forces applied. When using electromagnets, the actual excitation force is usually far from sinusoidal when the input signal is sinusoidal and, similarly, for the air jet it is difficult to generate a well-defined harmonic or other known force. Measurement of the forces is usually achieved indirectly, by measuring the reaction force of the fixed part of the excitation device against the supporting frame. This measurement demands zero deflection of the exciter body and in order to ensure reliable force measurement it is usually necessary to measure both the reaction force and the motion of the exciter itself in order to compensate for the inertia forces which such motion will generate.

It is possible to envisage the application of excitation using an on-board actuator device, such as a piezoelectric crystal. Such a device would have the advantage that the excitation would be applied in the rotating axes coordinate set, and would thus enable FRFs to be obtained with matching measurements of response, also made in the rotating-axes coordinates.

3.3. RESPONSE MEASUREMENT ON ROTATING STRUCTURES

3.3.1. *General comments*

In parallel with the difficulties of exciting a rotating structure, there are also problems associated with measuring the required responses. Once again, transducers to perform these measurements can be on-board, rotating with the test structure, or fixed in space. Strain gauges are frequently used to measure stress levels on rotating blades, wheels and rotors although the use of slip rings or telemetry devices is necessary in these cases to extract the signals for analysis. Alternatively, stationary transducers can be used such as the conventional capacitance or inductance pickups which are used in non-rotating applications as well. There are a number of new optical techniques which are particularly well suited to use in rotating machinery applications - laser holography and laser Doppler velocimetry. We shall discuss some of the features of the latter of these two devices as these are quite widely available.

3.3.2. *Laser Doppler velocimeters*

The laser Doppler velocimeter (LDV) is an instrument which is capable of measuring the instantaneous velocity of a point on the surface of a structure, measuring along the line of sight between the LDV and the measurement point. The instrument has a wide working range of amplitudes and frequencies and is capable of measuring on moving surfaces. It has been used to measure the flexural vibrations of a rotating disc on the same test rig as illustrated earlier (Figure 5). The more advanced LDVs have a scanning capability which means that they can be programmed to make measurements either at a sequence of pre-determined points or at a continuously changing point. This facility is achieved by driving the two mirrors that provide the direction of the measurement laser beam. In normal scanning mode, the SLDV is used to measure at a large number of points (DOFs) in a short time but in the application to rotating machinery testing, the scanning capability can be linked to the rotation of the rotating testpiece in such a way that the measurement point is locked onto a selected point on the disc or rotor. The synchronisation of the rotating shaft with the position of the measurement point is achieved using an encoder on the shaft to drive the mirrors in such a way that the measurement point rotates at exactly the same rate as the shaft. This, in effect, provides a non-contact transducer capable of measuring in the rotating-axes coordinate set, and works without the practical difficulties associated with slip rings or telemetry.

This device is likely to become much more widely available and further developments of its applications to modal testing will undoubtedly follow. One which has already been demonstrated for rotating machinery is the possibility of performing a non-synchronous rotating scan around the rotating disc. This is perhaps the most general measurement of all: with the fixed point in space as one limiting case (scan rate = 0) and rotating with the disc (scan rate = spin speed) as the other.

4. Modal identification for rotating structures

4.1. GENERAL

In this last section it is appropriate to add some comments about the final stage of a modal test: that of extracting the underlying modal properties of the tested structure, in order to construct the required mathematical model. Most of the differences and difficulties associated with modal analysis of rotating machinery lie in the basic theory and in some of the practicalities of performing tests on moving objects. However, there remain some difficulties at the modal identification stage and these are discussed below.

One requirement which must be observed carefully is the fact that true FRF data (as required for modal analysis) will only be obtained if the excitation and the response are measured in the same axes set (fixed or rotating). Problems will arise if, for example, the excitation is applied and measured in the fixed-axes set of coordinates and the response is measured in the rotating-axes coordinates. We have already seen how different frequencies are present in these signals, when described in terms of coordinates in the different axes sets. The complications which will result from this effect are quite extensive, and it is not possible to do more here than to alert the user to the scale of the problem. Considerable care must be taken when seeking to use the multi-frequency response functions which ensue from such a test where the two parameters of excitation and response are not measured in the same axes frame.

4.2. BASIC FORMULATION OF FRFS FOR ROTATING STRUCTURES

It has already been shown earlier in the paper that the basic formula for an FRF of a rotating structure will either be of the same essential form as those of stationary structures (if the equations of motion all have constant coefficients – LTI systems) or will be complicated by the periodic coefficients in the more general case. As it is the case that most practical structures can be described by the LTI type of expressions, then the basic formula used as the basis of curve-fitting to extract modal parameters can be used in the case of FRFs obtained on rotating structures.

When using simple single-FRF modal analysis procedures, then direct application to the measured FRFs is acceptable. However, if global modal analysis algorithms are to be used, then it is important to note that the measured FRFs for a rotating structure do not usually obey the reciprocity properties which are familiar in stationary structures. In some parameter extraction routines such reciprocity is assumed; in others it is not. Care must be taken to ensure that the appropriate type of algorithm is used.

4.3 SELECTION OF FRFS TO CONSTRUCT A MATHEMATICAL MODEL

In general, the system dynamics matrices for a rotating structure are not symmetric and this demands that twice as many FRFs be measured and analysed in these cases than are necessary for stationary structures. In principle, at least one column and one row from the complete FRF matrix are necessary in order to furnish sufficient information to define the modes of the structure. This requirement is in addition to all the usual considerations that must be made about which row and/or column to measure so that all the modes are properly represented and visible in the final model. All in all, these requirements place major demands on a would-be modal test, especially as it is particularly difficult to measure both a row and a column of the FRF matrix using the same excitation device.

In some cases, the lack of symmetry in the system matrices actually takes the form of skew symmetry, and in these cases, a relaxation of the demanding requirements for the FRFs to be measured is available. If the matrices are symmetric or skew symmetric, then the right-hand and the left-hand eigenvectors (whose separate existence is the direct cause of the double number of FRFs to be measured) are simply complex conjugates of each other, and it is sufficient to determine just one set. In these cases, then, it is possible to revert to the conventional practice of measuring the elements on just one row or one column of the FRF matrix – a much more practicable proposition.

5. Concluding remarks

In this paper we have sought to introduce the reader to the extensive subject of modal analysis applied to rotating structures, particularly those found in rotating machinery. This is considered to be an important area in modal analysis and one which demands proper development and exploitation.

The basis of the underlying theory of modal analysis for rotating structures has been presented and set in the context necessary for the application of modal testing techniques. Particular attention has been drawn to the choice of axes set as well as to a number of features which are common only to rotating components.

An introduction to some of the recent developments on the practical side has also been presented, with particular mention of the use of active magnetic bearings (AMBs) as excitation devices and scanning laser Doppler velocimeters (SLDVs) as response measuring devices, both highly suitable for rotating machinery applications.

6. Bibliography

1. Forch, P., Gahler, C. and Nordmann, R. (1996), Modal testing in rotating machiery using active magnertic bearings, Japan.
2. Irretier, H. and Reuter, F. (1995), Experimental modal analysis of rotating disk systems, *ASME Design Engineering Conference; DE-Vol. 84-2, Vol 3 Part B.*
3. Bucher, I., Ewins, D. J. and Robb, D. (1996), Modal testing of rotating structures: difficulties, assumptions and practical approach, *IMechE 6th Conference on Vibrations in Rotating Machines*, Oxford, UK.

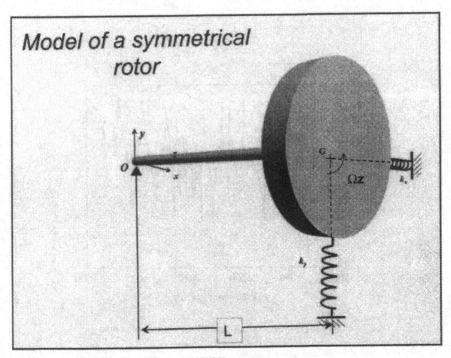

Figure 1 2DOF rotating system

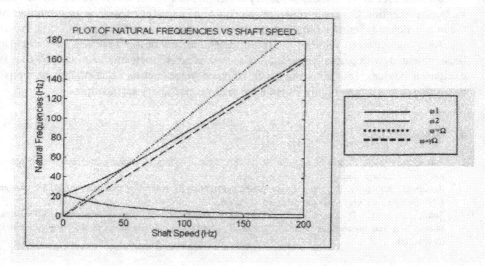

Figure 2 Natural frequencies of 2DOF rotating system

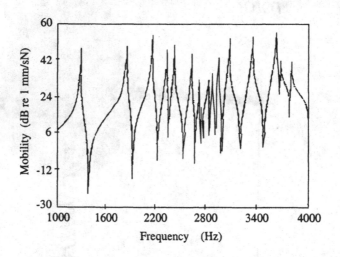

Figure 3 Typical FRF at a point on the rim of a disc

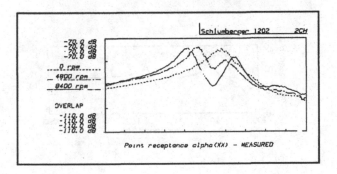

Figure 4 Excitation of rotor through auxiliary bearing plus FRFs

568

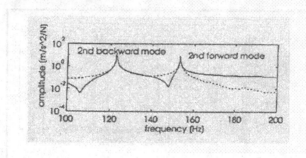

Figure 5 Test rig using AMB exciters for modal testing of rotors, plus FRF

NONLINEARITY IN MODAL ANALYSIS

G. R. TOMLINSON
Department of Mechanical Engineering,
University of Sheffield, UK

1. Introduction

Nonlinearity is a topic which is difficult to avoid as all structures encountered in practice are nonlinear to some degree, the nonlinearity often being a function of factors such as boundary/initial conditions, material properties, previous history, excitation levels etc. Nonlinearity can occur in a global (eg material nonlinearity) or a local sense (eg joints/interfaces).

In modal testing one is usually concerned with detecting and identifying nonlinearity and the question "is the structure under test linear or nonlinear" frequently arises and is a challenging one.

In an attempt to provide an answer, this paper gives an introduction to the techniques which have been successfully applied to the detection and identification of nonlinearity in dynamic testing and modal analysis. The approach taken is pragmatic in that it is more often the case to experience nonlinearity during testing than to have detailed a-priori knowledge of its existence/character. However, there are examples where one encounters structures or components which rely upon nonlinear behaviour to meet the required specification such as shock absorbers, nonlinear vibration absorbers, many damping elements (hysteretic and viscous), displacement limiting/impact devices to name but a few.

It is impossible to cover in detail all of the methods which fall into the category of applied nonlinear methods and thus the emphasis will be on those which have shown the most potential in terms of implementation and usefulness. A comprehensive bibliography is given at the end of the paper for the interested reader in terms of time domain methods and frequency domain methods.

2. Detection of nonlinearity

Every modal test should include a check of whether or not the structure is nominally linear, weakly nonlinear or strongly nonlinear. The distinction between these categories is somewhat open to interpretation and on how the results are to be used. For example, if the results are to be employed as a validation check of a linear FE analysis, then the results should satisfy the linearity checks. If they are to be used to characterise the

J.M.M. Silva and N.M.M. Maia (eds.), Modal Analysis and Testing, 569–597.
© 1999 *Kluwer Academic Publishers. Printed in the Netherlands.*

structure, deviation from linear behaviour may be essential information and the type of nonlinearity may be important, requiring identification. Linear behaviour is uniquely defined but nonlinear behaviour is not, thus the definition of weak/strong nonlinearity is non-unique. In this paper we can assume that a weak nonlinear structure is one in which no bifurcations occur in the response even though there may be some distortion of the FRFs or harmonics generated.

Linearity is defined by the principal of Superposition:

$$a x_1(t) + b x_2(t) \rightarrow a y_1(t) + b y_2(t) \tag{1}$$

This is usually too cumbersome to apply in practice as the above should be validated for a range of constants a and b and inputs x_1 and x_2. In practise a subset of this is commonly applied, namely a test for Homogeneity:

$$a x(t) \rightarrow a y(t) \tag{2}$$

This is normally carried out by comparing the FRFs for different levels of the input, i.e. if the FRF is nominally invariant when the input is say, at least doubled, then it is a good indicator that the system is linear. In addition, it is good practice to test for receprocity which is another indicator that if this holds, linearity is nominally ensured. Examples of Homogeneity and Reciprocity tests are shown in Figure 1.

The form of the excitation plays an important role in that sinusoidal excitation is the optimum type for use with the Homogeneity check; random excitation tends to linearise the FRF (see reference [74]) and impulse excitation has a high crest factor which tends to make it sensitive to the location of the input. It should be noted that a structure ought to be tested over its operating range in terms of excitation and response levels (if possible) as nonlinearity such as dry friction can be very significant at low excitation levels (creating an artificially stiff system) and its effect considerably reduced as the excitation level excees the friction force level.

Another simple measure which is used in the case of random or impact excitation during modal testing is the Conference function. This is a good indicator of measurement quality but it is not necessarily a good indicator of nonlinearity and should be used with caution.

More sophisticated methods have been widely used to detect nonlinearity which normally require a little more computational effort than the methods described above; these being the Hilbert transform, Correlation functions/NARMAX, Higher Order FRFs, Statistical Moments, Bi/Tri Spectra; the bibliography references these.

3. Typical effects of nonlinearity seen in dynamic testing

The most common cause of structural nonlinearity is due to tests on structures which result in large displacements. The simplest beam structure in the laboratory can be made to display a degree of nonlinearity if subjected to large displacement behaviour. The nonlinearity frequency arises from either a stiffness (hardening or softening) or a damping effect, sometimes both. A simple method of calculating the response characteristics of single degree of freedom systems is to use the method of Harmonic Balance, as described in references [2] and [43]. This provides some insight into nonlinear behaviour which can be used as a basis for subsequent identification work. The linearised FRF is written as,

$$H(\omega) = \left[\left(K_{eq} - m\omega^2\right) + iC_{eq}\right]^{-1}$$

For nonlinear elements the equivalent stiffness K_{eq} and damper C_{eq}, can be obtained from the Harmonic Balance method and these are shown in Table 1 for stiffness, damping and friction nonlinear elements:

Table 1

Nonlinearity	K_{eq}	C_{eq}	Notes
Cubic Stiffness: Hardening	$k\left(1+\dfrac{3}{4}\alpha A^2\right)$	—	A is the peak response amplitude
Softening	$k\left(1+\dfrac{3}{4}\alpha A^2\right)$	—	α is the nonlinear coefficient
Coulomb Damping	—	$\dfrac{4F}{\pi A\omega}$	F is the friction force level
Velocity Squared Damping	—	$\dfrac{8}{3}\alpha A\omega$	α is the nonlinear coefficient

Examples of how these affect the FRF when sinusoidal excitation is used are shown in Figure 2. In Figure 3 the effect of different types of excitation on the same nonlinearity can be seen.

In many cases the distortion caused can be used to classify the type of nonlinearity and this is exactly how, for example, the Hilbert transform is used. Using numerical simulation of single degree of freedom systems incorporating common nonlinearities and taking the Hilbert transforms of their FRFs, obtained via harmonic excitation, a catalogue of 'distortions' can be created and used in the identification process.

An example of the effectiveness of the Hilbert transform can be seen in Figure 4 where the Nyquist plots of systems with stiffness and damping nonlinearities can be uniquely identified by their Hilbert transforms.

Several methods of taking the Hilbert transform have been developed but the most efficient in terms of modal testing uses the Fourier transform, see reference [73].

The Hilbert transform has also been applied to transient data and, combined with the Gabor transform, has been successfully applied to structures with coupled modes as described in reference [75].

4. Restoring force surface method

In modal testing the process of extracting the modal parameters usually assumes the structure is linear. If this is not the case then the steps that are often employed are to detect and possibly identify the type of nonlinearity and then obtain a model of the nonlinearity. The techniques discussed above can be employed to detect and in some cases identify the nonlinearity. However, one of the most effective methods of identifying almost arbitrary nonlinear systems is the Restoring Force Surface method (see references [15 - 18]) and much of the remaining part of this paper will be devoted to this.

The features that make this method attractive are:

- It can be applied to single and multi degree of freedom systems.
- A direct visual representation of the nonlinearity is available.
- One can obtain a model of the nonlinearity.
- A measure of the significance and confidence limits of the parameters can be obtained.

Initially the discussion is restricted to structures with well separated modes such that they can be considered as systems with one degree of freedom.

$$m \ddot{y} + f(y, \dot{y}) = x(t) \tag{3}$$

m	mass of the system
$x(t)$	time dependent external force
$\ddot{y}$	acceleration response
$\dot{y}$	velocity response
y	displacement response

$f(y, \dot{y})$ is the restoring force which acts to return the system to equilibrium when disturbed. In general this will be a nonlinear function of the displacement y and velocity $\dot{y}$.

4.1. FORMING THE RESTORING FORCE SURFACE

One measures the signals $x(t)$ and $\ddot{y}(t)$ at regular time intervals, giving x_i, $\ddot{y}_i$ for i = 1, ..., N where

$$x_i = x(t_i) = x(t_0 + i \Delta t)$$
$$\ddot{y}_i = \ddot{y}(t_i) = \ddot{y}(t_0 + i \Delta t) \tag{4}$$

and Δt is the sampling interval.

For each sampling instant one has:

$$f_i(y_i, \dot{y}_i) = x_i - m \ddot{y}_i \tag{5}$$

If the mass m is known all quantities on the right-hand side of the equation are known. Thus, the value of the restoring force f_i is known at each sampling instant.

Integrating the acceleration data provides the velocity and displacement at each sampling instant, $\dot{y}_i$ and y_i.

The end result of this procedure is a sequence of triplets,

$$y_i, \quad \dot{y}_i, \quad f_i = x_i - m\ddot{y}_i \qquad i=1,...,N \qquad (6)$$

The first two numbers specify a point in the phase-plane, the third (i.e. f_i) specifies the restoring force value corresponding to that state of the system.

If one regards $f(y,\dot{y})$ as a surface over the phase plane, the procedure above has specified the height of the surface over an irregularly spaced set of points.

The restoring force surface values are specified over a regular grid in order to plot the surface. This can be carried out by using a 'near neighbour' averaging method, as shown in Figure 5.

Figure 6 shows the results of applying this method to a system with a nonlinear stiffness function where the nonlinearity is clearly seen as a function of the displacement only i.e. there is no variation in the surface parallel to the velocity axis.

4.2. TEST CONFIGURATIONS

One drawback of the procedure is that an estimate of the mass is required (see equation (6)). Errors in the mass can cause distortions in the force surface. The following test configurations remove the need for a mass estimate.

4.2.1. *Transmissibility measurements*

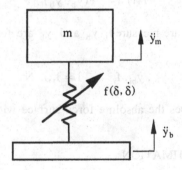

$\ddot{y}_m$ = acceleration of mass

$\ddot{y}_b$ = acceleration of base

$\ddot{\delta}$ = $\ddot{y}_m - \ddot{y}_b$ = relative acceleration

The equation of motion is:

$$m\ddot{y}_m + f(\delta,\dot{\delta}) = 0 \qquad (7)$$

The relative acceleration $\ddot{\delta}$ is integrated to form $\dot{\delta}$ and δ, the relative velocity and displacement. Notice that the mass m appears in the equation of motion as a scaling factor for $\ddot{y}_m$. One is now free to set the mass scale m = 1 and form the set of triplets.

$$\left(\delta_i, \ \dot\delta_i, \ f_i = -\ddot y_m\right) \qquad\qquad i = 1, ..., N \qquad\qquad (8)$$

One forms a surface exactly as before, the surface is correct up to an overall height scale. The type of nonlinearity is, of course, still represented faithfully. If an estimate of the mass should become available, one can correct the scale of the surface.

4.2.2. *Blocked force measurements*

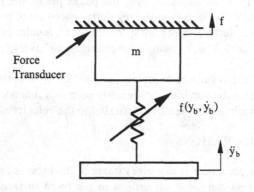

Here the mass does not accelerate, the equation of 'motion' is:

$$f(t) \ = \ f(y_b, \dot y_b) \qquad\qquad (9)$$

The f_i and $\ddot y_{b_i}$ values are measured, $\dot y_b$ and y_b are formed by integration. One obtains the triplets

$$y_{bi}, \ \dot y_{bi}, \ f_i \qquad i = 1, ..., N \qquad\qquad (10)$$

This set of triplets gives the absolute force surface without the need for a mass estimate.

4.3. PARAMETER ESTIMATION

Least squares methods allow one to obtain the equations of motion from the measured time data. Suppose one has a linear system:

$$m\ddot y + c\dot y + ky \ = \ x \qquad\qquad (11)$$

If $\ddot y_i, \dot y_i, y_i$ and x_i are sampled as before for $i = 1, ..., N$, then:

$$m\ddot y_i + c y_i + k y_i \ = \ x_i + \varepsilon_i \quad \text{for} \ i = 1, ..., N \qquad\qquad (12)$$

where ε_i is measurement noise. This set of n equations can be written as a matrix equation:

$$\begin{bmatrix} \ddot{y}_1 & \dot{y}_1 & y_1 \\ \vdots & \vdots & \vdots \\ \ddot{y}_N & \dot{y}_N & y_N \end{bmatrix} \begin{Bmatrix} m \\ c \\ k \end{Bmatrix} = \begin{Bmatrix} x_1 \\ \vdots \\ x_N \end{Bmatrix} + \begin{Bmatrix} \varepsilon_1 \\ \vdots \\ \varepsilon_N \end{Bmatrix} \tag{13}$$

or

$$[y]\,\{\beta\} = \{x\} + \{\varepsilon\} \tag{14}$$

The best (least-squares) estimate of the parameter vector $\{\beta\}$ is given by

$$\{\beta\} = \left([y]^T [y]\right)^{-1} [y]^T \{x\} \tag{15}$$

One can also fit a more general model if one allows for nonlinear systems,

$$f(y,\dot{y}) = \sum_i \sum_j a_{ij}\, y^i\, \dot{y}^j \tag{16}$$

The equation of motion is now,

$$m\ddot{y} + \sum_i \sum_j a_{ij}\, y^i\, \dot{y}^j = x \tag{17}$$

and one estimates the parameters m, a_{ij} $(i = 1,...,j = 1,...N)$

This model allows one to identify systems with polynomial stiffness, polynomial damping or even with cross product terms.

4.4. MDOF SYSTEMS

One can model an N degree of freedom system by a set of N masses, each connected to all other masses and to ground e.g. for N = 3

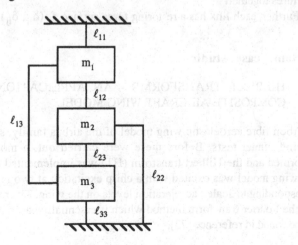

The force between mass m_1 and m_3, say, acts through the link ℓ_{13} and has the form

$$f_{13} = f_{13}(\delta_{13}, \dot{\delta}_{13})$$

where δ_{13} is the relative displacement of the masses.

The equation of motion for the mass m_i is therefore

$$m_i \ddot{y}_i + \sum_j f_{ij}(\delta_{ij}, \dot{\delta}_{ij}) = x_i(t) \tag{18}$$

where the index j runs over all masses connected to mass i, $x_i(t)$ is (if it exists) the external force at m_i .

As in the case of SDOF systems one can assume a model of the form

$$m_i \ddot{y}_i + \sum_j \sum_m \sum_n a_{(ij)_{mn}}(\delta_{ij})^m (\dot{\delta}_{ij})^n = x_i \tag{19}$$

If all the $\ddot{y}_i$, $\dot{y}_i$ etc. are measured, one can form the δ_{ij}'s and estimate parameters for the model above. This gives the equation of motion of m_i .

Note that if $x_i = 0$ (the system is excited at another point) m_i becomes a scaling factor in the equation above and cannot be estimated. One sets $m_i = 1$ in this case.

Because f_{ij} appears in the equation for m_i and m_j, if one estimates scaled parameters for both, one can give both equations a common scale by say, setting the linear parts of f_{ij} and f_{ji} to be equal.

As all masses are connected to all other masses, the common scale can be transferred to all equations which have $x_i = 0$. If only say $x_1 \neq 0$, one can obtain all parameters except those for f_{11} up to one scale.

Fitting a model to the $m_i \ddot{y}_i$ equation which has $x_i = 0$ fixes the scale.

This procedure allows one to identify each equation of motion separately.

Each mass with its set of links is a sub-structure which can be identified independently of the others. The structure can be identified by connecting the sub-structures together.

Further, each link has a restoring force surface $f_{ij}(\delta_{ij}, \dot{\delta}_{ij})$ which can be plotted.

5. Some case studies

5.1. HILBERT TRANSFORMS - AN APPLICATION TO A MODEL OF A COMPOSITE AIRCRAFT WING MODEL

A carbon fibre aeroelastic wing model of the airbus family; shown in Figure 7 was used in wind tunnel tests. Before these were carried out, a modal test of the model was performed and the Hilbert transform (HT) was implemented as a check for nonlinearity. The wing model was excited with a chirp excitation at two levels, a low and a high level corresponding to scaled acceleration levels on the wing. The method of calculating the HT was the Fourier transform method which is essentially based on the following, full details can be found in reference [73].

Let G(ω) be a measured FRF,

$$G(\omega) = Re\ G(\omega) + i\ Im\ G(\omega) \tag{20}$$

Taking a forward Fourier transform,

$$\mathcal{F}\{G(\omega)\} = H(\omega) = A(\omega) + i\ \hat{A}(\omega)$$

where $A(\omega)$ and $\hat{A}(\omega)$ are called pseudo spectra.

Now, $\hat{A}(\omega) = A(\omega)\left(-i\operatorname{sgn}(\omega)\right)$. Therefore,

$$H(\omega) = A(\omega)\left[1 + \operatorname{sgn}(\omega)\right] \tag{21}$$

i.e.,

$$
\begin{aligned}
H(\omega) &= 2\,G(\omega), \quad \omega > 0 \\
&= 2\,G(\omega), \quad \omega = 0 \\
&= 0, \qquad\quad \omega < 0
\end{aligned}
\tag{22}
$$

Equation (22) shows that the analytic signal can be calculated from only the positive spectrum of G(ω). Now,

$$
\begin{aligned}
\mathcal{F}^{-1}\{H(\omega)\} &= Re\,G(\omega) + HT\,\{Re\,G(\omega)\} \\
&= Re\,G(\omega) + i\,Im\,G(\omega)
\end{aligned}
\tag{23}
$$

Figure 8 shows the bode and Nyquist plots of the FRF and its HT for the wing model at a low level of chirp excitation. The close overlay implies that the system is linear. Figure 9 shows the effect of using a high level of chirp excitation. The mode at 129 Hz clearly shows a mismatch between the FRF and its HT. The distortion is typical of a damping nonlinearity. Further checks revealed that the model had a delamination near the trailing edge of the wing, which actually extended to the visible edge of the wing

5.2. IDENTIFICATION OF A BEAM STRUCTURE WITH NONLINEAR DAMPING VIA THE RSF METHOD

This case study concerns the use of the Restoring Force Surface (RSF) method.

Figure 10(a) shows the beam model and Figure 10(b) the schematic of how a nonlinear damper was introduced.

The system was treated as a 3 degree of freedom structure, the coordinates being the three point mass locations. The excitation used was a band limited random input at coordinate 3. Using the methodology described in section 4.4. the accelerations were used to obtain the relative velocities and displacements and the corresponding force surfaces. Figure 11 shows these for links ℓ_{21}, ℓ_{22}, ℓ_{23}. The ℓ_{22} force surface clearly displays nonlinearity in the velocity plane whilst the other two force surfaces indicate linear behaviour. This is the correct identification as the only nonlinear element occurs at the second coordinate and is linked to ground.

6. Conclusions

This paper has described several approaches for dealing with nonlinearity in modal testing. As far as the author is aware, to date there is no single method that can apply to all cases and it is clear that the choice of detection/identification method depends on the requirements of the test. The various methods can be ranked in terms of simplicity of use and their complementary to a modal test. A method which ranks high in the author's experience is the use of the Restoring Force Surface method because of its generality regarding the range of excitation types that can be used, the ability to handle weak and strong non-linearities and the fact that a predictive model can be obtained.

However, the choice of a particular method is analogous to the choice of FE codes, the more one uses a particular code/method, the more one gains experience and confidence with that method. Future techniques are likely to include a combination of methods and to take advantage of developments in neural networks, genetic algorithms and wavelets for analysing data and modelling complex phenomena where already these have seen applications in normal mode testing and structural health monitoring.

7. Bibliography

1. Material Nonlinearity in Vibration Problems, *ASME 71*, 1985.
2. Schmidt, G., Tondl, A. (1986), Nonlinear Vibration, *CUP*.
3. White, R. G. (1973), Effects of nonlinearity due to large deflections in the derivation of frequency response data from the impulse response of structures, *Jnl Sound and Vibration, 29 (3)*.
4. Wang, J. (1990), Dynamic modelling of joints, *Doctoraatsthesis*, Faculteit Toegepaste Weten Schappen, KUL, Belgium.
5. Bowden, M., Dugundji, J. (1990), Joint-damping and nonlinearity in the dynamics of space structures, *AIAA Jnl, 28(4)*.
6. Crawley, E. F. and O'Donnell, K. J. (1986), Identification of nonlinear system parameters in joints using the force state mapping technique, *AIAA Paper 86-1013*, 659-667.
7. Gaul, L., Bohlen, S. (1987), Identification of nonlinear structural joint models and implementation in discretised structure models, *Proc. 11th ASME Biennial Conf. on Mech Vibration and Noise*, Boston.
8. Tomlinson, G. R. (1987), A theoretical and experimental study of the force characteristics from electrodynamic exciters on linear and nonlinear systems, *Proc. 5th IMAC*, Imperial College, London, 1479-1487.
9. DTA Handbook on Nonlinearity in Dynamic Testing, *Vol 4*, DTA Cranfield University, 1994.
10. Tomlinson, G. R. (1979), Force distortion in resonance testing of structures with electrodynamic exciters, *Journal of Sound and Vibration, 63(3)*, 227-350.
11. Chen, S., Billings, S. A. (1989), Representations of nonlinear systems: the NARMAX model, *Int. Jnl Control, 49*, 1013-1032.
12. Tsang, K. M., Billings, S. A. (1992), Reconstruction of linear and nonlinear continuous time models for discrete time sampled data systems, *J. Mech. Syst. and Sig. Proc., 6*, 69-84.
13. Chen, Q., Tomlinson, G. R. (1992), Application of a series model for the identification of nonlinear stiffness and friction damped systems, *17th ISMA*, KUL Belgium.
14. Chen, Q., Tomlinson, G. R. (1993), The identification of systems with combined friction and stiffness nonlinearities using a time series model, *Acc. for publication in Journal of Vibrations and Acoustics, ASME*.
15. Worden, K., Tomlinson, G. R. (1988), Identification of linear/nonlinear restoring force surfaces in single and multi-mode systems, *Proc. 3rd Int. Conf. on Recent Advances in Structural Dynamics*, Southampton, 299-308.
16. Al-Hadid, M. A. (1989), Identification of nonlinear systems using the force-state mapping technique, *Ph.D. Thesis*, Univ. of London.

17. Worden, K. (1989), Parametric and non-parametric identification of nonlinearity in structural dynamics, *Ph.D. Thesis*, Heriot-Watt University, Edinburgh.

18. Wright, J. R., Hadid, M. A. (1991), Sensitivity of the force-state mapping approach to measurement errors, *Int. Journal of Anal. and Exp. Modal Analysis, 6*, 89-103.

19. Wyckaert, K. (1992), Development and evaluation of detection and identification schemes for the nonlinear dynamical behaviour of mechanical structures, *Ph.D. Thesis*, KUL Belgium.

20. Chouchai, T., Vinh, T. (1986), Analysis of nonlinear structures by impact testing and higher order transfer functions, *Proc. 4th IMAC*, Los Angeles.

21. Frachebourg, A. (1991), Identification of nonlinearities: application of the Volterra model to discrete MDOF systems, *Proc. Florence Modal Anal. Conf.*, Florence.

22. Gifford, S. (1989), Volterra series analysis of nonlinear structures, *Ph.D. Thesis*, Heriot-Watt University, Edinburgh.

23. Schetzen, M. (1980), The Volterra And Wiener Theories Of Nonlinear Systems, *J. Wiley and Sons Inc.*

24. Marmarelis, V. Z. and Marmarelis, P. Z. (1978), Analysis of physiological systems: white noise approach, *Plenum Press*, New York.

25. Lee, Y. W. and Schetzen, M. (1965), Measurement of the Wiener kernels of a nonlinear system by cross-correlation, *Int. Journal of Control, 2*, 237-254.

26. Gifford, S. J., Tomlinson, G. R. (1989), Recent advances in the application of functional series to nonlinear structures, *Journal of Sound and Vibration, 135 (2)*, 289-317.

27. Voon, W. S. F., Billings, S. A. (1983), Structure detection and model validity tests in the ident. of nonlinear systems, *IEE Proc. Pt D, 130*, 193-199.

28. Billings, S. A. and Fadzil, M. B. (1985), The practical identification of systems with nonlinearities, *Proc. IFAC Syst. Ident. and Param, Estimation*, York.

29. Billings, S. A., and Tsang, K. I. M. (1990), Spectral analysis of block structured nonlinear systems, *Mech. Syst. and Sig. Proc., 4*, 117-130.

30. Staszewski, W. (1994), The application of time variant analysis to gearbox fault detection, *University of Manchester.*

31. Staszewski, W., Tomlinson, G. R. (1994), Application of the Wavelet transform to the fault detection in a spur gear, *Mech. Syst. and Sig. Proc., 8 (3)*, 289-307.

32. Chui, C. K. (1992), An Introduction To Wavelets, *Academic Press, San Diego.*

33. Ruskai, N. D., Charles and Bartlett (eds.) (1992), Wavelets And Their Applications, Boston.

34. Newland, D. E. (1993), An introduction to random vibrations, spectral and wavelet analysis, *Longman Scientific and Technical*, UK.

35. Daubechies, I. (1991), Ten Lectures on Wavelets, *CBMS-NSF, Series in Applied Mathematics, Siam.*

36. Daubechies, I. (1991), The wavelet transform: A method for time-frequency localization, in Simon Haykin (ed.), *Advances In Spectrum Analysis And Array Processing, Vol. 1, Prentice Hall Advanced Reference Series on Engineering*, New Jersey.

37. Worden, K., Tomlinson, G. R. (1994), Modelling and classification of nonlinear systems using neural networks-I Simulation, *Mech. Syst. and Sig. Proc, 8 (3)*, 319-356.

38. Billings, S. A. et al (1991), Properties of neural networks with applications to modelling nonlinear dynamical systems, *Int. Jnl of Control, 55*, 193-224.

39. Masri, S. F. *et al* (1992), Structure unknown nonlinear dynamical systems: identification through neural networks, *Smart Matls and Structures, 1*, 45-46.

40. Navenda, K. S. and Parthasarathy, K. (1990), Identification and control of dynamical systems using neural networks, *IEEE Trans on Neural Networks, 1*, 4-27.

41. Brancaleoni, F., Spina, D., Valente, C. (1993), Damage assessment from the dynamic response of deteriorating structures, *Safety Evaluation based on Identification Approaches*, pub. Vieweg, ISBN 3-528-06535-4.

42. Mohammad, K. S., Tomlinson, G. R. (1989), A simple method of accurately determining the damping in nonlinear structures, *Proc. 7th IMAC, Las Vegas*, 1336-1346.

43. Tomlinson, G. R. (1985), Vibration analysis and identification of nonlinear structures, *Short Course Notes, University of Manchester.*

44. Vinh, T. (1986), Etudes des Structures nonlineaires, *Session de perfectionment: Dynamique des structure - Institut Superieur des Materiaux et de la Construction Mecanique*, St. Ouen, Paris.

580

45. Cartmell, M. (1990), Introduction to Linear, Paramatric and Nonlinear Vibrations, *Chapt. 5, Chapman and Hall.*

46. Tomlinson, G. R. (1986), Detection, identification and quantification of nonlinearity in modal analysis - a review, *Proc. 4th IMAC, LA*, 837-843.

47. He, J. (1987), Identification of structural dynamic characteristics, *Ph.D. Thesis*, Imperial College, London.

48. Ruder, M. (1983), Identification of the dynamic characteristics of a simple system with quadratic damping, *Série de Mécanique Appliquée, 28 (4)*, 439 - 446.

49. Peccate, L. (1984), Reciprocals of mobility and receptance - application to nonlinear systems, *Int. Report*, University of Manchester.

50. Rauch, A.(1992), Coherence, - a powerful estimator of nonlinearity, theory and application, *Proc. 10th IMAC*, San Diego, 784-795.

51. Tomlinson, G. R. (1987), Developments in the use of the Hilbert transform for detecting and quantifying nonlinearity associated with frequency response functions, *Mech. Syst. and Sig. Proc, 1 (2)*, 151-171.

52. Worden, K. and Tomlinson, G. R. (1991), The high frequency behaviour of frequency response functions and their effect on the Hilbert transform, *Proc. 9th IMAC*, Orlando.

53. King, N. and Tomlinson, G. R. (1993), Automated nonlinearity detection in aircraft ground vibration testing, *Proc. Int. Forum on Aeroelasticity and Struct. Dynamics, Strasbourg, Vol 2*, 767-786.

54. King, N. (1994), Detection of structural nonlinearity using Hilbert transform procedures, *Ph.D. Thesis to be submitted*, Engineering Department, University of Manchester.

55. Storer, D. and Tomlinson, G. R (1993), Recent developments in the measurement and interpretation of higher order transfer functions from nonlinear structures, *Mech. Syst and Sig. Proc., 7 (2)*, 173-189.

56. Storer, D. (1991), Dynamic analysis of nonlinear structures using higher order frequency response functions, *Ph.D. Thesis*, Engineering Department, University of Manchester.

57. Lee, G. M. and Tomlinson, G. R. (1994), A study of the convergence characteristics of the Volterra Series applied to the Duffing oscillator with a harmonic input, *to be published, Journal of Sound and Vibration.*

58. Tomlinson, G. R. (1991), The use of higher order frequency response functions in nonlinear structural dynamics, *Proc. 4th Int. Conf. on Recent Advances in Struct. Dynamics*, University of Southampton, 43 - 57.

59. Billings, S. A. and Tsang, K. M. (1989), Spectral analysis for nonlinear systems, Parts I and II, *Mech. Syst. and Sig. Proc., 3 (4).*

60. Worden, K., Manson, G. and Tomlinson, G. R. (1994), Pseudo fault induction in engineering structures, *Proc. Royal Soc., Part A*, LONDON, UK, 193-229.

61. Choi, D. N. *et al* (1990), Bispectral identification of nonlinear mode interaction, *Proc. 8th IMAC*, Orlando.

62. Brillinger D. R. (1970), The identification of polynomial systems by means of higher order spectra, *Journal of Sound and Vibration, 12 (3)*, 301-313.

63. Brillinger, D. R. and Rosenblatt, M. (1967), Spectral Analysis Of Time Series, *Ed. B. Harris, Wiley.*

64. Bendat, J. S. (1990), Nonlinear System Analysis and Identification, *John Wiley.*

65. Drazin, P. (1993), Nonlinear Systems, *Cambridge University Press.*

66. Moon, F. C. (1987), Chaotic Vibrations, *John Wiley Interscience.*

67. Nayfeh, A. H., Mook, D. T. (1979), Nonlinear Oscillations, *John Wiley and Sons*, London and New York.

68. Schetzen, M. (1980), The Volterra and Wiener Theories of Nonlinear Systems, *John Wiley Interscience*, New York.

69. Schmidt, G., Tondl, A. (1986), Nonlinear Vibrations, *Cambridge University Press.*

70. Starjinski, V. (1980), Methods Appliquees en Theorie des Oscillations Non Linearities, *Ed. MIR*, Moscow.

71. Thompson, J. M. T., Steward, H. B. (1986), Nonlinear Dynamics and Chaos, *John Wiley and Sons*, London and New York.

72. Cartmell, M. (1990), Introduction To Linear, Parametric And Nonlinear Vibrations, *Chapman and Hall*, London.

73. King, N. E. (1994), Detection of nonlinearity using Hilbert transform procedures, *Ph.D. Thesis*, University of Manchester.

74. Manson, G. (1996), Analysis of nonlinear mechanical systems using the Volterra series, *Ph.D. Thesis*, University of Manchester.
75. Spina, D., Valente, C., Tomlinson, G. R. (1996), A new procedure for detecting nonlinearity from transient data using the Gabor transform, *Journal of Nonlinear Dynamics, Vol. 11, No. 3*, 235-254.

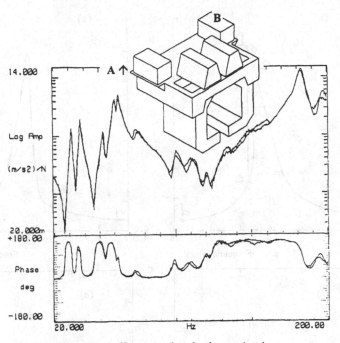

Figure 1 a) Homogeneity check at point A

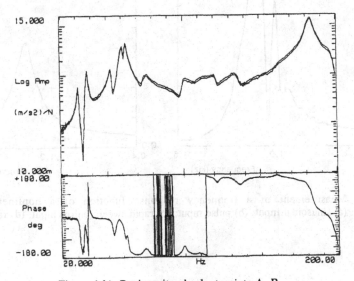

Figure 1 b) Reciprocity check at points A, B

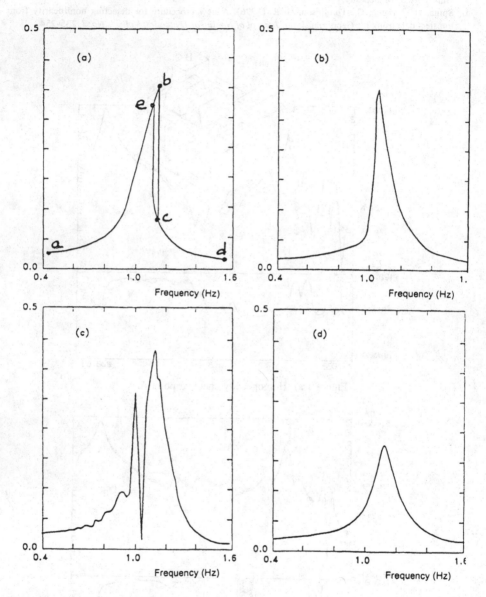

Figure 2   Measurement of a frequency response function of a nonlinear oscilator:
(a) sinusoidal input; (b) pulse input; (c) rapid sweep (chirp) input; (d) random input.

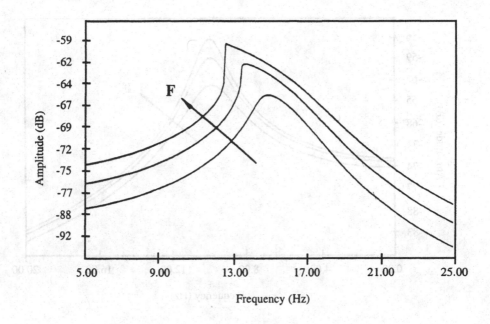

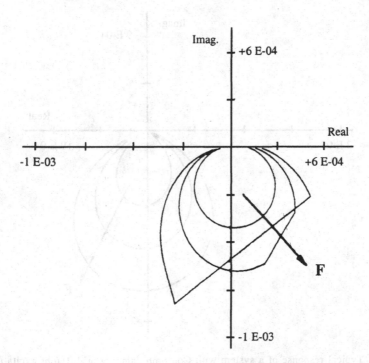

Figure 3 a) Typical response of a system with softening cubic stiffness at various excitation
levels (F)

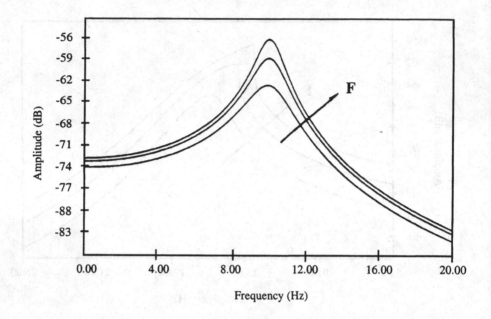

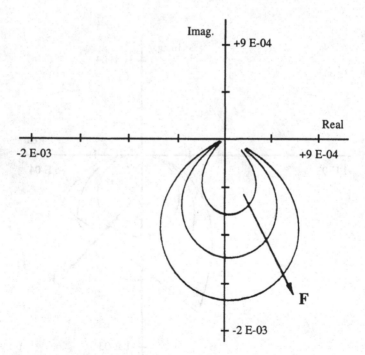

Figure 3 b) Typical response of a system with Coulomb damping at different excitation levels (F)

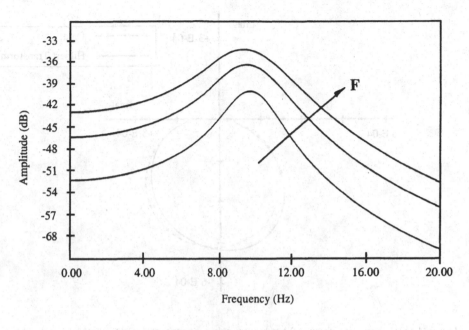

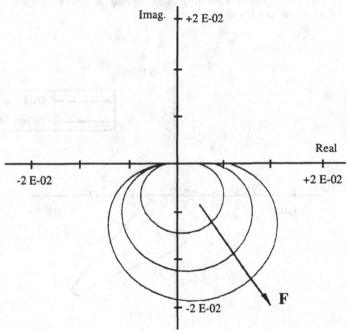

Figure 3 c) Typical response of a system with velocity squared damping at different excitation
levels (F)

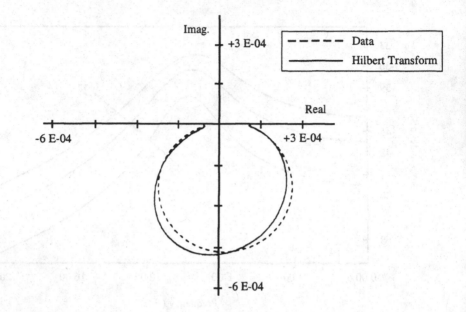

Figure 4 a) Hilbert transform of hardening cubic spring at low excitation level where it behaves near to a linear one

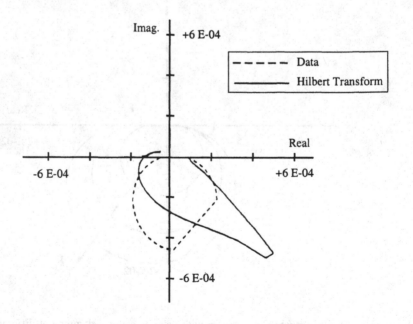

Figure 4 b) Hilbert transform of softening cubic spring

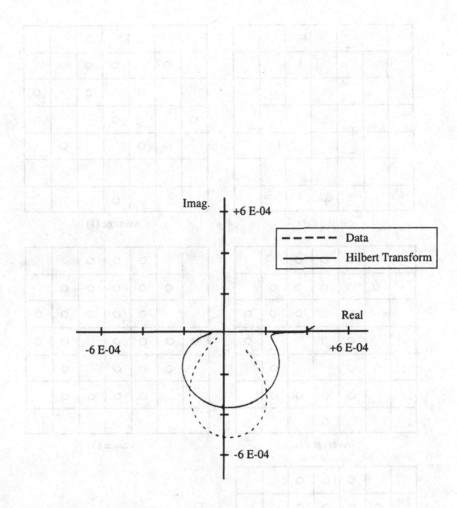

Figure 4 c) Hilbert transform of Coulomb friction at low excitation level where strong nonlinearity is detected

588

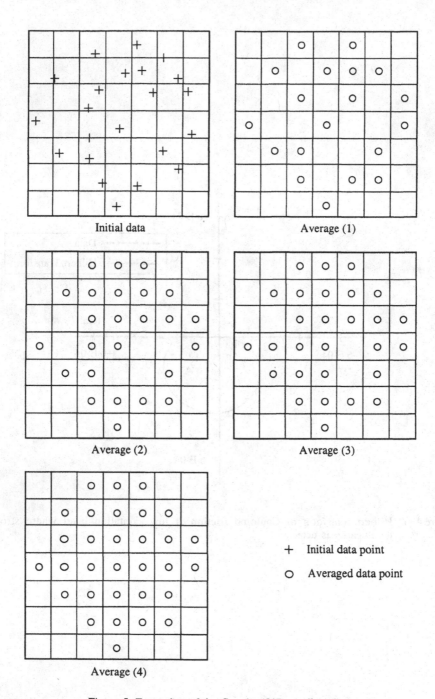

Initial data

Average (1)

Average (2)

Average (3)

Average (4)

+   Initial data point

o   Averaged data point

Figure 5 Formation of the Crawley/O'Donnell surface

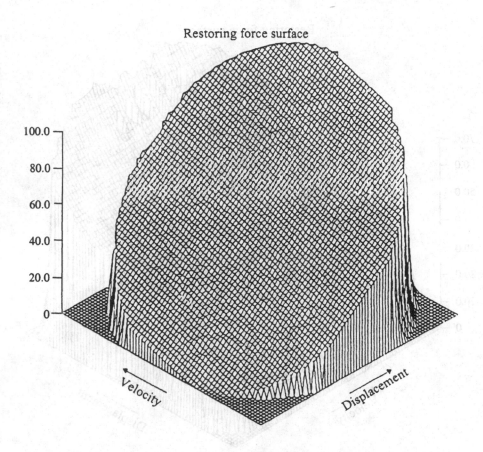

Figure 6 a) Estimated restoring force surface for a linear system

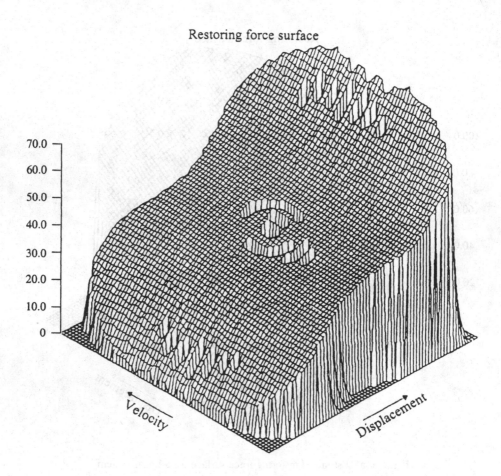

Figure 6 b) Estimated restoring force surface for a Duffing oscillator system

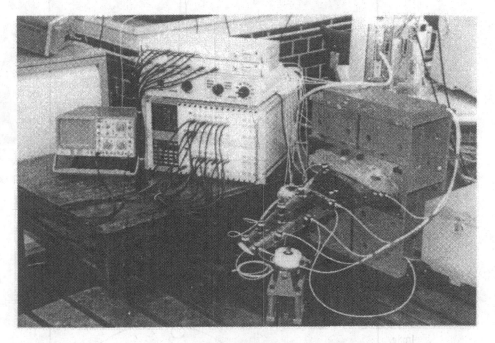

Figure 7 Experimental aeroelastic wing model

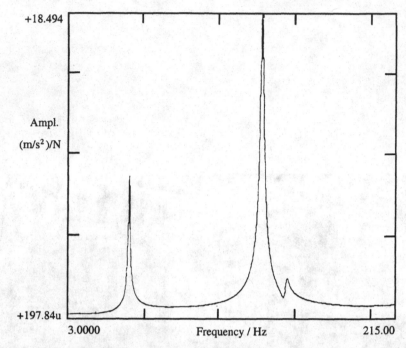

Figure 8 a) Bode amplitude plot of FRF and Hilbert transform — low level chirp excitation

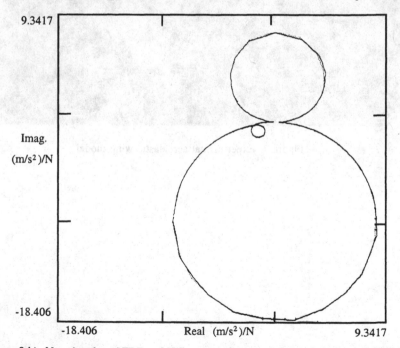

Figure 8 b) Nyquist plot of FRF and Hilbert transform — low level chirp excitation

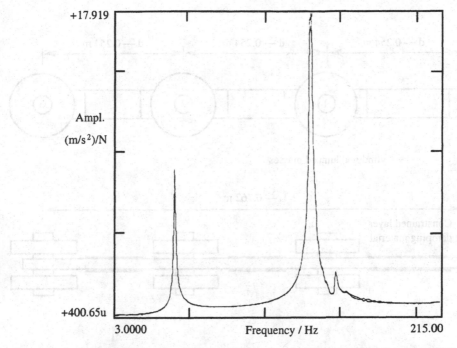

Figure 9 a) Bode amplitude plot of FRF and Hilbert transform — high level chirp excitation

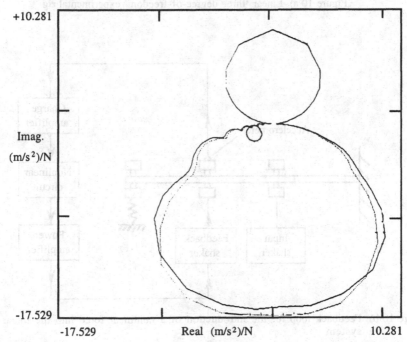

Figure 9 b) Nyquist plot of FRF and Hilbert transform — high level chirp excitation

594

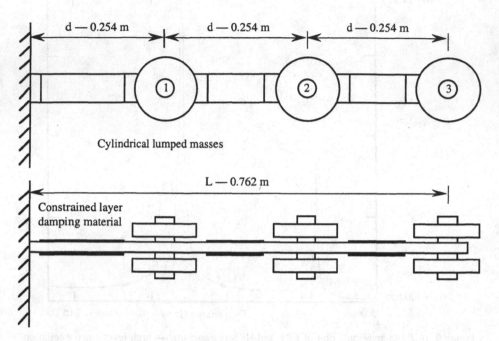

Cylindrical lumped masses

Constrained layer
damping material

Figure 10 a) Linear 'three degree-of-freedom' experimental rig

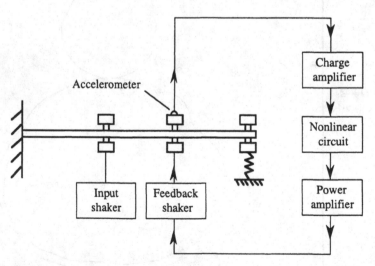

Figure 10 b) Feedback loop for the introduction of a nonlinear force into the linear 3 DOF
system

Restoring force surface

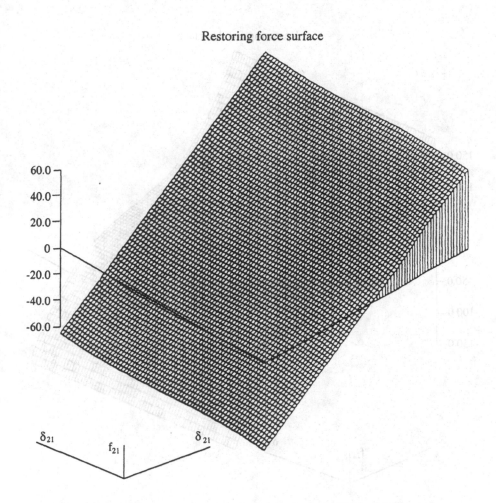

Figure 11 a) Restoring force surface for link ℓ_{21} in the nonlinear 3 DOF system

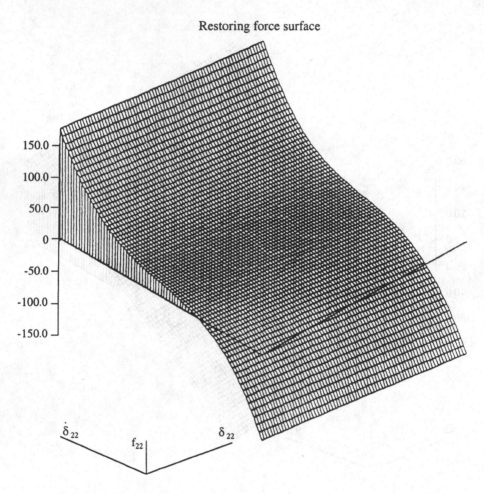

Figure 11 b) Restoring force surface for link ℓ_{22} in the nonlinear 3 DOF system

Restoring force surface

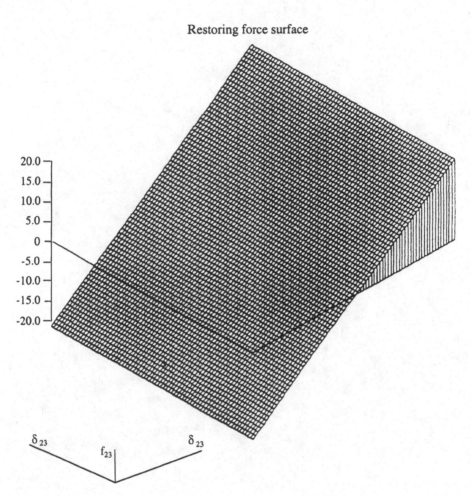

Figure 11 c) Restoring force surface for link ℓ_{23} in the nonlinear 3 DOF system